PARTICLES AND FIELDS

PARTICLES AND FIELDS

Eighth Mexican School

Oaxaca, México November 1998

EDITORS

Juan Carlos D'Olivo
ICN-UNAM, México

Gabriel López Castro
CINVESTAV, México

Myriam Mondragón
IF-UNAM, México

AIP CONFERENCE PROCEEDINGS 490

American Institute of Physics

Melville, New York

Editors:

Juan Carlos D'Olivo
Instituto de Ciencias Nucleares
Universidad Nacional Autonóma de México
Apdo. Postal 70-543
04510 México, D.F.
MÉXICO

E-mail: dolivo@nuclecu.unam.mx

Gabriel López Castro
Centro de Investigación y Estudios Avanzados del IPN
Apdo. Postal 14-740
07000 México, D.F.
MÉXICO

E-mail: glopez@fis.cinvestav.mx

Myriam Mondragón
Depto. de Física Teórica
Instituto de Física
Universidad Nacional Autonóma de México
Apdo. Postal 20-364
01000 México, D.F.
MÉXICO

E-mail: myriam@ft.ifisicacu.unam.mx

L.C. Catalog Card No. 99-067150
ISBN 1-56396-895-9
ISSN 0094-243X
DOE CONF- 981188

Printed in the United States of America

CONTENTS

Preface

The VIII Mexican School on Particles and Fields took place November 20-28, 1998, in the city of Oaxaca de Juárez, México. The School continued with the tradition of promoting High Energy Physics in Mexico, and it was aimed particularly at students and young researchers. To fulfill this purpose worldwide experts were brought together to give a series of courses and plenary lectures. There were also research seminars given by the participants to discuss their own work. This volume contains the written version of the material presented at the School. Parallel to the courses and seminars a series of lectures for the general public were given at the Universidad Autónoma Benito Juárez de Oaxaca (UABJO), by some of the invited speakers.

The School was attended by about 120 physicists, half of which were students. It was organized by the Particles and Fields Division of the Mexican Physical Society, and was sponsored by several institutions (listed separately). We are especially grateful for the warm hospitality and generosity offered by Dr. Leticia Mendoza, Chancellor of the UABJO.

We are also grateful to our colleagues in the Organizing Committee, A. Ayala, A. Rosado, and M. Ruíz-Altaba, for all their help. Special thanks go to our Conference Secretary Verónica Riquer, and to Martha Alonso and Trinidad Ramírez for their organizing efficiency. Last but not least, we thank warmly all the speakers for their excellent lectures.

Juan Carlos D'Olivo
ICN-UNAM

Gabriel López Castro
Cinvestav

Myriam Mondragón
IF-UNAM

Acknowledgments

The VIII-EMPC was sponsored by the following institutions:

- Centro de Investigación y de Estudios Avanzados
  - Dirección General
  - Departamento de Física (Zacatenco)
- Centro Latinoamericano de Física
- Centro Latinoamericano de Física, México
- Consejo Nacional de Ciencia y Tecnología, México
- European Laboratory for Particle Physics (CERN)
- Secretaría de Turismo, Estado de Oaxaca
- Universidad Autónoma Benito Juárez de Oaxaca
- Universidad Nacional Autónoma de México
  - Coordinación de la Investigación Científica
  - Dirección General de Asuntos del Personal Académico
  - Coordinación de Programas Académicos, Secretaría General
  - Dirección General de Intercambio Académico
  - Instituto de Ciencias Nucleares
  - Instituto de Física
  - Facultad de Ciencias

Organizing Committee

AYALA, Alejandro, ICN-UNAM
D'OLIVO, Juan Carlos, ICN-UNAM
LÓPEZ CASTRO, Gabriel, Cinvestav
MONDRAGÓN, Myriam, IF-UNAM
ROSADO, Alfonso, IF-BUAP
RUIZ-ALTABA, Martí, IF-UNAM

COURSES AND LECTURES

Lattice QCD

Rajan Gupta

Group T-8, Mail Stop B-285, Los Alamos National Laboratory
Los Alamos, NM 87545, U. S. A
E-mail: rajan@lanl.gov

INTRODUCTION

A large part of my lectures were devoted to the formulation of lattice QCD. A more detailed account of these, than presented at this school, is available from the e-print archive [1]. So, the only topic that I will cover in this writeup is renormalization on the lattice and the procedure for taking the continuum limit. These topics were presented in a different way from that in [1] and many participants expressed interest in seeing the alternate form written. The details of the lattice approach to extracting the quark masses and the hadron spectrum are also given in [1]. A status report on these calculations can be found in [2–4]. I was, contrary to plans, unable to cover the calculations of matrix elements using Lattice QCD and their impact on the Standard Model phenomenology. I have, therefore, included a list of references that should provide an interested reader a good starting point to learn about these calculations.

FORMULATION OF LATTICE QCD

1. Discretizing QCD on the lattice [1].
2. Improvement of lattice QCD [1,5–8].
3. Calculation of correlation functions and extraction of physical quantities like masses and decay constants from them [1].
4. Renormalization of lattice operators. Perturbative and non-perturbative renormalization [9,10].
5. Renormalization constants using external quark and gluon states in a fixed gauge [11,12].
6. Ward identities on the lattice and their use in non-perturbative determination of renormalization constants and the improvement coefficients [5,13–18]
7. Chiral symmetry, Ginsparg-Wilson condition, domain wall and overlap fermions [19–21]

CP490, *Particles and Fields: Eighth Mexican School,* edited by J. C. D'Olivo et al.

RENORMALIZATION ON THE LATTICE

The input parameters in simulations of lattice QCD are the gauge coupling constant g and the quark masses m_i. The results of the measurements are dimensionless numbers; all quantities with dimensions of energy are measured in units of the lattice spacing a. For example, lattice results for the mass of a hadron is the number $M_{hadron}(g, m_i)a(g)$. In practice the results also depend on the lattice size L, details of the lattice action for gauge fields and fermions, and on the number of flavors of dynamical quarks and their masses. I will first discuss an idealized calculation to explain renormalization and then reintroduce these systematic effects to explain how they are removed in order to get physical results.

First let me, for simplicity, assume that there are no finite volume corrections ($L = \infty$); no discretization errors (the lattice action is quantum perfect); and the quark masses have been tuned to their physical values. Then the results of simulations at some g will be the set of numbers

$$\begin{aligned} M_\pi^{latt} &= M_\pi a \\ M_\rho^{latt} &= M_\pi a \\ M_N^{latt} &= M_N a \\ M_\Delta^{latt} &= M_\Delta a \quad \ldots, \end{aligned} \tag{1}$$

which are the masses of the hadrons measured in units of a. We now renormalize the theory by demanding, for example, that $M_N = 938$ MeV at these simulation parameters. This renormalization condition uniquely fixes the lattice scale

$$a = \frac{M_N^{latt}}{938\text{MeV}} = 0.213\ M_N^{latt}\ \text{fermi} \tag{2}$$

for the input g.

We now repeat the calculation at a number of values of g and find the corresponding a

$$\begin{aligned} g_1^2 &\Longleftrightarrow a_1 = 0.213\ M_N^{latt}(1)\ \text{fermi} \\ g_2^2 &\Longleftrightarrow a_2 = 0.213\ M_N^{latt}(2)\ \text{fermi} \\ g_3^2 &\Longleftrightarrow a_3 = 0.213\ M_N^{latt}(3)\ \text{fermi} \\ &\vdots \\ &\vdots \\ g_n^2 &\Longleftrightarrow a_n = 0.213\ M_N^{latt}(n)\ \text{fermi}. \end{aligned} \tag{3}$$

From this data we can calculate the non-perturbative β-function

$$\frac{\partial g}{\partial \log a} = \beta_{NP}(g) \tag{4}$$

by taking a numerical derivative. If QCD is asymptotically free then this $\beta_{NP}(g)$ will coincide with the 2-loop perturbative β-function

$$\beta_{2-loop}(g) = \beta_0 g^3 + \beta_1 g^5 \tag{5}$$

as $g \to 0$. At large g the two will differ; the difference being the higher corrections in $\beta_{2-loop}(g)$.

For g sufficiently small that $\beta_{2-loop}(g)$ and $\beta_{NP}(g)$ agree, we can construct

$$\frac{1}{a_i} e^{-1/2\beta_0 g_i^2(a_i)} \left(\beta_0 g_i^2(a_i)\right)^{-\beta_1/2\beta_0^2} \equiv \frac{1}{a} f(g) = \Lambda_{QCD}^{(N)}, \tag{6}$$

using the above mentioned results for g_i and the corresponding a_i. This relation defines the fundamental non-perturbative scale of QCD, $\Lambda_{QCD}^{(N)}$, with the renormalization condition $M_N = 938$ MeV. Under the idealized assumptions made above, the Λ_{QCD} calculated with other renormalization conditions like $M_\rho = 770$ MeV will be the same, and $f(g)$ is a universal scaling function. Note that in Eq. 6 I have used the 2-loop β-function because corrections to the 1-loop expression do not vanish as $g \to 0$.

Thus, picking an input bare coupling g simply specifies the lattice spacing a. The physics obtained from a simulation at any g is the same. The renormalized coupling is fixed by the renormalization condition $M_N = 938$ MeV and the scale at which it is defined is fixed by the lattice spacing a. If QCD is the correct theory of nature then lattice simulations at each g will reproduce the complete hadron spectrum and all other properties of strong interactions. In this idealized case one can work on any coarse lattice and still get continuum physics.

It is important to note that unlike perturbative calculations, all results of lattice simulations are finite numbers –– there are no infinities. Only in the limit $a = 0$ do correlation lengths $\xi/a = 1/Ma$ diverge because the units of measurement $a \to 0$. Physical quantities are finite and well-defined. In other words the lattice with a finite spacing a simply serves as an non-perturbative ultra-violet regulator with cutoff $1/a$.

Now, let me start relaxing the assumptions. Simulations are done on finite lattices. This has two consequences: (i) the spectrum in a finite periodic box is modified from the continuum one, and (2) the allowed momentum states are discrete, $p = 2\pi n/La$ with $n = 0, 1, \ldots L$. Lüscher [22] showed that for sufficiently large periodic lattices the distortion in the spectrum is exponentially small. Our current estimates show that for quarks of mass $m_s/4$ the finite volume errors on lattices larger than 3 fermi are negligible, while for physical u and d quark masses one needs lattices of size 7 fermi. Current simulations use lattices that are ≈ 3 fermi and investigate quark masses $\geq m_s/4$. Consequently, from now on it is implicit that when I say lattice results, I am referring to state-of-the-art calculations in which finite volume corrections are negligible. The second issue, discrete momentum spectrum, is relevant in the calculation of form factors (semi-leptonic and rare kaon decays) and here the understanding of errors is still evolving.

The process of formulating QCD on the lattice introduces discretization errors. One reason is that derivatives in the action are replaced by finite differences,

$$\partial_x f(x) \longrightarrow \frac{f(x+a)-f(x-a)}{2a} = \frac{1}{2a}(e^{a\partial_x} - e^{-a\partial_x})f(x)\,. \tag{7}$$

This leads to errors that are proportional to the lattice spacing. The other reason is quantum corrections. Depending on the level of improvement (the order of the leading correction in the action and operators as a power in a), any quantity measured on the lattice will differ from its value in the continuum. For example

$$M_N^{\mathrm{lattice}} a(g) = M_N^{\mathrm{continuum}} a(g) \;+\; O(a)\,. \tag{8}$$

To remove these errors we extrapolate in a results obtained at a number of values of g (see the connection between g and a in Eq. 3) to $a \to 0$ using functional forms incorporating the expected leading discretization errors. Another way to state this is: as one simulates the theory on coarser and coarser lattices, the discretization errors increase. To get continuum field theory results one, therefore, has to either use higher order approximations to the derivatives (improved actions) or take the continuum limit by extrapolation. A consequence of these discretization errors is that different renormalization schemes are no longer equivalent —– the difference again being $O(a)$. These differences are killed by the extrapolation to $a = 0$.

The most serious approximation in lattice simulations is including dynamical quarks in the update of background gauge configurations. Most calculations so far have been done in the quenched approximation ($n_f = 0$) which corresponds to neglecting the momentum dependence of vacuum polarization. (This neglect of the momentum dependence is implicit in the statement often made that the coupling in the quenched and full theories can be matched at one value of the scale. Unfortunately, this scale is not *a priori* known and is not unique. It depends on the typical momentum characterizing a given physical quantity and could be highly non-perturbative, $<< 1$ GeV.) The quenched approximation is expected to be good to $\sim 10\%$ for many observables, thus quenched calculations have phenomenological relevance in addition to developing the methodology. For a more detailed discussion of the quenched approximation see [1] and references therein. Dynamical simulations have just begun and the first explorations are still an approximation to the real world as they include only two degenerate flavors of sea quarks with masses $\gtrsim m_s/4$. Each such theory ($n_f = 0, 2, 3, \ldots$ or eventually the real one with physical masses for u, d, and s) will have its own unique continuum limit. (The quenched theory is not unitary, so strictly speaking one cannot think of it as a mathematically rigorous field theory!) Numerical simulations of Lattice QCD will quantify how physics changes with these parameters.

All lattice results therefore have a dependence on additional free parameters in the simulations. These are the number and masses of sea quarks included in the generation of background gauge configurations. Physical results can be obtained from these lattice simulations via three extrapolations. The first extrapolation, called the chiral extrapolation, is in the quark masses. We extrapolate to the physical values using expressions derived from chiral perturbation theory as we expect these to be very reliable for at least $m \leq m_s/8$. This extrapolation is inflicted upon us because the computational cost goes up as a high power, $\sim 1/m^{4-5}$, with decreasing quark mass. At each a, the values of bare quark masses are fixed by requiring that M_π, M_{K^*}, M_D, M_B take on the experimental values after the lattice scale has been fixed using say M_ρ or M_N. (One could also fix these by minimizing the χ^2 for the whole spectrum.) It is worth pointing out that it is very hard, in practice, to take into account isospin breaking effects in lattice simulations, *i.e.* $m_u \neq m_d$, as in that case one simultaneously needs to take into account electromagnetic effects. Since both these effects introduce corrections of only a few MeV, it is common practice to ignore them for the present. Thus, current simulations use a common light quark mass $(m_u + m_d)/2$ for the u and d flavors and fix this using M_{π^+}.

The second extrapolation is to the continuum limit, $a = 0$, for fixed number of flavors and fixed sea quark masses to remove the discretization errors. The last extrapolation is in the number of flavors and masses of sea quarks to get the physical results. Depending on the nature of, and numerical control over, the various systematic errors in the lattice data, the order of the first two extrapolations can be interchanged.

Finally, the status of various calculations, and their impact on phenomenology, can be obtained from the list of review or state-of-the-art articles provided below.

PHYSICS OF HEAVY QUARKS

1. f_B, $B^0 - \overline{B^0}$ mixing, and CP violation [23–25].
2. Semi-leptonic form factors in the decays $D \to K(K^*)l\nu_l$, $B \to \pi(\rho)l\nu_l$, and $B \to D(D^*)l\nu_l$ [23].
3. Form factors in rare decays $B \to K^*\gamma$ [23].
4. Isgur-Wise function [23].
5. Operator product expansion and the heavy quark effective theory (HQET) [26].
6. Discussion of power corrections and renormalons [27–29].
7. Parameters of the Heavy Quark Effective Theory: $\overline{\Lambda}$, λ_1, and λ_2 [30,31].

NON-LEPTONIC DECAYS

1. Effective weak Hamiltonian for kaon decays and CP violation [25,32].
2. $K^0 - \overline{K^0}$ mixing (B_K) and CP violation [25,33–36].
3. $\Delta I = 1/2$ rule and B parameters for strong penguin operators [37,38].
4. B_7, B_8, the mass of the strange quark, and their implications for ϵ'/ϵ [25,33–35,39,40].
5. Direct evaluation of $K^+ \to \pi^+\pi$ matrix elements on the lattice. Reconciling lattice results for the $\Delta I = 3/2$ amplitude and B_K [41,42].

6. Matrix elements of 4-fermion operators that arise in the study of the lifetimes of heavy particles and in supersymmetric extensions of the Standard Model [34,43–45]

STRUCTURE FUNCTIONS

Quark and gluon distributions (structure functions) in hadrons are measured in deep inelastic lepton hadron scattering. It is possible to compute moments of these structure functions on the lattice. For recent attempts see Ref. [46].

ACKNOWLEDGEMENTS

It is a great pleasure to thank the organizers Alejandro Ayala, Juan Carlos D'Olivo, and Myriam Mondragon for a very stimulating school, and Piotr Kielanowski for inviting me to give the lectures. I thank all the students for their help and enthusiasm in making it a very memorable experience.

REFERENCES

1. R. Gupta, Lectures at the 1997 Les Houches Summer School "Probing the Standard Model of Particle Interactions", hep-lat/9806023.
2. T. Bhattacharya and R. Gupta, Nucl. Phys. (Proc. Suppl.) **B63** (1998) 95.
3. R. Kenway, Proceedings of Lattice 98, hep-lat/9810054.
4. R. Burkhalter, Proceedings of Lattice 98, hep-lat/9810043.
5. M. Lüscher, Lectures at the 1997 Les Houches Summer School, hep-lat/9802029.
6. K.Symanzik, Nucl. Phys. **B226**(1983) 187,205.
7. M. Alford, T. Klassen, G. P. Lepage, Nucl. Phys. **B496** (1997) 377.
8. P. Hasenfratz, hep-lat/9803027.
9. P. Lepage and P. Mackenzie, Phys. Rev. **D48** (1993) 2250.
10. G. Martinelli, *et al.*, Phys. Lett. **B411** (1997) 141.
11. G. Martinell, C. Pittori, C. Sachrajda, M. Testa, A. Vladikas, Nucl. Phys. **B445** (1995) 81
12. A. Donini, G. Martinelli, C. Sachrajda, M. Talevi, and A. Vladikas, Phys. Lett. **B360** (1995) 83.
13. M. Bochicchio *et al.*, Nucl. Phys. **B262** (1985) 331.
14. M. Lüscher, S. Sint, R. Sommer, P. Weisz, Nucl. Phys. **B478** (1996) 365.
15. M. Lüscher, S. Sint, R. Sommer, P. Weisz, and U. Wolff, Nucl. Phys. **B491** (1997) 323.
16. M. Lüscher, S. Sint, R. Sommer, H. Wittig, Nucl. Phys. **B491** (1997) 344.
17. G. Martinelli, G.C. Rossi, C. Sachrajda, S. Sharpe, M. Talevi, and M. Testa, Phys. Lett. **B411** (1997) 141.
18. T. Bhattacharya, S. Chandrasekharan, R. Gupta, W. Lee, and S. Sharpe, hep-lat/9810018.

19. M. Lüscher, Phys. Lett. **428B** (1998) 342, and hep-lat/9811032
20. D. Kaplan, Phys. Lett. **288B** (1992) 342; Y. Shamir, Nucl. Phys. (Proc. Suppl.) **B53** (1997) 664; Nucl. Phys. (Proc. Suppl.) **B47** (1996) 212.
21. H. Neuberger and R. Narayanan, Nucl. Phys. **B443** (1995) 305; R. Narayanan, hep-lat/9707035.
22. M. Lüscher, *Selected Topics In Lattice Field Theory*, Lectures at the 1988 Les Houches Summer School, North Holland, 1990.
23. J. Flynn and C. Sachrajda, hep-lat/9710057, hep-lat/9710080.
24. T. Draper, Plenary talk at LATTICE 98, hep-lat/9810032.
25. A. Buras, Proceedings of the 1997 Les Houches Summer School, North Holland, 1999. hep-ph/9806471.
26. M. Neubert, Phys. Rep. **245** (1994) 259.
27. M. Beneke, hep-ph/9807443.
28. G. Martinelli and C. Sachrajda, Phys. Lett. **B354** (1995) 423.
29. G. Martinelli and M. Neubert, C. Sachrajda, Nucl. Phys. **B461** (1996) 238.
30. V. Gimenez, G. Martinelli, and C. Sachrajda, Nucl. Phys. **B486** (1997) 227.
31. V. Gimenez, G. Martinelli, and C. Sachrajda, Phys. Lett. **B393** (1997) 124.
32. M. Ciuchini, E. Franco, G. Martinelli, and L. Silvestrini, Phys. Lett. **B380** (1996) 353.
33. L. Conti, A. Donini, V. Gimenez, G. Martinelli, M. Talevi, and A. Vladikas, Phys. Lett. **B421** (1998) 273.
34. R. Gupta, T. Bhattacharya, S. Sharpe, Phys. Rev. **D55** (1997) 4036.
35. G. Kilcup, R. Gupta, S. Sharpe, Phys. Rev. **D57** (1997) 1654.
36. JLQCD Collaboration, S. Aoki, *et al.*, Phys. Rev. Lett. **80** (1998) 5271.
37. C. Dawson, G. Martinelli, G.C. Rossi, C. Sachrajda, S. Sharpe, M. Talevi, and M. Testa, Nucl. Phys. **B514** (1998) 313.
38. D. Pekurovsky and G. Kilcup, hep-lat/9812019.
39. M. Ciuchini, E. Franco, G. Martinelli, L. Reina, and L. Silvestrini Z. Phys. **C68** (1995) 239.
40. R. Gupta, Proceedings of ORBIS SCIENTIAE 1997-II, PHYSICS OF MASS, Miami, Florida, Dec 12-15, 1997, hep-ph/9801412.
41. M. Golterman and Ka Chun Leung, Phys. Rev. **D56** (1997) 2950.
42. JLQCD Collaboration, S. Aoki, *et al.*, Phys. Rev. **D58** (1998) 054503.
43. M. Ciuchini, E. Franco, G. Martinelli, A. Masiero, and L. Silvestrini, Phys. Rev. Lett. **79** (1997) 978.
44. C. Allton, L. Conti, A. Donini, V. Gimenez, L. Giusti, G. Martinelli, M. Talevi, and A. Vladikas, hep-lat/9806016.
45. G. Martinelli, Plenary Talk at Lattice 98, hep-lat/9810013.
46. C. Best, M. Göckeler, R. Horsley, H. Perlt, P. Rakow, A. Schdfer, G. Schierholz, A. Schiller, S. Schramm, Nucl. Phys. Proc. Suppl. **63** (1998) 236; M. Göckeler, R. Horsley, H. Perlt, P. Rakow, G. Schierholz, A. Schiller, P. Stephenson, hep-ph/9711245. M. Gčkeler, R. Horsley, E.-M. Ilgenfritz, H. Perlt, P. Rakow, G. Schierholz, A. Schiller Phys. Rev. **D53** (1996) 2317.

Electroweak Precision Data and Precision Tests of the SM and MSSM

W. Hollik

Institut für Theoretische Physik, Universität Karlsruhe
D-76128 Karlsruhe, Germany

Abstract. These lectures give an overview of the status of radiative correction calculations to provide accurate theoretical predictions for testing the standard model (SM) in present and future experiments. The status of the standard model is discussed in the light of the recent electroweak precision data for the vector boson masses and for the Z resonance observables. As a theoretically well motivated example of physics beyond the SM, the minimal supersymmetric standard model (MSSM) is considered, reviewing precision tests of the MSSM and precise calculations for the mass of the lightest Higgs boson.

INTRODUCTION

The e^+e^- colliders LEP and the SLC, in operation since summer 1989, have collected an enormous amount of electroweak precision data on Z and W bosons [1,2]. The W boson properties have in parallel been determined at the $p\bar{p}$ collider Tevatron with a constant increase in accuracy [2,3]; after the discovery of the top quark there [4], its mass has been measured [5] with a precision of better than 3%, to 173.8 ± 5.0 GeV. The ongoing experiments at LEP 2 and the near-future Tevatron upgrade will also, in the coming years, support us with further increases in precision, in particular on the mass of the W and the top, and the SLC might continue to improve the impressive accuracy already obtained in the electroweak mixing angle. This stimulating experimental program together with the theoretical activities to provide accurate predictions from the standard model have formed the era of electroweak precision tests and will keep it alive also in the next years.

The standard theory of the electroweak interaction is a gauge-invariant quantum field theory with the symmetry group SU(2)$\times$U(1) spontaneously broken by the Higgs mechanism. The possibility to perform perturbative calculations for observable quantities in terms of a few input parameters is substantially based on the renormalizability of this class of theories [6]. A certain set of input parameters has to be taken from experiment. In the electroweak standard model essentially three free parameters are required to describe the gauge bosons $\gamma, W^{\pm}, Z$, and their

CP490, *Particles and Fields: Eighth Mexican School,* edited by J. C. D'Olivo et al.

interactions with the fermions. For a comparison between theory and experiment, hence, three independent experimental input data are required. The most natural choice consists of the electromagnetic fine structure constant α, the muon decay constant (Fermi constant) G_μ, and the mass of the Z boson, which has meanwhile been measured with the same accuracy as the Fermi constant [1,2]. Other measurable quantities are predicted in terms of the input data. Each additional precision experiment, which allows the detection of small deviations from the lowest-order predictions, can be considered a test of the electroweak theory at the quantum level. In the Feynman graph expansion of the scattering amplitude for a given process the higher-order terms show up as diagrams containing closed loops. The renormalizability of the standard model ensures that it retains its predictive power also in higher orders. The higher-order terms are the quantum effects of the electroweak theory. They are complicated in their concrete form, but they are finally the consequence of the basic Lagrangian with a simple structure. The quantum corrections (radiative corrections) contain the self-coupling of the vector bosons as well as their interactions with the Higgs field and the top quark, and provide the theoretical basis for electroweak precision tests. Assuming the validity of the standard model, the presence of the top quark and the Higgs boson in the loop contributions to electroweak observables allows an indirect probe of their mass ranges from comparison with precision data.

The generation of high-precision experiments hence imposes stringent tests on the standard model. A primordial step strengthening our confidence in the standard model has been the discovery of the top quark at the Tevatron [4], at a mass that agrees with the mass range obtained indirectly, through the radiative corrections. Moreover, with the top mass as an additional precise experimental data point one can now fully exploit the virtual sensitivity to the Higgs mass.

The experimental sensitivity in the electroweak observables, at the level of the quantum effects, requires the highest standards on the theoretical side as well. A sizeable amount of work has contributed, over the recent years, to a steadily rising improvement of the standard model predictions, pinning down the theoretical uncertainties to the level required for the current interpretation of the precision data. The availability of both highly accurate measurements and theoretical predictions, at the level of 0.1% precision and better, provides tests of the quantum structure of the standard model, thereby probing its still untested scalar sector, and simultaneously accesses alternative scenarios such as the supersymmetric extension of the standard model.

The lack of direct signals from new physics beyond the standard model makes the high-precision experiments a unique tool also in the search for indirect effects: through possible deviations of the experimental results from the theoretical predictions of the minimal standard model. Since such deviations are expected to be small, it is decisive to have the standard loop effects in the precision observables under control.

These lectures contain a discussion of the theoretical developments for testing the electroweak theory, the status of the standard model in view of the recent precision data, and the implications for the Higgs boson. Precision tests of the MSSM are also discussed, as well as a precise prediction for the mass of the lightest Higgs boson in the MSSM.

STATUS OF PRECISION CALCULATIONS

Radiative corrections entries

The possibility of performing precision tests is based on the formulation of the standard model as a renormalizable quantum field theory preserving its predictive power beyond tree-level calculations. With the experimental accuracy being sensitive to the loop-induced quantum effects, also the Higgs sector of the standard model is being probed. The higher-order terms induce the sensitivity of electroweak observables to the top and Higgs mass m_t, M_H and to the strong coupling constant α_s.

The calculation of electroweak observables in higher orders requires the concept of renormalization to get rid of the divergences in the Feynman integral evaluation and to define the physical input parameters. In QED and in the electroweak theory the classical Thomson scattering and the particle masses set natural scales where the parameters $e = \sqrt{4\pi\alpha}$ and the electron, muon, ... masses can be defined. In the electroweak standard model a distinguished set for parameter renormalization is given in terms of e, M_Z, M_W, M_H, m_f with the masses of the corresponding particles. The finite parts of the counter terms are fixed by the renormalization conditions that the propagators have poles at their physical masses, and e becomes the $ee\gamma$ coupling constant in the Thomson limit of Compton scattering. This electroweak on-shell scheme, the extension of the familiar QED renormalization, has been used in many practical applications [7]- [17]. The mass of the Higgs boson, as long as it is experimentally unknown, is treated as a free input parameter. In practical calculations, the W mass is replaced by G_μ as an input parameter by using relation (18).

Instead of the set e, M_W, M_Z as basic free parameters other renormalization schemes make use of α, G_μ, M_Z [18] or perform the loop calculations in the $\overline{MS}$ scheme [19]- [22]. Other schemes applied in the past utilize the parameters α, G_μ, $\sin^2\theta_W$, with the mixing angle deduced from neutrino-electron scattering [23], or the concept of effective running couplings [24,25].

Before predictions can be made from the theory, a set of independent parameters has to be taken from experiment. For practical calculations the physical input quantities α, G_μ, M_Z, m_f, M_H, α_s are commonly used to fix the free parameters of the standard model. Differences between various schemes are formally of higher order than the one under consideration. The study of the scheme dependence of the perturbative results, after improvement by resummation of the leading terms, allows us to estimate the missing higher-order contributions.

Related to charge and mass renormalization, there occur two sizeable effects in the electroweak loops that deserve a special discussion:

(i) Charge renormalization and light fermion contribution:

Charge renormalization introduces the concept of electric charge for real photons ($q^2 = 0$) to be used for the calculation of observables at the electroweak scale set by M_Z. Hence the difference

$$\mathrm{Re}\,\hat{\Pi}^\gamma(M_Z^2)) = \mathrm{Re}\,\Pi^\gamma(M_Z^2) - \Pi^\gamma(0) \tag{1}$$

of the photon vacuum polarization is a basic entry in the predictions for electroweak precision observables. The purely fermionic contributions correspond to standard QED and do not depend on the details of the electroweak theory. They are conveniently split into a leptonic and a hadronic contribution

$$\mathrm{Re}\,\hat{\Pi}^\gamma(M_Z^2)_{\mathrm{ferm}} = \mathrm{Re}\,\hat{\Pi}^\gamma_{\mathrm{lept}}(M_Z^2) + \mathrm{Re}\,\hat{\Pi}^\gamma_{\mathrm{had}}(M_Z^2)\,, \tag{2}$$

where the top quark is not included in the hadronic part (5 light flavours); it yields a small non-logarithmic contribution

$$\hat{\Pi}^\gamma_{\mathrm{top}}(M_Z^2) \simeq \frac{\alpha}{\pi}\,Q_t^2\,\frac{M_Z^2}{5\,m_t^2} \simeq 0.6\cdot 10^{-4}\,. \tag{3}$$

The quantity

$$\begin{aligned}\Delta\alpha &= \Delta\alpha_{\mathrm{lept}} + \Delta\alpha_{\mathrm{had}} \\ &= -\,\mathrm{Re}\,\hat{\Pi}^\gamma_{\mathrm{lept}}(M_Z^2) - \mathrm{Re}\,\hat{\Pi}^\gamma_{\mathrm{had}}(M_Z^2)\end{aligned} \tag{4}$$

corresponds to a QED-induced shift in the electromagnetic fine structure constant

$$\alpha \to \alpha(1+\Delta\alpha)\,, \tag{5}$$

which can be resummed according to the renormalization group accommodating all the leading logarithms of the type $\alpha^n \log^n(M_Z/m_f)$. The result can be interpreted as an effective fine structure constant at the Z mass scale:

$$\alpha(M_Z^2) = \frac{\alpha}{1-\Delta\alpha}\,. \tag{6}$$

It corresponds to a resummation of the iterated 1-loop vacuum polarization from the light fermions to all orders.

$\Delta\alpha$ is an input of crucial importance because of its universality and of its remarkable size of $\sim 6\%$. The leptonic content can be directly evaluated in terms of the known lepton masses, yielding at one loop order:

$$\Delta\alpha_{\text{lept}} = \sum_{\ell=e,\mu,\tau} \frac{\alpha}{3\pi}\left(\log\frac{M_Z^2}{m_\ell^2} - \frac{5}{3}\right) + O\left(\frac{m_\ell^2}{M_Z^2}\right). \tag{7}$$

The 2-loop correction has been known already for a long time [27], and also the 3-loop contribution is now available [28], yielding altogether

$$\begin{aligned}\Delta\alpha_{\text{lept}} &= 314.97687 \cdot 10^{-4} = \\ &[314.19007_{1-\text{loop}} + 0.77617_{2-\text{loop}} + 0.0106_{3-\text{loop}}] \cdot 10^{-4}\,.\end{aligned} \tag{8}$$

For the light hadronic part, perturbative QCD is not applicable and quark masses are not available as reasonable input parameters. Instead, the 5-flavour contribution to $\hat{\Pi}^{\gamma}_{\text{had}}$ can be derived from experimental data with the help of a dispersion relation

$$\Delta\alpha_{\text{had}} = -\frac{\alpha}{3\pi} M_Z^2 \,\text{Re} \int_{4m_\pi^2}^{\infty} \text{d}s' \frac{R^{\gamma}(s')}{s'(s' - M_Z^2 - i\varepsilon)} \tag{9}$$

with

$$R^{\gamma}(s) = \frac{\sigma(e^+e^- \to \gamma^* \to \text{hadrons})}{\sigma(e^+e^- \to \gamma^* \to \mu^+\mu^-)}$$

as an experimental input quantity in the problematic low energy range.

Integrating by means of the trapezoidal rule (averaging data in bins) over e^+e^- data for the energy range below 40 GeV and applying perturbative QCD for the high-energy region above, the expression (9) yields the value [29,30]

$$\Delta\alpha_{\text{had}} = -0.0280 \pm 0.0007\,, \tag{10}$$

which agrees with another independent analysis [31] with a different error treatment. Because of the lack of precision in the experimental data a large uncertainty is associated with the value of $\Delta\alpha_{\text{had}}$, which propagates into the theoretical error of the predictions of electroweak precision observables. Including additional data from τ-decays [32] yields about the same result with a slightly improved uncertainty. Recently other attempts have been made to increase the precision of $\Delta\alpha$ [33,35]– [37] by "theory-driven" analyses of the dispersion integral (9). The common basis is the application of perturbative QCD down to the energy scale given by the τ mass for the calculation of the quantity $R^{\gamma}(s)$ outside the resonances. Those calculations were made possible by the recent availability of the quark-mass-dependent $O(\alpha_s^2)$ QCD corrections [38] for the cross section down to close to the thresholds for b and c production. [A first step in this direction was done in [39] in the massless approximation.] In order to pin down the error, two different strategies are in use: the application of the method developed in [36] for minimizing the impact of data from less reliable regions, done in [33], and the rescaling of data in the open charm region of 3.7–5 GeV from PLUTO/DASP/MARKII, for the purpose of normalization to agree with perturbative QCD, done in [35]. The results obtained for $\Delta\alpha_{\text{had}}$ are very similar:

$$0.02763 \pm 0.00016 \quad \text{ref [33]}$$
$$0.02777 \pm 0.00017 \quad \text{ref [35]}$$

In [37] the $\overline{MS}$ quantity $\hat{\alpha}(M_Z)$ has been derived with the help of an unsubtracted dispersion relation in the $\overline{MS}$-scheme, yielding a comparable error. The history of the determination of the hadronic vacuum polarization is visualized in Figure 1.

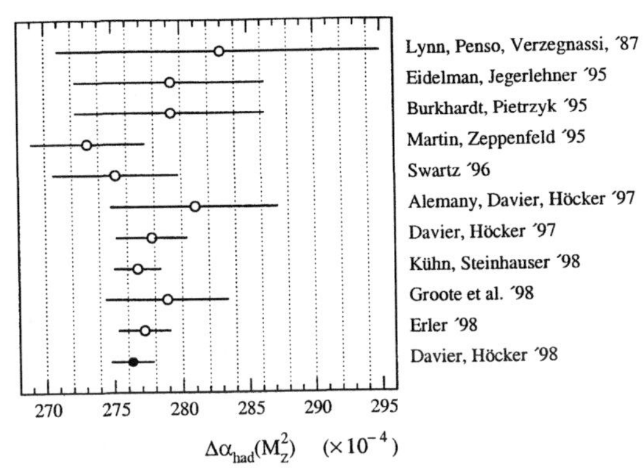

FIGURE 1. Various determinations of $\Delta\alpha_{\text{had}}$ (from ref [34]).

The basic assumption in the theory-driven approach, the validity of perturbative QCD and quark-hadron duality, is supported by the following empirical observations:

– The strong coupling constant $\alpha_s(m_\tau)$ determined from hadronic τ decays shows good agreement with $\alpha_s(M_Z)$ determined from Z-peak observables when the renormalization group evolution of α_s in perturbative QCD is imposed to run α_s from m_τ to the Z-mass scale.

– Non-perturbative contributions in $R^\gamma(s)$, parametrized in terms of condensates of quarks, gluons and of vacuum expectation values of higher-dimensional operators in the operator product expansion [40] can be probed by comparing spectral moments of $R^\gamma_{\text{exp}}(s)$ with the corresponding expressions involving the theoretical R^γ. It has been shown from fitting a set of moments that the non-perturbative contributions are negligibly small [33].

– Recent preliminary measurements of R^γ at BES at 2.6 and 3.3 GeV show values slightly lower than the previous data [41,2], better in alignment with the expectations from perturbative QCD.

Although the error in the QCD-based evaluation of $\Delta\alpha_{\text{had}}$ is considerably reduced, it should be kept in mind that the conservative estimate in Eq. (10) is independent of theoretical assumptions on QCD at lower energies and thus less sensitive to potential systematic effects [42].

(ii) Mixing angle renormalization and the ρ-parameter:

The ρ-parameter, originally defined as the ratio of the neutral to the charged current strength in neutrino scattering [43], is unity in the standard model at the tree level, but gets a deviation $\Delta\rho$ from 1 by radiative corrections. The dominating universal part has its origin in the renormalization of the relation between the gauge boson masses and the electroweak mixing angle. This relation is modified in higher orders according to

$$\sin^2\theta_W = 1 - \frac{M_W^2}{M_Z^2} + \frac{M_W^2}{M_Z^2}\Delta\rho + \cdots \tag{11}$$

The main contribution to the universal ρ-parameter

$$\rho = \frac{1}{1-\Delta\rho} \tag{12}$$

is from the (t,b) doublet [44], at the present level calculated as follows:

$$\Delta\rho = 3x_t \cdot [1 + x_t\,\rho^{(2)} + \delta\rho_{\rm QCD}] \tag{13}$$

with

$$x_t = \frac{G_\mu m_t^2}{8\pi^2\sqrt{2}}\,. \tag{14}$$

The electroweak 2-loop part [45,46] is described by the function $\rho^{(2)}(M_H/m_t)$, and $\delta\rho_{\rm QCD}$ is the QCD correction to the leading $G_\mu m_t^2$ term [47,48]

$$\delta\rho_{\rm QCD} = -\frac{\alpha_s(\mu)}{\pi}c_1 + \left(\frac{\alpha_s(\mu)}{\pi}\right)^2 c_2(\mu) \tag{15}$$

with

$$c_1 = \frac{2}{3}\left(\frac{\pi^2}{3}+1\right) = 2.8599$$

and the 3-loop coefficient [48] $c_2(\mu)$, which amounts to

$$c_2 = -14.59 \quad \text{for } \mu = m_t \text{ and 6 flavours}$$

with the on-shell top mass m_t. This reduces the scale dependence of ρ significantly and hence is an important entry to decrease the theoretical uncertainty of the standard model predictions for precision observables.

There is also a Higgs contribution to $\Delta\rho$, which, however, is not UV-finite by itself when derived from only the diagrams involving the physical Higgs boson. The M_H-dependence for large Higgs masses M_H is only logarithmic in 1-loop order [49]; the 2-loop contribution [50] shows a dependence $\sim M_H^2$ for large values of the Higgs mass. In the limit $\sin^2\theta_W \to 0, M_Z \to M_W$, where the $U(1)_Y$ is switched off, one finds $\Delta\rho_{\rm H} = 0$. This is the consequence of the global $SU(2)_R$ symmetry of the Higgs Lagrangian ('custodial symmetry'), which is broken by the $U(1)_Y$ group. Thus, $\Delta\rho_{\rm H}$ is a measure of the $SU(2)_R$ breaking by the weak hypercharge.

The vector boson mass correlation

The interdependence between the gauge boson masses is established through the accurately measured muon lifetime or, equivalently, the Fermi coupling constant G_μ. Originally, the μ-lifetime τ_μ has been calculated within the framework of the effective 4-point Fermi interaction. Beyond the well-known 1-loop QED corrections [51], the 2-loop QED corrections in the Fermi model have been calculated quite recently [52], yielding the expression (the error in the 2-loop term is from the hadronic uncertainty)

$$\frac{1}{\tau_\mu} = \frac{G_\mu^2 m_\mu^5}{192\pi^3}\left(1 - \frac{8m_e^2}{m_\mu^2}\right) \cdot \tag{16}$$
$$\cdot \left[1 + 1.810\,\frac{\alpha}{\pi} + (6.701\,\pm 0.002)\left(\frac{\alpha}{\pi}\right)^2\right].$$

This formula is the defining equation for G_μ in terms of the experimental μ-lifetime. Owing to the presence of order-dependent QED corrections, the numerical value of the Fermi constant changes after the second-order term is included. Compared with the value given in the 1998 report of the Particle Data Group [53], the latest value is now smaller by $2 \cdot 10^{-10}$ GeV^{-2}, namely [52]

$$G_\mu = (1.16637 \pm 0.00001) \cdot 10^{-5}\,\text{GeV}^{-2}\,, \tag{17}$$

where also the error has been reduced by a factor of about 1/2.

In the standard model, G_μ can be calculated including quantum corrections in terms of the basic standard model parameters, thereby separating off all diagrams that correspond to the QED corrections in the Fermi model. This yields the correlation between the masses M_W, M_Z of the vector bosons, expressed in terms of α and G_μ; in 1-loop order it is given by [9]:

$$\frac{G_\mu}{\sqrt{2}} = \frac{\pi\alpha}{2s_W^2 M_W^2}[1 + \Delta r(\alpha, M_W, M_Z, M_H, m_t)]\,. \tag{18}$$

with $\quad s_W^2 = 1 - M_W^2/M_Z^2$.
The decomposition

$$\Delta r = \Delta\alpha - \frac{c_W^2}{s_W^2}\,\Delta\rho^{(1)} + (\Delta r)_{\text{rem}} \tag{19}$$

separates the leading fermionic contributions $\Delta\alpha$ and $\Delta\rho$(1-loop). All other terms are collected in the remainder part $(\Delta r)_{\text{rem}}$, the typical size of which is of order ~ 0.01.

The presence of large terms in Δr requires the consideration of effects higher than 1-loop. The modification of Eq. (18) according to

$$1 + \Delta r \to \frac{1}{(1 - \Delta\alpha) \cdot (1 + \frac{c_W^2}{s_W^2} \Delta\rho) - (\Delta r)_{\text{rem}}}$$
$$\equiv \frac{1}{1 - \Delta r} \tag{20}$$

accommodates the following higher-order terms (Δr in the denominator is an effective correction including higher orders):

(i) the leading log resummation [54] of $\Delta\alpha$: $1 + \Delta\alpha \to (1 - \Delta\alpha)^{-1}$;

(ii) the resummation of the leading m_t^2 contribution [55] in terms of $\Delta\rho$ in Eq. (13). Beyond the QCD higher-order contributions through the ρ-parameter, the complete $O(\alpha\alpha_s)$ corrections to the self energies are available [56,57]. All these higher-order terms contribute with the same positive sign to Δr. Non-leading QCD corrections to Δr of the type

$$\Delta r_{(bt)} = 3x_t \left(\frac{\alpha_s}{\pi}\right)^2 \left(a_1 \frac{M_Z^2}{m_t^2} + a_2 \frac{M_Z^4}{m_t^4}\right)$$

are also available [58].

(iii) With the quantity $(\Delta r)_{\text{rem}}$ in the denominator, non-leading higher-order terms containing mass singularities of the type $\alpha^2 \log(M_Z/m_f)$ from light fermions are incorporated [59].

(iv) The subleading $G_\mu^2 m_t^2 M_Z^2$ contribution of the electroweak 2-loop order [60] in an expansion in terms of the top mass. This subleading term turned out to be sizeable, about as large as the formally leading term of $O(m_t^4)$ via the ρ-parameter. In view of the present and future experimental accuracy it constitutes a non-negligible shift in the W mass.

Meanwhile exact results have been derived for the Higgs-dependence of the fermionic 2-loop corrections in Δr [61], and comparisons were performed with those obtained via the top mass expansion [62]. Differences in the values of M_W of several MeV (up to 8 MeV) are observed when M_H is varied over the range from 65 GeV to 1 TeV.

Figure 2 shows the Higgs-mass dependence of the two-loop corrections to Δr associated with the t/b doublet, with $\Delta\alpha$, and with the light fermion terms not in $\Delta\alpha$, together with the leading m_t^4-term, which constitutes a very poor approximation.

Pure fermion-loop contributions (n fermion loops at n-loop order) have also been investigated [62,63]. In the on-shell scheme, explicit results have been worked out up to 4-loop order, which allows an investigation of the validity of the resummation (20) for the non-leading 2-loop and higher-order terms. It was found that numerically the resummation (20) works remarkably well, within 2 MeV in M_W.

Observables at the Z resonance

Measurements of the Z line shape in $e^+e^- \to f\bar{f}$ yield the parameters M_Z, Γ_Z, and the partial widths Γ_f or the peak cross section

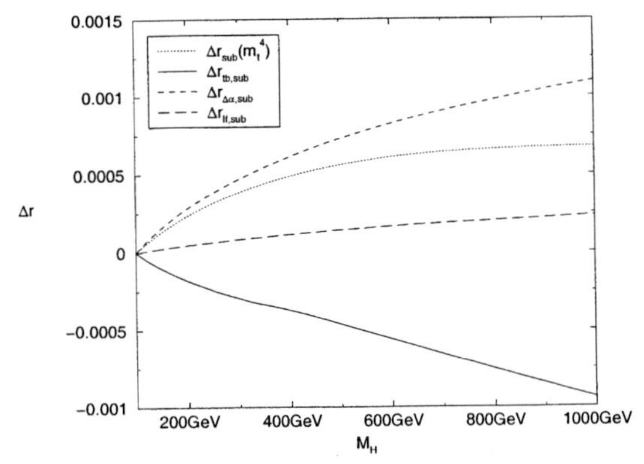

FIGURE 2. Higgs mass dependence of fermionic contributions to Δr at the two-loop level (from [62]). The different curves show the various contributions: light fermions via $\Delta\alpha$ ($\Delta r_{\Delta\alpha}$), residual light-fermion contribution not in $\Delta\alpha$ ($\Delta r_{\rm lf}$), the contribution from the (tb) doublet ($\Delta r_{\rm tb}$), and the approximation of the (tb) two-loop contribution by the term proportional to m_t^4. Displayed in each case is the difference $\Delta r(M_H) - \Delta r(100 \text{ GeV})$.

$$\sigma_0^f = \frac{12\pi}{M_Z^2} \cdot \frac{\Gamma_e \Gamma_f}{\Gamma_Z^2} \,. \tag{21}$$

Angular distributions and polarization measurements of the final fermions yield forward–backward and polarization asymmetries. Whereas M_Z is used as a precise input parameter, together with α and G_μ, the width, partial widths and asymmetries allow comparisons with the predictions of the standard model. The predictions for the partial widths as well as for the asymmetries can conveniently be calculated in terms of effective neutral current coupling constants for the various fermions.

Effective Z boson couplings: The effective couplings follow from the set of 1-loop diagrams without virtual photons, the non-QED or weak corrections. These weak corrections can conveniently be written in terms of fermion-dependent overall normalizations ρ_f and effective mixing angles s_f^2 in the NC vertices (see e.g. [64]):

$$\begin{aligned} J_\nu^{\rm NC} &= \left(\sqrt{2}G_\mu M_Z^2\right)^{1/2} \left(g_V^f\,\gamma_\nu - g_A^f\,\gamma_\nu\gamma_5\right) \\ &= \left(\sqrt{2}G_\mu M_Z^2 \rho_f\right)^{1/2} \left((I_3^f - 2Q_f s_f^2)\gamma_\nu - I_3^f \gamma_\nu\gamma_5\right) . \end{aligned} \tag{22}$$

ρ_f and s_f^2 contain universal parts, e.g. from the ρ-parameter via

$$\rho_f = \frac{1}{1-\Delta\rho} + \cdots, \quad s_f^2 = s_W^2 + c_W^2\,\Delta\rho + \cdots \tag{23}$$

with $\Delta\rho$ from Eq. (13) and non-universal parts that explicitly depend on the type of the external fermions.

The subleading 2-loop corrections $\sim G_\mu^2 m_t^2 M_Z^2$ for the leptonic mixing angle [60] s_ℓ^2 have also be obtained in the meantime, as well as for ρ_ℓ [65].

Meanwhile exact results have been derived for the Higgs-dependence of the fermionic 2-loop corrections in s_ℓ^2 [62,63], and comparisons were performed with those obtained via the top mass expansion [62]. Differences in the values of s_ℓ^2 can amount to $0.8 \cdot 10^{-4}$ when M_H is varied over the range from 100 GeV to 1 TeV.

For the b quark coupling to the Z boson, not only the universal contribution through the ρ-parameter but also the non-universal parts have a strong dependence on m_t, resulting from virtual top quarks in the vertex corrections. The difference between the d and b couplings can be parametrized in the following way

$$\rho_b = \rho_d(1+\tau)^2, \quad s_b^2 = s_d^2(1+\tau)^{-1}, \tag{24}$$

with the quantity

$$\tau = \Delta\tau^{(1)} + \Delta\tau^{(2)}$$

calculated perturbatively, including the complete 1-loop order term [66] with x_t from Eq. (14):

$$\Delta\tau^{(1)} = -2x_t - \frac{G_\mu M_Z^2}{6\pi^2\sqrt{2}}(c_W^2+1)\log\frac{m_t}{M_W} + \cdots, \tag{25}$$

and the leading electroweak 2-loop contribution of $O(G_\mu^2 m_t^4)$ [46,67]

$$\Delta\tau^{(2)} = -2\,x_t^2\,\tau^{(2)}, \tag{26}$$

where $\tau^{(2)}$ is a function of M_H/m_t with $\tau^{(2)} = 9 - \pi^2/3$ for small M_H.

Asymmetries and mixing angles: The effective mixing angles are of particular interest, since they determine the on-resonance asymmetries via the combinations

$$A_f = \frac{2g_V^f g_A^f}{(g_V^f)^2 + (g_A^f)^2}, \tag{27}$$

namely

$$A_{\rm FB} = \frac{3}{4}A_e A_f, \quad A_\tau^{\rm pol} = A_\tau, \quad A_{\rm LR} = A_e\,. \tag{28}$$

Measurements of the asymmetries hence are measurements of the ratios

$$g_V^f/g_A^f = 1 - 2Q_f s_f^2 \tag{29}$$

or the effective mixing angles, respectively.

Z width and partial widths: The total Z width Γ_Z can be calculated essentially as the sum over the fermionic partial decay widths. Expressed in terms of the effective coupling constants, they read up to second order in the fermion masses:

$$\Gamma_f = \Gamma_0 \left[(g_V^f)^2 + (g_A^f)^2 \left(1 - \frac{6m_f^2}{M_Z^2}\right)\right] \cdot \left(1 + Q_f^2 \frac{3\alpha}{4\pi}\right) + \Delta\Gamma_{\rm QCD}^f$$

with

$$\Gamma_0 \; = \; N_C^f \, \frac{\sqrt{2}G_\mu M_Z^3}{12\pi}, \qquad N_C^f = 1 \text{ (leptons)}, \;\; = 3 \text{ (quarks)}.$$

The QCD correction for the light quarks with $m_q \simeq 0$ is given by

$$\Delta\Gamma_{\rm QCD}^f \; = \; \Gamma_0 \left((g_V^f)^2 + (g_A^f)^2 \right) \cdot K_{\rm QCD} \tag{30}$$

with [68]

$$K_{\rm QCD} \; = \; \frac{\alpha_s}{\pi} + 1.41 \left(\frac{\alpha_s}{\pi}\right)^2 - 12.8 \left(\frac{\alpha_s}{\pi}\right)^3 - \frac{Q_f^2}{4} \frac{\alpha\alpha_s}{\pi^2} \, .$$

For b quarks the QCD corrections are different, because of finite b mass terms and to top-quark-dependent 2-loop diagrams for the axial part:

$$\Delta\Gamma_{\rm QCD}^b = \Delta\Gamma_{\rm QCD}^d \; + \; \Gamma_0 \left[(g_V^b)^2 \, R_V \; + \; (g_A^b)^2 \, R_A \right] \, . \tag{31}$$

The coefficients in the perturbative expansions

$$R_V = c_1^V \frac{\alpha_s}{\pi} + c_2^V \left(\frac{\alpha_s}{\pi}\right)^2 + c_3^V \left(\frac{\alpha_s}{\pi}\right)^3 + \cdots,$$
$$R_A = c_1^A \frac{\alpha_s}{\pi} + c_2^A \left(\frac{\alpha_s}{\pi}\right)^2 + \cdots$$

depending on m_b and m_t, are calculated up to third order in α_s, except for the m_b-dependent singlet terms, which are known to $O(\alpha_s^2)$ [69,70]. For a review of the QCD corrections to the Z width, see [71].

The partial decay rate into b-quarks, in particular the ratio $R_b = \Gamma_b/\Gamma_{\rm had}$, is an observable of special sensitivity to the top quark mass. Therefore, beyond the pure QCD corrections, also the 2-loop contributions of the mixed QCD–electroweak type, are important. The QCD corrections were first derived for the leading term of $O(\alpha_s G_\mu m_t^2)$ [72] and were subsequently completed by the $O(\alpha_s)$ correction to the $\log m_t/M_W$ term [73] and the residual terms of $O(\alpha\alpha_s)$ [74].

In the same spirit, also the complete 2-loop $O(\alpha\alpha_s)$ to the partial widths into the light quarks have been obtained, beyond those that are already contained in the factorized expression (30) with the electroweak 1-loop couplings [75]. These "non-factorizable" corrections yield an extra negative contribution of $-0.55(3)$ MeV to the total hadronic Z width (converted into a shift of the strong coupling constant, they correspond to $\delta\alpha_s = 0.001$). In summary, the 2-loop corrections of $O(\alpha\alpha_s)$ to the electroweak precision observables are by now completely under control.

Radiation of secondary fermions through photons from the primary final state fermions can yield another sizeable contribution to the partial Z widths; however, this is compensated by the corresponding virtual contribution through the dressed photon propagator in the final-state vertex correction for sufficiently inclusive final states, i.e. for loose cuts to the invariant mass of the secondary fermions [76].

Accuracy of the standard model predictions

For a discussion of the theoretical reliability of the standard model predictions, one has to consider the various sources contributing to their uncertainties:

Parametric uncertainties result from the limited precision in the experimental values of the input parameters, essentially $\alpha_s = 0.119 \pm 0.002$ [53], $m_t = 173.8 \pm 5.0$ GeV [5], $m_b = 4.7 \pm 0.2$ GeV, and the hadronic vacuum polarization as discussed in section 2.1. The conservative estimate of the error in Eq. (10) leads to $\delta M_W = 13$ MeV in the W-mass prediction, and $\delta \sin^2\theta = 0.00023$ common to all of the mixing angles.

The uncertainties from the QCD contributions can essentially be traced back to those in the top quark loops in the vector boson self-energies. The knowledge of the $O(\alpha_s^2)$ corrections to the ρ-parameter and Δr yields a significant reduction; they are small, although not negligible (e.g. $\sim 3 \cdot 10^{-5}$ in s_ℓ^2).

The size of unknown higher-order contributions can be estimated by different treatments of non-leading terms of higher order in the implementation of radiative corrections in electroweak observables ('options') and by investigations of the scheme dependence. Explicit comparisons between the results of 5 different computer codes based on on-shell and $\overline{MS}$ calculations for the Z-resonance observables are documented in the "Electroweak Working Group Report" [64] in ref [26]. The inclusion of the non-leading 2-loop corrections $\sim G_\mu^2 m_t^2 M_Z^2$ reduce the uncertainty in M_W below 10 MeV and in s_ℓ^2 below 10^{-4}, typically to $\pm 4 \cdot 10^{-5}$ [82].

PRECISION DATA AND THE STANDARD MODEL

We now confront the standard model predictions for the discussed set of precision observables with the most recent sample of experimental data [1,2]. In table 1 the standard model predictions for Z-pole observables and the W mass are put together for the best fit input data set, given in (33). The experimental results on the Z observables are from LEP and the SLC, the W mass is from combined LEP and $p\bar{p}$ data. The leptonic mixing angle determined via $A_{\rm LR}$ by the SLD experiment [77] and the s_ℓ^2 average from LEP:

$$s_e^2(A_{\rm LR}) = 0.23109 \pm 0.00029, \quad s_\ell^2(\text{LEP}) = 0.23189 \pm 0.00024$$

TABLE 1. Precision observables: experimental results from combined LEP and SLD data for Z observables and combined $p\bar{p}$ and LEP data for M_W, together with the standard model predictions for the best fit, Eq. (33).

Observable	Exp.	SM best fit
M_Z (GeV)	91.1867 ± 0.0019	91.1865
Γ_Z (GeV)	2.4939 ± 0.0024	2.4956
σ_0^{had} (nb)	41.491 ± 0.058	41.476
R_{had}	20.765 ± 0.026	20.745
R_b	0.21656 ± 0.00074	0.2159
R_c	0.1732 ± 0.0048	0.1722
$A_{\rm FB}^{\ell}$	0.01683 ± 0.00096	0.0162
$A_{\rm FB}^{b}$	0.0990 ± 0.0021	0.1029
$A_{\rm FB}^{c}$	0.0709 ± 0.0044	0.0735
A_b	0.867 ± 0.035	0.9347
A_c	0.647 ± 0.040	0.6678
ρ_ℓ	1.0041 ± 0.0012	1.0051
s_ℓ^2	0.23157 ± 0.00018	0.23155
M_W (GeV)	80.39 ± 0.06	80.372

have come closer to each other in their central value; owing to their smaller errors, however, they still differ by 2.8 standard deviations.

Table 1 contains the combined LEP/SLD value. ρ_ℓ and s_ℓ^2 are the leptonic neutral current couplings in Eq. (22), derived from partial widths and asymmetries under the assumption of lepton universality.

Note that the experimental value for ρ_ℓ points at the presence of genuine electroweak corrections by 3.5 standard deviations. In s_ℓ^2 the presence of purely bosonic radiative corrections is clearly established when the experimental result is compared with a theoretical value containing only the fermion loop corrections, an observation that has been persisting already for several years [78]. The deviation from the standard model prediction in the quantity R_b has been reduced below one standard deviation by now. Other small deviations are observed in the asymmetries: the purely leptonic $A_{\rm FB}$ is slightly higher than the standard model predictions, and $A_{\rm FB}$ for b quarks is lower. Whereas the leptonic $A_{\rm FB}$ favours a very light Higgs boson, the b quark asymmetry needs a heavy Higgs.

The effective mixing angle is an observable most sensitive to the mass M_H of the Higgs boson. Since a light Higgs boson corresponds to a low value of s_ℓ^2, the strongest upper bound on M_H is from $A_{\rm LR}$ at the SLC [77]. The inclusion of the two-loop electroweak corrections $\sim m_t^2$ from [60] yields a sizeable positive contribution to s_ℓ^2, see Figure 3. The inclusion of this term hence strengthens the upper bound on M_H.

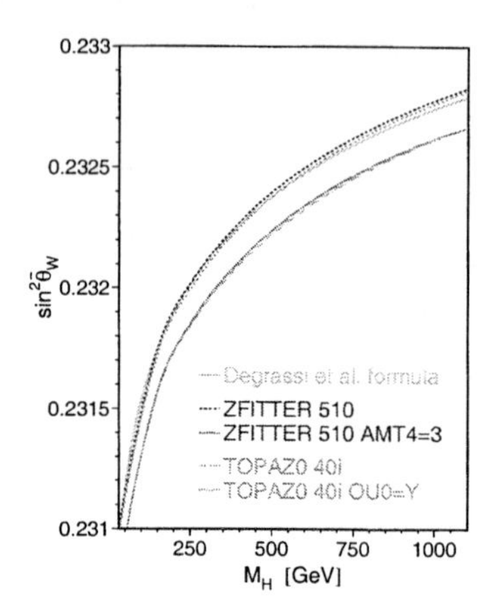

FIGURE 3. Higgs mass dependence of s_ℓ^2 with and without the electroweak 2-loop term $\sim m_t^2$, comparison of ZFITTER and TOPAZ0 codes. The lower sample of curves is without, the upper sample with the 2-loop term (figure prepared by C. Pauss).

The W mass prediction in table 1 is obtained from Eq. (18) (including the higher-order terms) from M_Z, G_μ, α and M_H, m_t. The present experimental value for the W mass from the combined LEP 2, UA2, CDF and D0 results is in best agreement with the standard model prediction.

The quantity s_W^2 resp. the ratio M_W/M_Z can indirectly be measured in deep-inelastic neutrino–nucleon scattering. The average from the experiments CCFR, CDHS and CHARM with the recent NUTEV result [79]

$$s_W^2 = 1 - M_W^2/M_Z^2 = 0.2255 \pm 0.0021 \tag{32}$$

for $m_t = 175$ GeV and $M_H = 150$ GeV corresponds to $M_W = 80.25 \pm 0.11$ GeV and is hence consistent with the direct vector boson mass measurements and with the standard theory.

Standard model global fits: The FORTRAN codes ZFITTER [80] and TOPAZ0 [81] have been updated by incorporating all the recent precision calculation results that were discussed in the previous section. Comparisons have shown good agreement between the predictions from the two independent programs [82]. Global fits of the standard model parameters to the electroweak precision data done by the Electroweak Working Group [1] are based on these recent versions. Including m_t and M_W from the direct measurements in the experimental data set, together with s_W^2 from neutrino scattering, the standard model parameters for the best fit result are:

	Measurement	Pull
m_Z [GeV]	91.1867 ± 0.0021	.09
Γ_Z [GeV]	2.4939 ± 0.0024	-.80
σ^0_{hadr} [nb]	41.491 ± 0.058	.31
R_e	20.765 ± 0.026	.66
$A^{0,e}_{fb}$	0.01683 ± 0.00096	.73
A_e	0.1479 ± 0.0051	.25
A_τ	0.1431 ± 0.0045	-.79
$\sin^2\theta^{lept}_{eff}$	0.2321 ± 0.0010	.53
m_W [GeV]	80.37 ± 0.09	-.01
R_b	0.21656 ± 0.00074	.90
R_c	0.1735 ± 0.0044	.29
$A^{0,b}_{fb}$	0.0990 ± 0.0021	-1.81
$A^{0,c}_{fb}$	0.0709 ± 0.0044	-.58
A_b	0.867 ± 0.035	-1.93
A_c	0.647 ± 0.040	-.52
$\sin^2\theta^{lept}_{eff}$	0.23109 ± 0.00029	-1.65
$\sin^2\theta_W$	0.2255 ± 0.0021	1.06
m_W [GeV]	80.41 ± 0.09	.43
m_t [GeV]	173.8 ± 5.0	.54
$1/\alpha^{(5)}(m_Z)$	128.878 ± 0.090	.00

FIGURE 4. Experimental results and pulls from a standard model fit (from ref [1,2]). pull = obs(exp)-obs(SM)/(exp.error).

$$
\begin{aligned}
m_t &= 171.1 \pm 4.9\,\text{GeV} \\
M_H &= 76^{+85}_{-47}\,\text{GeV} \\
\alpha_s &= 0.119 \pm 0.003\,.
\end{aligned}
\tag{33}
$$

The upper limit to the Higgs mass at the 95% C.L. is $M_H < 262$ GeV, where the theoretical uncertainty is included. Thereby the hadronic vacuum polarization in Eq. (10) has been used (solid line in Figure 5). With the theory-driven result on $\Delta\alpha_{\text{had}}$ of ref [33] one obtains [1] $M_H = 92^{+64}_{-41}$ (dashed line). The 1σ upper bound on M_H is influenced only marginally. The reason is that simultaneously with the error reduction the central value of M_H is shifted upwards (see Figure 5). Another recent analysis [83] (for earlier studies see [84,85]) based on the data set of summer 1998 yields a Higgs mass $M_H = 107^{+67}_{-45}$ GeV. About one half of the difference with (33) can be ascribed to the use of $\alpha(M_Z)$ of ref [37], which is very close to the value in ref [33,35]; the residual shift might be interpreted as due to different renormalization schemes and different treatments of α_s.

With an overall $\chi^2/\text{d.o.f.} = 15/15$ the quality of the fit is remarkably high. As can be seen from Figure 4, the deviation of the individual quantities from the standard model best-fit values are below 2 standard deviations.

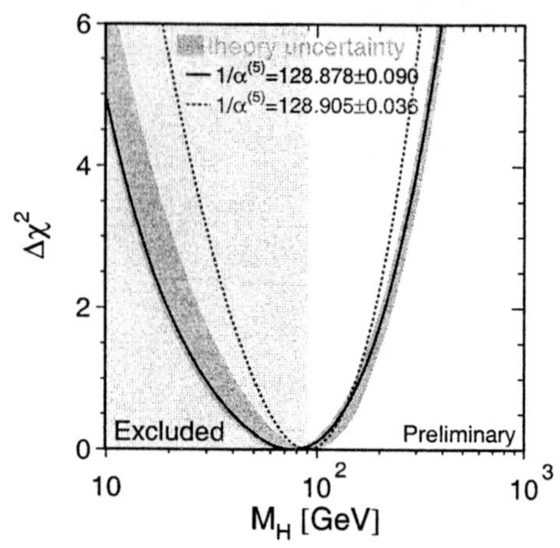

FIGURE 5. Higgs mass dependence of χ^2 in the global fit to precision data (from ref [1,2]). The shaded band displays the error from the theoretical uncertainties obtained from various options in the codes ZFITTER and TOPAZ0.

Compared with the results from 1997, the central value for the Higgs mass has moved to lower values and the error has been decreased. The Higgs mass bounds follow from the χ^2 distribution shown in Figure 5. The shift in the central value illustrates the effect of the inclusion of the electroweak two-loop contribution by Degrassi et al. [60], which was not implemented in the codes for the earlier analyses. Since it increases the prediction for s_ℓ^2 (Figure 3) for a given Higgs mass, the allowed values of M_H are shifted accordingly downwards.

The second observation is the decrease of the error, which besides the experimental improvements results from the reduction of the theoretical uncertainties of pure electroweak origin. The shaded band around the solid line in Figure 5 is the influence of the various 'options' (see section 2.4) in the codes ZFITTER and TOPAZ0 after the implementation of the 2-loop electroweak terms $\sim m_t^2$. It is thus the direct continuation of the error estimate done in the previous study [64]. Compared with the width of the uncertainty band in 1997 the shrinking is evident.

On the other hand, the remaining theoretical uncertainty associated with the Higgs mass bounds should be taken very seriously. The effect of the inclusion of the next-to-leading term in the m_t-expansion of the electroweak 2-loop corrections in the precision observables has shown to be sizeable, at the upper margin of the estimate given in [64]. It is thus not guaranteed that the subsequent subleading terms in the m_t-expansion are indeed smaller in size. Also the variation of the M_H-dependence at different stages of the calculation, as discussed in the previous subsections, indicate the necessity of more complete results at two-loop order. Having in mind also the variation of the Higgs mass bounds under the fluctuations of the experimental data [2], the limits for M_H derived from the analysis of electroweak data in the frame of the standard model still carry a noticeable uncertainty. Nevertheless, as a central message, it can be concluded that the indirect determination of the Higgs mass range has shown that the Higgs is light, with its mass well below the non-perturbative regime.

THE STANDARD HIGGS SECTOR

The minimal model with a single scalar doublet is the simplest way to implement the electroweak symmetry breaking. The experimental result that the ρ-parameter is very close to unity is a natural feature of models with doublets and singlets. In the standard model, the mass M_H of the Higgs boson appears as the only additional parameter beyond the vector boson and fermion masses. M_H cannot be predicted but has to be taken from experiment. The present lower limit (95% C.L.) from the search at LEP [86] is 89 GeV. Indirect determinations of M_H from precision data have already been discussed in section 3. The indirect mass bounds react sensitively to small changes in the input data, which is a consequence of the logarithmic dependence of electroweak precision observables. As a general feature, it appears that the data prefer a light Higgs boson.

There are also theoretical constraints on the Higgs mass from vacuum stability and absence of a Landau pole [87–89], and from lattice calculations [90]. Explicit perturbative calculations of the decay width for $H \to W^+W^-, ZZ$ in the large-M_H limit in 2-loop order [92] have shown that the 2-loop contribution exceeds the 1-loop term in size (same sign) for $M_H > 930$ GeV (Figure 6). This result is confirmed by the calculation of the next-to-leading order correction in the $1/N$ expansion, where the Higgs sector is treated as an $O(N)$ symmetric σ-model [93]. A similar increase of the 2-loop perturbative contribution with M_H is observed for the fermionic decay [94] $H \to f\bar{f}$, but with opposite sign leading to a cancellation of the one-loop correction for $M_H \simeq 1100$ GeV (Figure 6). The requirement of applicability of perturbation theory therefore puts a stringent upper limit on the Higgs mass. The indirect Higgs mass bounds obtained from the precision analysis show, however, that the Higgs boson is well below the mass range where the Higgs sector becomes non-perturbative. The lattice result [91] for the bosonic Higgs decay in Figure 6 for $M_H = 727$ GeV is not far from the perturbative 2-loop result. The difference may at least partially be interpreted as missing higher-order terms.

The behaviour of the quartic Higgs self-coupling λ, as a function of a rising energy scale μ, follows from the renormalization group equation with the β-function dominated by λ and the top quark Yukawa coupling g_t contributions:

$$\beta_\lambda = 24\,\lambda^2 + 12\,\lambda\,g_t^2 - 6\,g_t^4 + \cdots \qquad (34)$$

In order to avoid unphysical negative quartic couplings from the negative top quark contribution, a lower bound on the Higgs mass is derived. The requirement that the Higgs coupling remains finite and positive up to a scale Λ yields constraints

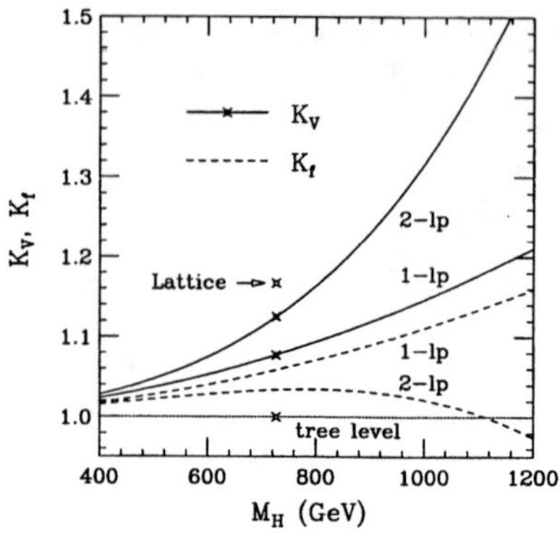

FIGURE 6. Correction factors for the Higgs decay widths $H \to VV$ $(V = W, Z)$ and $H \to f\bar{f}$ in 1- and 2-loop order (from ref [95])

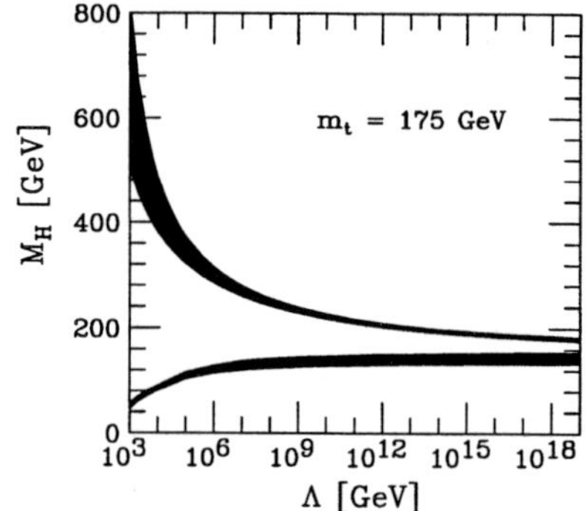

FIGURE 7. Theoretical limits on the Higgs boson mass from the absence of a Landau pole and from vacuum stability (from ref [89])

on the Higgs mass M_H, which have been evaluated at the 2-loop level [88,89]. These bounds on M_H are shown in Figure 7 as a function of the cut-off scale Λ up to which the standard Higgs sector can be extrapolated, for $m_t = 175$ GeV and $\alpha_s(M_Z) = 0.118$. The allowed region is the area between the lower and the upper curves. The bands indicate the theoretical uncertainties associated with the solution of the renormalization group equations [89]. It is interesting to note that the indirect determination of the Higgs mass range from electroweak precision data via radiative corrections is compatible with a value of M_H where Λ can extend up to the Planck scale.

PRECISION TESTS OF THE MSSM

Among the extensions of the standard model, the MSSM is the theoretically favoured scenario as the most predictive framework beyond the standard model. A

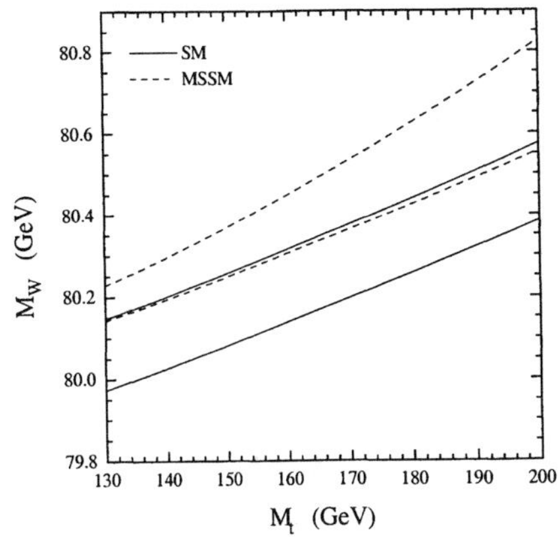

FIGURE 8. The W mass range in the standard model (—) and the MSSM (- - -). Bounds are from the non-observation of Higgs bosons and SUSY particles at LEP2.

definite prediction of the MSSM is the existence of a light Higgs boson with mass below ~ 135 GeV [107–109,112]. The detection of a light Higgs boson at LEP could be a significant hint for supersymmetry.

The structure of the MSSM as a renormalizable quantum field theory allows a similarly complete calculation of the electroweak precision observables as in the standard model in terms of one Higgs mass (usually taken as the CP-odd 'pseudoscalar' mass M_A) and $\tan\beta = v_2/v_1$, together with the set of SUSY soft-breaking parameters fixing the chargino/neutralino and scalar fermion sectors. It has been known for quite some time [96] that light non-standard Higgs bosons as well as light stop and charginos predict larger values for the ratio R_b [97,99,105]. Complete 1-loop calculations are available for Δr and for the Z boson observables [98,99,105].

So far, the direct search of SUSY particles at colliders has not been successful, and under some assumptions one can only set lower bounds on their masses. An alternative way to probe SUSY is to search for the virtual effects of the additional particles. One can use the high-precision measurements of electroweak observables to search for the quantum effects of the SUSY particles. In the MSSM one finds two types of potentially large effects: The first possibility is that charginos and scalar top quarks are light enough to affect the decay width of the Z boson into b–quarks; however, for masses beyond the LEP2 or Tevatron reach, these effects become too small to be observable. The second possibility is the contribution of the top and bottom squark loops to the electroweak gauge–boson self–energies: if there is a large splitting between the masses of $\tilde{b}_L$ and $\tilde{t}_L$, the contribution will grow with the mass of the heaviest scalar quark and can be sizable. This contribution enters the electroweak observables via the ρ parameter, which influences the normalization of the NC Z couplings, the effective electroweak mixing angle at the Z–boson resonance and the $W - Z$ mass correlation. Recently the 2-loop α_s corrections have been computed [100], which can amount to 30% of the 1-loop $\Delta\rho_{\tilde{b}\tilde{t}}$.

Figure 8 displays the range of predictions for M_W in the minimal model and in the MSSM. It is thereby assumed that no direct discovery has been made at LEP 2. As can be seen, precise determinations of M_W and m_t can become decisive for the separation between the models.

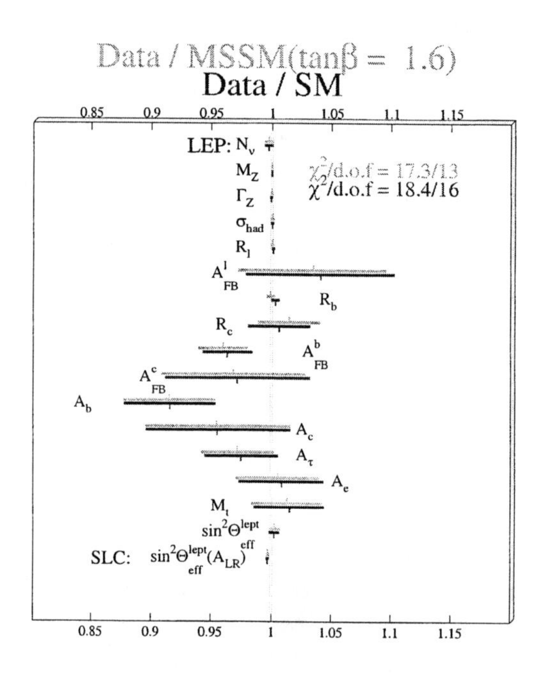

FIGURE 9. Best fits in the SM and in the MSSM, normalized to the data. Error bars are those from data. (Updated from ref [85] by U. Schwickerath).

As the standard model, the MSSM yields a good description of the precision data. A global fit [85] to all electroweak precision data, including the top mass measurement, shows that the χ^2 of the fit is slightly better than in the standard model; but, owing to the larger numbers of parameters, the probability is about the same as for the standard model (Figure 9).

The virtual presence of SUSY particles in the precision observables can be exploited also in the other way of constraining the allowed range of the MSSM parameters. Since the quality of the standard model description can be achieved only for those parameter sets where the standard model with a light Higgs boson is approximated, deviations from this scenario result in a rapid decrease of the fit quality. An analysis of the precision data in this spirit can be found in ref [101].

THE MASS OF THE LIGHTEST MSSM HIGGS BOSON

The search for the lightest Higgs boson provides a direct and very stringent test of SUSY. A precise prediction for the mass of the lightest Higgs boson in terms of the relevant SUSY parameters hence is crucial in order to determine the discovery and exclusion potential of LEP2 and the upgraded Tevatron and also for physics at the LHC, where a high-precision measurement of the mass of this particle might be possible.

In the MSSM, the mass m_h of the lightest Higgs boson is restricted at the tree level to be smaller than the Z-boson mass. This bound, however, is strongly affected by the inclusion of radiative corrections. The dominant one-loop corrections arise form the top and scalar-top sector via terms of the form $G_F m_t^4 \ln(m_{\tilde{t}_1} m_{\tilde{t}_2}/m_t^2)$ [102]. They increase the predicted values of m_h and yield an upper bound of about 150 GeV. These results have been improved by performing a complete one-loop calculation in the on-shell scheme, which takes into account the contributions of all sectors of the MSSM [103–105]. Beyond one-loop order renormalization group (RG) methods have been applied in order to obtain leading logarithmic higher-order contributions [106–109], and a diagrammatic calculation of the dominant two-loop contributions in the limiting case of vanishing $\tilde{t}$-mixing and infinitely large M_A and $\tan\beta$ has been carried out [110].

Recently a Feynman-diagrammatic calculation of the leading two-loop corrections of $\mathcal{O}(\alpha\alpha_s)$ to the masses of the neutral $\mathcal{CP}$-even Higgs bosons has been performed [111]. They have been combined with the complete one-loop diagrammatic calculation [112] to obtain in this way the currently most precise prediction for m_h within the Feynman-diagrammatic approach for arbitrary values of the parameters of the Higgs and scalar top sector of the MSSM. Further refinements concerning the leading two-loop Yukawa corrections of $\mathcal{O}(G_F^2 m_t^6)$ [107] and of leading QCD corrections beyond two-loop order are also included. The results will be discussed in this section.

In the MSSM two Higgs doublets are needed, decomposed as follows:

$$\begin{aligned} H_1 &= \begin{pmatrix} H_1^1 \\ H_1^2 \end{pmatrix} = \begin{pmatrix} v_1 + (\phi_1^0 + i\chi_1^0)/\sqrt{2} \\ \phi_1^- \end{pmatrix}, \\ H_2 &= \begin{pmatrix} H_2^1 \\ H_2^2 \end{pmatrix} = \begin{pmatrix} \phi_2^+ \\ v_2 + (\phi_2^0 + i\chi_2^0)/\sqrt{2} \end{pmatrix}. \end{aligned} \qquad (35)$$

The Higgs sector can be described with the help of two independent parameters (besides the gauge couplings g, g'), usually chosen as $\tan\beta = v_2/v_1$ and the mass M_A of the $\mathcal{CP}$-odd A boson. The $\mathcal{CP}$-even neutral mass eigenstates h^0, H^0 are obtained from $\phi_{1,2}^0$ by a rotation yielding the tree-level masses $m_{h,H,\text{tree}}$.

In the Feynman-diagrammatic approach the one-loop corrected Higgs masses are derived by finding the poles of the $h - H$-propagator matrix whose inverse is given by

$$\begin{pmatrix} q^2 - m^2_{H,\text{tree}} + \hat{\Sigma}_H(q^2) & \hat{\Sigma}_{hH}(q^2) \\ \hat{\Sigma}_{hH}(q^2) & q^2 - m^2_{h,\text{tree}} + \hat{\Sigma}_h(q^2) \end{pmatrix}, \tag{36}$$

where the $\hat{\Sigma}$ denote the full one-loop contributions to the renormalized Higgs-boson self-energies, which are taken from the complete one-loop on-shell calculation of [104]. The agreement with the result obtained in [103] is better than 1 GeV for almost the whole MSSM parameter space.

As mentioned above the dominant contribution arises from the $t-\tilde{t}$-sector. The current eigenstates of the scalar quarks, $\tilde{q}_L$ and $\tilde{q}_R$, mix to give the mass eigenstates $\tilde{q}_1$ and $\tilde{q}_2$. The non-diagonal entry in the scalar quark mass matrix is proportional to the mass of the quark and reads for the $\tilde{t}$-mass matrix

$$m_t M_t^{LR} = m_t(A_t - \mu \cot\beta\,). \tag{37}$$

Due to the large value of m_t mixing effects have to be taken into account. Diagonalizing the $\tilde{t}$-mass matrix one obtains the eigenvalues $m_{\tilde{t}_1}$ and $m_{\tilde{t}_2}$ and the $\tilde{t}$ mixing angle $\theta_{\tilde{t}}$.

The leading two-loop corrections have been obtained in [111] by calculating the $\mathcal{O}(\alpha\alpha_s)$ contribution of the $t-\tilde{t}$-sector to the renormalized Higgs-boson self-energies at zero external momentum from the Yukawa part of the theory. At the two-loop level the matrix (36) consists of the renormalized Higgs-boson self-energies

$$\hat{\Sigma}_s(q^2) = \hat{\Sigma}_s^{(1)}(q^2) + \hat{\Sigma}_s^{(2)}(0), \quad s = h, H, hH, \tag{38}$$

where the momentum dependence is neglected only in the two-loop contribution. The Higgs-boson masses at the two-loop level are obtained by determining the poles of the matrix Δ_{Higgs} in Eq. (36).

Two further steps of refinement have been implemented into the prediction for m_h, which are shown separately in the plots below. The leading two-loop Yukawa correction of $\mathcal{O}(G_F^2 m_t^6)$ is taken over from the result obtained by renormalization group methods [107]. The second step of refinement concerns leading QCD corrections beyond two-loop order, taken into account by using the $\overline{MS}$ top mass, $\overline{m}_t = \overline{m}_t(m_t) \approx 166.5$ GeV, for the two-loop contributions instead of the pole mass, $m_t = 175$ GeV. In the $\tilde{t}$ mass matrix, however, the pole mass is used as an input parameter. Only when performing the comparison with the RG results, m_t is replaced by $\overline{m}_t$ in the $\tilde{t}$ mass matrix for the two-loop result, since in the RG results the running masses appear everywhere. This three-loop effect gives rise to a shift up to 1.5 GeV in the prediction for m_h.

For the quantitative illustration, two representative values for $\tan\beta$ (favored by SUSY-GUT scenarios) are chosen: $\tan\beta = 1.6$ for the $SU(5)$ scenario and $\tan\beta = 40$ for the $SO(10)$ scenario; other parameters are $\alpha_s(m_t) = 0.1095$, and $m_t = 175$ GeV. For the figures below, furthermore $M = 400$ GeV ($M \equiv M_2$ is the soft SUSY breaking gaugino mass parameter), $M_A = 500$ GeV, and $m_{\tilde{g}} = 500$ GeV as typical values. The scalar top masses and the mixing angle are derived from

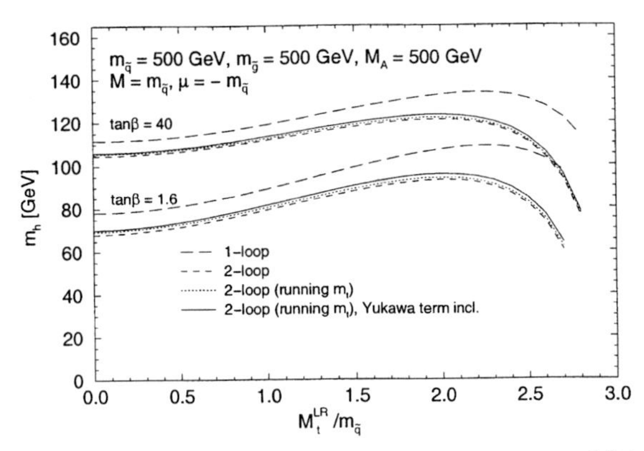

FIGURE 10. One- and two-loop results for m_h as a function of $M_t^{LR}/m_{\tilde{q}}$ for two values of $\tan\beta$.

the parameters $M_{\tilde{t}_L}$, $M_{\tilde{t}_R}$ and M_t^{LR} of the $\tilde{t}$ mass matrix. In the figures below, the assumption $m_{\tilde{q}} \equiv M_{\tilde{t}_L} = M_{\tilde{t}_R}$ is made.

The plot in Fig. 10 shows the result for m_h obtained from the diagrammatic calculation of the full one-loop and leading two-loop contributions. The two steps of refinement discussed above are shown in separate curves. For comparison the pure one-loop result is also given. The results are plotted as a function of $M_t^{LR}/m_{\tilde{q}}$, where $m_{\tilde{q}}$ is fixed to 500 GeV. The qualitative behavior is the same as in [111], where the result containing only the leading one-loop contribution (and without further refinements) was shown. The two-loop contributions give rise to a large reduction of the one-loop result of 10–20 GeV. The two steps of refinement both increase m_h by up to 2 GeV. A minimum occurs for $M_t^{LR} = 0$ GeV which we refer to as 'no mixing' (different from section 1). A maximum in the two-loop result for m_h is reached for about $M_t^{LR}/m_{\tilde{q}} \approx 2$ in the $\tan\beta = 1.6$ scenario as well as in the $\tan\beta = 40$ scenario. This case we refer to as 'maximal mixing' (differently from section 1). The maximum is shifted compared to its one-loop value of about $M_t^{LR}/m_{\tilde{q}} \approx 2.4$. The two steps of refinement have only a negligible effect on the location of the maximum.

The results of the diagrammatic calculation, including the higher-order terms discussed above, have been implemented into the FORTRAN code *FeynHiggs* [113]; it is available via its WWW-page
`http://www-itp.physik.uni-karlsruhe.de/feynhiggs`.

We now turn to the comparison of the diagrammatic results with the predictions obtained via renormalization group methods. We first compare the diagrammatic result for the no-mixing case, including the refinement terms, with the RG results obtained in [108]. After the inclusion of the refinement terms the diagrammatic result for the no-mixing case agrees very well with the RG result. The deviation

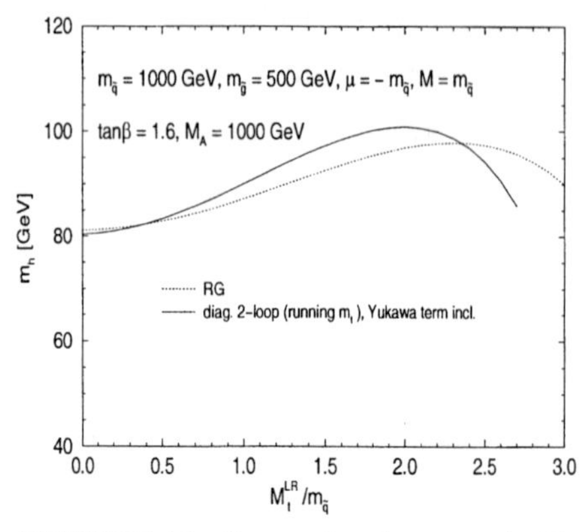

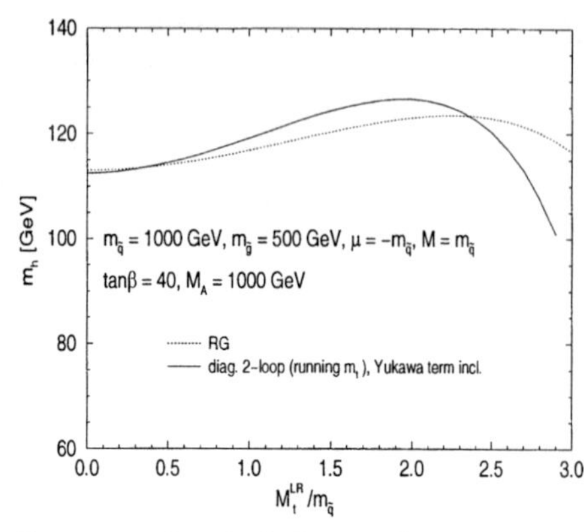

FIGURE 11. Comparison between the Feynman-diagrammatic calculations and the results obtained by renormalization group methods [108]. The mass of the lightest Higgs boson is shown for the two scenarios with $\tan\beta = 1.6$ and $\tan\beta = 40$ for increasing mixing in the $\tilde{t}$-sector and $m_{\tilde{q}} = M_A$.

between the results exceeds 2 GeV only for $\tan\beta = 1.6$ and $m_{\tilde{q}} < 150$ GeV. For smaller values of M_A the comparison for the no-mixing case looks qualitatively the same. For $\tan\beta = 1.6$ and values of M_A below 100 GeV slightly larger deviations are possible. Since the RG results do not contain the gluino mass as a parameter, varying $m_{\tilde{g}}$ gives rise to an extra deviation, which in the no-mixing case does not exceed 1 GeV. Varying the other parameters μ and M in general does not lead to a sizable effect in the comparison with the corresponding RG results.

Let us now consider the situation when mixing in the $\tilde{t}$ sector is taken into account. Comparing again the full one-loop result with the one-loop leading-log result used within the RG approach [109] yields good agreement for small mixing (only for values of M_A below 100 GeV and large mixing deviations of about 5 GeV occur). In Fig. 11 the diagrammatic result including the refinement terms is compared with the RG results [108] as a function of $M_t^{LR}/m_{\tilde{q}}$ for $\tan\beta = 1.6$ and $\tan\beta = 40$. For larger $\tilde{t}$-mixing, sizeable deviations between the diagrammatic and the RG results occur, which can exceed 5 GeV for moderate mixing and become very large for large values of $M_t^{LR}/m_{\tilde{q}}$. As already stressed above, the maximal value for m_h in the diagrammatic approach is reached for $M_t^{LR}/m_{\tilde{q}} \approx 2$, whereas the RG results have a maximum at $M_t^{LR}/m_{\tilde{q}} \approx 2.4$, i.e. at the one-loop value. Varying the value of $m_{\tilde{g}}$ in the Feynman-diagrammatic result leads to a larger effect than in the no-mixing case and shifts the diagrammatic result relative to the RG result within ± 2 GeV.

So far, the results of the diagrammatic on-shell calculation and the RG methods have been compared in terms of the parameters $M_{\tilde{t}_L}$, $M_{\tilde{t}_R}$ and M_t^{LR} of the $\tilde{t}$ mixing matrix, since the available numerical codes for the RG results [108,109] are given in terms of these parameters. However, since the two approaches rely on different renormalization schemes, the meaning of these (non-observable) parameters is not precisely the same in the two approaches starting from two-loop order. Indeed, assuming fixed values for the physical parameters $m_{\tilde{t}_1}$, $m_{\tilde{t}_2}$, $\theta_{\tilde{t}}$ and deriving the corresponding values of the parameters $M_{\tilde{t}_L}$, $M_{\tilde{t}_R}$, M_t^{LR} in the on-shell scheme as well as in the $\overline{MS}$ scheme, sizable differences occur between the values of the mixing parameter M_t^{LR} in the two schemes, while the parameters $M_{\tilde{t}_L}$, $M_{\tilde{t}_R}$ are approximately equal in the two schemes. Thus, part of the different shape of the curves in Fig. 11 may be attributed to a different meaning of the parameter M_t^{LR} in the on-shell scheme and in the RG calculation.

CONCLUSIONS

The experimental data for tests of the standard model have achieved an impressive accuracy. In the meantime, many theoretical contributions have become available to improve and stabilize the standard model predictions and to reach a theoretical accuracy clearly better than 0.1%.

The overall agreement between theory and experiment for the entire set of the precision observables is remarkable and instructively confirms the validity of the standard model. Fluctuations of data around the predictions are within two standard deviations, with no compelling evidence for deviations. Direct and indirect determinations of the top mass are compatible, and a light Higgs boson is clearly favoured by the analysis of precision data in the standard model context, which is far below the mass range where the standard Higgs sector becomes non-perturbative.

As a consequence of the high quality performance of the standard model, any kind of New Physics can only provoke small effects, at most of the size that is set by the radiative corrections. The MSSM, mainly theoretically advocated, is competitive to the standard model in describing the data with about the same quality in global fits. Since the MSSM predicts the existence of a light Higgs boson, the detection of a Higgs at LEP could be an indication of supersymmetry. The standard model can also accommodate such a light Higgs, but with the consequence that its validity cannot be extrapolated to energies much higher than the TeV scale.

ACKNOWLEDGEMENT

I want to express my gratitude to the organizers for the invitation, for the warm hospitality and the enjoyable atmosphere during the VIII Mexican School on Particles and Fields.

REFERENCES

1. The LEP Collaborations ALEPH, DELPHI, L3, OPAL, the LEP Electroweak Working Group and the SLD Heavy Flavor and Electroweak Groups, CERN-PPE/99-15 (1999)
2. D. Karlen, XXIX *International Conference on High Energy Physics*, Vancouver 1998 (to appear in the proceedings)
3. UA2 Collaboration, J. Alitti et al., *Phys. Lett.* B **276**, 354 (1992);
   CDF Collaboration, F. Abe et al., *Phys. Rev. Lett.* **65**, 2243 (1990); *Phys. Rev.* D **43**, 2070 (1991); *Phys. Rev. Lett.* **75**, 11 (1995); *Phys. Rev.* D **52**, 4784 (1995);
   D0 Collaboration, B. Abbott et al., *Phys. Rev. Lett.* **80**, 3008 (1998);
   P. Derwent, A. Kotwal, XXIX *International Conference on High Energy Physics*, Vancouver 1998 (to appear in the proceedings)
4. CDF Collaboration, F. Abe et al., *Phys. Rev. Lett.* **74**, 2626 (1995);
   D0 Collaboration, S. Abachi et al., *Phys. Rev. Lett.* **74**, 2632 (1995)
5. R. Partridge, XXIX *International Conference on High Energy Physics*, Vancouver 1998 (to appear in the proceedings)
6. G. 't Hooft, *Nucl. Phys.* B **33**, 173 (1971) and B **35**, 167 (1971)
7. G. Passarino, M. Veltman, *Nucl. Phys.* B **160**, 151 (1979)
8. M. Consoli, *Nucl. Phys.* B **160**, 208 (1979)
9. A. Sirlin, *Phys. Rev.* D **22**, 971 (1980);
   W. J. Marciano, A. Sirlin, *Phys. Rev.* D **22**, 2695 (1980);
   A. Sirlin, W. J. Marciano, *Nucl. Phys.* B **189**, 442 (1981)
10. D.Yu. Bardin, P.Ch. Christova, O.M. Fedorenko, *Nucl. Phys.* B **175**, 435 (1980) and B **197**, 1 (1982);
    D.Yu. Bardin, M.S. Bilenky, G.V. Mithselmakher, T. Riemann, M. Sachwitz, *Z. Phys.* C **44**, 493 (1989)
11. J. Fleischer, F. Jegerlehner, *Phys. Rev.* D **23**, 2001 (1981)
12. K.I. Aoki, Z. Hioki, R. Kawabe, M. Konuma,
    T. Muta, *Suppl. Prog. Theor. Phys.* **73**,1 (1982);
    Z. Hioki, *Phys. Rev. Lett.* **65**, 683 (1990), E: *ibidem* **65**, 1692 (1990); *Z. Phys.* C **49**, 287 (1991)
13. M. Consoli, S. LoPresti, L. Maiani, *Nucl. Phys.* B **223**, 474 (1983)
14. D.Yu. Bardin, M.S. Bilenky, G.V. Mithselmakher, T. Riemann, M. Sachwitz, *Z. Phys.* C **44**, 493 (1989)
15. M. Böhm, W. Hollik, H. Spiesberger, *Fortschr. Phys.* **34**, 687 (1986)
16. W. Hollik, *Fortschr. Phys.* **38**, 165 (1990)
17. M. Consoli, W. Hollik, F. Jegerlehner, in: *Z Physics at LEP 1*, eds. G. Altarelli, R. Kleiss and C. Verzegnassi, CERN 89-08 (1989)
18. G. Passarino, R. Pittau, *Phys. Lett.* B **228**, 89 (1989);
    V.A. Novikov, L.B. Okun, M.I. Vysotsky, *Nucl. Phys.* B **397**, 35 (1993)
19. G. Passarino, M. Veltman, *Phys. Lett.* B **237**, 537 (1990)
20. W.J. Marciano, A. Sirlin, *Phys. Rev. Lett.* **46**, 163 (1981);
    A. Sirlin, *Phys. Lett.* B **232**, 123 (1989)
21. G. Degrassi, S. Fanchiotti, A. Sirlin, *Nucl. Phys.* B **351**, 49 (1991)

22. G. Degrassi, A. Sirlin, *Nucl. Phys.* B **352**, 342 (1991)
23. M. Veltman, *Phys. Lett.* B **91**, 95 (1980);
    M. Green, M. Veltman, *Nucl. Phys.* B **169**, 137 (1980), E: *Nucl. Phys.* B **175**, 547 (1980);
    F. Antonelli, M. Consoli, G. Corbo, *Phys. Lett.* B **91**, 90 (1980);
    F. Antonelli, M. Consoli, G. Corbo, O. Pellegrino, *Nucl. Phys.* B **183**, 195 (1981)
24. D.C. Kennedy, B.W. Lynn, *Nucl. Phys.* B **322**, 1 (1989)
25. M. Kuroda, G. Moultaka, D. Schildknecht, *Nucl. Phys.* B **350**, 25 (1991)
26. *Reports of the Working Group on Precision Calculations for the Z Resonance*, CERN 95-03 (1995), eds. D. Bardin, W. Hollik, G. Passarino
27. G. Källén, A. Sabry, *K. Dan. Vidensk. Selsk. Mat.-Fys. Medd.* **29** (1955) No. 17
28. M. Steinhauser, *Phys. Lett.* B **429**, 158 (1998)
29. S. Eidelman, F. Jegerlehner, *Z. Phys.* C **67**, 585 (1995)
30. H. Burkhardt, B. Pietrzyk, *Phys. Lett.* B **356**, 398 (1995)
31. M.L. Swartz, *Phys. Rev.* D **53**, 5268 (1996)
32. R. Alemany, M. Davier, A. Höcker, *Eur. Phys. J.* C **2** (1998) 123
33. M. Davier, A. Höcker, *Phys. Lett.* B **419**, 419 (1998); hep-ph/9801361; *Phys. Lett.* B **435**, 427 (1998)
34. A. Höcker, XXIX *International Conference on High Energy Physics*, Vancouver 1998 (to appear in the proceedings)
35. J.H. Kühn, M. Steinhauser, *Phys. Lett.* B **437**, 425 (1998)
36. S. Groote, J. Körner, K. Schilcher, N.F. Nasrallah, *Phys. Lett.* B **440**, 375 (1998)
37. J. Erler, *Phys. Rev.* D **59** 054008 (1999)
38. A.H. Hoang, J.H. Kühn, T. Teubner, *Nucl. Phys.* B **452**, 173 (1995);
    K.G. Chetyrkin, J.H. Kühn, M. Steinhauser, *Phys. Lett.* B **371**, 93 (1996); *Nucl. Phys.* B **482**, 213 (1996); B **505**, 40 (1997);
    K.G. Chetyrkin, R. Harlander, J.H. Kühn, M. Steinhauser, *Nucl. Phys.* B **503**, 339 (1997)
39. A.D. Martin, D. Zeppenfeld, *Phys. Lett.* B **345**, 558 (1995)
40. E. Braaten, S. Narison, A. Pich, *Nucl. Phys.* B **373** (1992) 581
41. BES Collaboration, Z. Zhan, XXIX *International Conference on High Energy Physics*, Vancouver 1998 (to appear in the proceedings)
42. F. Jegerlehner, DESY 99-007, hep-ph/9901386
43. D. Ross, M. Veltman, *Nucl. Phys.* B **95**, 135 (1975)
44. M. Veltman, *Nucl. Phys.* B **123**, 89 (1977);
    M.S. Chanowitz, M.A. Furman, I. Hinchliffe, *Phys. Lett.* B **78**, 285 (1978)
45. J.J. van der Bij, F. Hoogeveen, *Nucl. Phys.* B **283**, 477 (1987)
46. R. Barbieri, M. Beccaria, P. Ciafaloni, G. Curci, A. Vicere, *Phys. Lett.* B **288**, 95 (1992); *Nucl. Phys.* B **409**, 105 (1993);
    J. Fleischer, F. Jegerlehner, O.V. Tarasov, *Phys. Lett.* B **319**, 249 (1993)
47. A. Djouadi, C. Verzegnassi, *Phys. Lett.* B **195**, 265 (1987)
48. L. Avdeev, J. Fleischer, S. M. Mikhailov, O. Tarasov, *Phys. Lett.* B **336**, 560 (1994); E: *Phys. Lett.* B **349**, 597 (1995);
    K.G. Chetyrkin, J.H. Kühn, M. Steinhauser, *Phys. Lett.* B **351**, 331 (1995)
49. M. Veltman, *Acta Phys. Polon.* B **8**, 475 (1977)

50. J.J. van der Bij, M. Veltman, *Nucl. Phys.* B **231**, 205 (1985)
51. R.E. Behrends, R.J. Finkelstein, A. Sirlin, *Phys. Rev.* **101**, 866 (1956);
T. Kinoshita, A. Sirlin, *Phys. Rev.* **113**, 1652 (1959)
52. T. van Ritbergen, R. Stuart, *Phys. Lett.* B **437**, 201 (1998), *Phys. Rev. Lett.* **82**, 488 (1999)
53. Particle Data Group, C. Caso et al., *Eur. Phys. J. C* **3**, 1 (1998)
54. W.J. Marciano, *Phys. Rev.* D **20**, 274 (1979)
55. M. Consoli, W. Hollik, F. Jegerlehner, *Phys. Lett.* B **227**, 167 (1989)
56. A. Djouadi, *Nuovo Cim.* A **100**, 357 (1988);
D. Yu. Bardin, A.V. Chizhov, Dubna preprint E2-89-525 (1989);
B.A. Kniehl, *Nucl. Phys.* B **347**, 86 (1990);
F. Halzen, B.A. Kniehl, *Nucl. Phys.* B **353**, 567 (1991) 567;
A. Djouadi, P. Gambino, *Phys. Rev.* D **49**, 3499 (1994)
57. B.A. Kniehl, J.H. Kühn, R.G. Stuart, *Phys. Lett.* B **214**, 621 (1988);
B.A. Kniehl, A. Sirlin, *Nucl. Phys.* B **371**, 141 (1992); *Phys. Rev.* D **47**, 883 (1993);
S. Fanchiotti, B.A. Kniehl, A. Sirlin, *Phys. Rev.* D **48**, 307 (1993)
58. K. Chetyrkin, J.H. Kühn, M. Steinhauser, *Phys. Rev. Lett.* **75**, 3394 (1995)
59. A. Sirlin, *Phys. Rev.* D **29**, 89 (1984)
60. G. Degrassi, P. Gambino, A. Vicini, *Phys. Lett.* B **383**, 219 (1996);
G. Degrassi, P. Gambino, A. Sirlin, *Phys. Lett.* B **394**, 188 (1997);
G. Degrassi, P. Gambino, M. Passera, A. Sirlin, *Phys. Lett.* B **418**, 209 (1998)
61. S. Bauberger, G. Weiglein, *Nucl. Instrum. Meth. A* **389**, 318 (1997); *Phys. Lett.* B **419**, 333 (1997)
62. G. Weiglein, *Acta Phys. Polon. B* **29**, 2735 (1998)
63. W. Hollik, B. Krause, A. Stremplat, G. Weiglein, to appear;
A. Stremplat, Diploma Thesis (Karlsruhe 1998)
64. D. Bardin et al., hep-ph/9709229, in: *Reports of the Working Group on Precision Calculations for the Z Resonance*, p. 7, CERN 95-03 (1995), eds. D. Bardin, W. Hollik, G. Passarino
65. G. Degrassi, P. Gambino, A. Sirlin, TUM-HEP-333/98 (to appear), and private communication
66. A.A. Akhundov, D.Yu. Bardin, T. Riemann, *Nucl. Phys.* B **276**, 1 (1986);
W. Beenakker, W. Hollik, *Z. Phys.* C **40**, 141 (1988);
J. Bernabeu, A. Pich, A. Santamaria, *Phys. Lett.* B **200**, 569 (1988)
67. A. Denner, W. Hollik, B. Lampe, *Z. Phys.* C **60**, 193 (1993)
68. K.G. Chetyrkin, A.L. Kataev, F.V. Tkachov, *Phys. Lett.* B **85**, 277 (1979);
M. Dine, J. Sapirstein, *Phys. Rev. Lett.* **43**, 668 (1979);
W. Celmaster, R. Gonsalves, *Phys. Rev. Lett.* **44**, 560 (1980);
S.G. Gorishny, A.L. Kataev, S.A. Larin, *Phys. Lett.* B **259**, 144 (1991);
L.R. Surguladze, M.A. Samuel, *Phys. Rev. Lett.* **66**, 560 (1991);
A. Kataev, *Phys. Lett.* B **287**, 209 (1992)
69. K.G. Chetyrkin, J.H. Kühn, *Phys. Lett.* B **248**, 359 (1990) and B **406**, 102 (1997);
K.G. Chetyrkin, J.H. Kühn, A. Kwiatkowski, *Phys. Lett.* B **282**, 221 (1992);
K.G. Chetyrkin, A. Kwiatkowski, *Phys. Lett.* B **305**, 285 (1993) and B **319**, 307 (1993)

70. B.A. Kniehl, J.H. Kühn, *Phys. Lett.* B **224**, 229 (1990); *Nucl. Phys.* B **329**, 547 (1990);
K.G. Chetyrkin, J.H. Kühn, *Phys. Lett.* B **307**, 127 (1993);
S. Larin, T. van Ritbergen, J.A.M. Vermaseren, *Phys. Lett.* B **320**, 159 (1994);
K.G. Chetyrkin, O.V. Tarasov, *Phys. Lett.* B **327**, 114 (1994)
71. K.G. Chetyrkin, J.H. Kühn, A. Kwiatkowski, in *Reports of the Working Group on Precision Calculations for the Z Resonance*, p. 175, CERN 95-03 (1995), eds. D. Bardin, W. Hollik, G. Passarino;
K.G. Chetyrkin, J.H. Kühn, A. Kwiatkowski, *Phys. Rep.* **277**, 189 (1996)
72. J. Fleischer, F. Jegerlehner, P. Rączka, O.V. Tarasov, *Phys. Lett.* B **293**, 437 (1992);
G. Buchalla, A.J. Buras, *Nucl. Phys.* B **398**, 285 (1993);
G. Degrassi, *Nucl. Phys.* B **407**, 271 (1993);
K.G. Chetyrkin, A. Kwiatkowski, M. Steinhauser, *Mod. Phys. Lett.* A **8**, 2785 (1993)
73. A. Kwiatkowski, M. Steinhauser, *Phys. Lett.* B **344**, 359 (1995);
S. Peris, A. Santamaria, *Nucl. Phys.* B **445**, 252 (1995)
74. R. Harlander, T. Seidensticker, M. Steinhauser, *Phys. Lett.* B **426**, 125 (1998)
75. A. Czarnecki, J.H. Kühn, *Phys. Rev. Lett.* **77**, 3955 (1996); E: **80**, 893 (1998)
76. A. Hoang, J.H. Kühn, T. Teubner, *Nucl. Phys.* B **455**, 3 (1995); **452**, 173 (1995)
77. SLD Collaboration, S. Fahey, XXIX *International Conference on High Energy Physics*, Vancouver 1998 (to appear in the proceedings)
78. P. Gambino, A. Sirlin, *Phys. Rev. Lett.* **73**, 621 (1994)
79. NuTev Collaboration, T. Bolton, XXIX *International Conference on High Energy Physics*, Vancouver 1998 (to appear in the proceedings)
80. D. Bardin et al., hep-ph/9412201
81. G. Montagna, O. Nicrosini, F. Piccinini, G. Passarino, hep-ph/9804211
82. D. Bardin, G. Passarino, hep-ph/9803425;
D. Bardin, M. Grünewald, G. Passarino, hep-ph/9902452
83. J. Erler, P. Langacker, hep-ph/9809352; hep-ph/9801422
84. K. Hagiwara, D. Haidt, S. Matsumoto, *Eur. Phys. J.* C **2**, 95 (1998);
J. Ellis, G.L. Fogli, E. Lisi, *Phys. Lett.* B **389**, 321 (1996); *Z. Phys.* C **69**, 627 (1996);
G. Passarino, *Acta Phys. Polon.* **28**, 635 (1997);
S. Dittmaier, D. Schildknecht, *Phys. Lett.* B **391**, 420 (1997);
S. Dittmaier, D. Schildknecht, G. Weiglein, *Phys. Lett.* B **386**, 247 (1996);
P. Chankowski, S. Pokorski, *Acta Phys. Polon.* **27**, 1719 (1996)
85. W. de Boer, A. Dabelstein, W. Hollik, W. Mösle, U. Schwickerath, *Z. Phys.* C **75**, 627 (1997)
86. D. Treille, XXIX *International Conference on High Energy Physics*, Vancouver 1998 (to appear in the proceedings)
87. L. Maiani, G. Parisi, R. Petronzio, *Nucl. Phys.* B **136**, 115 (1979);
N. Cabibbo, L. Maiani, G. Parisi, R. Petronzio, *Nucl. Phys.* B **158**, 259 (1979);
R. Dashen, H. Neuberger, *Phys. Rev. Lett.* **50**, 1897 (1983);
D.J.E. Callaway, *Nucl. Phys.* B **233**, 189;
M.A. Beg, C. Panagiotakopoulos, A. Sirlin, *Phys. Rev. Lett.* **52**, 883 (1984);

M. Lindner, *Z. Phys.* C **31**, 295 (1986)

88. M. Lindner, M. Sher, H. Zaglauer, *Phys. Lett.* B **228**, 139 (1989); G. Altarelli, G. Isidori, *Phys. Lett.* B **337**,141 (1994); J.A. Casas, J.R. Espinosa, M. Quiros, *Phys. Lett.* B **342**, 171 (1995) and B **382**, 374 (1996);
89. T. Hambye, K. Riesselmann, *Phys. Rev.* D **55**, 7255 (1997)
90. Kuti et al., *Phys. Rev. Lett.* **61**, 678 (1988); P. Hasenfratz et al., *Nucl. Phys.* B **317**, 81 (1989); M. Lüscher, P. Weisz, *Nucl. Phys.* B **318**, 705 (1989); M. Göckeler, H. Kastrup, T. Neuhaus, F. Zimmermann, *Nucl. Phys.* B **404**, 517 (1993)
91. M. Göckeler, H. Kastrup, J. Westphalen, F. Zimmermann, *Nucl. Phys.* B **425**, 413 (1994)
92. A. Ghinculov, *Nucl. Phys.* B **455**, 21 (1995); A. Frink, B. Kniehl, K. Riesselmann, *Phys. Rev.* D **54**, 4548 (1996)
93. T. Binoth, A. Ghinculov, J.J. van der Bij, *Phys. Rev.* D **57**, 1487 (1998); *Phys. Lett.* B **417**, 343 (1998)
94. L. Durand, B.A. Kniehl, K. Riesselmann, *Phys. Rev. Lett.* **72**, 2534 (1994); E: *ibidem* B **74**, 1699 (1995); A. Ghinculov, *Phys. Lett.* B **337**, 137 (1994); E: *ibidem* B **346**, 426 (1995); V. Borodulin, G. Jikia, *Phys. Lett.* B **391**, 434 (1997)
95. K. Riesselmann, hep-ph/9711456
96. A. Denner, R. Guth, W. Hollik, J.H. Kühn, *Z. Phys.* C **51**, 695 (1991); J. Rosiek, *Phys. Lett.* B **252**, 135 (1990; M. Boulware, D. Finnell, *Phys. Rev.* D **44**, 2054 (1991)
97. G. Altarelli, R. Barbieri, F. Caravaglios, *Phys. Lett.* B **314**, 357 (1993); C.S. Lee, B.Q. Hu, J.H. Yang, Z.Y. Fang, *J. Phys.* G **19**, 13 (1993); Q. Hu, J.M. Yang, C.S. Li, *Commun. Theor. Phys.* **20**, 213 (1993); J.D. Wells, C. Kolda, G.L. Kane, *Phys. Lett.* B **338**, 219 (1994); G.L. Kane, R.G. Stuart, J.D. Wells, *Phys. Lett.* B **354**, 350 (1995); M. Drees et al., *Phys. Rev.* D **54**, 5598 (1996)
98. P. Chankowski, A. Dabelstein, W. Hollik, W. Mösle, S. Pokorski, J. Rosiek, *Nucl. Phys.* B **417**, 101 (1994); D. Garcia, J. Solà, *Mod. Phys. Lett.* A **9**, 211 (1994)
99. D. Garcia, R. Jiménez, J. Solà, *Phys. Lett.* B **347**, 309 and 321 (1995); D. Garcia, J. Solà, *Phys. Lett.* B **357**, 349 (1995); A. Dabelstein, W. Hollik, W. Mösle, in *Perspectives for Electroweak Interactions in e^+e^- Collisions*, Ringberg Castle 1995, ed. B.A. Kniehl, World Scientific 1995 (p. 345); P. Chankowski, S. Pokorski, *Nucl. Phys.* B **475**, 3 (1996)
100. A. Djouadi, P. Gambino, S. Heinemeyer, W. Hollik, C. Jünger, G. Weiglein, *Phys. Rev. Lett.* **78**, 3626 (1997); *Phys. Rev.* D **57**, 4179 (1998)
101. J. Erler, D. Pierce, *Nucl. Phys.* B **526**, 53 (1998)
102. H. Haber, R. Hempfling, *Phys. Rev. Lett.* **66**, 1815 (1991); Y. Okada, M. Yamaguchi, T. Yanagida, *Prog. Theor. Phys.* **85**, 1 (1991);

J. Ellis, G. Ridolfi, F. Zwirner, *Phys. Lett.* B **257**, 83 (1991) and B **262**, 477 (1991); R. Barbieri, M. Frigeni, *Phys. Lett.* B **258**, 395 (1991)
103. P. Chankowski, S. Pokorski, J. Rosiek, *Nucl. Phys.* B **423**, 437 (1994)
104. A. Dabelstein, *Nucl. Phys.* B **456**, 25 (1995); *Z. Phys.* C **67**, 495 (1995)
105. J. Bagger, K. Matchev, D. Pierce, R. Zhang, *Nucl. Phys.* B **491**, 3 (1997)
106. J. Casas, J. Espinosa, M. Quirós, A. Riotto, *Nucl. Phys.* B **436**, 3 (1995), E: *ibid.* B **439**, 466 (1995)
107. M. Carena, J. Espinosa, M. Quirós, C. Wagner, *Phys. Lett.* B **355**, 209 (1995)
108. M. Carena, M. Quirós, C. Wagner, *Nucl. Phys.* B **461**, 407 (1996)
109. H. Haber, R. Hempfling, A. Hoang, *Z. Phys.* C **75**, 539 (1997)
110. R. Hempfling, A. Hoang, *Phys. Lett.* B **331**, 99 (1994)
111. S. Heinemeyer, W. Hollik, G. Weiglein, *Phys. Rev.* D **58**, 091701 (1998)
112. S. Heinemeyer, W. Hollik, G. Weiglein, *Phys. Lett.* B **440**, 296 (1998)
113. S. Heinemeyer, W. Hollik, G. Weiglein, hep-ph/9812320

Three Lectures on the Physics of Small x and High Gluon Density

Larry McLerran

Theoretical Physics Institute, University of Minnesota, Minneapolis, MN 55455

Abstract. In these lectures, I shall discuss small x physics and the consequences of the high gluon density which arises as x decreases. I argue that an understanding of this problem would lead to knowledge of the high energy asymptotics of hadronic processes. The high gluon density should allow a first principles computation of these asymptotics from QCD.

I LECTURE I: LOTS OF PROBLEMS

A Introduction

I think we all believe that QCD describes hadronic physics. It has been tested in a variety of environments. For high energy short distance phenomena, perturbative QCD computations successfully confront experiment. In lattice Monte-Carlo computations, one gets a successful semi-quantitative description of hadronic spectra, and perhaps in the not too distant future one will obtain precise quantitative agreement.

At present, however, all analytic computations and all precise QCD tests are limited to the small class of problems which correspond to short distance physics. Here there is some characteristic energy transfer scale E, and one uses asymptotic freedom,

$$\alpha_S(E) \rightarrow 0 \tag{1}$$

as $E \rightarrow \infty$

One question which we might ask is whether there are any non-perturbative "simple phenomena" which arise from QCD which are worthy of further effort. The questions I would ask before I would become interested in understanding such phenomena are

- Is the phenomenon simple and pervasive?
- Is it reasonably plausible that one can understand the phenomena from first principles, and compute how it would appear in nature?

CP490, *Particles and Fields: Eighth Mexican School,* edited by J. C. D'Olivo et al.

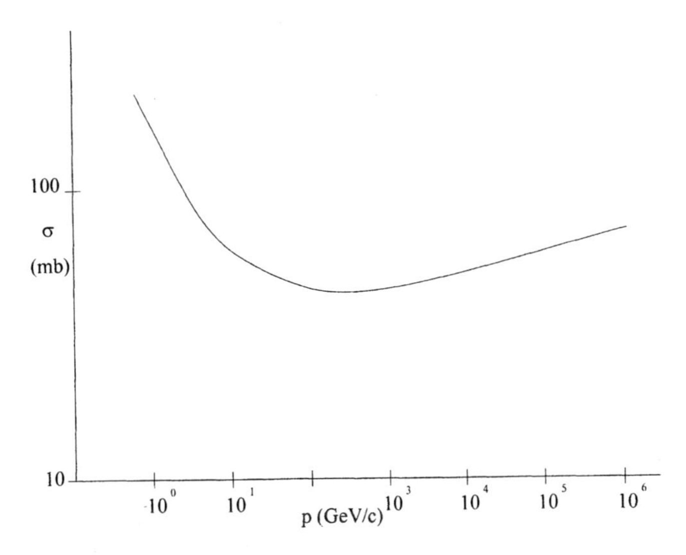

FIGURE 1. The total cross section for $p\bar{p}$ collisions

I will in this lecture try to explain a wide class of phenomena in QCD which are pervasive, and appear to follow simple patterns. I will then try to explain why I believe that these phenomena can be simply understood within QCD.

B Total Cross Sections at Asymptotic Energy

Computing total cross section as $E \to \infty$ is one of the great unsolved problems of QCD. Unlike for processes which are computed in perturbation theory, it is not required that any energy transfer become large as the total collision energy $E \to \infty$. Computing a total cross section for hadronic scattering therefore appears to be intrinsically non-perturbative. In the 60's and early 70's, Regge theory was extensively developed in an attempt to understand the total cross section. The results of this analysis were to my mind inconclusive, and certainly can not be claimed to be a first principles understanding from QCD.

The total cross section for $\bar{p}p$ collisions is shown in Fig. 1. Typically, it is assumed that the total cross section grows as $ln^2 E$ as $E \to \infty$. This is the so called Froisart bound which corresponds to the maximal growth allowed by unitarity of the S matrix. Is this correct? Is the coefficient of $ln^2 E$ universal for all hadronic precesses? Why is the unitarity limit saturated? Can we understand the total cross section from first principles in QCD? Is it understandable in weakly coupled QCD, or is it an intrinsically non-perturbative phenomenon?

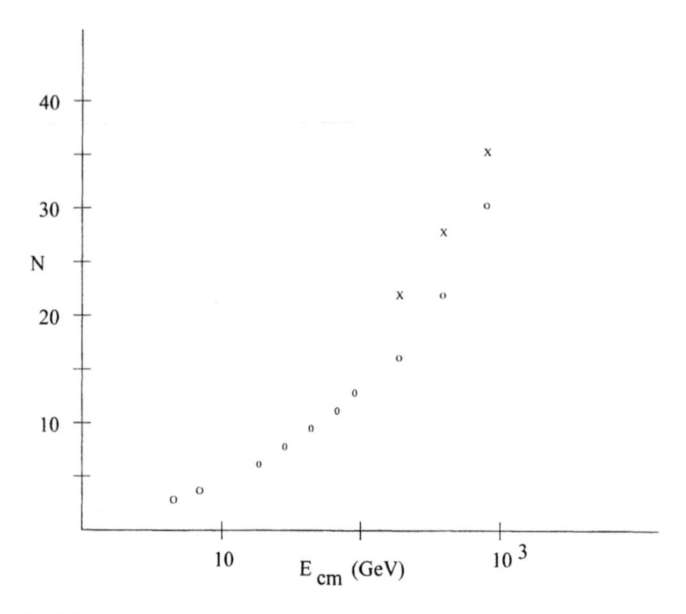

FIGURE 2. Multiplicity of produced particles in pp and $p\overline{p}$ collisions

C How Are Particle Produced in High Energy Collisions?

In Fig. 2, I plot the multiplicity of produced particles in pp and in $\overline{p}p$ collisions. The last three open circles correspond to the $\overline{p}p$ collisions with the multiplicity at zero energy subtracted. The remaining open circles to pp. The x's are $\overline{p}p$ collisions without the multiplicity at zero energy subtracted. Notice that the open circles fall on roughly the same curve. The implication is that whatever is causing the increase in multiplicity in these collisions may be from the same mechanism.

The obvious question is can we compute $N(E)$, the total multiplicity of produced particles as a function of energy?

At this point it is useful to develop some mathematical tools. I will introduce some useful kinematic variables: light cone coordinates. Let the light cone longitudinal momenta be

$$p^{\pm} = \frac{1}{\sqrt{2}}(E \pm p_z) \tag{2}$$

Note that the invariant dot product

$$p \cdot q = p_t \cdot q_t - p^+ q^- - p^- q^+ \tag{3}$$

and that

$$p^+ p^- = \frac{1}{2}(E^2 - p_z^2) = \frac{1}{2}(p_T^2 + m^2) = \frac{1}{2} m_T^2 \tag{4}$$

This equation defines the transverse mass m_T. (Please note that my metric is the negative of that conventionally used in particle physics. An unfortunate consequence of my education. Students, please feel free to convert everything to your favorite metric.)

Consider a collision in the center of mass frame as shown in Fig. 3. The right moving particle has $p_1^+ \sim \sqrt{2} \mid p_z \mid$ and $p_1^- \sim \frac{1}{2\sqrt{2}} m_T^2 / \mid p_z \mid$. For the colliding particles $m_T = m_{projectile}$, that is because the transverse momentum is zero, the transverse mass equals the particle mass For particle 2, we have $p_2^+ = p_1^-$ and $p_2^- = p_1^+$.

If we define the Feynman x of a produced pion as

$$x = p_\pi^+ / p_1^+ \tag{5}$$

then $0 \leq x \leq 1$. The rapidity of a pion is defined to be

$$y = \frac{1}{2} ln(p_\pi^+ / p_\pi^-) = \frac{1}{2} ln(2p^{+2} / m_T^2) \tag{6}$$

For pions, the transverse mass includes the transverse momentum of the pion.

The pion rapidity is always in the range $-y_{CM} \leq y \leq y_{CM}$ where $y_{CM} = ln(p^+ / m_{projectile})$ All the pions are produced in a distribution of rapidities within this range.

These definitions are useful, among other reasons, because of their simple properties under longitudinal Lorentz boosts: $p^\pm \to \kappa^{\pm 1} p^\pm$ where κ is a constant. Under boosts, the rapidity just changes by a constant. (Students, please check this relationship for momenta under boosts.)

A typical distribution of pions is shown in Fig. 4. It is convenient in the center of mass frame to think of the positive rapidity pions as somehow related to the right moving particle and the negative rapidity particles as related to the left moving particles. We define $x = p^+ / p^+_{projectile}$ and $x' = p^- / p^-_{projectile}$ and use x for positive rapidity pions and x' for negative rapidity pions.

Of course more than just pions are produced in high energy collisions. The variables we just presented easily generalize to these particles.

Several theoretical issues arise in multiparticle production. Can we compute dN/dy? or even dN/dy at $y = 0$? How does the average transverse momentum of produced particles $< p_T >$ behave with energy? What is the ratio of produced strange/nonstrange, and corresponding rations of charm, top, bottom etc at $y = 0$ as the center of mass energy approaches infinity?

Does multiparticle production as $E \to \infty$ at $y = 0$ become

- Simple?
- Understandable?
- Computable?

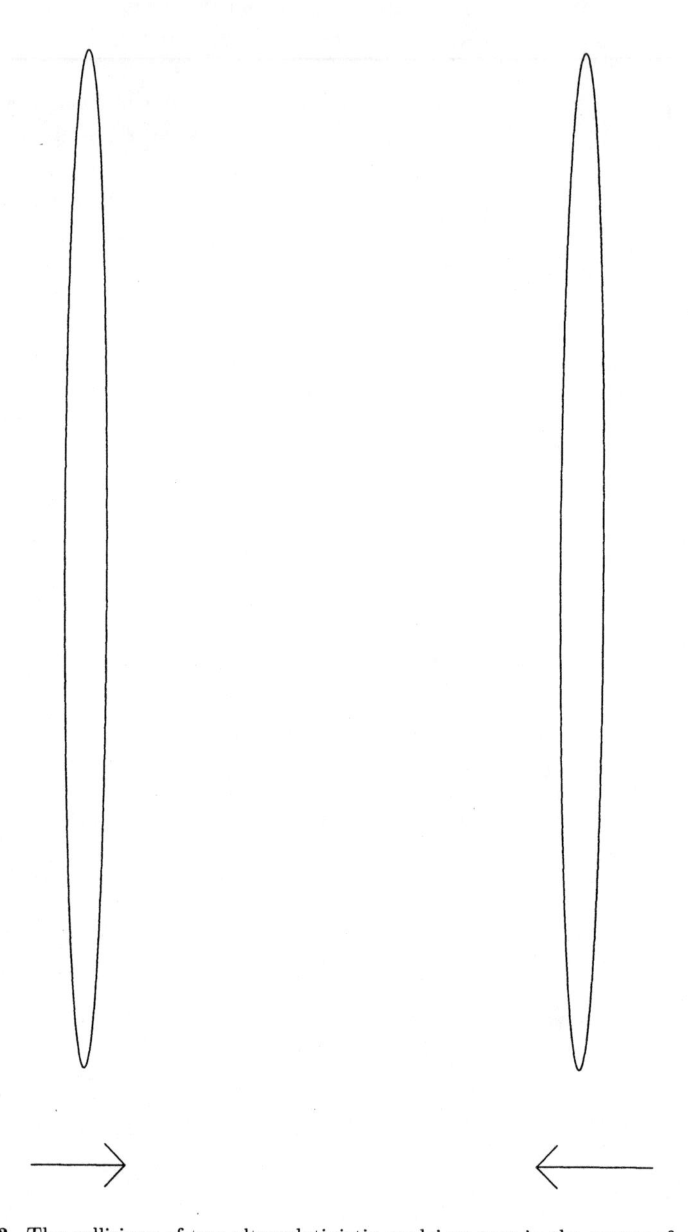

FIGURE 3. The collisions of two ultrarelativistic nuclei as seen in the center of mass frame

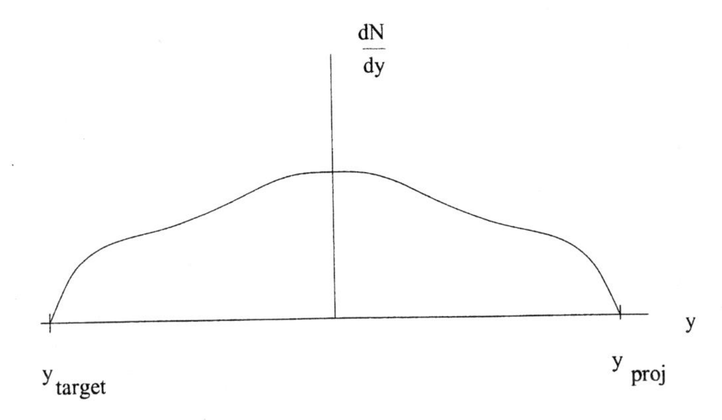

FIGURE 4. A typical pion rapidity distributions for hadronic collisions.

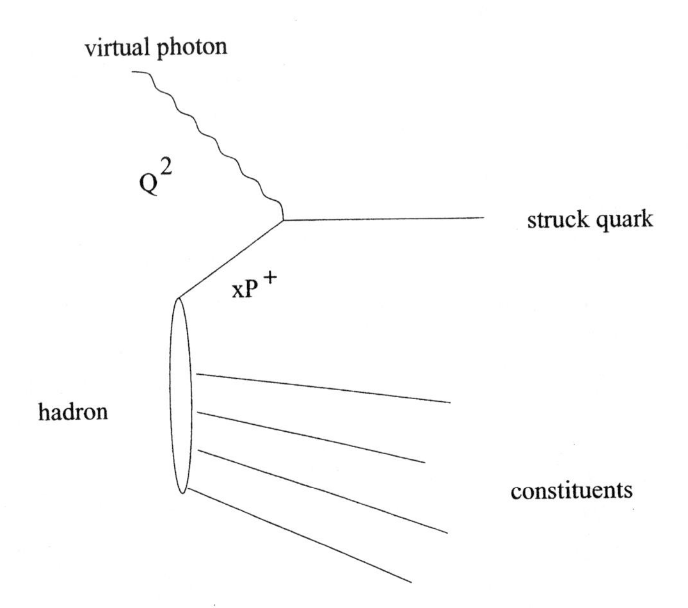

FIGURE 5. A cartoon of deep inelastic scattering.

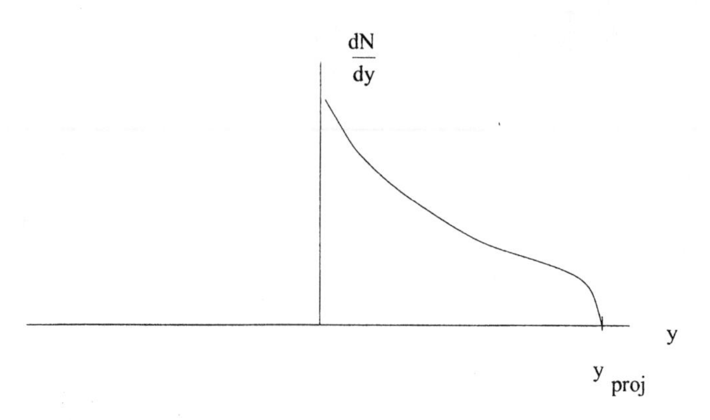

FIGURE 6. A rapidity distribution for gluons in the hadron wavefunction.

D Deep Inelastic Scattering

In Fig. 5, a cartoon of deep inelastic scattering is shown. Here an electron emits a virtual photon which scatters from a quark in a hadron. The momentum and energy transfer of the electron is measured, and the results of the break up are not. In these lectures, we cannot develop the theory of deep inelastic scattering. Suffice it to say, that this measurement is sufficient at large momenta transfer Q^2 to measure the distributions of quarks in a hadron.

To describe the quark distributions, it is convenient to work in a reference frame where the hadron has a large longitudinal momentum p^+_{hadron}. The corresponding light cone momentum of the constituent is $p^+_{constituent}$. We define $x = p^+_{constituent}/p^+_{hadron}$. (This x variable is equal to the Bjorken x variable, which can be defined in a frame independent way. In this frame independent definition, $x = Q^2/2p \cdot Q$ where p is the momentum of the hadronic target and Q is the momentum of the virtual photon. Students, please check that this is true.) The cross section which one extracts in deep inelastic scattering can be related to the distributions of quarks inside a hadron, dN/dx.

It is useful to think about the distributions as a function of rapidity. We define this for deep inelastic scattering as

$$y = y_{hadron} - ln(1/x) \tag{7}$$

and the invariant rapidity distribution as

$$dN/dy = xdN/dx \tag{8}$$

In Fig. 6, a typical dN/dy distribution for a constituent gluons of a hadron is shown. This plot is similar to the rapidity distribution of produced particles in deep inelastic scattering. The main difference is that we have only half of the plot, corresponding to the left moving hadron in a collision in the center of mass frame.

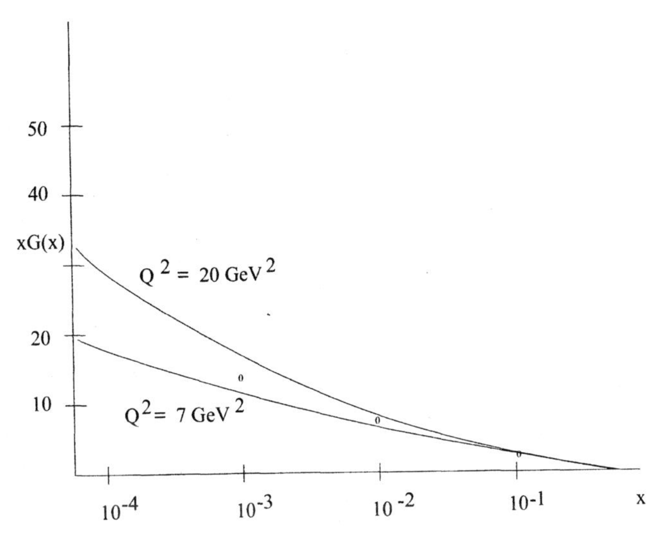

FIGURE 7. The Zeus data for the gluon structure functions. Error bars on the data are not shown, but are about 10 percent

We shall later argue that there is in fact a relationship between the structure functions as measured in deep inelastic scattering and the rapidity distributions for particle production. We will argue that the gluon distribution function is in fact proportional to the pion rapidity distribution.

The small x problem is that in experiments at Hera, the rapidity distribution function for quarks grows as the rapidity difference between the quark and the hadron grows. This growth appears to be more rapid than simply $\mid y_{proj} - y \mid$ or $(y_{proj} - y)^2$, and various theoretical models based on the original considerations of Lipatov and colleagues suggest it may grow as an exponential in $\mid y_{proj} - y \mid$. [1] If the rapidity distribution grew at most as y^2, then there would be no small x problem. We shall try to explain the reasons for this later in this lecture.

In Fig. 7, the Zeus data for the gluon structure function is shown. [2] I have plotted the structure function for $Q^2 = 7\ GeV^2$ and $20\ GeV^2$. The structure function depends upon the resolution of the probe, that is Q^2. Note the rise of $xg(x)$ at small x, this is the small x problem. I have also plotted the total multiplicity of produced particles in pp and $\overline{p}p$ collisions in the open circles on the same plot. Here I have used that $y = log(E_{cm}/1\ GeV)$ for the pion production data. This is approximately the maximal value of rapidity difference between centrally produced pions and the projectile rapidity. The total multiplicity has been rescaled so that at small x, it matches the gluon structure functions. (Strictly speaking, we should have plotted the total multiplicity at $y = 0$, but this is hard to extract from the data. If the distribution is an exponential in rapidity, then up to a constant these would be proportional.). Observe that the qualitative similarity between the gluon

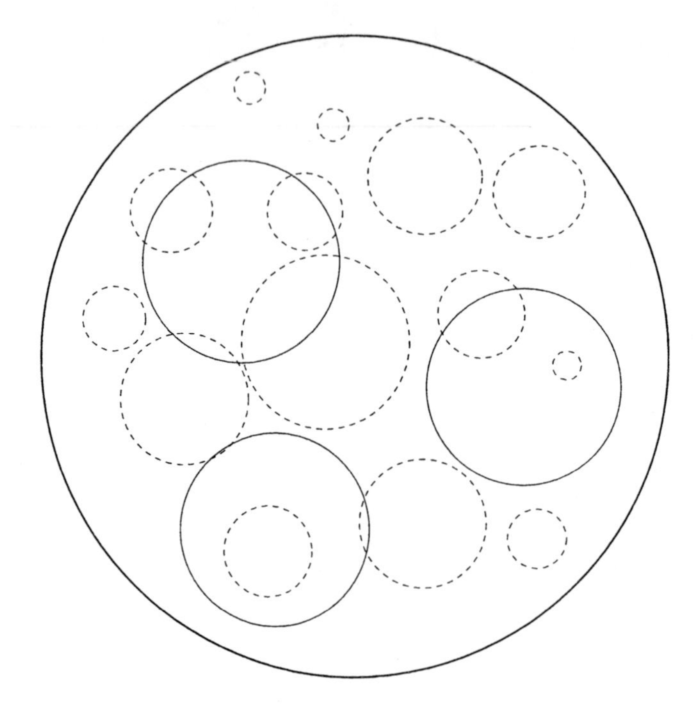

FIGURE 8. A picture of a hadron viewed head on.

structure function and the total multiplicity.

Why is the small x rise in the gluon distribution a problem? Consider Fig. 8, where we view hadron head on. [3] The constituents are the valence quarks, shown as solid circles, and the gluons and sea quarks shown as circles with dashed lines. As we add more and more constituents, the hadron becomes more and more crowded. If we were to try to measure these constituents with say an elementary photon probe, as we do in deep inelastic scattering, we might expect that the hadron would become so crowded that we could not ignore the shadowing effects of constituents as we make the measurement. (Shadowing means that some of the partons are obscured by virtue of having another parton in front of them. For hard spheres, for example, this would result in a decrease of the scattering cross section relative to what is expected from incoherent independent scattering.)

In fact, in deep inelastic scattering, we are measuring the cross section for a virtual photon γ^* and a hadron, $\sigma_{\gamma^* hadron}$. Making x smaller correspond to increasing the energy of the interaction (at fixed Q^2). An exponential growth in the rapidity corresponds to power law growth in $1/x$, which in turn implies power law growth with energy. This growth, if it continues forever, violates unitarity. The Froissart bound will allow at most $ln^2(1/x)$. (The Froissart bound is a limit on how rapidly a total cross section can rise. It follows from the unitarity of the scattering matrix.)

We shall later argue that in fact the distribution functions at fixed Q^2 do in fact saturate and cease growing so rapidly at high energy. The total number of gluons however demands a resolution scale, and we will see that the natural intrinsic scale is growing at smaller values of x, so that effectively, the total number of gluons within this intrinsic scale is always increasing. The quantity

$$\Lambda^2 = \frac{1}{\pi R^2} \frac{dN}{dy} \tag{9}$$

defines this intrinsic scale. Here πR^2 is the cross section for hadronic scattering from the hadron. For a nucleus, this is well defined. For a hadron, this is less certain, but certainly if the wavelngths of probes are small compared to R, this should be well defined. If

$$\Lambda^2 >> \Lambda_{QCD}^2 \tag{10}$$

as the Hera data suggests, then we are dealing with weakly coupled QCD since $\alpha_S(\Lambda) << 1$.

Even though QCD may be weakly coupled at small x, that does not mean the physics is perturbative. There are many examples of nonperturbative physics at weak coupling. An example is instantons in electroweak theory, which lead to the violation of baryon number. Another example is the atomic physics of highly charged nuclei. The electron propagates in the background of a strong nuclear Coulomb field, but on the other hand, the theory is weakly coupled and there is a systematic weak coupling expansion which allows for systematic computation of the properties of high Z (Z is the charge of the nucleus) atoms.

If the theory is local in rapidity, then the only parameter which can determine the physics at that rapidity is Λ^2. (Locality in rapidity means that there are not long range correlations in the hadronic wavefunction as a function of rapidity. In pion production, it is known that except for overall global conserved quantities such as energy and total charge, such correlations are of short range.) Note that if only Λ^2 determines the physics, then in an approximately scale invariant theory such as QCD, a typical transverse momentum of a constituent will also be of order Λ^2. If $\Lambda^2 >> 1/R^2$, where R is the radius of the hadron, then the finite size of the hadron becomes irrelevant. Therefore at small enough x, all hadrons become the same. The physics should only be controlled by Λ^2.

There should therefore be some equivalence between nuclei and say protons. When their Λ^2 values are the same, their physics should be the same. We can take an empirical parameterization of the gluon structure functions as

$$\frac{1}{\pi R^2} \frac{dN}{dy} \sim \frac{A^{1/3}}{x^\delta} \tag{11}$$

where $\delta \sim .2 - .3$. This suggests that there should be the following correspondences:

- RHIC with nuclei $\sim$ Hera with protons

ZEUS 1995

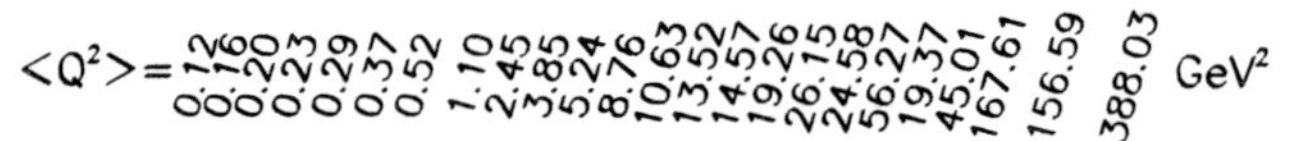

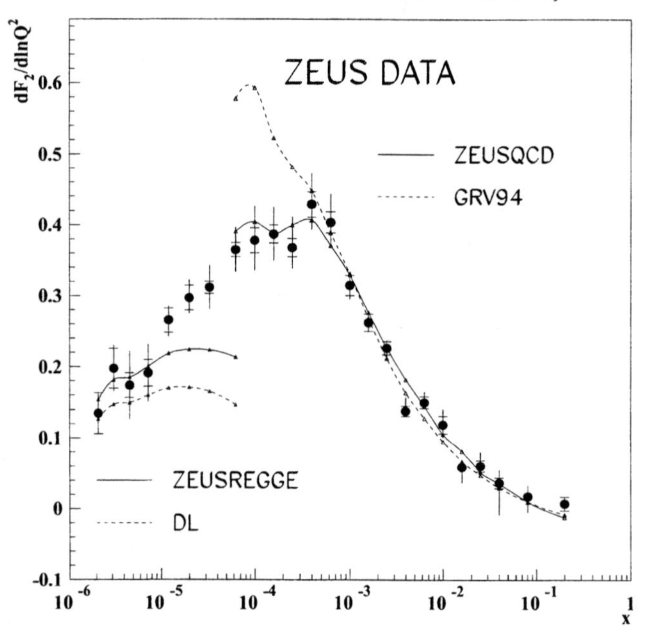

FIGURE 9. The Caldwell plot for the derivative of F_2

- LHC with nuclei $\sim$ Hera with nuclei

To get some rough idea of what the scales are which are important, consider the Caldwell plot of the Zeus data shown in Fig. 9. [2] The function $dF_2/dlnQ^2$, the derivative of the structure function F_2 is a function of both x and Q^2. The Caldwell plot takes a slice of this data in the Q^2, x plane. The important thing to observe is the qualitative change in the behavior of the function at $Q^2 \sim 3\ GeV^2$ at an $x \sim 10^{-4}$. Until we get to this value, the curve is adequately parameterized using DGLAP evolution equations and GRV parameterization of the gluon distribution functions. It is suggestive that the turnover has something to do with the physics of high gluon density. If so the typical scale associated with gluon momenta is rather large.

Since the physics of high gluon density is weak coupling we have the hope that we might be able to do a first principle calculation of

- the gluon distribution function
- the quark and heavy quark distribution functions
- the intrinsic p_T distributions quarks and gluons

We can also suggest a simple escape from unitarity arguments which suggest that the gluon distribution function must not grow at arbitrarily small x. The point is that at smaller x, we have larger Λ and correspondingly larger p_T. A typical parton added to the hadron has a size of order $1/p_T$. Therefore although we are increasing the number of gluons, we do it by adding in more gluons of smaller and smaller size. A probes of size resolution $\Delta x \geq 1/p_T$ at fixed Q will not see partons smaller than this resolution size. They therefore do not contribute to the fixed Q^2 cross section, and there is no contradiction with unitarity.

E Heavy Ion Collisions

In Fig. 10, the standard lightcone cartoon of heavy ion collisions is shown. [4] To understand the figure, imagine we have two Lorentz contracted nuclei approaching one another at the speed of light. Since they are well localized, they can be thought of as sitting at $x^\pm = 0$, that is along the light cone, for $t < 0$ At $x^\pm = 0$, the nuclei collide. To analyze this problem for $t \geq 0$, it is convenient to introduce a time variable which is Lorentz covariant under longitudinal boosts

$$\tau = \sqrt{t^2 - z^2} \tag{12}$$

and a space-time rapidity variable

$$\eta = \frac{1}{2} ln \left(\frac{t-z}{t+z} \right) \tag{13}$$

For free streaming particles

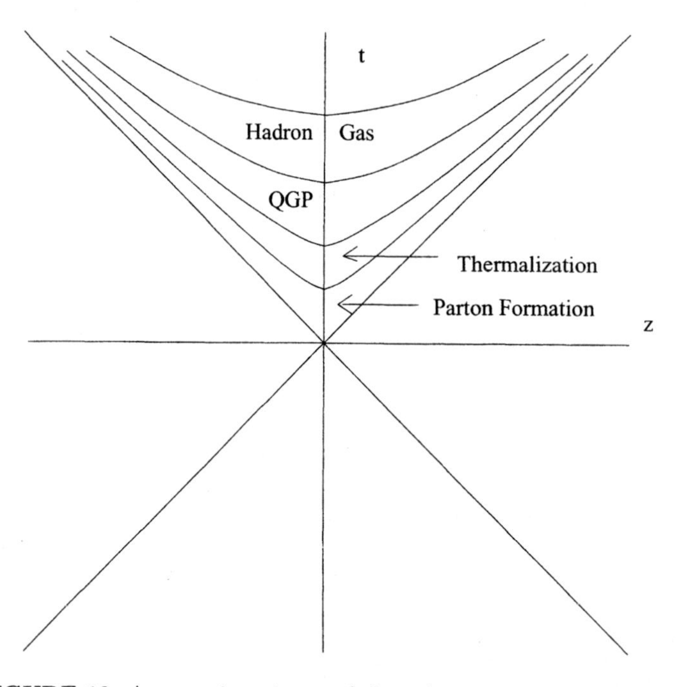

FIGURE 10. A space-time picture of ultrarelativistic nuclear collisions.

$$z = vt = \frac{p_z}{E} t \tag{14}$$

we see that the space-time rapidity equals the momentum space rapidity

$$\eta = y \tag{15}$$

If we have distributions of particles which are slowly varying in rapidity, it should be a good approximation to take the distributions to be rapidity invariant. This should be valid at very high energies in the central region. By the correspondence between space-time and momentum space rapidity, it is plausible therefore to assume that distributions are independent of η. Therefore distributions are the same on lines of constant τ, which is as shown in Fig. 10. At $z = 0$, $\tau = t$, so that τ is a longitudinally Lorentz invariant time variable.

We expect that at very late times, we have a free streaming gas of hadrons. These are the hadrons which eventually arrive at our detector. At some earlier time, these particle decouple from a dense gas of strongly interacting hadrons. As we proceed earlier in time, at some time there is a transition between a gas of hadrons and a plasma of quarks and gluons. This may be through a first order phase transition where the system might exist in a mixed phase for some length of time, or perhaps there is a continuous change in the properties of the system

At some earlier time, the quarks and gluons of the quark-gluon plasma are formed. This is at some time of the order of a Fermi, perhaps as small as .1 *Fermi*. As they form, the particles scatter from one another, and this can be described using the methods of transport theory. At some later time they have thermalized, and the system can be approximately described using the methods of perfect fluid hydrodynamics.

In the time between that for which the quarks and gluons have been formed and $\tau = 0$, the particles are being formed. This is where the initial conditions are made.

In various levels of sophistication, one can compute the properties of matter made in heavy ion collisions at times later than the formation time. The problems are understood in principle for $\tau \geq \tau_{formation}$ if perhaps not in fact. Very little is known about the initial conditions.

In principal, understanding the initial conditions should be the simplest part of the problem. At the initial time, the degrees of freedom are most energetic and therefore one has the best chance to understand them using weak coupling methods in QCD.

There are two separate classes of problems one has to understand for the initial conditions. First the two nuclei which are colliding are in single quantum mechanical states. Therefore for some early time, the degrees of freedom must be quantum mechanical. This means that

$$\Delta z \Delta p_z \geq 1 \tag{16}$$

Therefore classical transport theory cannot describe the particle down to $\tau = 0$ since classical transport theory assumes we know a distribution function $f(\vec{p}, \vec{x}, t)$, which is a simultaneous function of momenta and coordinates. This can also be understood as a consequence of entropy. An initial quantum state has zero entropy. Once one describes things by classical distribution functions, entropy has been produced. Where did it come from?

Another problem which must be understood is classical charge coherence. At very early time, we have a tremendously large number of particles packed into a longitudinal size scale of less than a fermi. This is due to the Lorentz contraction of the nuclei. We know that the particles cannot interact incoherently. For example, if we measure the field due to two opposite charge at a distance scale r large compared to their separation, we know the field fall as $1/r^2$, not $1/r$. On the other hand, in cascade theory, interactions are taken into account by cross sections which involve matrix elements squared. There is no room for classical charge coherence.

There are a whole variety of problems one can address in heavy ion collisions such

- What is the equation of state of strongly interacting matter?
- Is there a first order QCD phase transition?

These issues and others would take us beyond the scope of these lectures. The issues which I would like to address are related to the determination of the initial conditions, a problem which can hopefully be addressed using weak coupling methods in QCD.

F Universality

There are two separate formulations of universality which are important in understanding small x physics.

The first is a weak universality. This is the statement that physics should only depend upon the variable

$$\Lambda^2 = \frac{1}{\pi R^2}\frac{dN}{dy} \tag{17}$$

As discussed above, this universality has immediate experimental consequences which can be directly tested.

The second is a strong universality which is meant in a statistical mechanical sense. At first sight it appear a formal idea with little relation to experiment. If it is however true, its consequences are very powerful and far reaching. What we shall mean by strong universality is that the effective action which describes small x distribution function is critical and at a fixed point of some renormalization group. This means that the behavior of correlation functions is given by universal critical exponents, and these universal critical exponents depend only on general properties of the theory such as the symmetries and dimensionality.

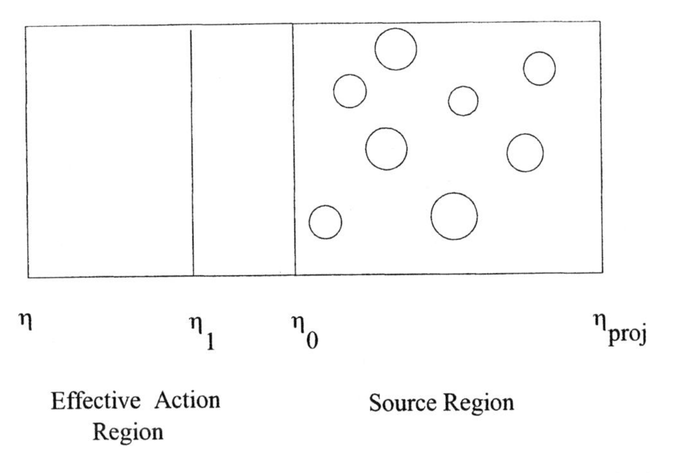

FIGURE 11. The space-time distribution function for glue inside of a hadron.

Since the correlation functions determine the physics, this statement says that the physics is not determined by the details of the interactions, only by very general properties of the underlying theory!

We can see how a renormalization group arises. In Fig 11, the space-time distribution of gluons is shown. The coordinate space rapidity is used. The effective action which we shall develop in later lectures is valid only for gluons with rapidity less than η_0. [5]- [8] Those at larger rapidity have been integrated out of the theory and appear only as recoilless sources of color charge for $\eta_0 \leq \eta \leq y_{proj}$.

The way a renormalization group is generated is by integrating out the gluon degrees of freedom in the range $\eta_1 \leq \eta \leq \eta_o$, to generate a new effective action for rapidity $\eta \leq \eta_1$. We can show that this results in new source strength of color charge, now in the range $\eta_1 \leq \eta \leq y_{frag}$, and in a modification of some of the coefficients of the effective action.

At high Q^2, the renormalization group analysis simplifies, and one can show that in various limits reduces to the BFKL or DGLAP analysis. The renormalization group equations at smaller q^2 become more complicated, and have yet to be written in explicit form and evaluated. This is in principle possible to do. [9]

An essential ingredient in this analysis is the appearance of an action and of a classical gluon field. We should expect the appearance of a classical gluon field when the phase space density of gluons becomes high. Gluons are after all bosons, and when the phase space density is large they should be described classically. This provides the essential difference between older renormalization group analysis which was formulated in terms of a distribution function. High density and its complications due to coherence require the introduction of a field. The new renormalization group analysis is phrased in terms of the effective action for the classical gluon field and jumps from an ordinary equation to a functional equation.

The classical gluon fields which we shall find are the non-abelian generalization of the Lienard-Wiechart potentials of electrodynamics. If we use the coordinate space variable x^- and realize that the source for the gluons arise from much smaller x^- than that at which we make the measurements, since they arise from gluons of much higher longitudinal momenta which are more Lorentz contracted, then the sources can be imagined as arising from a $\delta(x^-)$. The Lienard-Wiechart potentials also are proportional to $\delta(x^-)$, and so exist only in the sheet. The fields are also transversely polarized

$$B^i_a \perp E^i_a \perp \hat{z} \tag{18}$$

G Why an Effective Action?

The effective action formalism which will be advocated in the next lectures is very powerful. It is used to compute a gluon effective field. This field can be related to the wavefunction of the hadron.

This field allows one to generalize it from the original description of a single hadron, to collisions of hadrons and also to diffractive processes. We shall see that the effective action formulation is incredibly powerful.

The careful student might at this point be very worried: How has the problem been in any way simplified? We have just introduced a source, and to specify the field one has to specify the source. Moreover, such a specification is gauge dependent. What happens is amusing: We integrate over the all color orientations of the source. Gauge invariance is restored by the integration. The theory becomes specified by the local density of color charge squared. This is a gauge invariant quantity. It can be related to the gluon distribution functions, and is determined by the renormalization group equations.

II AN INTRODUCTION TO LIGHT CONE PHYSICS

This lecture will provide an introduction to light cone kinematics and quantization of field theory on the light cone. We will eventually use light cone methods to quantize QCD, using the light cone gauge.

Light cone coordinates are

$$x^\pm = \frac{1}{\sqrt{2}}(x^0 \pm x^3) \tag{19}$$

and momenta

$$p^\pm = \frac{1}{\sqrt{2}}(p^0 \pm p^3) \tag{20}$$

The invariant dot product is

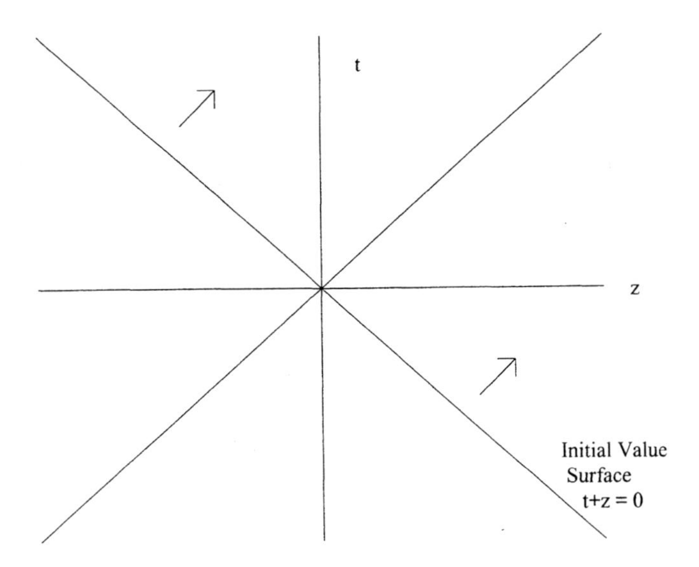

FIGURE 12. The initial data problem on the light cone.

$$p \cdot x = p_t \cdot x_t - p^+ x^- - p^- x^+ \tag{21}$$

where p_t and x_t are transverse coordinates. This implies that in this basis the metric is $g^{+-} = g^{-+} = -1$, $g^{ij} = \delta^{ij}$ where i, j refer to transverse coordinates. All other elements of the metric vanish.

An advantage of light cone coordinates is that if we do a Lorentz boost along the longitudinal direction with Lorentz gamma factor $\gamma = cosh(y)$ then $p^\pm \rightarrow e^{\pm y} p^\pm$

If we let x^+ be a time variable, we see that the variable p^- is to be interpreted as an energy. Therefore when we have a field theory, the component of the momentum operator P^- will be interpreted as the Hamiltonian. The remaining variables are to be thought of as momenta and spatial coordinates. In Fig. 12, there is a plot of the z, t plane. The line $x^+ = 0$ provides a surface where initial data might be specified. Time evolution is in the direction normal to this surface.

We see that an elementary wave equation

$$(p^2 + M^2)\phi = 0 \tag{22}$$

is particularly simple in light cone gauge. Since $p^2 = p_t^2 - 2p^+ p^-$ this equation is of the form

$$p^- \phi = \frac{p_t^2 + M^2}{2p^+} \phi \tag{23}$$

is first order in time. In light cone coordinates, the dynamics looks similar to that of the Schrodinger equation. The initial data to be specified is only the value of the field on the initial surface.

In the conventional treatment of the Klein-Gordon field, one must specify the field and its first derivative (the momentum) on the initial surface. In the light cone coordinate, the field is sufficient and the field momentum is redundant. This means that the field momentum will not commute with the field on the initial time surface!

Lets us work all this out with the example of the Klein Gordon field. The action for this theory is

$$S = -\int d^4x \left\{ \frac{1}{2}(\partial\phi)^2 + \frac{1}{2}M^2\phi^2 \right\} \tag{24}$$

The field momentum is

$$\Pi(x_t, x^-) = \frac{\delta S}{\delta \partial_+ \phi} = \partial_- \phi = \frac{\partial}{\partial x^-}\phi \tag{25}$$

Note that Π is a derivative of ϕ on the initial time surface. It is therefore not an independent variable, as would be the case in the standard canonical quantization of the scalar field.

We postulate the equal time commutation relation

$$[\Pi(x_t, x^-), \phi(y_t, y^-)] = -i\delta^{(3)}(x-y) \tag{26}$$

Here we time is $x^+ = y^+ = 0$ in both the the field and field momentum. We see therefore that

$$\partial_-[\phi(\vec{x}), \phi(\vec{y})] = -i\delta^{(3)}(x-y) \tag{27}$$

or

$$[\phi(x), \phi(y)] = -i\epsilon(x^- - y^-)\delta^{(2)}(x-y) \tag{28}$$

Here $\epsilon(v)$ is 1/2 for $v > 0$ and $-1/2$ for $v < 0$.

These commutation relations may be realized by the field

$$\begin{aligned} \phi(x) &= \int \frac{d^3p}{(2\pi)^3 2p^+} e^{ipx} a(p) \\ &= \int_{p^+>0} \frac{d^3p}{(2\pi)^3 2p^+} \left\{ e^{ipx} a(p) + e^{-ipx} a^\dagger(p) \right\} \end{aligned} \tag{29}$$

Using

$$[a(p).a^\dagger(q)] = 2p^+(2\pi)^3\delta^{(3)}(p-q) \tag{30}$$

the student can verify that the equal time commutation relations for the field are satisfied.

The quantity $1/p^+$ in the expression for the field in terms of creation and annihilation operators is singular when $p^+ = 0$. When we use a principle value prescription, we reproduce the form of the commutation relations postulated above with the factor of $\epsilon(x^- - y^-)$. Different prescriptions correspond to different choices for the inversion of $\frac{1}{\partial^-}$. One possible prescription is the Liebrandt-Mandelstam prescription $1/p^+ = p^-/(p^+p^- + i\epsilon)$. This prescription has some advantages relative to the principle value prescription in that it maintains causality at intermediate stages of computations and the principle value prescription does not. In the end, for physical quantities, the choice of prescription cannot result in different results. Of course, in some schemes the computations may become prohibitively difficult.

The student should now check that with the field above, the light cone Hamiltonian is

$$P^- = \int_{p^+>0} \frac{d^3p}{(2\pi)^3 2p^+} \frac{p_t^2 + M^2}{2p^+} a^\dagger(p) a(p) \tag{31}$$

as it must be.

In a general interacting theory, the Hamiltonian will of course be more complicated. The representation for the fields in terms of creation and annihilation operators will be the same as above. Note that all particles created by a creation operator have positive P^+. Therefore, since the vacuum has $P^+ = 0$, there can be no particle content to the vacuum. It is a trivial state. Of course this must be wrong since the physical vacuum must contain condensates such as the one responsible for chiral symmetry restoration. It can be shown that such non-perturbative condensates arise in the $P^+ = 0$ modes of the theory. We have not been careful in treating such modes. For perturbation theory, presumably to all orders, the above treatment is sufficient for our purposes.

A Light Cone Gauge QCD

In QCD we have a vector field A^μ_a. This can be decomposed into longitudinal and transverse parts as

$$A^\pm_a = \frac{1}{\sqrt{2}}(A^0_a \pm A^z) \tag{32}$$

and the transverse as lying in the tow dimensional plane orthogonal to the beam z axis. Light cone gauge is

$$A^+_a = 0 \tag{33}$$

In this gauge, the equation of motion

$$D_\mu F^{\mu\nu} = 0 \tag{34}$$

is for the + component

$$D_i F^{i+} - D^+ F^{-+} = 0 \tag{35}$$

which allows one to compute A^- in terms of A^i as

$$A^- = \frac{1}{\partial^{+2}} D^i \partial^+ A^i \tag{36}$$

This equation says that we can express the longitudinal field entirely in terms of the transverse degrees of freedom which are specified by the transverse fields entirely and explicitly. These degrees of freedom correspond to the two polarization states of the gluons.

We therefore have

$$A^i_a(x) = \int_{p^+>0} \frac{d^3p}{(2\pi)^3 2p^+} \left(e^{ipx} a^i_a(p) + e^{-ipx} a^{i\dagger}_a(p) \right) \tag{37}$$

where

$$[a^i_a(p), a^{j\dagger}_b(q)] = 2p^+ \delta_{ab} \delta^{ij} (2\pi)^3 \delta^{(3)}(p-q) \tag{38}$$

where the commutator is at equal light cone time x^+.

B Distribution Functions

We would like to explore some hadronic properties using light cone field operators. For example, suppose we have a hadron and ask what is the gluon content of that hadron. Then we would compute

$$\frac{dN_{gluon}}{d^3p} =< h \mid a^\dagger(p) a(p) \mid h > \tag{39}$$

The quark distribution for quarks of flavor i (for the sum of quarks and antiquarks) would be given in terms of creation and annihilation operators for quarks as

$$\frac{dN_i}{d^3p} =< h \mid \{ b^\dagger_i(p) b_i(p) + d^\dagger_i(p) d_i(p) \} \mid h > \tag{40}$$

where b corresponds to quarks and d to antiquarks. The creation and annihilation operators for quarks and gluons can be related to the quark coordinate space field operators by techniques similar to those above. The interested student should read the notes of Venugopalan for details. [10]

How would we begin computing such distribution functions? We will start with the example of a large nucleus, as this makes some issues conceptually simpler. We will then generalize to hadrons, where we shall see that the ideas presented here have a generalization.

For a large nucleus, we assume that the gluon distribution which we shall try to compute has longitudinal momentum soft compared to that of the valence quarks.

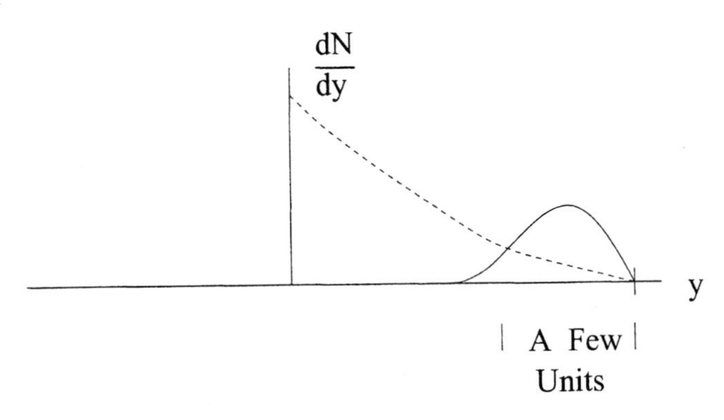

FIGURE 13. The rapidity distribution for quarks and gluons inside of a hadron.

Valence quarks have longitudinal momentum of the order of that of the nucleus, so this requirement is only that $x << 1$ In fact, we will require that $x << A^{-1/3}$. This is the requirement that in the frame where the longitudinal momentum of the gluons is zero, the nucleus has a Lorentz contracted size much less than the wavelength associated with the gluons transverse momentum $\lambda \sim 1/p_T$. This is the requirement that the gluon resolve the nucleus as a whole and is insensitive to the details of the nuclear structure (spatial distribution of valence quarks within the nucleus).

In Fig. 13, a rapidity distribution for gluons and for valence quarks is shown. We are interested in the region where the overlap between the quark and gluon rapidity distribution is small.

If the gluon phase space density is very large, the quantum gluons of the nucleus may be treated as classical fields. This may be true if

$$\Lambda^2 = \frac{1}{\pi R^2} \frac{dN_{glue}}{dy} \tag{41}$$

satisfies $\Lambda^2 >> \Lambda^2_{QCD}$. Certainly if the typical p_T of the gluons was of order Λ_{QCD} this would be true since the gluons would then be closely packed together. We shall see that this is true in fact for gluons with $p_T << \Lambda$ which for high gluons density can become very large.

If the valence quarks have a longitudinal momentum much larger than the typical gluon momentum then their typical interactions with these gluons should be characterized by the soft momentum scale (otherwise the soft gluons would not remain soft). In an emission of a gluon, the emitted gluon has momentum very small compared to the valence quark longitudinal momentum. Its velocity therefore is barely changed by this emission. The quarks are therefore recoilless sources of color charge for the gluon classical field.

The picture we have is therefore that of Fig. 14. The valence quarks are sources for the gluons and sit on a sheet of thickness infinitesimal compared to the typical

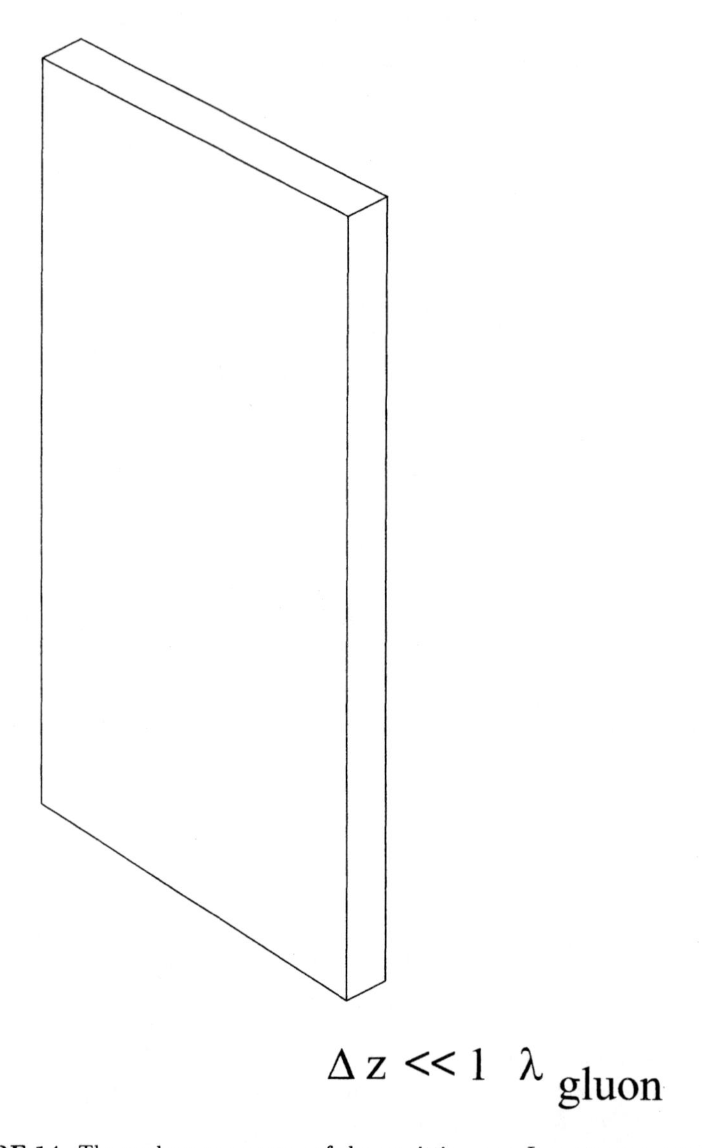

FIGURE 14. The nucleus as a source of charge sitting on a Lorentz contracted sheet.

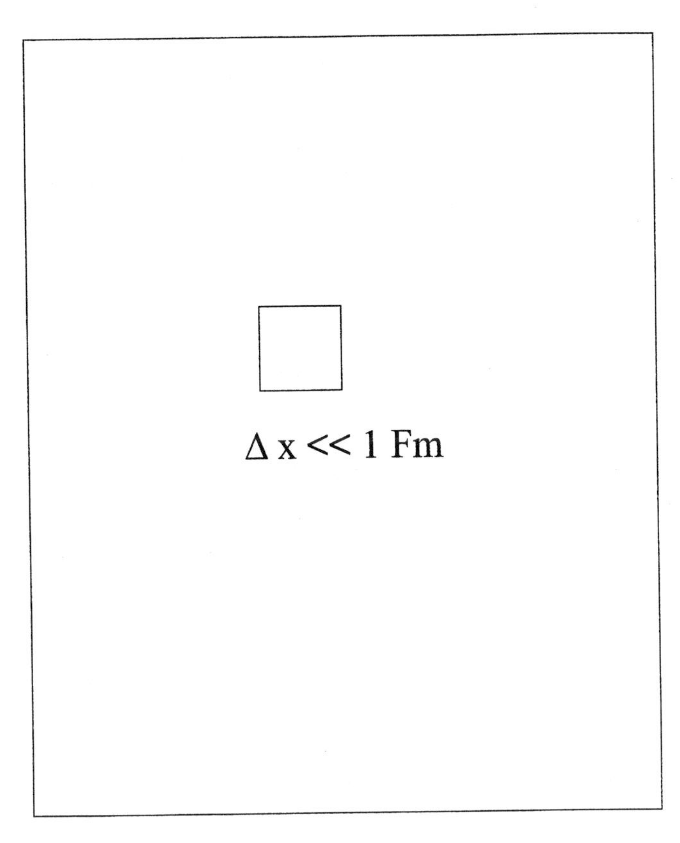

FIGURE 15. A head on picture of a nucleus.

wavelength associated with the gluon field. We shall further assume for simplicity that the transverse distribution of charge is uniform. (In fact, a better approximation is that the distribution of charge is slowly varying on the gluon wavelength scale of interest. This case can be computed directly from knowledge of the uniform transverse distribution case, The transverse size scale of variation is the nuclear radius, which is much larger than a fermi, so that this criteria is satisfied so long as $\Lambda >> \Lambda_{QCD}$, and the typical transverse momentum is of order Λ.)

In Fig. 15, a head on picture of the nucleus is shown. A small square is indicated which shows the transverse resolution size scale that we shall employ to probe the nucleus. In order that we can use weak coupling methods, we require that the transverse size be $\Delta x << 1Fm$.

In Fig. 16, the nucleus is shown along the beam axis. Where the tube generated by the square associated with Δx intersects a quark or gluons shown an x. (We use space-time rapidity variables to spread out the nucleus. The precise definition of this variable is below. For now, just imagine that we have chosen a longitudinal

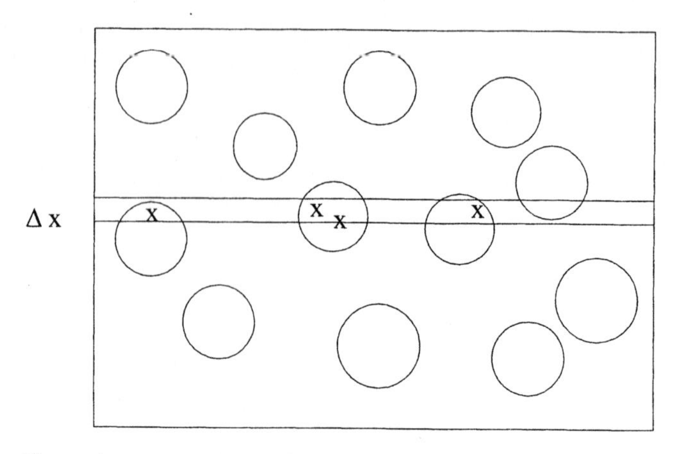

FIGURE 16. The nucleus as seen longitudinally with the longitudinal axis blown up by a choice of variable such as rapidity.

spatial variable which spreads out the Lorentz contracted hadron.) The physical extent in longitudinal spatial coordinate is of course small. If we require that $\Delta x << 1Fm$ and require that a tube intersects a quark or gluon from some hadron, then typically, that hadron will be far separated from any other hadron which has one of its quarks or gluons within the tube. The quarks and gluons within the tube are therefore uncorrelated in color. If we further require that the tube be large enough so that many quarks and gluons are in each tube, a situation always possible if we have large enough energy or a large enough nucleus, then the color total color charge associated with these quarks and gluons will be in a high dimensional representation, and can be treated classically. To see this, recall that in a high dimensional representation, $Q^2 >> Q \sim [Q,Q]$ so that commutators may be ignored.

For partons at the x values where we wish to compute the gluon distribution function, the higher rapidity sources sit on a sheet of infinitesimal thickness. The problem we must solve is to compute the typical correlation functions which give distribution functions for stochastic sources sitting on a sheet of infinite transverse extent traveling at the speed of light. Mathematically, this problem is similar to spin glass problems in condensed matter physics. Such problems typically have entirely nontrivial renormalization groups, and we will see that this is the case for our problem.

To understand a little better how the sources are distributed, it is useful to go inside the sheet. This is accomplished by introducing a space-time rapidity variable. Let us assume that the rapidity of the projectile is y_{proj} and its longitudinal momentum is P^+, we define

$$\eta = y_{proj} - ln(P^+_{proj} x^-) \tag{42}$$

We define the momentum space rapidity as

$$y = y_{proj} - ln(P^+_{proj}/p^+) \tag{43}$$

Here x^- is the coordinate of a source and p^+ its momentum. The previous definition of momentum space rapidity for produced particles was

$$y = \frac{1}{2} ln(p^+/p^-) = ln(p^+/m_t) \tag{44}$$

where $m_t = \sqrt{p_t^2 + m^2}$. This last expression is valid for particles on mass shell, whereas the other definition of momentum space rapidity works for constituents of the hadron wavefunction, which are of course not on mass shell. These definitions are equal within about a unit of rapidity for typical values of transverse momenta. In the wavefunction, the uncertainty principle relation gives $x^- p^+ \sim 1$, so that we see the space time rapidity is up to about a unit of rapidity the same as both momentum space values. We will therefore use these rapidities interchangeably in what follows.

We can now implement our static sources of charge by assuming the theory is defined by an ensemble of such charges:

$$Z = \int [d\rho] exp \left\{ -\frac{1}{2} \int_y^{y_{proj}} dy' d^2x_t \frac{1}{\mu^2(y')} \rho^2(y', x_t) \right\} \tag{45}$$

This ensemble gives

$$< \rho^a(y, x_T) \rho^b(y', x_t') >= \mu^2(y) \delta^{ab} \delta(y - y') \delta^{(2)}(x_t - x_t') \tag{46}$$

The parameter μ^2 therefore has the interpretation of a charge squared per unit rapidity

$$\mu^2(y) = \frac{1}{N_c^2 - 1} \frac{1}{\pi R^2} \frac{dQ^2(y)}{dy} \tag{47}$$

The total charge squared is

$$\chi(y) = \int_y^{y_{proj}} dy' \mu^2(y') \tag{48}$$

Since the sources are individual quarks and gluons, this can also be related directly to the total number of gluons and quarks contributing to the source as [11]

$$\chi(y) = \frac{1}{\pi R^2} \left(\frac{N_g}{2N_c} + \frac{N_c N_q}{N_c^2 - 1} \right) \int_x^1 dx G(x) \tag{49}$$

The factors above arise from computing the color charge squared of a singe quark or a single gluon.

We may now construct the non-abelian Lienard-Wiechart potentials generated by this distribution of sources. We must solve the equation

$$D_\mu F^{\mu\nu} = g^2 \delta^{\nu +} \rho(y, x_t) \tag{50}$$

To solve this equation, we look for a solution of the form

$$A^{\pm} = 0 \tag{51}$$

and

$$A^i = \frac{1}{i} U(y, x_t) \nabla^i U^\dagger(y, x_T) \tag{52}$$

We could equally well solve this equation in a gauge where A^+ is nonzero and all other components vanish. The student should check that the gauge transformation induced by U gives

$$\overline{A}^+ = \frac{1}{i} U^\dagger \partial^+ U \tag{53}$$

In this non-lightcone gauge, the field equations simplify and we get

$$\nabla_t^2 \overline{A}^+ = g^2 \rho(y, x_t) \tag{54}$$

Solving for U gives

$$U = exp\left\{ i \int_y^{y_{proj}} dy' \frac{1}{\nabla_t^2} \rho(y', x_t) \right\} \tag{55}$$

We have therefore constructed an explicit expression for the light cone field A^i in terms of the source ρ for arbitrary ρ! The system is integrable.

The structure of the fields strengths E and B which follow from this field strength is now easy to understand. The only nonzero longitudinal derivative is ∂^+. The field strength F_{ij} where both ij are transverse vanishes since the filed looks like a pure gauge transformation in the two dimensional space. It is in fact a pure gauge everywhere but in the sheet where the charge sits. Therefore the only nonvanishing field strength is F^{i+}. This gives $E \perp B \perp \vec{z}$, that is the fields are transversely polarized. This is the precise analog of the Lienard-Wiechart potentials of electrodynamics.

III LECTURE 3: THE ACTION AND RENORMALIZATION GROUP

The effective action for the theory we have described must be gauge invariant and properly describe the dynamics in the presence of external sources, up to an overall gauge transformation which is constant in $x^{\pm}$. (The lack of precise gauge invariance arises from the the desire to define the intrinsic transverse momentum of gluon distribution functions. Although when used in computations of gauge dependent quantities, the gauge dependence disappears, it is useful to not have the full gauge invariance in intermediate steps of computation.) The student should verify, that consistency with the Yang-Mills equations in the presence of an external source requires that

$$D_\mu J^\mu = 0 \tag{56}$$

For a source of the type we have here,

$$J^\mu_a = \delta^{\mu a}\delta(x^-)\rho^a(x_t) \tag{57}$$

This is an approximation we make when we describe the source on scales much larger than the spatial extent of the source. To properly regularize the delta function, we need to spread the source out in x^- as was done in the previous section. We find that the action is

$$\begin{aligned} S = &-\frac{1}{4}\int d^4x F^a_{\mu\nu}F^{\mu\nu}_a \\ &+\frac{i}{N_c}\int d^2x_t dx^- \delta(x^-)\rho^a(x_t) trT^a exp\left\{i\int_{-\infty}^{\infty} dx^+ T\cdot A^-(x)\right\} \end{aligned} \tag{58}$$

In this equation, the matrix T is in the adjoint representation of the gauge group. This is required for reality of the action. The student should minimize this action to get the Yang-Mills equations, identify the current, and show that the current is covariantly conserved.

This action is gauge invariant under gauge transformations which are required to be periodic in the time x^+. This is a consequence of the gauge invariance of the measure of integration over the sources ρ. This will be taken as a boundary condition upon the theory. In general if we had not integrated over sources, one could not define a gauge invariant source, as gauge rotations would change the definition of the source. Here because the source is integrated over in a gauge invariant way, the problem does not arise.

In the most general gauge invariant theory which we can write down is generated from

$$Z = \int [d\rho]e^{-F[\rho]}\int [dA]e^{iS[A,\rho]} \tag{59}$$

This is a generalization of the Gaussian ansatz described in the previous lecture. It allows for a slightly more complicated structure of stochastic variation of the sources. The Gaussian ansatz can be shown to be valid when the evaluating structure functions at large transverse momenta.

$$F_{Gaussian}[\rho] = \frac{1}{2\chi}\int d^2x_t\rho^2(x_t) \tag{60}$$

This theory is a slight modification of what we described in the first lecture. We have here assumed that our theory is an effective theory valid only in a limited range of rapidity much less than the rapidity of the source. In this restricted range, the structure of the source in rapidity cannot be important, and therefore we couple only to the total charge seen at the rapidity of interest. The local charge density as a function of rapidity is replaced by the total charge at rapidities greater than that at which we measure the field. The scale of fluctuation of the source is instead of the local charge squared per unit are per unit rapidity $\mu^2(y)$ becomes replaces by the scale of fluctuation of the total charge. The student should prove that

$$\chi = \int_y^{y_{proj}} dy' \mu^2(y') \tag{61}$$

To fully determine F in the above equation demands a full solution of the renormalization group equations of the theory. This has yet to be done. We shall confine our attention in most of the analysis below to the Gaussian ansatz. It is remarkable that within this simple ansatz for F, most of the general feature which a full treatment should generate, such as proper unitary behavior of deep inelastic scattering, arises in a natural way.

Let us turn our attention for the moment to the gluon distribution function

$$(2\pi)^3 2p^+ \frac{dN}{d^3p} =< a^\dagger(p) a(p) > \tag{62}$$

Using the results of the previous lecture, you should prove that

$$\frac{dN}{d^3p} = \frac{2p^+}{(2\pi)^3} \sum_{i,a} D^{ii}_{aa}(p, -p) \tag{63}$$

Here a is a color index and i is a transverse index associated with the gluon field. D is the gluon propagator in the external field

$$D^{\mu\nu}_{ab}(p, q) =< A^\mu_a(p) A^\nu_b(q) > \tag{64}$$

(In field theory language, D is the propagator in the external field including both connected and disconnected pieces.)

In lowest order, the A in the expression for D is simply the external Lienard-Wiechart potential. Recall that

$$A^i = -iU\nabla^i U^\dagger \tag{65}$$

where the U's are the explicit functions of the source ρ computed in the previous lecture.

The expectation value above may be computed using the Gaussian weight function. We need the propagator $1/\nabla_t^4$ to perform this calculation. The student should try to do this computation (it is equivalent to normal ordering exponentials), and

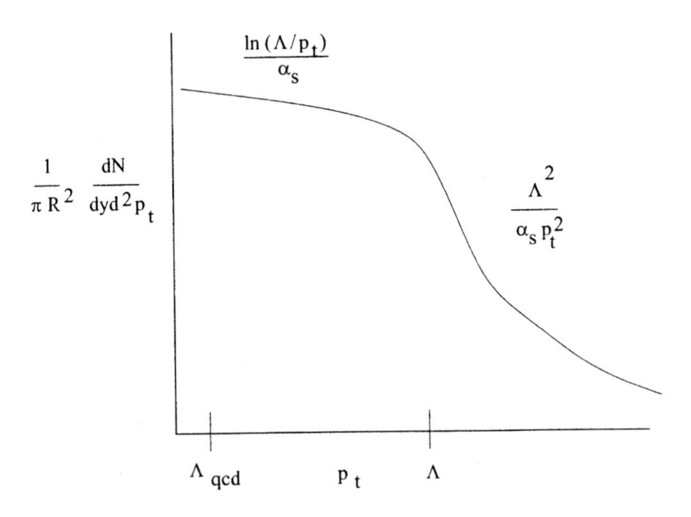

FIGURE 17. The intrinsic transverse momentum dependence of the gluon distribution function.

if it is too hard, refer to the paper of Jalilian-Marian et. al, where it is worked out in some detail. [7] The result is

$$\frac{dN}{d^2p_T} \sim \int d^2z e^{-ip_Tz_T} \frac{4(N_c^2-1)}{N_c z_T^2} \theta(1 - z_t\Lambda_{QCD}) \left\{1 - (z_t^2\Lambda_{QCD}^2)^{2\pi\alpha_s^2\chi z_t^2}\right\} \tag{66}$$

This results in the form for the gluon distribution function as shown in Fig. 17. At $p_t >> \Lambda$ where $\Lambda = \alpha_s\sqrt{\chi}$, the

$$\frac{1}{\pi R^2}\frac{dN}{dyd^2p_t} \sim \Lambda^2/\alpha_s p_t^2 \tag{67}$$

This is because this part of the distribution can be thought of as arising from bremstrahlung from the sources of the gluon field at high rapidities. As $p_t \leq \Lambda$, the gluon distribution function goes to a slowly varying function of p_T, which in the Gaussian ansatz is

$$\frac{1}{\pi R^2}\frac{dN}{dyd^2p_t} \sim \frac{1}{\alpha_s} ln(\Lambda/p_t) \tag{68}$$

In general, we expect saturation in this region, for arguments which will be presented in the next few pages. This means we get a slowly varying function of p_T, which by dimensional arguments, is therefore a slowly varying function of Λ

The only place that any non-trivial rapidity dependence enters the problem is through Λ. This dependence can be found from the renormalization group analysis of the effective action. Therefore in the bremstrahlung region, the dependence on Λ^2 is linear, and in the saturation region it is very weak.

Let us assume that when we go beyond the Gaussian ansatz, the general features of saturation presented above remain. We will show that this results in a reasonable physical picture for small x physics, and solve the unitarity puzzle outlined in the first lecture. First consider the gluon distribution function. The total number of gluons which can be measured at some scale size larger than a resolution size scale $1/Q^2$ is

$$xG(x,Q^2) = \int_0^{Q^2} \frac{d^2p_T}{(2\pi)^2} \frac{dN}{dyd^2p_T} \tag{69}$$

For $Q^2 >> \Lambda^2$, the integral is dominated by the bremstrahlung tail, and $G \sim R^2\Lambda^2$. (The gluon distribution is proportional to the charge per unit area times the area.) In our random walk scenario, the effective charge squared must be proportional to the length of the random walk, R, so that $G \sim R^3$ which for a nucleus is proportional to $A^{1/3}$. This is what we expect at large Q^2, except the reasoning is a little different than usual. The standard argument would have been that at large Q^2, the degrees of freedom in a nucleus for example should act incoherently, and we should have the gluon distribution functions proportional to A. Here we have random fields generating precisely the same A dependence!

When $Q^2 \leq \Lambda^2$, the integral is dominated by the saturation region. Here $G \sim R^2Q^2$, and the gluons can be thought of as arising from the surface of the hadron. Again this is consistent with what is expected from phenomenology. The gluons are so soft that they cannot see the entire hadron, only its surface.

The effect of saturation for unitarization of deep inelastic scattering can also be easily understood. Suppose we are at some fixed $Q^2 >> \Lambda^2(y)$. As y decreases corresponding to going to smaller x,, the gluon distribution function increases as $\Lambda^2(y)$. When x becomes so small that we get into the saturation region, the linear Λ dependence is weakened (in the Gaussian ansatz it is logarithmic in Λ), and the structure function stops growing. We expect that the number of gluons of size smaller than our resolution scale have ceased to grow, and the cross section should become slowly varying. This is in spite of the fact that Λ continues to grow!

The physics is again simple to understand: At small x we are indeed adding more and more gluons to the hadron wavefunction. These gluons are smaller as their inverse size scales as gluon density,

$$1/r^2 \sim \frac{1}{\pi R^2} \frac{dN}{dy} \tag{70}$$

They do not contribute to a fixed Q^2 cross section when they become smaller than the resolution size scale. What saves unitarity is p_t broadening.

A Deep Inelastic Scattering

In the previous discussion, we were concerned with the gluon distribution function. In deep inelastic scattering, we measure the quark distribution functions. How are these related?

In deep inelastic scattering we measure

$$\begin{aligned} W^{mu\nu} &= \frac{1}{2\pi} Im \int d^4x e^{iqx} < P_{had} \mid T(J^\mu(x) J^\nu(0) \mid P_{had} > \\ &= \frac{1}{M_{had}} \left\{ - \left(g^{\mu\nu} - \frac{q^\mu q^\nu}{q^2} \right) F_1 + \right. \\ & \left. \left(p^\mu - q^\mu \frac{p \cdot q}{q^2} \right) \left(p^\nu - q^\nu \frac{p \cdot q}{q^2} \right) \frac{1}{p \cdot q} F_2 \right\} \end{aligned} \tag{71}$$

We must be able to compute the imaginary part of the current-current correlation function. This is simply vacuum polarization in the presence of the Lienard-Wiechart potentials. In fact, it is straightforward to compute the propagators in these background fields. [13] The reason why it is simple is because the background field is basically gauge transformations in two regions of spaces separated by a surface of discontinuity.

This propagator may be then used to compute the vacuum polarization. The computation can be done so far as to relate explicitly the structure functions to correlation functions of exponentials involving ρ and expressions very similar to those for the gluon distribution function result. These results are currently being applied to the study of deep inelastic scattering to see whether the effects of saturation might be seen experimentally.

B Renormalization Group and How It Works

In the above computation of the gluon propagator, one used the classical effective action to compute the Lienard-Wiechart potentials. These classical fields were then used to compute the propogator. What about the quantum corrections?

The classical field contribution is shown in Fig 18. The x in the figure marks the position of the source. In Fig 19, the contribution arising by inserting the lowest order contribution to the connected piece of the gluon propogator is shown. Here there is a loop digram and the dashed line with the solid ellipse in it represents the gluon propogator to all orders in the strength of the external field. In Fig. 20, the piece where a quantum loop correction to one of the classical fields is shown.

When the quantum corrections are computed, one gets a correction to the classical field of order $\alpha_s ln(x_{cutoff}/x)$, where x is the value for gluon distribution function, and x_{cutoff} is the maximum x appropriate for the effective action. Although the coupling is small, the logarithm can become big if we go to x values far below the cutoff. The overall theory should remain valid however, the problem is that the classical approximation is no longer good.

The way to fix this up is to use the renormalization group. Suppose we want a new effective theory below x_{new}, and we require that $x_{new} << x_{cutoff}$ but that $\alpha_s ln(x_{cutoff}/x_{new}) << 1$. This means that the quantum corrections are small in the range $x_{new} << x << x_{cutoff}$, and can be handled perturbatively. We proceed by

FIGURE 18. The classical field contribution to the gluon distribution function.

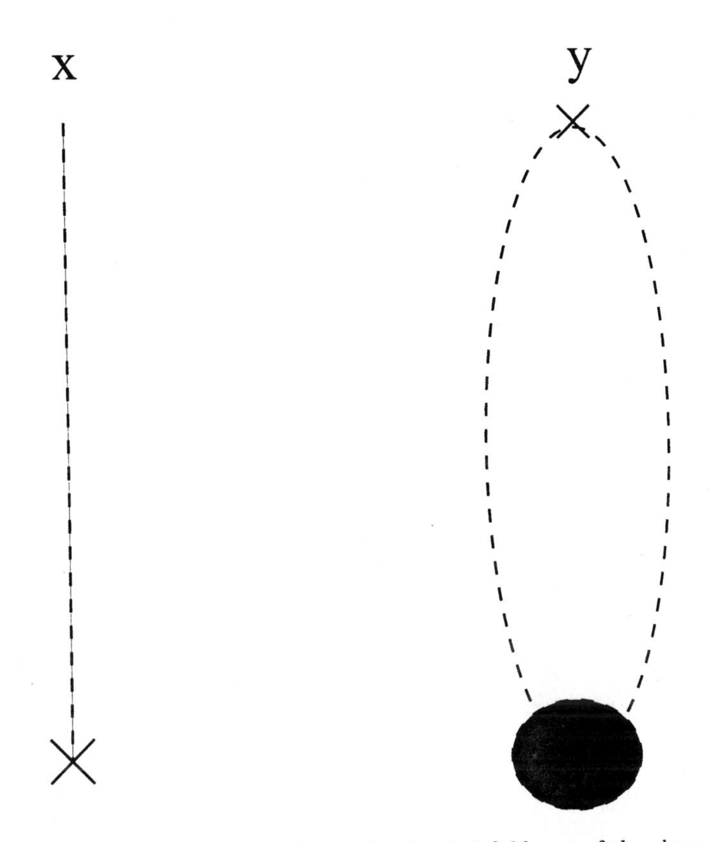

FIGURE 19. The first quantum correction to the classical field part of the gluon distribution function.

FIGURE 20. The quantum propogator correction to the gluon distribution function.

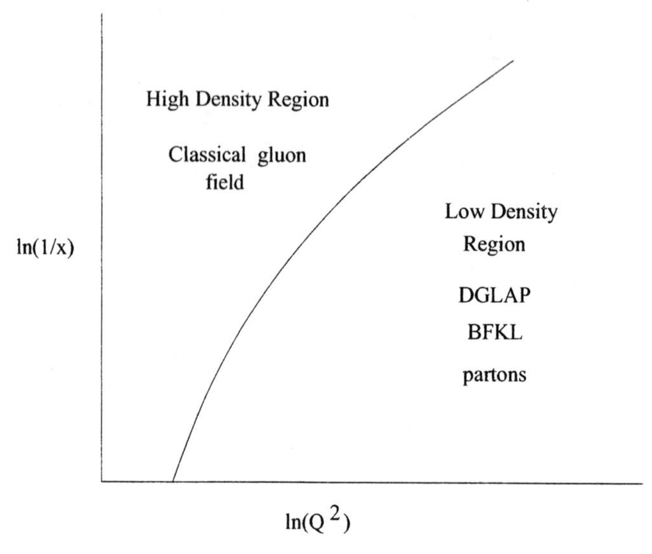

FIGURE 21. The $ln(1/x)$ and $ln(Q^2)$ plane and various regions of high and low parton density.

integrating out the degrees of freedom in this intermediate range of x, to generate a new effective theory in the region $x << x_{new}$.

In this process, the one can show that the only thing that changes in the effective action formalism is the weight function $F[\rho]$ for fluctuations in the color source . Effectively, we are changing what we call source for the gluon field, and trading it for what we call the gluon field. This renormalization group procedure is that of Wilson-Kadanoff. One can derive the form of the renormalization group equations and they can be written explicitly in the low gluon density region. One can prove that here the function F is Gaussian, and that in appropriate kinematic limits, one reproduces the BFKL equation, the DGLAP equation and their non-linear generalizations to first order in the non-linearities. Work is currently in progress to get the explicit form of these equations in the high density region, and to solve them.

The various regions of high and low density are shown in the plot of Fig. 21. If we are in the low density region, we evolve in x by the BFKL equation and in Q^2 by the DGLAP equation. Hopefully a full solution to the problem will lead to an effective action which is at a fixed point of the renormalization group. In this case, at very small x, the form for F will simplify, and all correlation functions will have universal critical exponents. If this is true, the dynamics of high energy scattering for all hadrons becomes the same, and presumably has a simple structure.

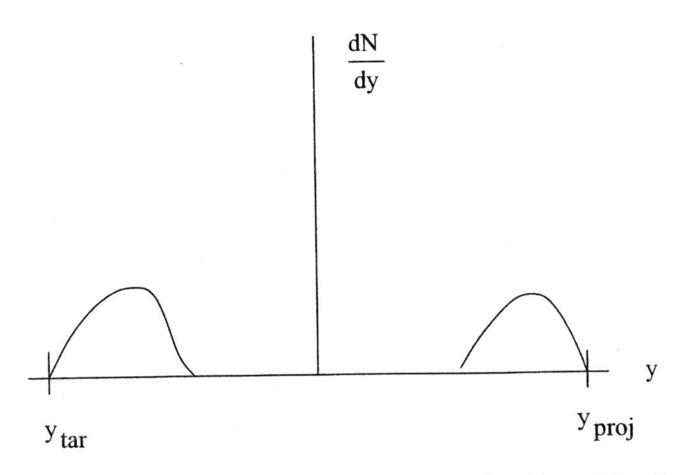

FIGURE 22. The rapidity distribution of particles produced in a diffractive event.

C Other Directions

In Raju Venugopalan's talk, you heard about using these techniques to describe high energy hadron-hadron collisions. [14]- [16] I have nothing to add to his talk, except that this shows that there is a direct relationship between hadron-hadron collisions and deep inelastic scattering.

A third class of phenomena is diffraction. In Fig. 22, a rapidity distribution of particle for a diffractive event is shown. There are two clusters of produced particles. In deep inelastic scattering these could be say the clump associated with a quark-anti-quark jet produced by the virtual photon, and a clump associated with the fragmentation of the target. If we restrict our attention to the class of diffractive phenomenon where the target does not fragment, that is scatters elastically, one can prove the following: [17]- [19]

- Deep inelastic scattering is given by computing the amplitude for vacuum polarization for the electromagnetic current in the presence of the non-abelian classical field, squaring and then averaging over color.
- Diffractive events are given by computing the amplitude in the presence of the non-abelian color field, averaging over color, and then squaring.

We have argued the first case earlier in the lecture. To understand how the second case works, note that only the diagrams of Fig. 23 survive averaging over color before squaring the amplitude. In this figure a virtual photon produces a pair of quark jets. These jets exchange gluons (dash dot line) with the source which comprise the hadron. For sources are far away in rapidity from the jet, and therefore the longitudinal momentum transfer is very small. (The sources are not kicked by the jet. They do not change their velocity.) Moreover the color averaging

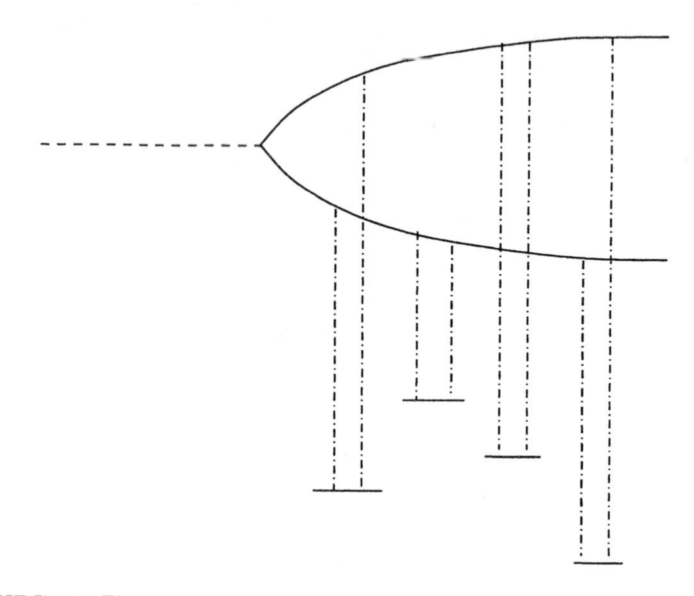

FIGURE 23. The non-zero contributions to the amplitude after color averaging.

and translational invariance guarantees that the transverse momentum given to the quarks after two scatterings vanishes.

So if we imagine a hadron and its Fock space constituents, they are unchanged by the scatterings shown in Fig. 23. Therefore the initial hadron state projects on to the hadron after scattering, in the amplitude, with weight very close to one, that is the hadron has scattered elastically (although many of its constituents have scattered.)

The amplitude shown in Fig. 23 therefore describes the rapidity distribution of particles shown in Fig. 22 only if the hadron does not fragment, that is, maintains its identity as a hadron. Therefore only one particle emerges on the hadron side of the collision. It may be possible to generalize these considerations to fully diffractive processes where the hadron is allowed to fragment.

IV ACKNOWLEDGMENTS

I thank my colleagues Alejandro Ayala-Mercado, Miklos Gyulassy, Yuri Kovchegov, Alex Kovner, Jamal Jalilian-Marian, Andrei Leonidov, Raju Venugopalan and Heribert Weigert with whom the ideas presented in this talk were developed. This work was supported under Department of Energy grants in high energy and nuclear physics DOE-FG02-93ER-40764 and DOE-FG02-87-ER-40328. I particularly thank Andrei Leonidov for a critical reading of the manuscript.

REFERENCES

1. E.A. Kuraev, L.N. Lipatov and Y.S. Fadin, *Zh. Eksp. Teor. Fiz* **72**, 3 (1977) (*Sov. Phys. JETP* **45**, 1 (1977)); I.A. Balitsky and L.N. Lipatov, *Sov. J. Nucl. Phys.* **28** 822 (1978); G. Altarelli and G. Parisi, *Nucl. Phys.* **B126** 298 (1977); Yu.L. Dokshitser, *Sov.Phys.JETP* **46** 641 (1977).
2. J. Breitweg et. al., *Eur. Phys. J.* **67**, 609 (1999) and references therein.
3. L. V. Gribov, E. M. Levin and M. G. Ryskin, *Phys. Rep.* **100** 1 (1983).
4. J. D. Bjorken, *Phys. Rev.* **D27**, 140 (1983).
5. L. McLerran and R. Venugopalan, *Phys. Rev.* **D49** 2233 (1994); **D49** 3352 (1994).
6. Y. Kovchegov, *Phys. Rev.* **D54**, 5463 (1996); **D55**, 5445 (1997).
7. J. Jalilian-Marian, A. Kovner, L. McLerran and H. Weigert, *Phys. Rev.* **D55** 5414 (1997);
8. J. Jalilian-Marian, A. Kovner, A. Leonidov and H. Weigert *Nucl. Phys.* **B504** 415 (1997); *Phys. Rev.* **D59** 014014 (1999); J. Jalilian-Marian, A. Kovner and H. Weigert, *Phys. Rev.* **D59** 014015 (1999).
9. E. Iancu, A. Leonidov and L. McLerran, work in progress.
10. R. Venugopalan nucl-th/9808023.
11. M. Gyulassy and L. McLerran, *Phys. Rev.* **C56** 2219 (1997).
12. Al Mueller *Nucl. Phys.* **B307**, 34 (1988); **B355**, 115 (1990).
13. L. Mclerran and R. Venugopalan, *Phys. Rev.* **D59** 094002 (1999).
14. A. Kovner, L. McLerran, and H. Weigert *Phys. Rev.* **D52**, 3809 (1995); 6231 (1995).
15. A. Krasnitz and R. Venugopalan hep-ph/9808332, hep-ph/9809433.
16. S. Bass, B. Muller and W. Poschl, nucl-th/9808011.
17. A. Hebecker, *Nucl. Phys.* **B505**, 349 (1997).
18. W. Buchmuller, T. Gehrmann and A. Hebecker, *Nucl. Phys.* **B537**, 477 (1999).
19. L. Mclerran and Y. Kovchegov hep-ph/9903246

Neutrino Physics

R. D. Peccei

Department of Physics and Astronomy, UCLA, Los Angeles, CA 90095-1547

Abstract. These lectures describe some aspects of the physics of massive neutrinos. After a brief introduction of neutrinos in the Standard Model, I discuss possible patterns for their masses. In particular, I show how the presence of a large Majorana mass term for the right-handed neutrinos can engender tiny neutrino masses for the observed neutrinos. If neutrinos have mass, different flavors of neutrinos can oscillate into one another. To analyze this phenomena, I develop the relevant formalism for neutrino oscillations, both in vacuum and in matter. After reviewing the existing (negative) evidence for neutrino masses coming from direct searches, I discuss evidence for, and hints of, neutrino oscillations in the atmosphere, the sun, and at accelerators. Some of the theoretical implications of these results are emphasized. I close these lectures by briefly outlining future experiments which will shed further light on atmospheric, accelerator and solar neutrino oscillations. A pedagogical discussion of Dirac and Majorana masses is contained in an appendix.

I NEUTRINOS IN THE STANDARD MODEL

Neutrinos play a special role in the $SU(2) \times U(1)$ electroweak theory. While the left-handed neutrinos are part of $SU(2)$ doublets

$$L_i = \begin{pmatrix} \nu_{\ell_i} \\ \ell_i \end{pmatrix}_{\rm L} , \qquad \ell_i = \{e, \mu, \tau\} , \tag{1}$$

the right-handed neutrinos are $SU(2)$ singlets. Since the electromagnetic charge and the $U(1)$ hypercharge differ by the value of the third component of weak isospin

$$Q = T_3 + Y , \tag{2}$$

one sees that the right-handed neutrinos $(\nu_{\ell_i})_{\rm R}$ carry **no** $SU(2) \times U(1)$ quantum numbers. The above has two important experimental implications:

i) The neutrinos seen experimentally are those produced by the weak interactions. These neutrinos are purely left-handed.

ii) Because one cannot infer the existence of right-handed neutrinos from weak processes, the presence of $(\nu_{\ell_i})_{\rm R}$ can only be seen indirectly, most likely through the existence of neutrino masses. However, it should be noted that neutrino masses do not necessarily imply the existence of right-handed neutrinos.

CP490, *Particles and Fields: Eighth Mexican School,* edited by J. C. D'Olivo et al.

Before discussing issues connected with neutrino masses, it is useful to summarize what we know about the left-handed neutrinos from their weak interactions. In the electroweak theory these neutrinos couple to the Z-boson in a universal fashion

$$\mathcal{L}_{Z\nu\bar{\nu}} = \frac{e}{2\cos\theta_W \sin\theta_W} Z^\mu [J_\mu^{\rm NC}]_\nu \ , \tag{3}$$

where

$$[J_\mu^{\rm NC}]_\nu = \sum_i (\bar{\nu}_{\ell_i})_{\rm L} \gamma_\mu (\nu_{\ell_i})_{\rm L} \ . \tag{4}$$

Precision studies of the Z-line shape allow the determination of the number of neutrino species i, since as this number increases so does the Z total width. Each neutrino type (provided its mass $m_{\nu_i} \ll \frac{M_Z}{2}$) contributes the same amount to the Z-width [1]

$$\Gamma(Z \to \nu_{\ell_i}\bar{\nu}_{\ell_i}) = \frac{\sqrt{2} G_F M_Z^3}{24\pi}\rho \ . \tag{5}$$

Here G_F is the Fermi constant as determined in μ-decay [2] and ρ is related to the axial coupling of the charged leptons to the Z-boson: $\rho = (2g_A^\ell)^2$. Using [1]

$$g_A^\ell = -0.50102 \pm 0.00030 \ , \tag{6}$$

which is the average of the results obtained by the four LEP collaborations and SLD, one has numerically

$$\Gamma_\nu \equiv \Gamma(Z \to \nu_{\ell_i}\bar{\nu}_{\ell_i}) = (167.06 \pm 0.22) \ {\rm MeV} \ . \tag{7}$$

Using the above, the number of different neutrino species, N_ν, can then be derived from the precision measurements of the Z total width and of its partial width into hadrons and leptons using the (obvious) equation

$$\Gamma_{\rm tot} = \Gamma_{\rm had} + 3\Gamma_{\rm lept} + N_\nu \Gamma_\nu \ . \tag{8}$$

From a fit of the Z-line shape for $e^+e^- \to \mu^+\mu^-$ and $e^+e^- \to$ hadrons at LEP one can extract very accurate values for $\Gamma_{\rm tot}$, $\Gamma_{\rm lept}$ and $\Gamma_{\rm had}$:

$$\begin{aligned} \Gamma_{\rm tot} &= (2.4939 \pm 0.0024) \ {\rm GeV} \\ \Gamma_{\rm lept} &= (83.90 \pm 0.1) \ {\rm MeV} \\ \Gamma_{\rm had} &= (1.7423 \pm 0.0023) \ {\rm GeV} \ . \end{aligned} \tag{9}$$

Whence it follows that the, so-called, invisible width is

$$\Gamma_{\rm inv} = N_\nu \Gamma_\nu = (499.9 \pm 3.4) \ {\rm MeV} \tag{10}$$

and thus one deduces for N_ν—the number of neutrino species:

$$N_\nu = 2.992 \pm 0.020 \ . \tag{11}$$

This result strongly supports the notion, expressed in Eq. (1), that there are only 3 generations of leptons.

It turns out that one can get a more accurate value for N_ν by using other information derivable from the Z-line shape. The cross-section for $e^+e^- \to$ hadrons can be expressed in terms of three factors: [3] a peak cross-section

$$\sigma_o = \frac{12\pi \Gamma_{\text{lept}} \Gamma_{\text{had}}}{\Gamma^2_{\text{tot}} M_Z^2} \ , \tag{12}$$

a Breit-Wigner factor

$$\text{BW}(s) = \frac{s \ \Gamma^2_{\text{tot}}}{(s - M_Z^2)^2 + s^2 \Gamma^2_{\text{tot}}/M_Z^2} \ , \tag{13}$$

and a computable initial state bremsstrahlung correction $(1 - \delta_{\text{QED}}(s))$, with

$$\sigma_{\text{had}} = \sigma_o \ \text{BW}(s)(1 - \delta_{\text{QED}}(s)) \ . \tag{14}$$

The value of σ_o extracted from an analysis of the cross-section for $e^+e^- \to$ hadrons at LEP [1]

$$\sigma_o = (41.491 \pm 0.058) \text{ nb} \tag{15}$$

can be combined with the LEP results for the ratio of hadronic to leptonic partial widths of the Z

$$R_\ell = \frac{\Gamma_{\text{had}}}{\Gamma_{\text{lept}}} = 20.765 \pm 0.026 \tag{16}$$

to deduce, with a little bit of theoretical input, a value for N_ν. This value, as we will see, is slightly more accurate than that given in Eq. (11).

Using Eqs. (8), (10), (12), and (16), one can write

$$N_\nu = \frac{\Gamma_{\text{inv}}}{\Gamma_\nu} = \frac{\Gamma_{\text{lept}}}{\Gamma_\nu} \left\{ \frac{\Gamma_{\text{tot}}}{\Gamma_{\text{lept}}} - \frac{\Gamma_{\text{had}}}{\Gamma_{\text{lept}}} - 3 \right\} = \frac{\Gamma_{\text{lept}}}{\Gamma_\nu} \left\{ \sqrt{\frac{12\pi R_\ell}{\sigma_o M_Z^2}} - R_\ell - 3 \right\} \ . \tag{17}$$

In the Standard Model, the ratio $\Gamma_{\text{lept}}/\Gamma_\nu$ is very accurately known:

$$\left. \frac{\Gamma_\nu}{\Gamma_{\text{lept}}} \right|_{\text{SM}} = 1.991 \pm 0.001 \ . \tag{18}$$

Using this value in Eq. (17), along with the experimentally determined Z mass $M_Z = (91.1867 \pm 0.0021)$ GeV and the values of σ_o and R_ℓ measured at LEP, gives

$$N_\nu = 2.994 \pm 0.011 \ ; \qquad \Gamma_{\text{inv}} = (500.1 \pm 1.9) \text{ MeV} \ . \tag{19}$$

These values are consistent with those in Eqs. (10) and (11), but are about a factor of two more accurate.

There is an analogous equation to Eq. (3) describing the coupling of the $W^\pm$ boson to the leptonic charged currents. Again only left-handed neutrinos are involved. One has

$$\mathcal{L}_{W\ell\nu_\ell} = \frac{e}{\sqrt{2}\sin\theta_W}\{W_+^\mu J_{\mu-}^{\rm lept} + W_-^\mu J_{\mu+}^{\rm lept}\} , \tag{20}$$

where

$$J_{\mu-}^{\rm lept} = (J_{\mu+}^{\rm lept})^\dagger = \sum_i \bar{\ell}_{i\rm L}\gamma_\mu \nu_{\ell_i\rm L} . \tag{21}$$

The states that appear in Eq. (21) in general are not mass eigenstates, since mass generation can mix leptons of the same charge among each other. Nevertheless, one can always diagonalize the charged lepton mass matrix by a by-unitary transformation of the left- and right-handed charged lepton fields:

$$\ell_{\rm L} = U^\ell \tilde{\ell}_{\rm L} ; \qquad \ell_{\rm R} = V^\ell \tilde{\ell}_{\rm R} . \tag{22}$$

After this transformation, the charged currents in Eq. (21) read

$$J_{\mu-}^{\rm lept} = (J_{\mu+}^{\rm lept})^\dagger = \sum_{ij} \overline{\tilde{\ell}}_{i\rm L}\gamma_\mu (U^\ell)^\dagger_{ij}\nu_{\ell_j\rm L} = \sum_i \overline{\tilde{\ell}}_{i\rm L}\gamma_\mu \tilde{\nu}_{\ell_i\rm L} , \tag{23}$$

where

$$\tilde{\nu}_{\ell\rm L} = (U^\ell)^\dagger \nu_{\ell L} . \tag{24}$$

Note that because U^ℓ is unitary, the neutral current $[J_\mu^{\rm NC}]_\nu$ is the same whether it is expressed in terms of $\nu_{\ell\rm L}$ or $\tilde{\nu}_{\ell\rm L}$. Conventionally, the states $\tilde{\nu}_{\ell_i\rm L}$ are called **weak interaction eigenstates**, since they are produced in the decay of a W^+ boson in association with a physically charged lepton $\tilde{\ell}_i$ of definite mass. These states, of course, are also pair produced by the Z-boson. For ease of notation, in what follows I will drop the tilde on both $\tilde{\ell}_{i\rm L}$ and $\tilde{\nu}_{\ell_{i\rm L}}$ with the understanding that the states now called $\nu_{\ell_i\rm L}$ are those produced by the weak interactions—they are the weak interaction eigenstates. Similarly, the charged leptons ℓ_i are the states associated with a diagonal mass matrix

$$M_\ell = \begin{pmatrix} m_e & & \\ & m_\mu & \\ & & m_\tau \end{pmatrix} . \tag{25}$$

II PATTERNS OF NEUTRINO MASSES

With these preliminaries underway, I want now to examine in a bit of detail the possible patterns of neutrino masses. To do so, it is useful to first review how fermion masses originate in field theory. The mass term with which everybody is acquainted with is one involving a fermion ψ and its conjugate $\bar{\psi} = \psi^\dagger \gamma^0$:

$$\mathcal{L}_{\text{mass}} = -m\bar{\psi}\psi = -m(\overline{\psi_{\text{L}}}\psi_{\text{R}} + \overline{\psi_{\text{R}}}\psi_{\text{L}}) \ , \tag{26}$$

with ψ_{L}, ψ_{R} being the usual projections

$$\psi_{\text{L}} = \frac{1}{2}(1-\gamma_5)\psi \ ; \qquad \psi_{\text{R}} = \frac{1}{2}(1+\gamma_5)\psi \ . \tag{27}$$

This term, obviously, conserves fermion number

$$\psi \to e^{i\alpha}\psi \ ; \qquad \bar{\psi} \to e^{-i\alpha}\bar{\psi} \tag{28}$$

and gives equal mass for particles and antiparticles

$$m_{\bar{\psi}} = m_{\psi} = m \ . \tag{29}$$

For particles carrying any $U(1)$ quantum number, like electromagnetic charge, it is clear that $\mathcal{L}_{\text{mass}}$ is the **only** possible mass term, since to preserve these $U(1)$ quantum numbers one needs always to have particle-antiparticle interactions.

Neutrinos, however, provide an interesting exception. Because neutrinos do not have electromagnetic charge, it is possible to contemplate other types of mass terms for them besides the particle-antiparticle term given in Eq. (26). These other neutrino mass terms, contain two neutrino (or two antineutrino) fields. Hence they violate fermion number (and in some cases $SU(2)\times U(1)$), but otherwise are allowed by Lorentz invariance.

As we discuss in more detail in Appendix A, one can write three different types of mass terms for neutrinos:

$$\begin{aligned}\mathcal{L}^{\nu}_{\text{mass}} = & -[\overline{\nu_{\text{R}}} m_D \nu_{\text{L}} + \overline{\nu_{\text{L}}} m_D^\dagger \nu_{\text{R}}] - \frac{1}{2}[\overline{\nu_{\text{R}}}\tilde{C} m_S \overline{\nu_{\text{R}}}^T + \nu_{\text{R}}^T \tilde{C} m_S^\dagger \nu_{\text{R}}] \\ & - \frac{1}{2}[\nu_{\text{L}}^T \tilde{C} m_T \nu_{\text{L}} + \overline{\nu_{\text{L}}}\tilde{C} m_T^\dagger \overline{\nu_{\text{L}}}^T] \ .\end{aligned} \tag{30}$$

Here the mass matrices m_D, m_S, m_T are Lorentz scalars. However, their presence is only possible as a result of different symmetry breakdowns. Specifically, m_D conserves fermion number, but violate $SU(2) \times U(1)$ since it does not transform as an $SU(2)$ doublet. This fermion number conserving mass is often called a Dirac mass. Thus, in a happy confluence of notation, m_D can stand both for a Dirac mass and a doublet mass. Both m_S and m_T violate fermion number by two units and are known as Majorana masses. Because m_S couples ν_{R} with itself, clearly it is an

$SU(2) \times U(1)$ invariant. This is not the case for m_T, which violates $SU(2) \times U(1)$ because it does not transform as an $SU(2)$ triplet.

The matrix $\tilde{C}$ which enters in the Majorana mass terms in Eq. (30) is there to preserve Lorentz invariance. Appendix A contains a detailed discussion of this point, along with a pedagogical review of how one constructs 4-spinors starting from 2-dimensional Weyl spinors. I note here only that $\tilde{C}$ is **not** to be confused with the matrix C connected with how Dirac fields transform under charge conjugation. [4][1] Under the charge conjugation operator $U(C)$ a Dirac field ψ is transformed into its Hermitian conjugate $\psi^\dagger$:

$$U(C)\psi U(C)^{-1} = C\psi^\dagger(x) . \tag{31}$$

The matrix C is necessary to insure the invariance of the Dirac equation under charge conjugation and obeys the restriction

$$C\gamma_\mu^* C^{-1} = -\gamma_\mu . \tag{32}$$

In general, C depends on the γ-matrix basis used. In the Majorana basis, where the γ-matrices are purely imaginary, then $C = 1$. At any rate, the matrix $\tilde{C}$ appearing in Eq. (30) is related to C by [5]

$$\tilde{C} = C\gamma^{oT} . \tag{33}$$

It is easy to check that instead of (32) $\tilde{C}$ obeys

$$\tilde{C}\gamma_\mu^T \tilde{C}^{-1} = -\gamma_\mu . \tag{34}$$

The reason the matrix $\tilde{C}$ appears in Eq. (30) is because it relates the, so-called, charge conjugate field ψ^c to $\bar{\psi}$ rather than to $\psi^\dagger$. In view of the way the charge conjugation operator acts on the fermion field ψ (see Eq. (31)), it is natural to define the charge conjugate field ψ^c as

$$\psi^c(x) = C\psi^\dagger(x) . \tag{35}$$

Now $\bar{\psi} = \psi^\dagger\gamma^0$, so one also has that

$$\psi^c(x) = C\gamma^{0T}\bar{\psi}^T(x) = \tilde{C}\bar{\psi}^T(x) . \tag{36}$$

So $\tilde{C}$, indeed, serves to relate $\bar{\psi}$ to ψ^c.

---

1) Unfortunately, the distinction between C and $\tilde{C}$ is often blurred in the literature.

III THE SEE-SAW MECHANISM

Eq. (30) displays the most general neutrino mass term, involving three distinct mass matrices m_D, m_S and m_T. If there are only three flavors of neutrinos these are 3×3 matrices. For the moment this is what we shall assume, but we shall return to this point later on.

One can write Eq. (30) in a more symmetrical way by replacing the transposed fields in this equation by the charge conjugate field. Recall that [cf. Eq. (35)]

$$\psi^c = \tilde{C}\bar{\psi}^T \; ; \quad \overline{\psi^c} = \psi^T \tilde{C} \; . \tag{37}$$

Hence, for example, one can write[2]

$$\overline{\nu_{\rm R}}\nu_{\rm L} = -\nu_{\rm L}^T \overline{\nu_{\rm R}}^T = \nu_{\rm L}^T \tilde{C}\tilde{C}\overline{\nu_{\rm R}}^T = \overline{\nu_{\rm L}^c}\nu_{\rm R}^c \; . \tag{38}$$

Thus

$$\overline{\nu_{\rm R}}\nu_{\rm L} = \frac{1}{2}[\overline{\nu_{\rm R}}\nu_{\rm L} + \overline{\nu_{\rm L}^c}\nu_{\rm R}^c] \; . \tag{39}$$

Using these equations, $\mathcal{L}^\nu_{\rm mass}$ can be written in the following compact way:

$$\mathcal{L}^\nu_{\rm mass} = -\frac{1}{2}\left[(\overline{\nu_{\rm L}^c} \; \overline{\nu_{\rm R}}) \begin{pmatrix} m_T & m_D^T \\ m_D & m_S \end{pmatrix} \begin{pmatrix} \nu_{\rm L} \\ \nu_R^c \end{pmatrix} \right] + \text{h.c.} \tag{40}$$

For 3 generations of neutrinos, the six mass eigenstates m_i are the eigenvalues of the 6×6 matrix

$$M = \begin{pmatrix} m_T & m_D^T \\ m_D & m_S \end{pmatrix} . \tag{41}$$

Because M is not necessarily Hermitian, its diagonalization necessitates a bi-unitary transformation

$$U_{\rm R}^\dagger M U_{\rm L} = M_{\rm diag} \; , \tag{42}$$

where $U_{\rm L}$ and $U_{\rm R}$ are 6×6 unitary matrices. This diagonalization is accomplished by a basis change on the original neutrino fields

$$\psi_{\rm L} = \begin{pmatrix} \nu_{\rm L} \\ \nu_{\rm R}^c \end{pmatrix} \; ; \quad \psi_{\rm R} = \begin{pmatrix} \nu_{\rm L}^c \\ \nu_{\rm R} \end{pmatrix} \tag{43}$$

to a new set of fields $\eta_{\rm L}$ and $\eta_{\rm R}$ defined by the equations:

$$\psi_{\rm L} = U_{\rm L}\eta_{\rm L} \; ; \quad \psi_{\rm R} = U_{\rm R}\eta_{\rm R} \; . \tag{44}$$

---

[2] The minus sign in the second term below comes from Fermi statistics.

It is useful to consider the simple, but physically interesting case, [6] of just one family of neutrinos. Further, let us imagine $m_T = 0$ and $m_S \gg m_D$. The 2×2 matrix M in this case reads simply

$$M = \begin{pmatrix} 0 & m_D \\ m_D & m_S \end{pmatrix} . \tag{45}$$

This matrix has two eigenvalues, given approximately by m_S and $-m_D^2/m_S$. That is, in this case the spectrum splits into a very heavy neutrino of (approximate) mass m_S and a very light neutrino of (approximate) mass m_D^2/m_S.[3] This, so called, **see-saw mechanism** is very suggestive. It is natural to expect that m_D should be of the order of the charged lepton mass, corresponding to the neutrino in question: $m_D \sim m_\ell$. Then the spectrum of leptons has a natural hierarchy:

$$(m_\nu)_{\text{light}} \sim m_\ell \left(\frac{m_\ell}{m_S}\right) \ll m_\ell \ll (m_\nu)_{\text{heavy}} \sim m_S . \tag{46}$$

So, if there is a large mass scale associated with the right-handed neutrinos (the mass scale m_S, which is **not** constrained by the scale of $SU(2) \times U(1)$ breaking, since it is an $SU(2) \times U(1)$ singlet) one readily understands why neutrino masses could be so much lighter than the corresponding charged lepton masses.

The matrix M in the simple 2×2 example above is diagonalized (approximately) by the orthogonal matrix

$$U = \begin{pmatrix} 1 & m_D/m_S \\ -m_D/m_S & 1 \end{pmatrix} . \tag{47}$$

The two neutrino mass eigenstates are then

$$\eta_{\text{L}} \equiv \begin{pmatrix} \eta_1 \\ \eta_2 \end{pmatrix}_{\text{L}} = \begin{pmatrix} 1 & -m_D/m_S \\ m_D/m_S & 1 \end{pmatrix} \begin{pmatrix} \nu_{\text{L}} \\ \nu_{\text{R}}^c \end{pmatrix} \tag{48}$$

$$\eta_{\text{R}} \equiv \begin{pmatrix} \eta_1 \\ \eta_2 \end{pmatrix}_{\text{R}} = \begin{pmatrix} 1 & -m_D/m_S \\ m_D/m_S & 1 \end{pmatrix} \begin{pmatrix} \nu_{\text{L}}^c \\ \nu_{\text{R}} \end{pmatrix} . \tag{49}$$

I note that the mass eigenstates η_1 and η_2 are Majorana (self-conjugate) states

$$\eta_1 = \eta_{1\text{L}} + \eta_{1\text{R}} = (\nu_{\text{L}} + \nu_{\text{L}}^c) - \frac{m_D}{m_S}(\nu_{\text{R}}^c + \nu_{\text{R}}) = \eta_1^c \tag{50}$$

$$\eta_2 = \eta_{2\text{L}} + \eta_{2\text{R}} = (\nu_{\text{R}}^c + \nu_{\text{R}}) + \frac{m_D}{m_S}(\nu_{\text{L}} + \nu_{\text{L}}^c) = \eta_{\text{L}}^c . \tag{51}$$

The ν_{L} state which enters in the weak interactions, for all practical purposes is, essentially $\eta_{1\text{L}}$. That is, it is the state associated with the light neutrino eigenstate $(m_1 \simeq m_D^2/m_S)$:

---

3) For fermion fields, the sign of the mass term is irrelevant since it can be changed by a chiral transformation $\psi_{\text{R}} \to \exp\left[i\frac{\pi}{2}\right]\psi_{\text{R}}$; $\psi_{\text{L}} \to \exp\left[-i\frac{\pi}{2}\right]\psi_{\text{L}}$ which leaves the rest of the Lagrangian invariant.

$$\nu_L = \eta_{1L} + \frac{m_D}{m_S}\eta_{2L} \; . \tag{52}$$

The right-handed neutrino ν_R, on the other hand, is essentially the heavy neutrinos eigenstate η_{2R} $(m_2 \simeq m_S)$:

$$\nu_R = \eta_{2R} - \frac{m_D}{m_S}\eta_{1R} \; . \tag{53}$$

This simple example can be easily generalized to the 3×3 case of interest. Again, if the matrix m_T is negligible (i.e. if its eigenvalues are negligibly small), then the neutrino mass matrix M takes the approximate form

$$M = \begin{pmatrix} 0 & m_D^T \\ m_D & m_S \end{pmatrix} \; . \tag{54}$$

Provided the eigenvalues of m_S are large compared to those of m_D, then again the spectrum separates into a light and heavy neutrino sector. The light neutrinos have a 3×3 mass matrix

$$(M_\nu)_{\text{light}} = m_D^T m_S^{-1} m_D \; , \tag{55}$$

while the heavy neutrino mass eigenstates are the eigenstates of the 3×3 matrix

$$(M_\nu)_{\text{heavy}} = m_S \; . \tag{56}$$

The see-saw mechanism, in my view, is the only natural way to understand eV neutrino masses. Let me expand a bit on this point. Simce m_S is an $SU(2)\times U(1)$ invariant parameter, there are no constraints on it. On the other hand, as we discussed earlier, both m_D and m_T can only originate **after** $SU(2)\times U(1)$ breaking. The Yukawa interaction, of ν_R with a left-handed doublet $L = \begin{pmatrix} \nu_\ell \\ \ell \end{pmatrix}_L$ via a Higgs doublet $\Phi = \begin{pmatrix} \phi^0 \\ \phi^- \end{pmatrix}$

$$\mathcal{L}_{\text{Yukawa}} = -\Gamma \bar{\nu}_R \Phi L + \text{h.c.} \; , \tag{57}$$

leads to a Dirac mass

$$m_D = \Gamma\langle\phi^0\rangle \; . \tag{58}$$

Since $\langle\phi^0\rangle$ is fixed by the scale of the $SU(2) \times U(1)$ breakdown:

$$\langle\phi^0\rangle = \frac{1}{(\sqrt{2}\,G_F)^{1/2}} \sim 180 \text{ GeV} \; , \tag{59}$$

to get m_D to have a value in the eV range requires that $\Gamma \sim 10^{-11}$!

The situation is not much less artificial in the case of m_T. In this case, to get a non-zero value for m_T it is necessary to introduce a Higgs triplet field $\vec{\Delta}$. This field can couple to $L \otimes L$ so that if, indeed, $\vec{\Delta}$ gets a VEV one can generate a triplet mass m_T. In detail, the triplet coupling involving $\vec{\Delta}$ has the form

$$\mathcal{L}_{\text{triplet}} = -\frac{1}{2}\{\Gamma_T L^T \tilde{C}\vec{\tau} \cdot \vec{\Delta} L\} + \text{h.c.} \,, \tag{60}$$

where Γ_T is an unknown coupling constant. When the neutral component of $\vec{\Delta}$, Δ^o, gets a vacuum expectation, then $\mathcal{L}_{\text{triplet}}$ generates a mass term for ν_{L}:

$$\mathcal{L}^{\nu}_{\text{mass}} = -\frac{1}{2}\Gamma_T \langle \Delta^o \rangle \{\nu_{\text{L}}^T \tilde{C} \nu_{\text{L}}\} + \text{h.c.} \tag{61}$$

and $m_T = \Gamma_T \langle \Delta^o \rangle$. The only real constraint on $\langle \Delta^o \rangle$ comes from precision measurements of the ρ parameter, typifying the NC to CC ratio. Experimentally [1] one finds

$$\rho_{\text{exp}} = 1.00412 \pm 0.00124 \,. \tag{62}$$

The presence of the triplet Higgs interaction modifies the ρ parameter from unity at the tree level and one has: [7]

$$\rho = 1 - 2\left(\frac{\langle \Delta^o \rangle}{\langle \phi^0 \rangle}\right)^2 + \text{rad. corr.} \tag{63}$$

Using the error on ρ in Eq. (62) as an estimate of the size of $\langle \Delta^o \rangle$ implies that $\langle \Delta^o \rangle \leq 4$ GeV. So, also in this case, if $\langle \Delta^o \rangle$ is near this limit to get neutrino masses in the eV range one needs a Yukawa strength of order $\Gamma_T \sim 10^{-9}$. If $\langle \Delta^o \rangle << \langle \phi^0 \rangle$ then Γ_T can be larger, but one is left to explain the reason for the doublet-triplet VEV hierarchy.

Elementary Higgs triplets do not emerge very naturally in models. [4] However, one can always get an **effective** triplet out of two Higgs doublets:

$$\vec{\Delta} \sim \Phi^T C \vec{\tau} \Phi \tag{64}$$

where C is an appropriate charge conjugation matrix. Effective L-violating interactions involving pairs of doublet Higgs fields arise quite naturally in Grand Unified Theories [9] as dimension 5 terms:

$$\mathcal{L}^{d=5}_{\text{eff}} = \frac{g}{2\Lambda}(L^T \tilde{C} \vec{\tau} L) \cdot (\Phi^T C \vec{\tau} \Phi) + \text{h.c.} \tag{65}$$

In the above, Λ is a scale associated with the GUT breakdown scale and g is a coupling constant. Clearly the above interaction gives

---

4) An exception is provided by left-right symmetric models where triplets have often been considered to give the requisite symmetry breaking. [8]

$$m_T = \frac{g\langle\phi^0\rangle^2}{\Lambda} \; . \tag{66}$$

Since $\langle\phi^0\rangle \sim 10^2$ GeV, with $g \sim O(1)$, one gets $m_T \sim 10^{-2}$ eV for scales $\Lambda \sim 10^{15}$ GeV, which are typical of GUTs. Note that the above formula for m_T is quite similar in spirit to the see-saw expression for light neutrinos

$$(m_\nu)_{\rm light}^{\rm see-saw} \sim \frac{m_D^2}{m_S} \sim \frac{\langle\phi^0\rangle^2}{m_S} \; , \tag{67}$$

since $m_D \sim \langle\phi^0\rangle$. In either case, new physics at a large scale (either a large $\nu_{\rm R}$ mass scale m_S or the GUT scale Λ) produces a light neutrino. It is clearly more appealing physically to have light neutrinos be the result of new physics at high scales, rather than simply as a result of some Yukawa coupling being unnaturally small.

IV NEUTRINO OSCILLATIONS IN VACUUM

If neutrinos have mass then, in general, the neutrinos produced by the weak interactions (weak interaction eigenstates) are not states of definite mass (mass eigenstates). In the basis where the charged lepton mass matrix is diagonal [c.f. Eq. (25)], it follows from Eq. (23) that the neutrino weak interaction eigenstates are fixed by the corresponding lepton produced in the associated weak process. That is, the piece of the weak current $J_{\mu-}^{\rm lept}$ involving the charged lepton $\ell_i = \{e, \mu, \tau\}$ will **always** involve the corresponding neutrino weak interaction eigenstate $\nu_{\ell_i} = \{\nu_e, \nu_\mu, \nu_\tau\}$. These, left-handed, neutrino weak interaction eigenstates are superpositions of neutrino mass eigenstates ν_i:

$$\nu_{\ell_j} = \sum_i U_{\ell_j i}\nu_i \; . \tag{68}$$

The matrix $U_{\ell_j i}$, in general, is a 3×6 matrix. However, if the see-saw mechanism is operative, one expects that the contributions of **superheavy** neutrinos in Eq. (68) should be negligible. Then, to a very good approximation, the matrix $U_{\ell_j i}$ is a 3×3 unitary matrix

$$U_{\ell_j i}U^*_{\ell_k i} = \delta_{\ell_j \ell_k} \; . \tag{69}$$

For the moment I will restrict myself to the case when Eq. (69) holds. Furthermore, I will discuss the phenomenology of neutrino oscillations in the simple case of just 2 flavors of neutrinos, since this is how most of the data is usually presented. However, the formalism which we will develop can be generalized straightforwardly to three families of light neutrinos. [10] For definitiveness, let us consider then just ν_e and ν_μ weak interaction eigenstates. In this case, Eq. (68) reads, using a convenient quantum mechanical notation,

$$|\nu_e\rangle = \cos\theta|\nu_1\rangle + \sin\theta|\nu_2\rangle$$
$$|\nu_\mu\rangle = -\sin\theta|\nu_1\rangle + \cos\theta|\nu_2\rangle \ . \tag{70}$$

The mass eigenstates $|\nu_i\rangle$ have a time evolution which just follows from the Schrödinger equation:

$$|\nu_i(t)\rangle = e^{-iE_it}|\nu_i(0)\rangle \ ; \quad E_i = \sqrt{\vec{p}^2 + m_i^2} \ . \tag{71}$$

Because $m_1 \neq m_2$, it is easy to see that the weak interaction eigenstate ν_e produced at $t = 0$ evolves in time into a superposition of ν_e and ν_μ states. Taking by definition $|\nu_i\rangle \equiv |\nu_i(0)\rangle$, it follows that

$$\begin{aligned} |\nu_e(t)\rangle &= \cos\theta e^{-iE_1t}|\nu_1\rangle + \sin\theta e^{-iE_2t}|\nu_2\rangle \\ &= \left[\cos^2\theta e^{-iE_1t} + \sin^2\theta e^{-iE_2t}\right]|\nu_e\rangle + \left[\cos\theta\sin\theta(e^{-iE_2t} - e^{-iE_1t})\right]|\nu_\mu\rangle \\ &\equiv A_{ee}(t)|\nu_e\rangle + A_{e\mu}(t)|\nu_\mu\rangle \ . \end{aligned} \tag{72}$$

Using the above, one can compute immediately the probabilities that at time t the state $\nu_e(t)$ is either a ν_e or a ν_μ weak interaction eigenstate:

$$P(\nu_e \to \nu_e; t) = |A_{ee}(t)|^2 = 1 - \frac{1}{2}\sin^2 2\theta[1 - \cos(E_2 - E_1)t] \tag{73}$$
$$P(\nu_e \to \nu_\mu; t) = |A_{e\mu}(t)|^2 = \frac{1}{2}\sin^2 2\theta[1 - \cos(E_2 - E_1)t] \ . \tag{74}$$

Since the masses of neutrinos are small compared to the momentum, one can write

$$E_i \simeq |p| + \frac{m_i^2}{2|p|} \ ; \qquad t \simeq L \ , \tag{75}$$

where L is the distance travelled by the neutrinos in a time t. Using the above, one can write, for instance,

$$P(\nu_e \to \nu_\mu; L) = \frac{1}{2}sin^2 2\theta\left[1 - \cos\frac{\Delta m^2}{2|p|}L\right] = \sin^2 2\theta \sin^2\frac{\Delta m^2 L}{4|p|} \ , \tag{76}$$

where $\Delta m^2 = m_2^2 - m_1^2$. Numerically, it turns out that

$$\frac{\Delta m^2 L}{4|p|} \simeq 1.27\frac{\Delta m^2(\mathrm{eV}^2)L(m)}{|p|(\mathrm{MeV})} \ . \tag{77}$$

Recapitulating, for the case of 2 neutrino species, one gets the following formula quantifying the probability that a weak interaction eigenstate neutrino (ν_e) has oscillated to other weak interaction eigenstate neutrino (ν_μ) after traversing a distance L:

$$P(\nu_e \to \nu_\mu; L) = \sin^2 2\theta \sin^2 \left[\frac{1.27\Delta m^2(\text{eV}^2)L(m)}{|p|(\text{MeV})} \right] . \tag{78}$$

The probability that no such oscillation took place, of course, is just

$$P(\nu_e \to \nu_e; L) = 1 - P(\nu_e \to \nu_\mu; L) . \tag{79}$$

These probabilities depend on two factors: (i) a mixing angle factor $\sin^2 2\theta$ and (ii) a kinematical factor which depends on the distance travelled, on the momentum of the neutrinos, as well as on the difference in the squared mass of the two neutrinos. Obviously, for oscillations to be important the mixing factor $\sin^2 2\theta$ should be of $O(1)$. However, large mixing is not enough. It is also important that the kinematical factor $\Delta m^2(\text{eV}^2)L(m)/|p|(\text{MeV}) \gtrsim O(1)$, so that the second oscillatory factor in Eq. (78) can be significant.

It is useful to develop the formalism a bit more for future use. The probability amplitudes $A_{ee}(t)$ and $A_{e\mu}(t)$ of Eq. (72) can be recognized as matrix elements of a 2×2 matrix e^{-iHt} defined by

$$e^{-iHt} = U \; e^{-iH_{\text{diag}}t} \; U^\dagger , \tag{80}$$

where

$$U = \begin{pmatrix} \cos\theta & \sin\theta \\ -\sin\theta & \cos\theta \end{pmatrix} \tag{81}$$

is the mixing matrix of the weak interaction eigenstates and

$$H_{\text{diag}} = \begin{pmatrix} E_1 & 0 \\ 0 & E_2 \end{pmatrix} . \tag{82}$$

One has

$$A_{ee}(t) = [e^{-iHt}]_{11} ; \qquad A_{e\mu}(t) = [e^{-iHt}]_{12} . \tag{83}$$

Using the fact that $E_i = |p| + m_i^2/2|p|$, it proves convenient to separate H_{diag} into two different pieces:

$$H_{\text{diag}} = \left(|p| + \frac{m_1^2 + m_2^2}{4|p|} \right) \begin{pmatrix} 1 & 0 \\ 0 & 1 \end{pmatrix} + \frac{\Delta m^2}{4|p|} \begin{bmatrix} -1 & 0 \\ 0 & 1 \end{bmatrix} . \tag{84}$$

The first piece above, because it is proportional to the unit matrix, gives an overall phase factor which is irrelevant for calculating the neutrino oscillation probabilities. Hence, effectively, one can replace

$$H_{\text{diag}} \to H_o = -\frac{\Delta m^2}{4|p|}\sigma_3 . \tag{85}$$

In view of (85), it is convenient to define the 2×2 Hamiltonian matrix H_{vac} by

$$
\begin{aligned}
H_{\text{vac}} = U\ H_o\ U^\dagger &= \frac{\Delta m^2}{4|p|}\begin{pmatrix} -\cos 2\theta & \sin 2\theta \\ \sin 2\theta & \cos 2\theta \end{pmatrix} \\
&= \frac{\Delta m^2}{4|p|}\{\sin 2\theta\sigma_1 - \cos 2\theta\sigma_3\}\ . \qquad (86)
\end{aligned}
$$

Then, effectively,

$$
A_{ee}(t) = [e^{-iH_{\text{vac}}t}]_{11}\ ; \qquad A_{e\mu}(t) = [e^{-iH_{\text{vac}}t}]_{12}\ . \qquad (87)
$$

Just as the above coefficients describe the time evolution of a state that started at $t = 0$ as a ν_e weak interaction eigenstate [cf. Eq. (72)], one can define coefficients

$$
A_{\mu e}(t) = [e^{-iH_{\text{vac}}t}]_{21}\ ; \qquad A_{\mu\mu}(t) = [e^{-iH_{\text{vac}}t}]_{22} \qquad (88)
$$

which will detail the time evolution of a state which started out at $t = 0$ as a ν_μ weak interaction eigenstate:

$$
|\nu_\mu(t)\rangle = A_{\mu e}(t)|\nu_e\rangle + A_{\mu\mu}(t)|\nu_\mu\rangle\ . \qquad (89)
$$

It is easy to deduce from these considerations that the 2×2 Hamiltonian H_{vac} is just the Hamiltonian which enters in the Schrödinger equation for $|\nu_e(t)\rangle$ and $|\nu_\mu(t)\rangle$:

$$
i\frac{\partial}{\partial t}\begin{bmatrix} |\nu_e(t)\rangle \\ |\nu_\mu(t)\rangle \end{bmatrix} = H_{\text{vac}}\begin{bmatrix} |\nu_e(t)\rangle \\ |\nu_\mu(t)\rangle \end{bmatrix}\ . \qquad (90)
$$

V NEUTRINO OSCILLATIONS IN MATTER

When neutrinos propagate in matter, a subtle but important effect takes place which alters the ways in which neutrinos oscillate into one another. The origin of this effect, which is known as the MSW effect for the initials of the physicists who first discussed it, [11] is connected to the fact that the electron neutrinos can interact in matter also through charged current interactions. While all neutrino species have the same interactions in matter due to the neutral currents, the ν_e weak interaction eigenstates, because of their charged current interactions, as they propagate in matter experience a slightly different index of refraction than the ν_μ weak interaction eigenstates (and the ν_τ weak interaction eigenstates). This different index of refraction for ν_e alters the time evolution of the system from what it was in vacuum.

Let us again consider the two-neutrino case. The relative index of refraction between ν_e and ν_μ is the result of the difference between the forward scattering amplitudes for ν_e and ν_μ, caused by the charged current interactions of the ν_e. In detail, one has [11]

$$1 - n_{\rm rel} = -\frac{2\pi N_e}{|p|^2}\left[A(0)|_{\nu_e e} - A(0)|_{\nu_\mu e}\right] \tag{91}$$

where N_e is the electron density and $A(0)$ is the forward scattering amplitude. The contribution of the neutral current interactions cancels in Eq. (91), while the charged current contribution to $A(0)|_{\nu_e e}^{\rm CC}$ gives [11]

$$1 - n_{\rm rel} = \frac{\sqrt{2}G_F\, N_e}{|p|} . \tag{92}$$

One can use Eq. (92) and the formalism we developed at the end of the last section to study the evolution of neutrinos in matter. The relevant Hamiltonian now is

$$H = H_{\rm vac} + |p|(1 - n_{\rm rel})\begin{bmatrix} 1 & 0 \\ 0 & 0 \end{bmatrix} . \tag{93}$$

Again, because relative phases are irrelevant, we can subtract from the above a term proportional to the identity. This yields the following effective Hamiltonian describing the propagation of neutrinos in matter:

$$\begin{aligned} H_{\rm matter} &= H_{\rm vac} + |p|\frac{(1 - n_{\rm rel})}{2}\begin{bmatrix} 1 & 0 \\ 0 & -1 \end{bmatrix} \\ &= \frac{\Delta m^2}{4|p|}\sin 2\theta\sigma_1 - \left(\frac{\Delta m^2}{4|p|}\cos 2\theta - \frac{G_F\, N_e}{\sqrt{2}}\right)\sigma_3 . \end{aligned} \tag{94}$$

Because of the term proportional to the electron density in Eq. (94), in matter it is no longer true that the eigenstates of $H_{\rm matter}$ are ν_1 and ν_2. Calling these matter eigenstates ν_1^M and ν_2^M, one has that

$$\begin{bmatrix} |\nu_e\rangle \\ |\nu_\mu\rangle \end{bmatrix} = \begin{pmatrix} \cos\theta_M & \sin\theta_M \\ -\sin\theta_M & \cos\theta_M \end{pmatrix}\begin{bmatrix} |\nu_1^M\rangle \\ |\nu_2^M\rangle \end{bmatrix} \equiv U_M \begin{bmatrix} |\nu_1^M\rangle \\ |\nu_2^M\rangle \end{bmatrix} \tag{95}$$

and

$$H_{\rm matter} = U_M H_{\rm matter}^{\rm diag} U_M^\dagger . \tag{96}$$

It is easy to check that $H_{\rm matter}^{\rm diag}$ is given by

$$H_{\rm matter}^{\rm diag} = -\sigma_3\left[\left(\frac{\Delta m^2}{4|p|}\cos 2\theta - \frac{G_F\, N_e}{\sqrt{2}}\right)^2 + \left(\frac{\Delta m^2}{4|p|}\sin 2\theta\right)^2\right]^{1/2} , \tag{97}$$

with the mixing angle in matter θ_M determined by the equation

$$\sin 2\theta_M = \frac{\frac{\Delta m^2}{4|p|}\sin 2\theta}{\left[\left(\frac{\Delta m^2}{4|p|}\cos 2\theta - \frac{G_F\ N_e}{\sqrt{2}}\right)^2 + \left(\frac{\Delta m^2}{4|p|}\sin 2\theta\right)^2\right]^{1/2}} \ . \tag{98}$$

The presence of the term proportional to the electron density gives rise to interesting resonance phenomena. [12] There is a **critical density** N_e^{crit}, given by

$$N_e^{\text{crit}} = \frac{\Delta m^2 \cos 2\theta}{2\sqrt{2}|p|G_F} \ , \tag{99}$$

for which the matter mixing angle θ_M becomes **maximal** ($\sin 2\theta_M \to 1$), irrespective of what the vacuum mixing angle θ is. If one is in such a medium, then $H_{\text{matter}}^{\text{diag}}$ reduces to

$$H_{\text{matter}}^{\text{diag}}\Big|_{N_e = N_e^{\text{crit}}} = -\frac{\Delta m^2}{4|p|}\sin 2\theta\sigma_3 \ . \tag{100}$$

The probability that a ν_e transmutes into a ν_μ after traversing a distance L in this medium is given by Eq. (78), with two differences. First, since we are in a medium $\sin 2\theta \to \sin 2\theta_M$. However, because the density is assumed to be the critical density, $\sin 2\theta_M \to 1$. Second, since $H_{\text{matter}}^{\text{diag}}$ in Eq. (100) differs from H_o by the replacement of $\Delta m^2 \to \Delta m^2 \sin 2\theta$, such a replacement also will enter in the kinematical factor in the probability formula. Hence, it follows that

$$P_{\text{matter}}\left(\nu_e \to \nu_\mu; L\right)\big|_{N_e = N_e^{\text{crit}}} = \sin^2\left(\frac{\Delta m^2}{4|p|}\sin 2\theta L\right) \ . \tag{101}$$

This formula shows that one can get **full conversion** of a ν_e weak interaction eigenstate into a ν_μ weak interaction eigenstate, provided that the length L and momentum $|p|$ satisfy the relation

$$\frac{\Delta m^2}{4|p|}\sin 2\theta L = \frac{n\pi}{2} \ ; \qquad n = 1, 2, \ldots \ . \tag{102}$$

There is a second interesting limit to consider. [12] This is when the electron density N_e is so large that it overwhelms the other terms in $H_{\text{matter}}^{\text{diag}}$. If $G_F\ N_e \gg \Delta m^2/2\sqrt{2}|p|$, then one has, approximately,

$$H_{\text{matter}}^{\text{diag}} = -\sigma_3 \frac{G_F\ N_e}{\sqrt{2}} \ . \tag{103}$$

In this limit, it is easy to check that $\sin 2\theta_M \to 0$; $\cos 2\theta_M \to -1$, so that $\theta_M \to \frac{\pi}{2}$. In this case, there are no oscillations in matter because $\sin 2\theta_M$ vanishes

$$P_{\text{matter}}\left(\nu_e \to \nu_\mu; L\right)\big|_{N_e \gg \frac{\Delta m^2}{2\sqrt{2}|p|G_F}} \to 0 \ . \tag{104}$$

This actually is immediate also since, in this limit, $H_{\rm matter}$ itself is diagonal

$$H_{\rm matter} = U_M H^{\rm diag}_{\rm matter} U^\dagger_M = \sigma_3 \frac{G_F\ N_e}{\sqrt{2}} \ . \tag{105}$$

Hence the Schrödinger equation (90), with $H_{\rm vac} \to H_{\rm matter}$, is diagonal and there can be no transitions. For future use I note that in this limit, since $\theta_M = \pi/2$, the ν_e weak interaction eigenstate in matter coincides with the state ν_2^M:

$$|\nu_e\rangle = \cos\theta_M |\nu_1^M\rangle + \sin\theta_M |\nu_2^M\rangle \underset{\theta_M=\pi/2}{\to} |\nu_2^M\rangle \ . \tag{106}$$

VI EVIDENCE FOR NEUTRINO MASSES

Most experiments searching for direct evidence for neutrino masses have, up to now, only set limits on these masses and the associated mixing angles. However, there are now both strong hints, and some real evidence, that neutrino masses really exist coming from neutrino oscillation experiments.

Most oscillation data is presented as an allowed region, or limits at some confidence level, in a $\Delta m^2 - \sin^2 2\theta$ plot. That is, experimentalists find it convenient to quantify their results using the formalism discussed in the last section, involving oscillations among two neutrino species ν_α and ν_β The oscillation probability formulas for $P(\nu_\alpha \to \nu_\beta; L)$ [cf. Eq. (78)] involves both the $\alpha - \beta$ mixing angle $\theta_{\alpha\beta}$, as well as a kinematical factor depending on the mass squared difference Δm^2 between the mass eigenstates in the two neutrino system. Because neutrino beams have a rather large energy spread, for $\Delta m^2 \gg |p|/L$ the kinematical oscillating factor in Eq. (78) averages to 1/2. This implies that, in general, the sensitivity to a signal for neutrino oscillations goes down to $\sin^2 2(\theta_{\alpha\beta})_{\rm min} \simeq 2P(\nu_\alpha \to \nu_\beta; L)$.

A Direct Mass Measurements

The classical way to try to infer a non-vanishing value for neutrino masses is by measuring β-decay spectra near their endpoint. The presence of neutrino masses alters the dependence of the measured intensity $I(T_e)$ on the electron kinetic energy T_e as it approaches the maximum energy release Q. One has [13]

$$I(T_e) = (Q - T_e) \sum_i |U_{ei}|^2 [(Q-T_e)^2 - m^2_{\nu_i}]^{1/2} \ . \tag{107}$$

If $m_{\nu_i} = 0$ then the intensity spectrum is quadratically dependent on the energy release $(Q - T_e)$. If neutrinos have mass, one has a spectrum distorsion and $\sqrt{I}$ is no longer linear in $(Q - T_e)$, but vanishes at some value of T_e less than the maximum energy released Q. These distorsions are best detected in β-decay spectra which have low Q values; an ideal candidate being Tritium where $Q = 18.6$ KeV.

TABLE 1. Neutrino Mass Limits from ^{3}He β-decay, from Ref. [14].

Experiment	"$m^2_{\nu_e}$"(eV^2)	"m_{ν_e}"(eV)
Tokyo	$-65 \pm 85 \pm 65$	< 13.1
Los Alamos	$-147 \pm 68 \pm 41$	< 9.3
Zürich	$-24 \pm 48 \pm 61$	< 11.7
Livermore	$-130 \pm 20 \pm 15$	< 7.0
Mainz	$-22 \pm 17 \pm 14$	< 5.6
Troitsk	$1.5 \pm 5.9 \pm 3.6$	< 3.9

Tritium β-decay experiments are sensitive to neutrino masses in the "few eV" range. Remarkably, most of the high precision experiments performed with tritium actually see an **excess** of events near the end-point, setting poorer limits than their theoretical sensitivity. [14] However, very recently, the Troitsk experiment [15] has been able to determine a very stringent result for the largest eigenvalue "m_{ν_e}" principally contributing to Tritium β-decay:

$$\text{“}m^2_{\nu_e}\text{”} = (1.5 \pm 5.9 \pm 3.6)\ \text{eV}^2\ ; \quad \text{“}m_{\nu_e}\text{”} < 3.9\ \text{eV}\ \ (90\%\ \text{C.L.}) \tag{108}$$

Table 1 gives a compilation of the existing β-decay results in Tritium and the corresponding limit for "m_{ν_e}".

Similar, but less accurate, kinematical bounds are also known for the largest eigenvalues principally contributing to decays involving ν_μ and ν_τ weak eigenstates. Denoting these eigenvalues, respectively, as "m_{ν_μ}" and "m_{ν_τ}", one finds the following results. From studying $\pi^+ \to \mu^+\nu_\mu$ decay at PSI [16] one has

$$\text{“}m^2_{\nu_\mu}\text{”} = (-0.016 \pm 0.028)\ \text{MeV}^2;\quad \text{“}m_{\nu_\mu}\text{”} < 170\ \text{KeV}\ (90\% C.L.)\ . \tag{109}$$

From studying the decay $\tau \to \nu_\tau + 5\pi$ at LEP [17] one has

$$\text{“}m_{\nu_\tau}\text{”} < 18.2\ \text{MeV}\ (95\%\ \text{C.L.})\ . \tag{110}$$

A different, and in some ways more interesting, limit on the neutrinos associated with the ν_e weak interaction eigenstate comes from neutrinoless double β-decay. This process, if it exists, violates lepton number. Thus it is only possible if neutrinos have a Majorana mass. Ordinary double β-decay $Z \to (Z+2) + 2e^- + 2\bar{\nu}_e$ conserves lepton number. However, in a double β-decay processes where no neutrinos are emitted $Z \to (Z+2) + 2e^-$, lepton number is violated. As shown schematically in Fig. 1, these processes can only occur if there is a neutrino-antineutrino transition engendered by the presence of a Majorana mass term.

There is now a variety of measurements of ordinary double β-decay, [18] but up to now there are only limits on neutrinoless double β-decay. [19] The half-life for these later processes is a measure of the neutrino Majorana mass associated with these decays:

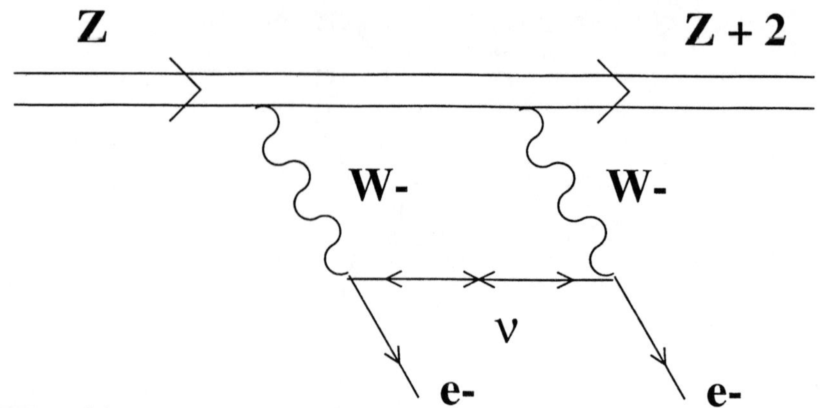

FIGURE 1. Schematic diagram which gives rise to neutrinoless double β-decay if neutrinos have a Majorana mass.

$$\left[T_{\frac{1}{2}}^{2\beta o\nu}\right]^{-1} \sim \langle m_{\nu_e}\rangle_{ee}^2 \,. \tag{111}$$

Here the mass $\langle m_{\nu_e}\rangle_{ee}$ is given by

$$\langle m_{\nu_e}\rangle_{ee} = \sum_i U_{ei}^2 m_{\nu_i}, \tag{112}$$

with m_{ν_i} being the neutrino mass eigenstates. Note that $\langle m_{\nu_e}\rangle_{ee}$ vanishes if the neutrinos are Dirac particles. That is, if neutrinos have only lepton number conserving Dirac masses.

This point can be appreciated readily by considering, for simplicity, the case of one neutrino species. In this case, if these neutrinos only have a Dirac mass m_D, the corresponding 2×2 neutrino mass matrix M has the form

$$M = \begin{pmatrix} 0 & m_D \\ m_D & 0 \end{pmatrix} \,. \tag{113}$$

This matrix is diagonalized to

$$M_{\rm diag} = \begin{pmatrix} -m_D & 0 \\ 0 & m_D \end{pmatrix} \tag{114}$$

by the orthogonal matrix

$$U = \begin{pmatrix} \frac{1}{\sqrt{2}} & \frac{1}{\sqrt{2}} \\ -\frac{1}{\sqrt{2}} & \frac{1}{\sqrt{2}} \end{pmatrix} \,. \tag{115}$$

Using Eq. (115), for this case one easily checks that $\langle m_{\nu_e}\rangle_{ee}$ vanishes. One has

TABLE 2. Bounds on neutrinoless double β-decays half lives and associated bounds on $\langle m_{\nu_e}\rangle_{ee}$, from Ref. [14].

Decay	$T_{\frac{1}{2}}^{2\beta o\nu}$ (Years)	$\langle m_{\nu_e}\rangle_{ee}$ (eV)
$^{76}Ge \to {}^{76}Se$	$> 5.7 \times 10^{25}$ (90% C.L.)	< 0.2 (90% C.L.)
$^{128}Te \to {}^{128}Xe$	$> 7.7 \times 10^{24}$ (68% C.L.)	< 1.1 (68% C.L.)
$^{130}Te \to {}^{130}Xe$	$> 5.6 \times 10^{22}$ (90% C.L¿)	< 3.0 (90% C.L.)
$^{136}Xe \to {}^{136}Ba$	$> 4.4 \times 10^{23}$ (90% C.L.)	< 2.3 (90% C.L.)

$$\langle m_{\nu_e}\rangle_{ee} = \left(\frac{1}{\sqrt{2}}\right)^2 (-m_D) + \left(\frac{1}{\sqrt{2}}\right)^2 (m_D) = 0 \ . \tag{116}$$

Table 2 reproduces a recent compilation of experimental results on neutrinoless double β-decay. [14] The most sensitive of these experiments involves the double β-decay of ^{76}Ge to ^{76}Se, with a half-life limit of over 10^{25} years. [20] The resulting bound on $\langle m_{\nu_e}\rangle_{ee}$ quoted is

$$\langle m_{\nu_e}\rangle_{ee} < 0.2 \text{ eV} \quad (90\% \text{ C.L.}) \ . \tag{117}$$

This bound, however, has probably a factor of two uncertainty due to uncertainties associated with calculating the nuclear matrix elements involved in the decay. [21]

B Cosmological Constraints

There are some indirect constraints on neutrino masses provided by cosmology. The most relevant is the constraint which follows from demanding that the energy density in neutrinos should not overclose the Universe. Neutrinos are thermal relics; they decoupled from the Universe's expansion when their interaction rate Γ fell below the Universe's expansion rate H. [22] In the usual Robertson-Walker expanding Universe, the rate of expansion $H \sim T^2/M_{\rm P}$, where $M_{\rm P} \sim 10^{19}$ GeV is the Planck mass and T is the Universe's temperature. Since

$$\Gamma = n_\nu \langle \sigma v \rangle \sim G_F^2 T^5 \ , \tag{118}$$

with G_F the Fermi constant, $G_F \sim 10^{-5}$ GeV^{-2}, decoupling occurs at a temperature T_D determined by setting $\Gamma \simeq H$. This gives

$$T_D \simeq \left(\frac{1}{G_F^2 M_{\rm P}}\right)^{1/3} \sim 1 \text{ MeV} \ . \tag{119}$$

Two cases are of interest. If neutrinos have a mass much less than T_D ($m_\nu \ll T_D$) they are **hot relics**. That is, they are relativistic at the time of decoupling. For hot relics, the density of neutrinos is comparable to that of photons at decoupling:

$n_\nu \sim n_\gamma|_{T_D}$. **Cold relics**, on the other hand, are neutrinos whose mass is much greater than T_D ($m_\nu \gg T_D$). In this case, at the time of decoupling the neutrino density n_ν, because of the Boltzmann factor, is much below that of the photons. Thus for cold relics, $n_\nu \ll n_\gamma|_{T_D}$.

In either case, one can compute the neutrino contribution to the energy density of the Universe. [22] This is simplest for the case of hot relics, since their number density essentially tracks the photon number density.[5] Thus the number density of neutrinos now is fixed by the measured temperature of the microwave background radiation:

$$n_\nu = \frac{3\zeta(3)}{2\pi^2} T_\nu^3 . \tag{120}$$

Hence the contribution of neutrinos to the present energy density of the Universe is

$$\rho_\nu = n_\nu \sum_i m_{\nu_i} . \tag{121}$$

It has become conventional to normalize all densities in terms of the Universe's closure density ρ_c:

$$\rho_c = \frac{3H_o^2}{8\pi G_N} \simeq 1.9 \times 10^{-29} h^2 \frac{\text{g}}{\text{cm}^3} \simeq 1.1 \times 10^4 h^2 \frac{\text{eV}}{\text{cm}^3} . \tag{122}$$

In the above H_o is the Hubble constant and h is a measure of its uncertainty. One finds

$$H_o = 100h \frac{\text{Km}}{\text{sec Mpsec}} \tag{123}$$

with [24]

$$h = 0.65 \pm 0.1 . \tag{124}$$

Defining

$$\Omega_\nu = \frac{\rho_\nu}{\rho_c} , \tag{125}$$

then, for hot relics, one has

$$\Omega_\nu^{\text{Hot}} = \frac{\sum_i m_{\nu_i}}{92 \text{ eV } h^2} . \tag{126}$$

---

5) Because of photon reheating at the time of recombination (and a small statistical difference because neutrinos are fermions and photons are bosons), the neutrino temperature T_ν is not quite the same as the photon temperature T_γ. One finds $T_\nu = \left(\frac{4}{11}\right)^{1/3} T_\gamma$. [23]

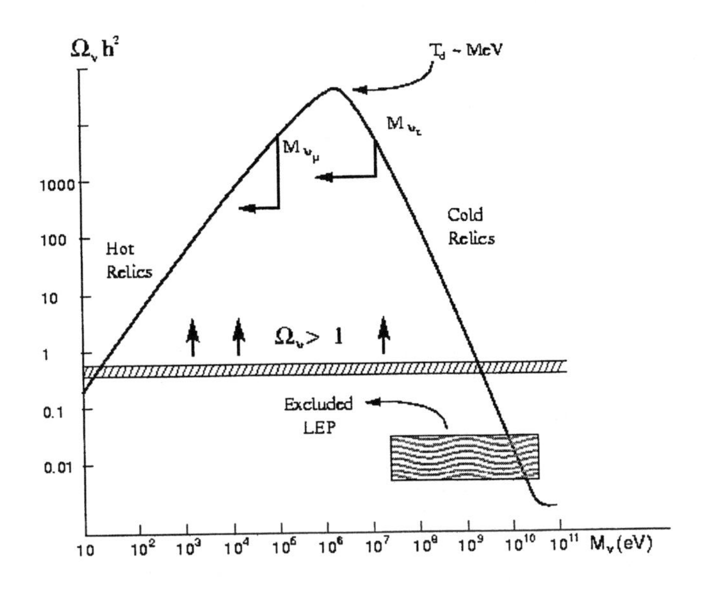

FIGURE 2. Plot of $\Omega_\nu h^2$ as a function of the neutrino (sum) mass.

It follows from the above that if the sum of neutrino masses $\sum_i m_{\nu_i} \simeq 30$ eV, then neutrinos would close the Universe. Because we know that the Universe is not very far from closure density, if neutrinos are hot relics the sum of their masses cannot be much above 30 eV. Thus, although direct bounds allows for a presence of a "ν_μ" neutrino with mass "m_{ν_μ}" less than 170 KeV, cosmology forbids neutrinos to have masses as large as that. [6]

When neutrino masses are above $T_D \sim$ MeV, then the simple formula given in Eq. (126) no longer applies. Nevertheless, it is still possible to compute Ω_ν taking into account now of the appropriate Boltzmann factor. Figure 2, adapted from [22], plots $\Omega_\nu h^2$ as a function of the neutrino (sum) mass. This quantity rises linearly with mass up to $m_\nu \sim 1$ MeV and then drops rather rapidly. Cosmology allows neutrino masses for which $\Omega_\nu \leq 1$. So, as mentioned above, "m_{ν_μ}" and "m_{ν_τ}" must really be well below their kinematical bounds. On the other hand, we note that our bounds for "m_{ν_e}" and $\langle m_{\nu_e} \rangle_{ee}$ lie in a cosmologically allowed region. In principle, cosmology also allows neutrinos to exist with masses greater than a few GeV, since these cold relics give $\Omega_\nu \leq 1$. However, as we discussed earlier, neutrino counting at LEP excludes additional neutrinos besides ν_e, ν_μ and ν_τ, with mass $m_\nu < M_Z/2$. This exclusion region is also indicated in Fig. 2.

---

6) These cosmological bounds can be avoided if the massive neutrinos were unstable and had a sufficiently short lifetime. [25]

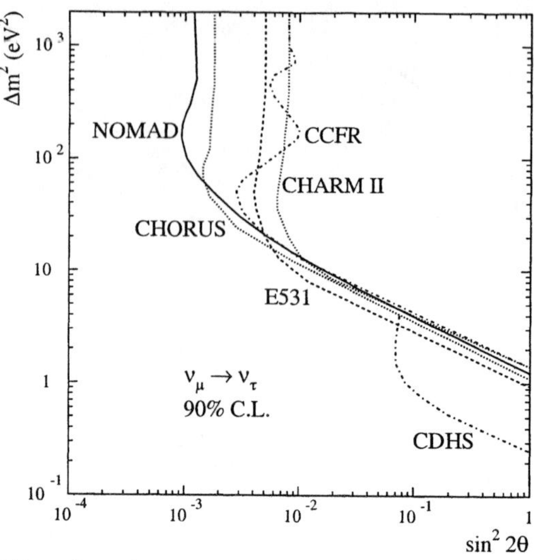

FIGURE 3. Bounds on $\nu_\mu \to \nu_\tau$ oscillations for large Δm^2, from [27].

C Accelerator limits and hints for neutrino masses

Two experiments at CERN, using ν_μ beams of average neutrino energies $\langle E_{\nu_\mu} \rangle \sim$ 15 GeV and typical decay lengths $L \sim 1$ Km, put strong limits on $\nu_\mu \to \nu_\tau$ neutrino oscillations for $\Delta m^2 \geq 10$ eV^2. In view of the discussion in the last subsection, this is a very interesting mass range to explore, since neutrinos with masses in this range could be of cosmological interest.

These two experiments use quite different techniques to detect ν_τ's. CHORUS [26] uses an emulsion target to try to detect the τ track produced in ν_τ charged current interactions. NOMAD, [27] on the other hand, uses drift chambers and kinematical techniques to detect a ν_τ signal. The result of both experiments are shown in Fig. 3, along with limits obtained by some earlier experiments. The CHORUS and NOMAD results for $\Delta m^2 \geq 10$ eV^2 exclude oscillations with mixing angles $\sin^2 2\theta_{\mu\tau} \geq 10^{-3}$ at the 90% C.L., improving previous limits by about a factor of 5.

Because NOMAD has a very good electron identification, this experiment is also able to set a strong limit for $\nu_\mu \to \nu_e$ oscillations. For $\Delta m^2 \geq 10$ eV^2, one excludes oscillations with mixing angles $\sin^2 2\theta_{\mu e} > 2 \times 10^{-3}$ at 90% C.L. Finally, because the ν_μ beam at CERN has about a 1% ν_e admixture, both experiments are also able to exclude $\nu_e \to \nu_\tau$ oscillations for $\Delta m^2 \geq 10$ eV^2, but now only for mixing angles $\sin^2 2\theta_{e\tau} \geq 10^{-1}$, at 90% C.L.

The situation regarding neutrino oscillations is much less clear cut in the region

$\Delta m^2 \leq 10$ eV^2. Here there are limits from past accelerator and reactor searches for oscillations and recent bounds from the KARMEN experiment. [28]. However, there is also some evidence for $\nu_\mu \rightarrow \nu_e$ oscillations coming from the LSND experiment. [29] Let me begin by briefly detailing the results of this last experiment first. LSND studies neutrinos originating from pions produced at rest at the LAMPF beam stop by an 800 MeV proton beam. These neutrinos, which are produced in the chain $\pi^+ \rightarrow \mu^+\nu_\mu \rightarrow e^+\nu_e\bar{\nu}_\mu\nu_\mu$, have average momenta in the 30 to 50 MeV range. What LSND looks for is the oscillation of the $\bar{\nu}_\mu$ produced in μ^+ decay into a $\bar{\nu}_e$, using a delayed coincidence in a target 30 meters from the beam dump. If $\bar{\nu}_\mu \rightarrow \bar{\nu}_e$ oscillations take place, the $\bar{\nu}_e$ inverse β-decay in the target ($\bar{\nu}_e p \rightarrow e^+ n$) produces a prompt photon from e^+e^- annihilation, while the produced neutron gives a delayed photon, as a result of the process $np \rightarrow d\gamma$.

The LSND experiment observes an excess of $e^+\gamma$ coincidence events which, if interpreted as $\bar{\nu}_\mu \rightarrow \bar{\nu}_e$ oscillations, give a substantial **allowed region** in the $\Delta m^2 -$ $\sin^2 2\theta$ plane. However, as I mentioned above, other experiments performed in the past, [30] as well as the recent KARMEN experiment, [28] exclude almost all of this allowed region. Furthermore, new data from the KARMEN2 detector [31] which became available in summer 1998 appeared to exclude even the small remaining allowed region for LSND!

This rather confusing situation is displayed in Fig. 4. It was discussed in some detail in the summary talk of Janet Conrad at the 1998 Vancouver International Conference on High Energy Physics. [32] As one can see from Fig. 4, a combination of the BNL 776 data and the Bugey reactor data only leaves the region between 0.2 eV$^2 < \Delta m^2 < 4$ eV^2 as an "allowed" region for the LSND signal. However, this region is essentially excluded by the KARMEN2 data, if one uses the 90% C.L. bound from this experiment. However, this result itself is somewhat anomalous, since the 90% C.L. sensitivity for KARMEN2 is actually below the LSND signal.[7] Furthermore, the LSND experiment has also looked for $\nu_\mu \rightarrow \nu_e$ oscillations by studying ν_e quasielastic scattering events and the collaboration, again, find an excess of events. citeLSND2 If interpreted as resulting from oscillations, this additional signal gives a $\Delta m^2 - \sin^2 2\theta$ allowed region which is consistent with that obtained by the $\bar{\nu}_\mu \rightarrow \bar{\nu}_e$ analysis.

It is difficult to make strong statements at this stage. The best that one can say is that there are hints of $\nu_\mu \rightarrow \nu_e$ oscillations in the region 0.2 eV$^2 < \Delta m^2 < 4$ eV^2, with rather small mixing angles $\sin^2 2\theta_{\mu e} \sim 10^{-2}$.

---

[7]) The 90% C.L. for KARMEN2 of Fig. 4 uses data only from the initial part of their run-where no background events were seen, even though 3 events were expected. Additional data from KARMEN2 now appears to have the number of background events expected. [33] As a result, it looks like the full KARMEN2 results will probably be closer to the 90% C.L. sensitivity line in Fig. 4.

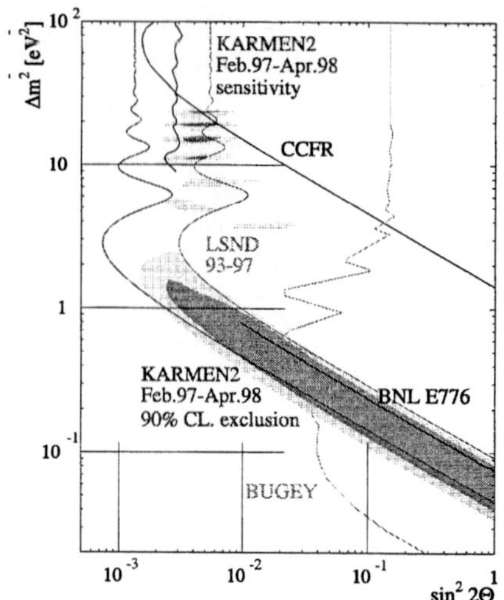

FIGURE 4. Summary of the experimental situation for $\bar{\nu}_\mu \to \bar{\nu}_e$ oscillations at the time of the Vancouver Conference, from Ref. [28].

D Atmospheric Neutrino Oscillations.

Large underground detectors, originally conceived to search for proton decay, are sensitive to the flux of neutrinos produced in the atmosphere. These neutrinos are mostly produced through the decay of pions, with the $\pi \to \mu \to e$ chain producing two ν_μ neutrinos and antineutrinos for each ν_e neutrino and antineutrino. One has known since the early 1990's that the observed flux of ν_μ's appeared to be much smaller than expected, with the ratio [35]

$$R = \frac{\left(\frac{\nu_\mu}{\nu_e}\right)_{\rm observed}}{\left(\frac{\nu_\mu}{\nu_e}\right)_{\rm expected}} \simeq 0.6 \; . \tag{127}$$

Although the anomalous ratio R could be the result of neutrino oscillations, strong evidence for neutrino oscillations only emerged in summer 1998 from the SuperKamiokande experiment. The SuperKamiokande collaboration [36] reported a pronounced zenith angle dependence for the flux of multi-GeV ν_μ neutrinos, but no such dependence from ν_e neutrinos. For neutrino energies in the multi-GeV range, the neutrino fluxes are not affected by geomagnetic effects in an asymmetric fashion. Thus one expects the observed neutrino signal to be **up-down symmetric**. As can be seen in Fig. 5, the SuperKamiokande data for multi-GeV ν_μ's is clearly up-down asymmetric. There are 139 up-going ν_μ compared to 256 down going events. The observed asymmetry

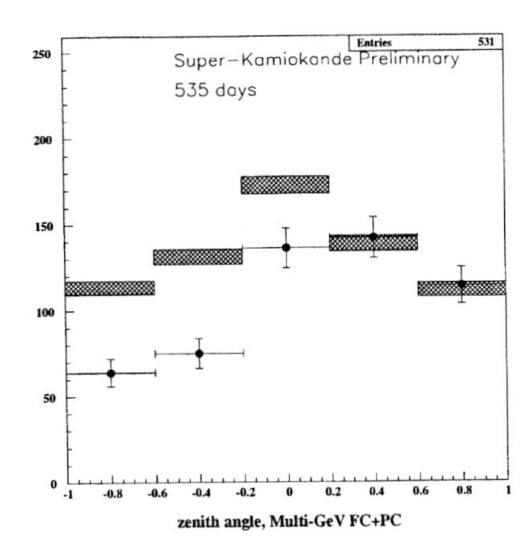

FIGURE 5. SuperKamiokande results on multi-GeV ν_μ events.

$$\left(\frac{U-D}{U+D}\right)_{\nu_\mu}^{\text{Multi-GeV}} = -0.296 \pm 0.048 \pm 0.010 \tag{128}$$

is a 6σ effect. The corresponding asymmetry for multi-GeV ν_e

$$\left(\frac{U-D}{U+D}\right)_{\nu_e}^{\text{Multi-GeV}} = -0.036 \pm 0.067 \pm 0.020 \tag{129}$$

is quite consistent with zero.

The SuperKamiokande collaboration [36] interprets these results as evidence for $\nu_\mu \to \nu_X$ oscillations, with ν_X some other neutrino species. This is most dramatically demonstrated in Fig. 6 where the ratio of data to Monte Carlo is plotted as a function of L/E_ν for both ν_e and ν_μ events. No L/E_ν dependence is seen in the ν_e data, but the ν_μ data drops down to a value of 1/2 for $L/E_\nu \geq 10^3$ Km/GeV. Recalling the simple 2-neutrino formula for the probability of oscillations [Eq. (78)], Fig. 6 suggest immediately that $\Delta m^2 \sim 10^{-3}$ eV^2 and that the mixing angle θ is near maximal. This is confirmed by a more detailed analysis, which for $\nu_\mu \to \nu_X$ oscillations gives $\sin^2 2\theta = 1$ and $\Delta m^2 = 2.2 \times 10^{-3}$ eV^2 as the best fit point.

It is unlikely, however, that the SuperKamiokande results are due to $\nu_\mu \to \nu_e$ oscillations (that is, that $\nu_X \equiv \nu_e$). First, the region in the $\Delta m^2 - \sin^2 2\theta$ plane

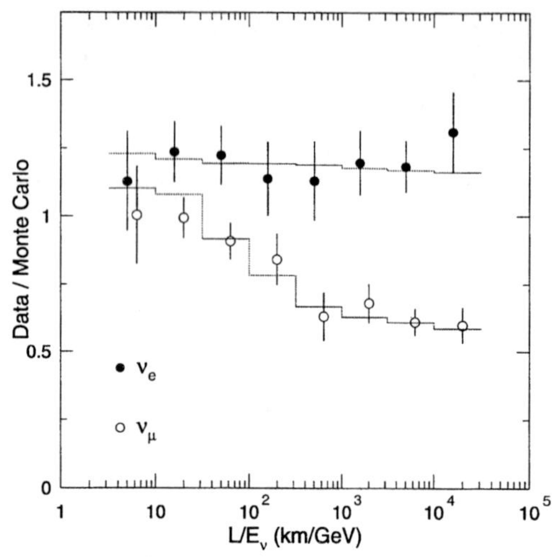

FIGURE 6. Plot of neutrino signals in the SuperKamiokande experiment as a function of L/E_ν.

favored by the SuperKamiokande results, is almost totally excluded already by the null results of the CH00Z reactor experiment [37] which looks at ν_e oscillations into another neutrino species. Furthermore, the up-down ratio (129) for the ν_e flux is more than 3σ away from what one would expect if one were dealing with $\nu_\mu \to \nu_e$ oscillations, where one expects

$$\left(\frac{U-D}{U+D}\right)^{\text{theory}}_{\nu_\mu \to \nu_e} = 0.205 \ . \tag{130}$$

If there are only three neutrino species, then most likely what is being seen in SuperKamiokande are $\nu_\mu \to \nu_\tau$ oscillations. However, at this stage, it is not possible to rule out the possibility that ν_X may be a sterile neutrino ν_s.[8]

The SuperKamiokande results [36] are consistent with previous Kamiokande results, [38] which also had indicated a (less pronounced) zenith angle dependence of the ν_μ flux. Although the $\Delta m^2 - \sin^2 2\theta$ regions for SuperKamiokande and Kamiokande do not appear to overlap much, the 90% C.L. region of SuperKamiokande is "more-significant", since the Kamiokande best fit has $\sin^2 2\theta = 1.35$. In fact, recent results presented by SuperKamiokande at DPF 99, [39] with more data collected, help span the gap, indicating even more clearly the consistency of all data with each other.

In addition, data from other underground experiments (Soudan [40] and MACRO [41]) as well as other phenomena—like the flux of upward going muons [42] produced by ν_μ interactions in the earth—when interpreted in a neutrino oscillation

---

[8] Sterile neutrinos are, by definition, $SU(2) \times U(1)$ singlets. Because they do not couple to the Z, they are not excluded by the neutrino counting results from LEP.

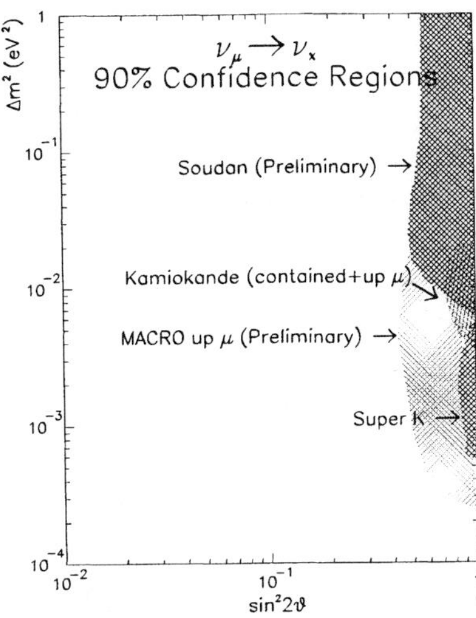

FIGURE 7. Evidence for atmospheric neutrino oscillations from all experiments, from Ref. [24].

framework, are totally consistent with the SuperKamiokande ν_μ zenith angle results. Fig. 7 summarizes all this information in one graph. This figure provides quite strong evidence in favor of neutrino masses and is probably the strongest evidence we have to date for physics beyond the Standard Model.

E Solar Neutrinos

The study of the solar neutrino flux was started in the early 1970's by Ray Davis and his group. [43] At present there are five different experiments which give information on solar neutrinos (Homestake, [44] Gallex, [45] SAGE, [46] Kamiokande [47] and SuperKamiokande [48]) and all five have some bearing on the issue of neutrino oscillations. In fact, roughly speaking, all five experiments see approximately half of the expected rate, as shown in Fig. 8. However, these experiments are sensitive to different parts of the solar neutrino spectrum, because the reactions they use to detect solar neutrinos in their detectors have different thresholds. SAGE and Gallex study the reaction $\nu_e + {}^{71}\mathrm{Ga} \rightarrow {}^{71}\mathrm{Ge} + e^-$, which has a threshold of 0.23 MeV. Homestake looks for the excitation of chlorine ($\nu_e + {}^{37}\mathrm{Cl} \rightarrow {}^{37}\mathrm{Ar} + e^-$) which has a 0.8 MeV threshold. The water Cerenkov detectors, Kamiokande and SuperKamiokande, study elastic $\nu_e e$ scattering and their threshold is in the neighborhood of 6.5 MeV.[9]

It has long been felt that the observed discrepancy between the neutrino signals detected and the expectations of the, so called, Standard Solar Model(SSM) [49]

9) SuperKamiokande is making strong efforts to move this threshold down to 5.5 MeV.

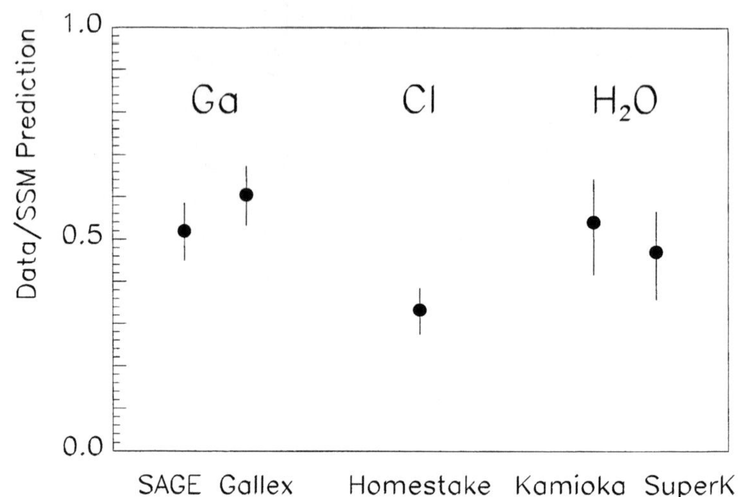

FIGURE 8. Rates seen by the diferent solar neutrino experiments, compared to the expectations of the Standard Solar Model. [49]

is not due to defects in this model but to the presence of some new physical phenomena. One of the principal arguments in favor of this latter solution to the solar neutrino puzzle, has to do with the details of the signal expected in each experiment. Because of the quite different threshold involved, each of the solar neutrino experiments, in fact, feels different pieces of the neutrino producing reactions in the solar cycle. For example, the Gallium experiments are the only ones which are sensitive to neutrinos originating in the *pp* cycle (the main solar cycle), with these neutrinos contributing about 50% of the expected rate. The Homestate detector mostly measures neutrinos from ^8B, although it is also sensitive to ^{7}Be neutrinos. Finally, because of their high threshold, the big water Cerenkov detectors only see Boron neutrinos. These circumstances make it difficult to argue for an astrophysical solution to the solar neutrino deficit. Much more natural is to imagine that this deficit arises as a result of neutrino oscillations.

There are two distinct neutrino oscillation solutions to the solar neutrino problem. Because roughly all experiments are reduced by about a factor of two from expectations, it is possible to fit the data by using vacuum neutrino oscillations $\nu_e \to \nu_X$. Clearly, for this fit one must appeal to large mixing angles and assume a tiny Δm^2. Since $E_\nu \sim$ MeV and the earth-sun distance $L \sim 10^{11}$ m, typically $\Delta m^2 \sim 10^{-11}$ eV^2. However, because the Homestake result is only about 30% of the predicted value, one has to fine-tune the parameters, so that only a few "just so" regions are favored. [50] A recent "just so" fit by Bahcall, Krastev, and Smirnov [51] is shown in Fig. 9.

In my opinion, much more interesting that the above "solution" is the possibility

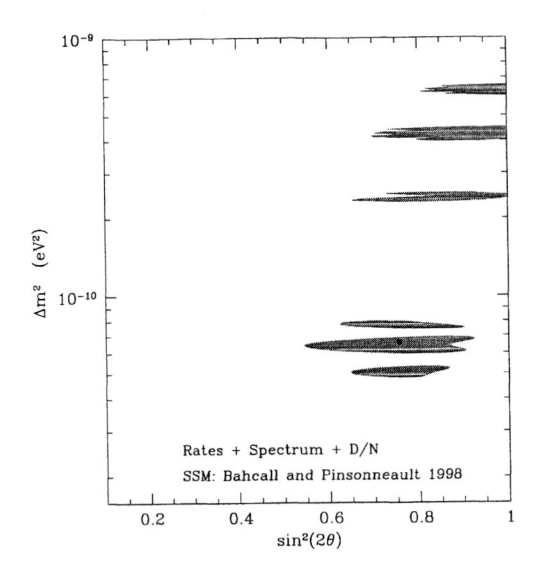

FIGURE 9. "Just so" solar neutrino fit, from [51].

that the solar neutrino results are a reflection of matter induced oscillations (the MSW effect we discuss in Section V). In the sun, the electron density to a good approximation, can be characterized by an exponential profile function [52]

$$N_e(r) = N_e(0)e^{-\frac{10r}{R_0}} \,. \tag{131}$$

The central density $N_e(0) \simeq 10^{26}\ \mathrm{cm}^{-3} \simeq 10^{12}\ (\mathrm{eV})^3$ is rather high and for appropriate values of Δm^2 and $\sin^2 2\theta$ can exceed the critical MSW density. For instance, for θ small, $\Delta m^2 \simeq 10^{-5}\ \mathrm{eV}^2$ and $|p| \sim 3$ MeV

$$N_e^{\mathrm{crit}} = \frac{\Delta m^2 \cos^2\theta}{\sqrt{2}|p|G_F} \sim 10^{11}\ (\mathrm{eV})^3 \,. \tag{132}$$

It follows from Eq. (132) that, for these parameters, ν_e's produced in the core of the sum (where $N_e(o) \gg N_e^{\mathrm{crit}}$) as they radiate outward go through a region with $N_e \sim N_e^{\mathrm{crit}}$ and can oscillate to ν_μ's (or other neutrino types) without paying a mixing angle penalty, since $\sin 2\theta_M|_{N_e^{\mathrm{crit}}} \to 1$.

The actual calculation of what happens in the sun is rather complicated, [12] since the density N_e changes along the neutrino trajectory. Since $N_e = N_e(t)$, the matter Hamiltonian of Eq. (94) is now **time dependent**:

$$H_{\mathrm{matter}}(t) = \frac{\Delta m^2}{4|p|}\sin 2\theta\sigma_1 - \left(\frac{\Delta m^2}{4|p|}\cos 2\theta - \frac{G_F}{\sqrt{2}}N_e(t)\right)\sigma_3 \,. \tag{133}$$

Although one can diagonalize this Hamiltonian, the resulting mixing angles and energies will be time dependent:

$$E_{1,2}^M(t) = \pm\left[\left(\frac{\Delta m^2}{4|p|}\cos 2\theta - \frac{G_F N_e(t)}{\sqrt{2}}\right)^2 + \left(\frac{\Delta m^2}{4|p|}\sin 2\theta\right)^2\right]^{1/2} \tag{134}$$

$$\tan 2\theta_M(t) = \frac{\frac{\Delta m^2}{4|p|}\sin 2\theta}{\frac{\Delta m^2}{4|p|}\cos 2\theta - \frac{G_F N_e(t)}{\sqrt{2}}} \ . \tag{135}$$

Because of this time dependence, it is no longer true that the states $|\nu_1^M(t)\rangle$ and $|\nu_2^M(t)\rangle$ are actual eigenstates. In fact, transitions can occur between these states. A simple calculation [12] shows that the states $|\nu_i^M(t)\rangle$ obey a coupled Schrödinger equation

$$i\frac{\partial}{\partial t}\begin{bmatrix} |\nu_1^M(t)\rangle \\ |\nu_2^M(t)\rangle \end{bmatrix} = \begin{bmatrix} E_1^M(t) & i\frac{\partial}{\partial t}\theta_M(t) \\ i\frac{\partial}{\partial t}\theta_M(t) & E_2^M(t) \end{bmatrix}\begin{bmatrix} |\nu_1^M(t)\rangle \\ |\nu_2^M(t)\rangle \end{bmatrix} \ . \tag{136}$$

If

$$|E_2^M(t) - E_1^M(t)| \gg \left|2\frac{\partial}{\partial t}\theta_M(t)\right| \tag{137}$$

then transiting between $|\nu_1^M(t)\rangle$ and $|\nu_2^M(t)\rangle$ will be relatively unimportant and one has an **adiabatic** situation. For an exponential density profile, Eq. (137) is satisfied at N_e^{crit} provided that [12]

$$\frac{\Delta m^2 \sin^2 2\theta}{2\cos 2\theta} \gg 2\times 10^{-8}\ (\text{eV})^2 \ . \tag{138}$$

For the adiabatic case, one can use our discussion of matter oscillations to give a qualitative picture of how the MSW mechanism could work in the sun. Because at the solar core $N_e(o) \gg N_e^{\text{crit}}$, according to Eq. (106) $|\nu_e\rangle \simeq |\nu_2^M(o)\rangle$. Because we are assuming adiabaticity, as the neutrinos diffuse out of the core of the sun, the state $|\nu_2^M(o)\rangle$ will evolve into $|\nu_2^M(t)\rangle$. That is, there are no transitions in the sun. Thus, when the neutrinos exit the sun, the state $|\nu_2^M(t_{\text{surface}})\rangle$ will just simply become $|\nu_2\rangle$. Because

$$\langle\nu_e|\nu_2\rangle = \sin\theta \ , \tag{139}$$

it follows that, in this case,

$$P_{\text{solar}}^{\text{adiabatic}}(\nu_e \to \nu_e; L) = \sin^2\theta \ . \tag{140}$$

A more careful analysis shows that there are actually two MSW solutions, one adiabatic and one non-adiabatic, [53] both having $\Delta m^2 \sim 10^{-5}$ eV^2. The adiabatic

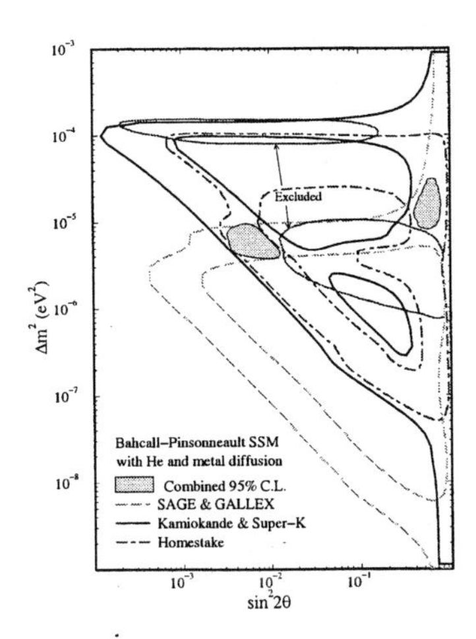

FIGURE 10. Regions in the the $\Delta m^2 - \sin^2 2\theta$ plane favored by the MSW explanations of the solar neutrino data, from Ref. [51].

solution has large mixing angles $\sin^2 2\theta \simeq 1$. Hence, according to Eq. (140), $P_{\rm solar}^{\rm adiabatic}(\nu_e \to \nu_e; L) \simeq 1/2$ so, indeed, roughly half the flux is lost. The non-adiabatic solution has $\sin^2 2\theta \sim 5 \times 10^{-3}$. Furthermore, a rather large range in the $\Delta m^2 - \sin^2 2\theta$ plane is eliminated by the absence of a day/night effect, which would be a sign of matter oscillations in the earth. The favored MSW regions are depicted in Fig. 10.

I want to close this brief discussion of solar neutrinos, and in particular of the MSW explanation of the solar data, by making a more quantitative remark. To fit the solar neutrino data using the MSW effect requires that the probability $P(\nu_e \to \nu_e; L)$ have considerable energy dependence. The required energy dependence is shown in Fig. 11. I indicate also in this figure at what energies the neutrinos produced in the various solar reactions are effective. One sees from Fig. 11 that essentially all *pp* neutrinos survive, the Berylium neutrinos disappear and the flux of Boron neutrinos is roughly halved. This reconciles nicely with what is seen in the data, as detailed in Table 3.

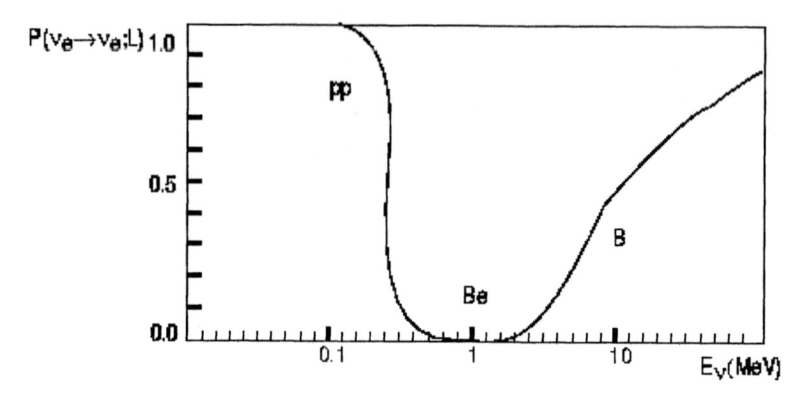

FIGURE 11. Energy dependence needed for $P(\nu_e \to \nu_e; L)$ to fit the solar neutrino data.[54]

Table 3. Summary of neutrino flux observations and expectations of solar neutrino experiments

^{37}Cl (13% Be; 80% B)
(2.56 ± 0.23) SNU Ref.[44] $\left[\text{predicted SSM}: \left(7.7^{+1.2}_{-1.0}\right) \text{SNU}\right]$

^{71}Ga (51% pp; 15% Be; 12% B)
(77.5 ± 7.7) SNU Ref.[45] [predicted SSM : (129 ± 8) SNU]
(66.6 ± 8.0) SNU Ref.[46]

Water $\check{C}$ (100% B)
2.80 ± 0.38 SNU Ref.[47] [predicted SSM : $\left(5.15^{+1.0}_{-0.7}\right)$ SNU]
2.44 ± 0.10 SNU Ref.[48]

VII THEORETICAL IMPLICATIONS

I summarize in Fig. 12 where one is at present on the issue of neutrino masses and mixings. This figure collects together all the neutrino oscillation evidence, as well as the hints for oscillations, which we have at the moment. As is clear from the figure there are three regions suggested in the $\Delta m^2 - \sin^2 2\theta$ plane. The strongest evidence is that for atmospheric neutrino oscillations coming from the SuperKamiokande zenith angle data. Here the suggested parameters are $(\Delta m^2) \sim 3 \times 10^{-3}$ eV^2, $\sin^2 2\theta \sim 1$ with $\nu_\mu \to \nu_X$ $(\nu_X \neq \nu_e)$. Solar neutrinos also are strongly suggestive of oscillations. Interpreting the data this way leads to $(\Delta m^2) \sim 10^{-5}$ eV^2, with $\sin^2 2\theta \sim 1$ or $\sin^2 2\theta \sim 5 \times 10^{-3}$, for MSW $\nu_e \to \nu_X$ oscillations, or $\Delta m^2 \sim 10^{-11}$ eV^2 and $\sin^2 2\theta \sim 1$, for "just-so" $\nu_e \to \nu_X$ oscillations. The weakest hint for oscillations probably is that of LSND, because of other contrary evidence. At

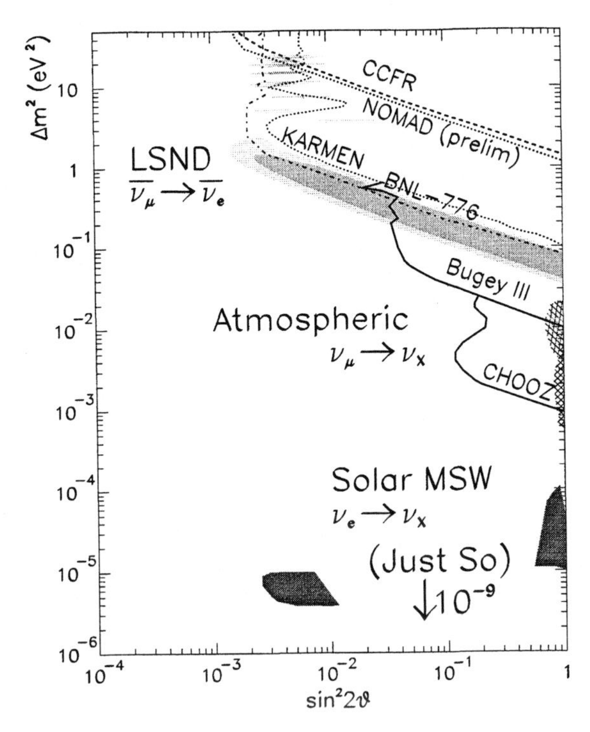

FIGURE 12. Summary of evidence of, and hints for, neutrino oscillations.[24]

any rate, the suggested region here is $\Delta m^2 \sim 5 \times 10^{-1}$ eV^2 and $\sin^2 2\theta \sim 10^{-2}$ for $\nu_e \to \nu_\mu$ oscillations.

Besides neutrino oscillation phenomena, there is really no other direct evidence for neutrino masses. However, both β-decay and neutrinoless double β-decay put rather strong bounds on neutrino masses connected to ν_e. From β-decay the largest neutrino mass obeys the bound: "m_{ν_e}" < 3.9 eV (90% C.L.). Double β-decay, bounds the Majorana mass $\langle m_{\nu_e} \rangle_{ee}$ even more strongly: $\langle m_{\nu_e} \rangle_{ee} < 0.2$ eV (90% C.L.). These bounds are interesting, since they are close to the kind of neutrino masses which could have substantial cosmological influence. Using the central value for the Hubble parameter [c.f. Eq.(124)], I note that $\sum_i m_{\nu_i} = 30$ eV, 6 eV, 2 eV correspond, respectively, to neutrinos closing the Universe, to neutrinos being 20% of the dark matter in the Universe (assuming $\Omega_M \simeq 1$), and to neutrinos being 20% of the dark matter, with $\Omega_M \simeq 0.3$. The last case is, perhaps, the one that is most cosmologically realistic. [55] This stresses the the importance of continuing the search for neutrino masses in the eV range.

Concentrating only on the SuperKamiokande evidence for neutrino masses already has important implications. Taking

$$\Delta m^2 = m_3^2 - m_2^2 \sim 3 \times 10^{-3} \text{ eV}^2 \tag{141}$$

gives one already a **lower bound** on some neutrino mass: $m_3 \geq 5 \times 10^{-2}$ eV. This mass value, in turn, gives a lower bound for the cosmological contribution of neutrinos

$$\Omega_\nu \geq \frac{m_3}{92 \text{ eV } h^2} \sim 1.5 \times 10^{-3} \; . \tag{142}$$

Although this number is far from that needed for closure of the Universe, I note that the Ω_ν of Eq. (142) is comparable to the contribution of luminous matter to the energy density of the Universe [56]

$$\Omega_{\text{luminous}} \sim (3 - 7) \times 10^{-3} \; . \tag{143}$$

So, from SuperKamiokande we learn that the neutrino contribution to the energy density of the Universe is the same as, in the words of Carl Sagan, that of "billions and billions of stars"!

For particle physics, a value of $m_3 \sim 5 \times 10^{-2}$ eV is also quite interesting. If we use either the simple see-saw formula of Eq. (55) or the GUT relation (66), the identification

$$m_3 = \begin{cases} \frac{[(m_D)_3]^2}{m_S} \sim \frac{m_t^2}{m_S} \\ m_T \simeq \frac{\langle\phi^0\rangle^2}{\Lambda} \end{cases} \tag{144}$$

give comparable values for m_S and Λ:

$$m_S \sim \Lambda \sim 10^{15} \text{ GeV} \; . \tag{145}$$

Of course, these values are only justifiable in specific models where one has a bit more control of other constants, which are taken above all to be of $O(1)$.

If one goes beyond the SuperKamiokande data, then many theoretical scenarios emerge. Unfortunately, in general, these scenarios mostly reflect the prejudices one has regarding the data. Nevertheless, it is useful to briefly discuss two differing broad theory scenario. In the first scenario, one assumes that all hints for oscillations seen are true. In the second, one disregards some oscillation hints. In most cases, the discarded data is that of LSND.

If one believes all hints for neutrino oscillations, since there are three different Δm^2 involved, the neutrino mass matrix M necessarily is a 4×4 matrix.[10] To get a 4×4 neutrino matrix one adds to the usual three neutrinos a sterile neutrino ν_s. Most 4 neutrino models attempt to fit all data, since this was after all the reason for introducing the fourth neutrino. The most promising scenario [58] has two pairs

[10] There have been attempts to "stretch" some of the data, so that all hints can be accounted for with only two different Δm^2. These attempts [57] seem rather forced to me.

of quasi-Dirac neutrinos split by a small mass difference.[11] The heaviest pair (m_2 and m_3) have masses of order 0.5 eV and contribute $\Omega_\nu \simeq 0.03$ to cosmology. The atmospheric neutrino oscillations involve this pair, so that $\Delta m^2_{23} \sim 3 \times 10^{-3}$ eV^2. The second pair (m_1 and m_4) are much lighter, with mass around $10^{-2} - 10^{-1}$ eV. Their mass difference $\Delta m^2_{14} \sim 10^{-5}$ eV^2 is what enters in solar neutrino oscillations. The LSND result is explained as an oscillation between the light pair and the heavy pair, with $\Delta m^2 \sim m_3^2 \sim 0.6$ eV^2. In this scheme, the solar neutrino oscillations involve oscillations of ν_e to a sterile neutrino, while the atmospheric neutrino oscillation is $\nu_\mu \to \nu_\tau$ and LSND $\nu_e \to \nu_\mu$. Although this scheme works phenomenologically, theoretically it is difficult to get light sterile neutrinos almost degenerate with ordinary neutrinos.

Different patterns arise if one is prepared to disregard some of the neutrino oscillation limits. If one disregards, in particular, the results from LSND then the CH00Z bound, [37] and the quite different mass squared differences involved in atmospheric and solar neutrino oscillations, suggest a very simple 3-neutrino mixing matrix. CH00Z suggests that $\theta_{13} \simeq 0^o$. On the other hand, atmospheric neutrino oscillations suggest $\theta_{23} \simeq 45^o$. Finally, depending on what solar neutrino oscillation solution one picks, the angle θ_{12} can either be large or small. Thus, neglecting possible CP violating phases, the neutrino mixing matrix looks like [59]

$$U \simeq \begin{bmatrix} 1 & 0 & 0 \\ 0 & \frac{1}{\sqrt{2}} & -\frac{1}{\sqrt{2}} \\ 0 & \frac{1}{\sqrt{2}} & \frac{1}{\sqrt{2}} \end{bmatrix} \begin{bmatrix} 1 & 0 & 0 \\ 0 & 1 & 0 \\ 0 & 0 & 1 \end{bmatrix} \begin{bmatrix} c_{12} & -s_{12} & 0 \\ s_{12} & c_{12} & 0 \\ 0 & 0 & 1 \end{bmatrix} \qquad (146)$$

$$= \begin{bmatrix} c_{12} & -s_{12} & 0 \\ \frac{s_{12}}{\sqrt{2}} & \frac{c_{12}}{\sqrt{2}} & -\frac{1}{\sqrt{2}} \\ \frac{s_{12}}{\sqrt{2}} & \frac{c_{12}}{\sqrt{2}} & \frac{1}{\sqrt{2}} \end{bmatrix},$$

where $c_{12} = \cos\theta_{12}$; $s_{12} = \sin\theta_{12}$. However, even with U of the above form, there are many open questions to answer. For instance, is maximal mixing ($\theta_{12} \simeq 45^o$) allowed? Are nearly degenerate neutrino masses ($m_1 \simeq m_2 \simeq m_3 \simeq 0.5$ eV) allowed? What neutrino mass matrix gives rise to this particular mixing matrix?

These questions cannot really be answered in a straightforwad manner, without making some more assumptions. There is really a lot of freedom. Given U and some assumptions for the neutrino mass spectrum $\{m_i\}$ then one can deduce a neutrino mass matrix M. However, recall from our discussion of the see-saw mechanism, that M itself depends on both the neutrino Dirac mass m_D and on the right-handed neutrino mass matrix m_S [cf. Eq. (55)]. Thus, to make progress, even "knowing" M one has to make some assumptions on m_D (or m_S) to learn something further. For instance, one could use GUTs which naturally ties the matrix m_D in Eq. (55) to the u-quark mass matrix. [60]

Some authors have preferred to focus on some simple structure for the 3×3 matrix M. [61] Two of these are particularly appealing. The first of these has

---

11) A quasi-Dirac neutrino pair reduces to a Dirac neutrino, as the mass difference between the pair vanishes.

a total degeneracy for the neutrinos, the other is Dirac-like with an additional massless neutrino. In the first pattern

$$M = m \begin{pmatrix} 1 & 0 & 0 \\ 0 & 0 & 1 \\ 0 & 1 & 0 \end{pmatrix} . \tag{147}$$

In this case, $\sum_i m_{\nu_i} = 3m$, so that cosmology impose a bound on m, depending on what one believes Ω_ν is. However, in the degenerate case, one has also that

$$\langle m_{\nu_e} \rangle_{ee} = \sum_i U_{ei}^2 m_{\nu_i} = m . \tag{148}$$

The double β-decay bound then tell us that $m < 0.2$ eV. If we push m to its upper bound, however, it is difficult to see what perturbation can then give $\Delta m^2_{\rm atmos} \simeq 3 \times 10^{-3}$ eV^2; $\Delta m^2_{\rm solar} \sim 10^{-5}$ eV^2.

The second simple pattern for neutrino masses has [62]

$$M = m \begin{pmatrix} 0 & 1 & 1 \\ 1 & 0 & 0 \\ 1 & 0 & 0 \end{pmatrix} \tag{149}$$

which has a degenerate pair and a zero eigenvalue. Note that this pattern conserves $L_e - L_\mu - L_\tau$. Since for this mass matrix $\Delta m^2_{\rm atmos} = m^2$, it follows that $m \sim 5 \times 10^{-2}$ eV. So in this case, neutrinos do not contribute much to the energy density of the Universe ($\Omega_\nu \simeq 3 \times 10^{-3}$). To get solar neutrino oscillations one has to introduce some perturbation on the mass matrix (149) that will split the massive degenerate states and give $\Delta m^2_{12} \sim 10^{-5}$ eV^2.

VIII FUTURE EXPERIMENTS

It seems pretty clear that progress in understanding what is going on in the neutrino sector can only come from further data. Fortunately, new data will be forthcoming in all the relevant Δm^2 regions. I want to end these lectures by briefly discussing these future experiments.

A Solar Neutrinos

SuperKamiokande will continue to take data in years to come, thus refining their present measurements of solar neutrinos. Furthermore, a real effort is taking place to lower the neutrino energy threshold further so as to be able to study the shape dependence of the signal as a function of E_ν. In addition to this continuing effort, relatively soon two other experiments will be coming on line which have considerable promise. The first of these is SNO (the Sudbury Neutrino Observatory [63]) which uses a Kiloton of D_2O. The advantage of having heavy water is that it allows SNO to study simultaneously both charged current and neutral current processes. The charged current process

$$\nu_e + d \to e^- + p + p \;, \tag{150}$$

like all charged current processes, is sensitive to whether oscillations have occurred or not. The neutral current disintegration of the deuteron, on the other hand, is insensitive to oscillations since it is the same for all neutrino species ν_X:[12]

$$\nu_X + d \to \nu_X + p + n \;. \tag{151}$$

Comparison of the rates for the two neutrino reactions(150) and (151) should help rule out possible astrophysical explanations for the solar neutrino puzzle. The SNO detector should begin taking data in 1999.

The second solar neutrino experiment of interest is Borexino. [64] This experiment is presently under construction at the Gran Sasso Laboratory and should be ready for data taking in 2001. Borexino uses 300 tons of scintillator, which has a relatively low threshold ($E_{\rm thr} > 340$ KeV). As a result, Borexino should be particularly sensitive to the $E_\nu = 862$ GeV neutrino line coming from ^{7}Be. Recalling Fig. 11, one sees that if the MSW explanation is correct, the solar neutrino signal in Borexino should be significantly below the theoretical expectations. Indeed, if there are no solar oscillations, Borexino is supposed to detect about 50 events/day, while if the MSW explanation is true, this number should go down to about 10 events/day.

B Atmospheric Neutrinos

Here again SuperKamiokande will continue to integrate data with time. However, the Δm^2 region, will also be probed more directly by using neutrino beams from accelerators. Three such long baseline experiments are in different stages of readiness. K2K, which uses a neutrino beam from KEK, aimed at SuperKamiokande 250 Km away, should shortly be operational. [65] MINOS, [66] in the Soudan Mine, is under construction and will be the target of a dedicated neutrino beam from Fermilab, 730 Km away. First data should become available around 2001-2002. Finally, a variety of proposals exist for experiments in the Gran Sasso Laboratory, which is 740 Km from CERN, to become targets of neutrino beams from CERN.

The main advantage that these long baseline experiments have over SuperKamiokande is that the neutrino beam used is well characterized, both in energy and in its time structure. Furthermore, these beams also have a higher intensity. So many possible systematic effects will be under better control. In the case of the higher energy Fermilab and CERN beams, it may also be possible to directly search and detect ν_τ's, if these neutrinos are produced in the oscillations.

---

12) This is only true for neutrinos whose neutral couplings to the Z are universal. It does not apply to sterile neutrinos.

C LSND Region

The $\Delta m^2 - \sin^2 2\theta$ region identified by the LSND experiment as potentially interesting also will be explored further. At Fermilab, there is an approved experiment, Mini BooNE, [67] which will run around 2001-2002, which should be about a factor of five more sensitive than LSND in a comparable kinematical region. With this sensitivity, it should be quite clear whether $\nu_\mu \to \nu_e$ oscillations with $\Delta m^2 \sim (0.1 - 1)$ eV^2 exist or not.

ACKNOWLEDGEMENTS

I am grateful to Juan Carlos D'Olivo and Myriam Mondragon for their wonderful hospitality in Oaxaca. I am also thankful to all the students at the VIII Escuela Mexicana de Particulas y Campos for their attention and enthusiasm. This work was supported in part by the Department of Energy inder contract No. DE-FG03-91ER40662, Task C.

APPENDIX A: DIRAC AND MAJORANA MASSES

To understand how Dirac and Majorana masses can arise, it is useful to review here some of the properties of the spinor representations of the Lorentz group. The Lorentz group, besides the well known vector and tensor representations has also spinor representations. It turns out that there are two inequivalent spinor representations. It is out of these two-dimensional spinors that one builds up the usual four-dimensional Dirac spinor ψ.

Under a Lorentz transformation, a vector field V^μ has the well known transformation

$$V^\mu \to V'^\mu = \Lambda^\mu{}_\nu V^\nu \ , \tag{1}$$

where the 4-dimensional representation matrices Λ obey the pseudo-orthogonality conditions

$$\eta_{\mu\nu} = \Lambda^\alpha{}_\mu \eta_{\alpha\beta} \Lambda^\beta{}_\nu \ , \tag{2}$$

involving the metric tensor

$$\eta_{\mu\nu} = \begin{bmatrix} -1 & & & \\ & 1 & & \\ & & 1 & \\ & & & 1 \end{bmatrix} . \tag{3}$$

Besides vector representations, the Lorentz group has two inequivalent spinor representation. The corresponding 2-dimensional Weyl spinors are conventionally denoted by ξ_a and $\dot{\xi}_a$, known as undotted and dotted spinors, respectively. Under Lorentz transformations they transform as

$$\xi_a \to \xi'_a = M_a{}^b \xi_b \tag{4}$$

$$\dot{\xi}_a \to \dot{\xi}'_a = M_a^{*b} \dot{\xi}_b \ . \tag{5}$$

The 2×2 matrices M and M^*, with det M = det $M^* = 1$, provide inequivalent representation of $SL(2, C)$. Obviously, from the above it follows that $\dot{\xi} \sim \xi^*$.

One can establish a relationship between the 2×2 matrices M and the 4×4 matrices Λ, since the vector field V^μ transforms as $V \sim \xi \otimes \dot{\xi}$. For these purposes, it is useful to define a set of four matrices $\sigma^\mu \equiv (1, \vec{\sigma})$, with $\vec{\sigma}$ being the usual Pauli matrices. The 2×2 matrix

$$V = \sigma^\mu \eta_{\mu\nu} V^\nu \equiv \sigma^\mu V_\mu \tag{6}$$

under a Lorentz transformation transforms as

$$V \to V' = M V M^\dagger = \sigma^\mu V'_\mu \ . \tag{7}$$

Using Eq. (A1), it follows that

$$\sigma^\mu_{ac} \Lambda_\mu{}^\nu = M_a{}^b \sigma^\nu_{bd} M_c^{*d} \ . \tag{8}$$

Because det $M = 1$, the analogue of the scalar product for vectors $V^\mu \eta_{\mu\nu} V^\nu \equiv V^\mu V_\mu$, for the spinors ξ and $\dot{\xi}$ leads to the following Lorentz scalars:

$$\xi_a \epsilon^{ab} \xi_b \equiv \xi_a \xi^b \ ; \qquad \dot{\xi}_a \epsilon^{ab} \dot{\xi}_b \equiv \dot{\xi}_a \dot{\xi}^b \tag{9}$$

where $\epsilon^{ab} = -\epsilon^{ba}$ and $\epsilon^{12} = 1$. Similarly, just as the contraction of a covariant and contravariant metric tensor gives the identity $[\eta_{\mu\rho}\eta^{\rho\nu} = \delta^\nu_\mu]$, one can define 2×2 antisymmetric ϵ-matrices, ϵ_{ab}, which obey

$$\epsilon_{ac} \epsilon^{cb} = \delta^b_a \ . \tag{10}$$

It follows that $\epsilon_{12} = -1$.

The usual 4-component Dirac spinor ψ is made up of a dotted and an undotted Weyl spinor:

$$\psi = \begin{pmatrix} \xi_a \\ \dot{\chi}^a \end{pmatrix} \tag{11}$$

In this, so called, Weyl-basis the Dirac γ-matrices γ^μ, which obey the anticommutation relations $\{\gamma^\mu, \gamma^\nu\} = -2\eta^{\mu\nu}$, take the form

$$\gamma^\mu = \begin{pmatrix} 0 & \sigma^\mu \\ \bar{\sigma}^\mu & 0 \end{pmatrix} \ . \tag{12}$$

Here $\bar{\sigma}^\mu = (1, -\vec{\sigma})$, so that in this basis

$$\gamma^0 = \begin{pmatrix} 0 & 1 \\ 1 & 0 \end{pmatrix} ; \quad \gamma^i = \begin{pmatrix} 0 & \sigma^i \\ -\sigma^i & 0 \end{pmatrix} \tag{13}$$

and

$$\gamma_5 = i\gamma^0\gamma^1\gamma^2\gamma^3 = \begin{pmatrix} -1 & 0 \\ 0 & 1 \end{pmatrix} . \tag{14}$$

It follows from the above that ξ_a and $\dot{\chi}^a$ are **chiral projections** of ψ:

$$\psi_{\rm L} = \frac{1}{2}(1-\gamma_5)\psi = \begin{pmatrix} \xi_a \\ 0 \end{pmatrix} ; \quad \overline{\psi_{\rm L}} = \bar{\psi}\frac{1}{2}(1+\gamma_5) = (0 \;\; \xi_a^*) \tag{15}$$

$$\psi_{\rm R} = \frac{1}{2}(1+\gamma_5)\psi = \begin{pmatrix} 0 \\ \dot{\chi}^a \end{pmatrix} ; \quad \overline{\psi_{\rm R}} = \bar{\psi}\frac{1}{2}(1-\gamma_5) = (\dot{\chi}^{a*} \; 0) . \tag{16}$$

Using these equations, it is easy to see that the Dirac mass term connects ξ with $\dot{\chi}$. Specifically, one has

$$\mathcal{L}_{\rm Dirac} = -m_D(\overline{\psi_{\rm L}}\psi_{\rm R} + \overline{\psi_{\rm R}}\psi_{\rm L}) = -m_D(\xi_a^*\dot{\chi}^a + \dot{\chi}^{a*}\xi_a) . \tag{17}$$

Recall, however, that dotted spinors are related to the complex conjugate of an undotted spinor. Choosing a phase convention where

$$\xi_a^* = \dot{\xi}_a ; \quad \dot{\chi}_a^* = \chi_a, \tag{18}$$

one can write the Dirac mass term simply as

$$\mathcal{L}_{\rm Dirac} = -m_D(\dot{\xi}_a\dot{\chi}^a + \chi^a\xi_a) . \tag{19}$$

In view of Eq. (A9) this term is obviously Lorentz invariant. However, Lorentz invariance does not require one to have two distinct Weyl spinors ξ and χ to construct a mass term. Majorana masses, basically, make use of this "simpler" option.

One can define a 4-component **Majorana spinor** in terms of the Weyl spinor ξ and its complex conjugate $\dot{\xi}$:

$$\psi_M = \begin{pmatrix} \xi_a \\ \dot{\xi}^a \end{pmatrix} . \tag{20}$$

Because $\dot{\xi}^a = \xi^{a*}$, effectively ψ_M has only one independent helicity projection. One can choose this projection to be, say, $(\psi_M)_{\rm L}$:

$$(\psi_M)_{\rm L} = \frac{1}{2}(1-\gamma_5)\psi_M = \begin{pmatrix} \xi_a \\ 0 \end{pmatrix} ; \quad \overline{(\psi_M)_{\rm L}} = \overline{\psi_M}\frac{1}{2}(1+\gamma_5) = (0 \;\; \dot{\xi}_a) . \tag{21}$$

One can construct $(\psi_M)_{\rm R}$ by using the charge conjugate matrix $\tilde{C}$. In the Weyl basis $\tilde{C}$ is given by

$$\tilde{C} = \begin{bmatrix} \epsilon_{ab} & 0 \\ 0 & \epsilon^{ab} \end{bmatrix} = \begin{bmatrix} 0 & -1 & 0 & 0 \\ 1 & 0 & 0 & 0 \\ 0 & 0 & 0 & 1 \\ 0 & 0 & -1 & 0 \end{bmatrix} . \tag{22}$$

Clearly

$$\begin{aligned} (\psi_M)_{\rm R} &= \begin{pmatrix} 0 \\ \dot{\xi}^a \end{pmatrix} = \begin{pmatrix} 0 \\ \epsilon^{ab}\dot{\xi}_b \end{pmatrix} = \tilde{C}\overline{(\psi_M)_{\rm L}}^T \\ \overline{(\psi_M)_{\rm R}} &= (\xi^a \ 0) = (\epsilon^{ab}\xi_b \ 0) = (\xi_b \epsilon_{ba} \ 0) = (\psi_M)_{\rm L}^T \tilde{C} . \end{aligned} \tag{23}$$

That is, $(\psi_M)_{\rm R}$ is the charge conjugate of $(\psi_M)_{\rm L}$ (c.f. Eq. (36)):

$$[(\psi_M)_{\rm L}]^c = (\psi_M)_{\rm R} . \tag{24}$$

Because of Eq. (A24) it follows that the Majorana spinor ψ_M obeys a constraint. It is **self-conjugate**:

$$\psi_M = \begin{pmatrix} \xi_a \\ \dot{\xi}^a \end{pmatrix} = (\psi_M)_{\rm L} + (\psi_M)_{\rm R} = (\psi_M)_{\rm L} + [(\psi_M)_{\rm L}]^c . \tag{25}$$

Hence,

$$\psi_M = [\psi_M]^c . \tag{26}$$

The Majorana mass term

$$\mathcal{L}_{\rm Majorana} = -\frac{1}{2} m_M \overline{\psi_M} \psi_M \tag{27}$$

involves a product of ξ with itself and $\dot{\xi}$ with itself

$$\mathcal{L}_{\rm Majorana} = -\frac{1}{2} m_M \left(\overline{(\psi_M)_{\rm L}}(\psi_M)_{\rm R} + \overline{(\psi_M)_{\rm R}}(\psi_M)_{\rm L} \right) = -\frac{1}{2} m_M (\dot{\xi}_a \dot{\xi}^a + \xi^a \xi_a) . \tag{28}$$

Eq. (A28) can also be written purely in terms of $(\psi_M)_{\rm L}$ by using the charge conjugation matrix $\tilde{C}$. Using Eq. (A23) one has also

$$\mathcal{L}_{\rm Majorana} = -\frac{1}{2} m_M \left(\overline{(\psi_M)_{\rm L}} \tilde{C} \overline{(\psi_M)_{\rm L}}^{\,T} + (\psi_M)_{\rm L}^T \tilde{C} (\psi_M)_{\rm L} \right) . \tag{29}$$

Equally well one can write this mass term entirely as a function of $(\psi_M)_{\rm R}$. One finds

$$\mathcal{L}_{\rm Majorana} = -\frac{1}{2} \left((\psi_M)_{\rm R}^T \tilde{C} (\psi_M)_{\rm R} + \overline{(\psi_M)_{\rm R}} \tilde{C} \overline{(\psi_M)_{\rm R}}^{\,T} \right) . \tag{30}$$

REFERENCES

1. A compilation of the latest electroweak data from LEP and the SLC is in the report of the LEP Electroweak Working Group, CERN-EP/99-15, February 1999.
2. Particle Data Group, C. Caso *et al.*, Europ. Phys. J. **C3** (1998) 1.
3. For a discussion, see for example, F. A. Berends *et al.*, in **Z Physics at LEP 1**, eds. G. Altarelli, R. Kleiss and C. Verzegnassi, Vol 1, p. 89, CERN report CERN 89-08, 1989.
4. See, for example, R. D. Peccei, in **Broken Symmetries**, eds. L. Mathelitsch and W. Plessas, Lecture Notes in Physics 521 (Springer Verlag, Berlin 1999).
5. See, for example, J. D. Bjorken and S. D. Drell, **Relativistic Quantum Fields** (Mc Graw Hill, New York 1965).
6. T. Yanagida, in Proceedings of the Workshop on the Unified Theories and Baryon Number in the Universe, Tsukuba, Japan 1979, eds. O. Sawada and A. Sugamoto, KEK Report No. 79-18; M. Gell-Mann, P. Ramond and R. Slansky in **Supergravity**, Proceedings of the Workshop at Stony Brook, NY, 1979, eds. P. van Nieuwenhuizen and D. Freedman (North-Holland, Amsterdam, 1979).
7. G. Gelmini and M. Roncadelli, Phys. Lett. **99B** (1981) 411.
8. See, for example, A. Masiero, R. N. Mohapatra and R. D. Peccei, Nucl. Phys. **B192** (1981) 66.
9. See, for example, S. Weinberg, in Proceedings of the XXIII International Conference on High Energy Physics, Berkeley, California 1986, ed. S. C. Loken (World Scientific, Singapore 1987).
10. Some recent reviews which discuss this material include, W. C. Haxton and B. R. Holstein, to appear in the American Journal of Physics, hep-ph/9905257; P. Fisher, B. Kayser and K. S. McFarland, to appear in the Annual Review of Nuclear and Particle Science, Vol. 49 (1999), hep-ph/9906244
11. S. P. Mikheyev and A. Y. Smirnov, Sov. J. Nucl. Phys. **42** (1985) 441; Nuovo Cimento **9C** (1986) 17; L. Wolfenstein, Phys. Rev. D**17** (1978) 2369.
12. See, for example, S. T. Petcov, in **Computing Particle Properties**, eds. H. Gausterer and C. Lang, Lecture Notes in Physics 512 (Springer Verlag, Berlin 1998).
13. For early work on neutrino mass limits, see for example, D. R. Hamilton, W. P. Alford and L. Gross, Phys. Rev. **92** (1953) 1521.
14. For a recent review see, for example, K. Zuber, Phys. Rept. **305** (1998) 295.
15. V. Lobashev, to be published in the Proceedings of the 17th International Workshop on Weak Interactions and Neutrinos WIN 99, Capetown, South Africa, January 1999; see also, V. Lobashev, Prog. Part. Nucl. Phys. **40** (1998) 337; for earlier results see, A.I. Belesev *et al.* , Phys. Lett. **B350** (1995) 263.
16. K. Assamagan *et al.* Phys. Rev. **D53** (1996) 6065.
17. ALEPH Collaboration, R. Barate *et al.*, Europ. Phys. J. **C2** (1998) 395.
18. The first laboratory meausurement of double β-decay in ^{82}Se is reported in S. R. Elliot, A.A. Hahn and M. Moe, Phys. Rev. Lett. **59** (1987) 1649.
19. For a recent review see, for example, A. Morales in the Proceedings of the 18th International Conference on Neutrino Physics and Astrophysics, Tokayama, Japan, June 1998.

20. Heidelberg- Moscow Collaboration, L. Baudis *et al.* hep-ex/9902014.
21. For a discussion see, P. Vogel and M. Moe, Ann. Rev. Nucl. Part. Phys. **44** (1994) 247. For a recent review see, A. Faessler and F. Simkovic, hep-ph/9901215.
22. See, for example, E. W. Kolb and M. Turner, **The Early Universe** (Addison-Wesley, Redwood City, California, 1990).
23. See, for example, S. Weinberg, **Gravitation and Cosmology** (Wiley, New York, 1972).
24. For a recent review see, for example, W. L. Freedman, to be published in the Proceedings of the Nobel Symposium on Particle Physics and the Universe, Haga Slott, Sweden, August 1988, astro-ph/9905222.
25. For a discussion see, for example, R. N. Mohapatra and P. Pal, **Massive Neutrinos in Physics and Astrophysics** (World Scientific, Singapore 1991).
26. CHORUS Collaboration, E. Eskut *et al.*, Phys. Lett. **B434** (1998) 205.
27. NOMAD Collaboration, P. Astier *et al.*, Phys. Lett. **B453** (1999) 169.
28. KARMEN Collaboration, B. Zeitnitz *et al.*, Prog. Part. Nucl. Phys. **32** (1994) 351; see also, K. Eitel, in the Proceedings of the 32nd Rencontres de Moriond, Electroweak Interactions and Unified Theories, Les Arcs, March 1997.
29. LSND Collaboration, C. Athanassopoulos *et al.*, Phys. Rev. Lett. **77** (1996) 3082; Phys. Rev. **C54** (1996) 2685.
30. BNL E776 Collaboration, L. Borodovsky *et al.*, Phys. Rev. Lett. **68** (1992) 274; Bugey Reactor Collaboration, B. Achkar *et al.* Nucl. Phys. **B434** (1995) 503.
31. KARMEN2 Collaboration, K. Eitel and B. Zeitnitz in the Proceedings of the 18th International Conference on Neutrino Physics and Astrophysics, Tokayama, Japan, June 1998.
32. J. Conrad, in the Proceedings of the XXIX International Conference on High Energy Physics, Vancouver, Canada 1998.
33. R. Maschuw, to be published in the Proceedings of the 17th International Workshop on Weak Interactions and Neutrinos WIN 99, Capetown, South Africa, January 1999.
34. LSND Collaboration, C. Athanassopoulos *et al.*, Phys. Rev. Lett. **81** (1998) 1774; Phys. Rev. **C58** (1998) 2511.
35. This effect was first reported by the deep underground water Cerenkov detectors, Kamiokande [K. S. Hirata *et al.*, Phys. Lett. B**205** (1988) 416, B**280** (1992) 145; Y. Fukuda *et al.*, Phys. Lett. B**335** (1994) 237] and IMB [D. Cooper *et al.*, Phys. Rev. Lett. **66** (1991) 2561; R. Becker-Szendy *et al.*, Phys. Rev. D**46** (1992) 3720].
36. SuperKamiokande Collaboration, Y Fukuda *et al.*, Phys. Rev. Lett. **81** (1998) 1562.
37. CH00Z Collaboration, M. Apollonio *et al.*, Phys. Lett. **B420** (1998) 397.
38. Kamiokande Collaboration, Y Fukuda *et al.*, Phys. Lett. **B335** (1994) 237.
39. M. Messier, to be published in the Proceedings of DPF99, Los Angeles, California, January 1999.
40. Soudan 2 Collaboration, W. W. M. Allison *et al.*, Phys. Lett. **B449** (1999) 137; H. Gallager, Ph. D. Thesis, University of Minnesota 1996.
41. MACRO Collaboration, M. Ambrosio *et al.*, Phys. Lett. **B434** (1998) 451.
42. SuperKamiokande Collaboration, Y Fukuda *et al.*, Phys. Rev. Lett. **82** (1999) 2644.
43. R. Davis Jr., D. S. Harmer, and K. C. Hoffman, Phys. Rev. Lett. **20** (1968) 1205.

44. Homestake Collaboration, R. Davis Jr., Prog. Part. Nucl. Phys. **32** (1994) 13.
45. Gallex Collaboration, W. Hampel *et al.*, Phys. Lett. **B447** (1999) 127.
46. SAGE Collaboration J. N. Abdurashitov *et al.*, Phys. Lett. **B328** (1994) 234.
47. Kamiokande Collaboration, Y Fukuda *et al.*, Phys. Rev. Lett. **77** (1996) 1683
48. SuperKamiokande Collaboration, Y Fukuda *et al.*, Phys. Rev. Lett. **81** (1998) 1158; Erratum **81** (1998) 4279.
49. For a review, see J. N. Bahcall in the Proceedings of the 18th International Conference on Neutrino Physics and Astrophysics, Tokayama, Japan, June 1998; see also, J. N. Bahcall, S. Basu and M. H. Pinsonneault, Phys. Lett. **B433** (1998) 1.
50. V. Barger, R. J. N. Phillips and K. Whisnant, Phys. Rev. **D43** (1991) 1110.
51. J. N. Bahcall, P. I. Krastev and A. Yu. Smirnov, Phys. Rev. **D58** (1998) 096016.
52. See, for example, J. N. Bahcall, **Neutrino Astrophysics** (Cambridge University Press, Cambridge 1989).
53. For representative fits see, for example, N. Hata and P. Langacker, Phys. Rev. **D56** (1997) 6107; J. N. Bahcall, P. I. Krastev and A. Yu. Smirnov, Phys. Rev. **D58** (1998) 096016.
54. Adapted from S. P. Rosen and J. M. Gelb, Phys. Rev. **D34** (1986) 969. For a more recent discussion see, for example, L. Wolfenstein, and P.I. Krastev, Phys. Rev. **D55** (1997) 4405; S.T. Petcov, Ref. [12].
55. For a recent review see, for example, M. S. Turner, to be published in The Proceedings of Particle Physics and the Universe (Cosmo-98), astro-ph/9904051.
56. For a discussion, see, for example, J. R. Primack, in **Particle Physics and Cosmology at the Interface**, eds. J. Pati, P. Ghose, and J. Maharana (World Scientific, Singapore, 1995).
57. Some examples include: R. P. Thun and S. McKee, Phys. Lett. **B439** (1998) 123; T. Teshima and T. Sakai, Prog. Theor. Phys. **101** (1999) 147.
58. For a general discussion see, for example, B.Kayser, Proceedings of the XXIX International Conference on High Energy Physics, Vancouver, Canada, July 1998. These models were first proposed by D. Caldwell and R. N. Mohapatra, Phys. Rev. **D48** (1993) 3259: see also, V. Barger, S. Pakvasa, T.J. Weiler and K. Whisnant, Phys. Lett. **B437** (1998) 107.
59. G. Altarelli and F. Feruglio, Phys. Lett. **B439** (1998) 112.
60. K.S. Babu, J. C. Pati and F. Wilczek, hep-ph/9812538.
61. R. Barbieri, L. J. Hall, G. L. Kane and G. G. Ross, hep-ph/9901228.
62. R. Barbieri, L. J. Hall and A. Strumia, Phys. Lett. **B445** (1999) 407.
63. For a discussion of the status of SNO see, for example, C. Okada, to be published in the Proceedings of DPF99, Los Angeles, California, January 1999.
64. For a recent discussion of Borexino see, G. Alimonti *et al.*, Nucl. Phys. Proc. Suppl. **32** (1998) 149.
65. For a recent discussion, see Y. Oyama, hep-ex/9803014.
66. E. Ables *et al.*, Fermilab proposal FERMILAB-P-875.
67. E. Chruch *et al.*, Fermilab proposal FERMILAB-P-898.

Stars as Particle-Physics Laboratories

Georg G. Raffelt

Max-Planck-Institut für Physik (Werner-Heisenberg-Institut)
Föhringer Ring 6, 80805 München, Germany

Abstract. Low-mass particles such as neutrinos, axions, other Nambu-Goldstone bosons, gravitons, and so forth are produced in the hot and dense interior of stars. Therefore, astrophysical arguments constrain the properties of these particles in ways which are often complementary to cosmological arguments and to laboratory experiments. The most important stellar-evolution arguments are explained and the resulting particle-physics limits are reviewed in the context of other information from cosmology and laboratory experiments.

I INTRODUCTION

Astrophysical and cosmological arguments and observations have become part of the main-stream methodology to obtain empirical information on existing or hypothetical elementary particles and their interactions. The "heavenly laboratories" are complementary to accelerator and non-accelerator experiments, notably at the "low-energy frontier" of particle physics, which includes neutrino physics, other weakly interacting low-mass particles such as the hypothetical axions, novel long-range forces, and so forth.

The present lectures are dedicated to stars as particle-physics laboratories, or more precisely, to what can be learned about weakly interacting low-mass particles from the observed properties of stars. The prime argument is that a hot and dense stellar plasma emits low-mass weakly interacting particles in great abundance. They subsequently escape from the stellar interior directly, without further interactions, and thus provide a local energy sink for the stellar medium. The astronomically observable impact of this effect provides some of the most powerful limits on the properties of neutrinos, axions, and the like.

Once the particles have escaped they can decay on their long way to Earth, allowing one to derive interesting limits on radiative decay channels from the absence of unexpected x- or γ-ray fluxes from the Sun or other stars.

Finally, the weakly interacting particles can be directly detected at Earth, thus far only the neutrinos from the Sun and supernova (SN) 1987A, allowing one to extract important information on their properties.

CP490, *Particles and Fields: Eighth Mexican School,* edited by J. C. D'Olivo et al.

The subject of "Stars as Particle-Physics Laboratories" is broader than what is covered here. I will not touch on the solar neutrino problem and its oscillation interpretation—this is a topic unto itself and has been extensively reviewed by other authors, for example [1–3].

The high densities encountered in neutron stars make them ideal for studies concerning novel phases of nuclear matter (e.g. meson condensates or quark matter), an area covered by two recent books [4,5]. Quark stars are also the subject of an older review [6] and are covered in the proceedings of two topical conferences [7,8].

Certain grand unified theories predict the existence of primordial magnetic monopoles. They would get trapped in stars and then catalyze the decay of nucleons by the Rubakov-Callan effect. The ensuing anomalous energy release is constrained by the properties of stars, in particular neutron stars and white dwarfs, a topic that has been reviewed a long time ago [9]. It was re-examined, and the limits improved, in the wake of the discovery of the faintest white dwarf ever which puts restrictive limits on an anomalous internal heat source [10].

Weakly interacting massive particles (WIMPs), notably in the guise of the supersymmetric neutralinos, are a prime candidate for the cosmic dark matter. Some of them would get trapped in stars, annihilate with each other, and produce a secondary flux of high-energy neutrinos. The search for such fluxes from the Sun and the center of the Earth by present-day and future neutrino telescopes is the "indirect method" to detect galactic particle dark matter, an approach which is competitive with direct laboratory searches—see [11] for a review.

Returning to the topics which are covered here, Sections II–IV are devoted to a discussion of the main stellar objects that have been used to constrain low-mass particles, viz. the Sun, globular-cluster stars, compact stars, and SN 1987A. In Sections V and VI the main constraints on neutrinos and axions are summarized. Section VII is given over to brief concluding remarks.

II THE SUN

A Basic Energy-Loss Argument

The Sun is the best-known single star and thus a natural starting point for our survey of astrophysical particle laboratories. It is powered by hydrogen burning which amounts to the net reaction $4p + 2e^- \to {}^4\mathrm{He} + 2\nu_e + 26.73$ MeV, giving rise to a measured ν_e flux which now provides one of the most convincing indications for neutrino oscillations [1–3]. Instead of neutrinos from nuclear processes we focus here on particle fluxes which are produced in thermal plasma reactions. The photo neutrino process $\gamma + e^- \to e^- + \nu\bar{\nu}$ is a case in point, or gravitons from electron bremsstrahlung. The solar energy loss from such standard processes is small, but it may be large for new particles. To be specific we consider axions (Sec. VI) which arise in a variety of reactions, and in particular by the Primakoff process where thermal photons mutate into axions in the electric field of the medium's charged particles (Fig. 1). In Sec. VI B 1 we will discuss direct search experiments for solar axions, while here we focus on what is the main topic of this review, the backreaction of a new energy loss on stars.

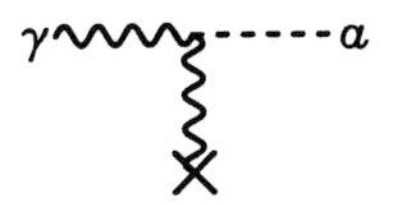

FIGURE 1. Primakoff production of axions in the Sun.

The Sun is a normal star which supports itself against gravity by thermal pressure, as opposed to degenerate stars like white dwarfs which are supported by electron degeneracy pressure. If one pictures the Sun as a self-gravitating monatomic gas in hydrostatic equilibrium, the "atoms" obey the virial theorem $\langle E_{\rm kin}\rangle = -\frac{1}{2}\langle E_{\rm grav}\rangle$. One consequence of this relationship is that extracting energy from such a system, i.e. reducing the total energy $\langle E_{\rm kin}\rangle + \langle E_{\rm grav}\rangle$, leads to contraction and to an *increase* of $\langle E_{\rm kin}\rangle$. Therefore, axion losses lead to contraction and heating. The nuclear energy generation rate scales with a high power of the temperature. Therefore, the heating implied by the new energy loss causes increased nuclear burning—the star finds a new equilibrium configuration where the new losses are compensated by an increased rate of energy generation.

The main lesson is that the new energy loss does not "cool" the star, it leads to heating and an increased consumption of nuclear fuel. The Sun, where energy is transported from the central nuclear furnace by radiation, actually overcompensates the losses and brightens, while it would dim if the energy transfer were by convection. Either behavior is understood by a powerful "homology argument" where the nonlinear interplay of the equations of stellar structure is represented in a simple analytic fashion [12].

The solar luminosity is well measured, yet this brightening effect is not observable because all else need not be equal. The present-day luminosity of the Sun depends on its unknown initial helium mass fraction Y; in a solar model it has to be adjusted such that $L_\odot = 3.85 \times 10^{33}$ erg s^{-1} is reproduced after 4.6×10^9 years of nuclear burning. For solar models with axion losses the required presolar helium abundance Y as a function of the axion-photon coupling constant $g_{a\gamma}$ is shown in Table 1. The axion luminosity L_a is also given as well as the central helium abundance Y_c, density ρ_c, and temperature T_c of the present-day Sun.

Even axion losses as large as $L_\odot$ can be accommodated by reducing the presolar helium mass fraction from about 27% to something like 23% [13,14]. The "standard Sun" has completed about half of its hydrogen-burning phase. Therefore, the anomalous energy losses cannot exceed something like $L_\odot$ or else the Sun could not have reached its observed age. Indeed, for $g_{10} = 30$ no consistent present-day Sun could be constructed for any value of Y [14]. The emission rate of other hypothetical particles would have a different temperature and density dependence than the Primakoff process, yet the general conclusion remains the same that a novel energy loss must not exceed about $L_\odot$.

TABLE 1. Solar-model parameters and relative detection rates in the Cl, Ga, and water neutrino observatories as a function of the axion-photon coupling constant $g_{10} \equiv g_{a\gamma}/10^{-10}\ \mathrm{GeV}^{-1}$ [13].

g_{10}	L_a $[L_\odot]$	Y	Y_c	ρ_c [g cm^{-3}]	T_c [10^7 K]	Cl	Ga	H_2O
0	0	0.266	0.633	153.8	1.563	1	1	1
4.5	0.04	0.265	0.641	158.0	1.575	1.07	1.16	1.20
10	0.20	0.257	0.679	177.5	1.626	1.45	2.2	2.4
15	0.53	0.245	0.751	218.3	1.722	2.5	6.0	6.7
20	1.21	0.228	0.914	324.2	1.931	6.4	20	23

This crude limit is improved by the solar neutrino flux which has been measured in five different observatories with three different spectral response characteristics, i.e. by the absorption on chlorine, gallium, and by the water Cherenkov technique. The axionic solar models produce larger neutrino fluxes; in Table 1 we show the expected detection rates for the Cl, Ga, and H_2O experiments relative to the standard case. For $g_{10} \lesssim 10$ one can still find oscillation solutions to the observed ν_e deficit, but larger energy-loss rates look excluded [13]. It appears safe to conclude that the Sun does not emit more than a few 10% of $L_\odot$ in new forms of radiation.

B Helioseismology

Over the past few years the precision measurements of the solar p-mode frequencies have provided a more reliable way to study the solar interior. For example, the convective surface layer is found to reach down to 0.710–0.716 $R_\odot$ [15], the helium content of these layers to exceed 0.238 [16]. Gravitational settling has reduced the surface helium abundance by about 0.03 so that the presolar value must have been at least 0.268, in good agreement with standard solar models. The reduced helium content required of the axionic solar models in Table 1 disagrees significantly with this lower limit for $g_{10} \geq 10$.

One may also invert the p-mode measurements to construct a "seismic model" of the solar sound-speed profile, e.g. [17]. All modern standard solar models agree well with the seismic model within its uncertainties (shaded band in Fig. 2) which mostly derive from the inversion method itself, not the measurements. The difference between the sound-speed profile of a standard solar model and those including axion losses are also shown in Fig. 2. For $g_{10} \geq 10$ the difference is larger than the uncertainties of the seismic model, implying a limit

$$g_{a\gamma} \lesssim 10 \times 10^{-10}\ \mathrm{GeV}^{-1}. \tag{1}$$

Other cases may be different in detail, but it appears safe to assume that any new energy-loss channel must not exceed something like 10% of $L_\odot$.

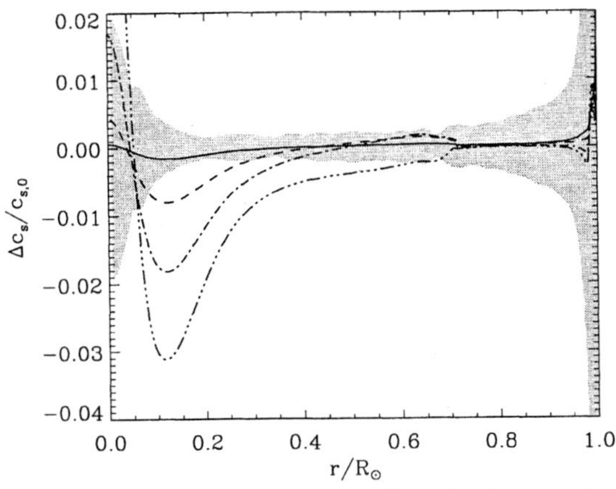

FIGURE 2. Fractional difference in sound-speed profiles of solar models with axion losses compared to the reference model [13]. The shaded area is the uncertainty of the seismic model [16]. The axion-photon coupling constant was g_{10}=4.5 (solid line), 10 (short-dashed), 15 (dash-dotted), 20 (dash-dot-dot-dotted).

C "Strongly" Interacting Particles

Thus far we have assumed that the new particles couple so weakly that they escape from the stellar interior without further interactions, in analogy to neutrinos or gravitons. They emerge from the entire stellar volume, i.e. their emission amounts to a local energy sink for the stellar plasma. But what if the particles interact so strongly that their mean free path is less than the solar radius?

The impact of such particles on a star compares to that of photons which are also "trapped" by their "strong" interaction. Their continuous thermal production and re-absorption amounts to the net effect of energy transfer from regions of higher temperature to cooler ones. In the Sun this radiative form of energy transfer is more important than conduction by electrons or convection, except in the outer layers. A particle which interacts more weakly than photons is more effective because it travels a larger distance before re-absorption—the ability to transfer energy is proportional to the mean free path. The properties of the Sun roughly confirm the standard photon opacities, whence a new particle would have to interact more strongly than photons to be allowed [18,19].

Therefore, contrary to what is sometimes stated in the literature, a new particle is by no means allowed just because its mean free path is less than the stellar dimensions. The impact of a new particle is maximal when its mean free path is of order the stellar radius. Of course, usually one is interested in very weakly interacting particles so that this point is mute.

III LIMITS ON STELLAR ENERGY LOSSES

A Globular-Cluster Stars

1 Evolution of Low-Mass Stars

Emitting new weakly interacting particles from stars modifies the time scale of evolution. For the Sun this effect is less useful to constrain particle emission than, say, the modified p-mode frequencies or the direct measurement of the neutrino fluxes. However, the observed properties of other stars provide far more restrictive limits on their evolutionary time scales so that anomalous modes of energy loss can be far more tightly constrained. We begin with globular-cluster stars which, together with SN 1987A, are the most successful example where astronomical observations provide nontrivial limits on the properties of elementary particles.

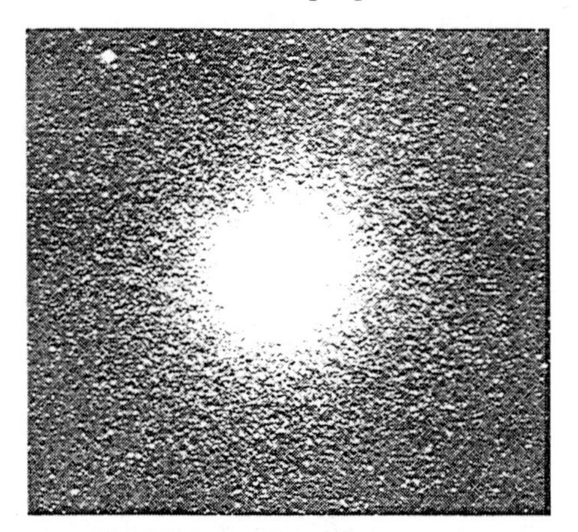

FIGURE 3. Globular cluster M3.

Our galaxy has about 150 globular clusters such as M3 (Fig. 3) which are gravitationally bound systems of up to a million stars. In Fig. 4 these stars are arranged according to their color or surface temperature (horizontal axis) and brightness (vertical axis), leading to a characteristic pattern which allows for quantitative tests of the stellar-evolution theory. Globular clusters are the oldest objects in the galaxy; the stars in a given cluster all formed at about the same time with essentially the same chemical composition, differing primarily in their mass. As more massive stars evolve faster, present-day globular-cluster stars are somewhat below[1] $1\,\mathcal{M}_\odot$ so that we are concerned with low-mass stars ($\mathcal{M} \lesssim 2\,\mathcal{M}_\odot$). Textbook expositions of stellar structure and evolution are [22,23].

---

[1] We always use the letter $\mathcal{M}$ to denote stellar masses with $1\,\mathcal{M}_\odot = 2 \times 10^{33}$ g the solar mass. The letter M is traditionally reserved for the absolute stellar brightness (in magnitudes or mag). The total or bolometric brightness is defined as $M_{\rm bol} = 4.74 - 2.5\,\log_{10}(L/L_\odot)$ with the solar luminosity $L_\odot = 3.85 \times 10^{33}$ erg s^{-1}.

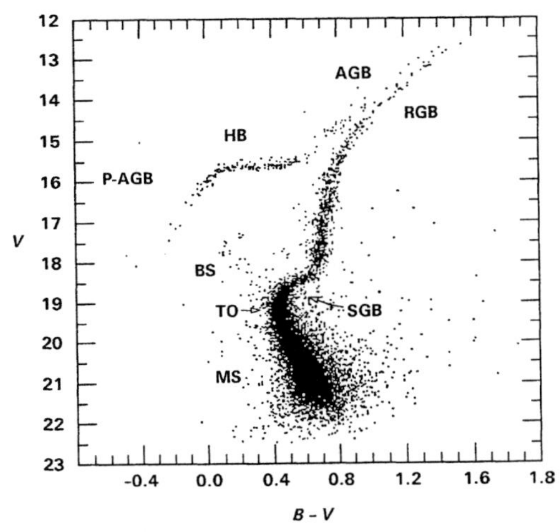

FIGURE 4. Color magnitude diagram for the globular cluster M3, based on the photometric data of 10,637 stars [20]. Vertically is the brightness in the visual (V) band, horizontally the difference between B (blue) and V brightness, i.e. a measure of the color and thus surface temperature, where blue (hot) stars lie toward the left. The classification for the evolutionary phases is as follows [21]. MS (main sequence): core hydrogen burning. BS (blue stragglers). TO (main-sequence turnoff): central hydrogen is exhausted. SGB (subgiant branch): hydrogen burning in a thick shell. RGB (red-giant branch): hydrogen burning in a thin shell with a growing core until helium ignites. HB (horizontal branch): helium burning in the core and hydrogen burning in a shell. AGB (asymptotic giant branch): helium and hydrogen shell burning. P-AGB (post-asymptotic giant branch): final evolution from the AGB to the white-dwarf stage.

Stars begin their life on the main sequence (MS) where they burn hydrogen in their center. Different locations on the MS in a color-magnitude diagram like Fig. 4 correspond to different masses, with more massive stars shining more brightly. When central hydrogen is exhausted the star develops a degenerate helium core, with hydrogen burning in a shell. Curiously, the stellar envelope expands, leading to a large surface area and thus a low surface temperature (red color)—they become "red giants." The luminosity is governed by the gravitational potential at the edge of the growing helium core so that these stars become ever brighter: they ascend the red-giant branch (RGB). The higher a star on the RGB, the more massive and compact its helium core.

It grows until about $0.5\,\mathcal{M}_\odot$ when it has become dense and hot enough to ignite helium. The ensuing core expansion reduces the gravitational potential at its edge and thus lowers the energy production rate in the hydrogen shell source, dimming these stars. Helium ignites at a fixed core mass, but the envelope mass differs due to varying rates of mass loss on the RGB, leading to different surface areas and thus surface temperatures. These stars thus occupy the horizontal branch (HB) in the color-magnitude diagram. In Fig. 4 the HB turns down on the left (blue color) where much of the luminosity falls outside the V filter; in terms of the total or "bolometric" brightness the HB is truly horizontal.

Finally, when helium is exhausted, a degenerate carbon-oxygen core develops, leading to a second ascent on what is called the asymptotic giant branch (AGB). These low-mass stars cannot ignite their carbon-oxygen core—they become white dwarfs after shedding most of their envelope.

The advanced evolutionary phases are fast compared with the MS duration which is about 10^{10} yr for stars somewhat below $1\,\mathcal{M}_{\odot}$. For example, the ascent on the upper RGB and the HB phase each take around 10^{8} yr. Therefore, the distribution of stars along the RGB and beyond can be taken as an "isochrone" for the evolution of a single star, i.e. a time-series of snapshots for the evolution of a single star with a fixed initial mass. Put another way, the number distribution of stars along the different branches are a direct measure for the duration of the advanced evolutionary phases. The distribution along the MS is different in that it measures the distribution of initial masses.

2 *Core Mass at Helium Ignition*

Anomalous energy losses modify this picture in measurable ways. We first consider an energy-loss mechanism which is more effective in the degenerate core of a red giant before helium ignition than on the HB so that the post-RGB evolution is standard. Since an RGB-star's helium core is supported by degeneracy pressure there is no feedback between energy-loss and pressure: the core is actually *cooled.* Helium burning $3^{4}\mathrm{He} \to {}^{12}\mathrm{C}$ depends very sensitively on temperature and density so that the cooling delays the ignition of helium, leading to a larger core mass $\mathcal{M}_{c}$, with several observable consequences.

First, the brightness of a red giant depends on its core mass so that the RGB would extend to larger luminosities, causing an increased brightness difference $\Delta M_{\mathrm{HB}}^{\mathrm{tip}}$ between the HB and the RGB tip. Second, an increased $\mathcal{M}_{c}$ implies an increased helium-burning core on the HB. For a certain range of colors these stars are pulsationally unstable and are then called RR Lyrae stars. From the measured RR Lyrae luminosity and pulsation period one can infer $\mathcal{M}_{c}$ on the basis of their so-called mass-to-light ratio A. Third, the increased $\mathcal{M}_{c}$ increases the luminosity of RR Lyrae stars so that absolute determinations of their brightness M_{RR} allow one to constrain the range of possible core masses. Fourth, the number ratio R of HB stars vs. RGB stars brighter than the HB is modified.

These observables also depend on the measured cluster metallicity as well as the unknown helium content which is usually expressed in terms of Y_{env}, the envelope helium mass fraction. Since globular clusters formed shortly after the big bang, their initial helium content must be close to the primordial value of 22–25%. Y_{env} should be close to this number because the initial mass fraction is somewhat depleted by gravitational settling, and somewhat increased by convective dredge-up of processed, helium-rich material from the inner parts of the star.

An estimate of $\mathcal{M}_c$ from a global analysis of these observables except A was performed in [25] and re-analysed in [24], A was used in [26], and an independent analysis using all four observables in [27]. In Fig. 5 we show the allowed core mass excess $\delta\mathcal{M}_c$ and envelope helium mass fraction Y_{env} from the analyses [24,27]; references to the original observations are found in these papers.

From Fig. 5 one concludes that within the given uncertainties the different observations overlap at the standard core mass ($\delta\mathcal{M}_c = 0$) and at an Y_{env} which is compatible with the primordial helium abundance. Of course, the error bands do not have a simple interpretation because they combine observational and estimated systematic errors, involving some judgement by the authors. The difference between the two panels of Fig. 5 gives one a sense of how sensitive the conclusions are to these more arbitrary aspects of the analysis. As a nominal limit it appears safe to adopt $|\delta\mathcal{M}_c| \lesssim 0.025$ or $|\delta\mathcal{M}_c|/\mathcal{M}_c \lesssim 5\%$; how much additional "safety-margin" one wishes to include is a somewhat arbitrary decision which is difficult to make objective in the sense of a statistical confidence level.

In [24] it was shown that this limit can be translated into an approximate limit on the average anomalous energy-loss rate ϵ_x of a helium plasma,

$$\epsilon_x \lesssim 10 \text{ erg g}^{-1}\text{ s}^{-1} \quad \text{at} \quad T \approx 10^8 \text{ K}, \quad \rho \approx 2\times10^5 \text{ g cm}^{-3}. \tag{2}$$

The density represents the approximate average of a red-giant core before helium ignition; the value at its center is about 10^6 g cm^{-3}. The main standard-model neutrino emission process is plasmon decay $\gamma \to \nu\bar{\nu}$ with a core average of about 4 erg g^{-1} s^{-1}. Therefore, Eq. (2) means that a new energy-loss channel must be less effective than a few times the standard neutrino losses.

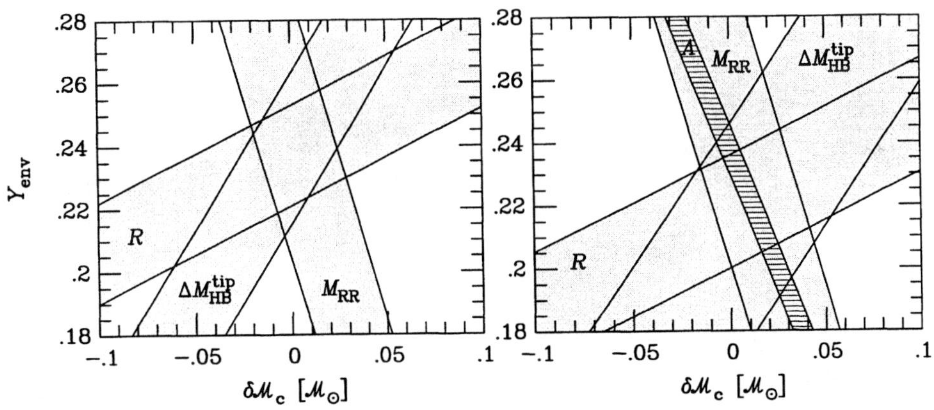

FIGURE 5. Allowed values for a core-mass excess at helium ignition $\delta\mathcal{M}_c$ and the envelope helium mass fraction Y_{env} of evolved globular-cluster stars. Left panel after [24], right panel [27]. The observables are the brightness difference $\Delta M_{\text{HB}}^{\text{tip}}$ between the HB and the RGB tip, the RR Lyrae mass-to-light ratio A, their absolute brightness M_{RR}, and the number ratio R between HB and RGB stars.

3 Helium-Burning Lifetime of Horizontal-Branch Stars

We now turn to an energy-loss mechanism which becomes effective in a non-degenerate medium, i.e. we imagine that the core expansion after helium ignition "switches on" an energy-loss channel that was negligible on the RGB. Therefore, the pre-HB evolution is taken to be standard. As in the case of the Sun (Sec. II A) there will be little change in the HB stars' brightness, rather they will consume their nuclear fuel faster and thus begin to ascend the AGB sooner. The net observable effect is a reduction of the number of HB relative to RGB stars.

From the measured HB/RGB number ratios in 15 globular clusters [28] and with plausible assumptions about the uncertainties of other parameters one concludes that the duration of helium burning agrees with stellar-evolution theory to within about 10% [24]. This implies that the new energy loss of the helium core should not exceed about 10% of its standard energy production rate. Therefore, the new energy-loss rate at average core conditions is constrained by [24]

$$\epsilon_x \lesssim 10 \text{ erg g}^{-1} \text{ s}^{-1} \quad \text{at} \quad T \approx 0.7 \times 10^8 \text{ K}, \quad \rho \approx 0.6 \times 10^4 \text{ g cm}^{-3}. \tag{3}$$

This limit is slightly more restrictive than the often-quoted "red-giant bound," corresponding to $\epsilon_x \lesssim 100$ erg g^{-1} s^{-1} at $T = 10^8$ K and $\rho = 10^4$ g cm^{-3}. It was based on the helium-burning lifetime of the "clump giants" in open clusters [29]. They have fewer stars, leading to statistically less significant limits. The "clump giants" are the physical equivalent of HB stars, except that they occupy a common location at the base of the RGB, the "red-giant clump."

4 Applications

After the energy-loss argument has been condensed into the simple criteria of Eqs. (2) and (3) it can be applied almost mechanically to a variety of cases. The main task is to identify the dominant emission process for the new particles and to calculate the energy-loss rate ϵ_x for a helium plasma at the conditions specified in Eqs. (2) or (3). The most important limits will be discussed in the context of specific particle-physics hypotheses in Secs. V and V. Here we just mention that these and similar arguments were used to constrain neutrino electromagnetic properties [25,26,29–34], axions [14,35–49], paraphotons [50], the photo production cross section on ^{4}He of new bosons [51,52], the Yukawa couplings of new bosons to baryons or electrons [53,54], and supersymmetric particles [55–57].

One may also calculate numerical evolution sequences including new energy losses [14,26,32,34,49,58]. Comparing the results from such studies with what one finds from Eqs. (2) and (3) reveals that, in view of the overall theoretical and observational uncertainties, it is indeed enough to use these simple criteria [24].

B White Dwarfs

White dwarfs are another case where astronomical observations provide useful limits on new stellar energy losses. These compact objects are the remnants of stars with initial masses of up to several $\mathcal{M}_\odot$ [23,59]. For low-mass progenitors the evolution proceeds as described in Sec. III A 1. When they ascend the asymptotic giant branch they eventually shed most of their envelope mass. The degenerate carbon-oxygen core, having reached something like $0.6\,\mathcal{M}_\odot$, never ignites. Its subsequent evolution is simply one of cooling, first dominated by neutrino losses throughout its volume, later by surface photon emission.

The cooling speed can be observationally infered from the "luminosity function," i.e. the white-dwarf number density per brightness interval. As white dwarfs are intrinsically dim they are observed only in the solar neighborhood, out to perhaps 100 pc (1 pc = 3.26 lyr) which is far less than the thickness of the galactic disk. The measured luminosity function (Fig. 6) reveals that there are few bright white dwarfs and many faint ones. The dotted line represents Mestel's cooling law [59,63], an analytic treatment based on surface photon cooling. The observed luminosity function dips at the bright end, a behavior ascribed to neutrino emission which quickly "switches off" as the star cools.

The luminosity function drops sharply at the faint end. Even the oldest white dwarfs have not yet cooled any further, implying that they were born 8–12 Gyr ago, in good agreement with the estimated age of the galaxy. Therefore, a novel cooling agent cannot be much more effective than the surface photon emission. This conclusion also follows from the agreement between the implied birthrate with independent estimates. The shape of the luminosity function can be deformed for an appropriate temperature dependence of the particle emission rate, e.g. enhancing

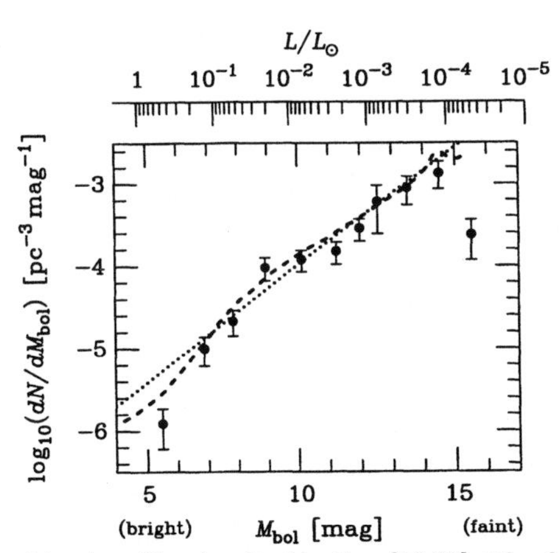

FIGURE 6. Observed white-dwarf luminosity function [60,61]. The dotted line represents Mestel's cooling law with a constant birthrate of 10^{-3} pc^{-3} Gyr^{-1}. The dashed line is from the cooling curve of a $0.6\,\mathcal{M}_\odot$ white dwarf which includes neutrino losses [62].

the "neutrino dip" at the bright end. Finally, white dwarfs in a certain range of surface temperatures are pulsationally unstable and are then called ZZ Ceti stars. The pulsation period of a few minutes depends on the luminosity, the period decrease thus on the cooling speed. For G117–B15A the period change was measured [64,65], implying a somewhat large cooling rate, yet better than a factor of 2 within the range of theoretical predictions.

White dwarfs were used to constrain the axion-electron coupling [66–71]. It was also noted that the somewhat large period decrease of G117–B15A could be ascribed to axion cooling [72]. Finally, a limit on the neutrino magnetic dipole moment was derived [70]. A detailed review of these limits is provided in [24]; they are somewhat weaker than those from globular-cluster stars, but on the same general level. Therefore, white-dwarf cooling essentially corroborates some of the globular-cluster limits, but does not improve on them.

IV SUPERNOVAE

A SN 1987A Neutrino Observations

When the star Sanduleak −69 202 exploded on 23 February 1987 in the Large Magellanic Cloud, a satellite galaxy of our Milky Way at a distance of about 50 kpc (165,000 lyr), it became possible for the first time to measure the neutrino emission from a nascent neutron star, turning this supernova (SN 1987A) into one of the most important stellar particle-physics laboratories [73–75].

A type II supernova explosion [76–81] is physically the implosion of an evolved massive star ($\mathcal{M} \gtrsim 8\,\mathcal{M}_\odot$) which has reached an "onion-skin structure" with several burning shells surrounding a degenerate iron core. It cannot gain further energy by fusion so that it becomes unstable when it has reached the limiting mass (Chandrasekhar mass) of 1–2 $\mathcal{M}_\odot$ that can be supported by electron degeneracy pressure. The ensuing collapse is intercepted when the equation of state stiffens at around nuclear density (3×10^{14} g cm^{-3}), corresponding to a core size of a few 10 km. At temperatures of tens of MeV this compact object is opaque to neutrinos. The gravitational binding energy of the newborn neutron star of about 3×10^{53} erg is thus radiated over several seconds from the "neutrino sphere." Crudely put, the collapsed SN core cools by thermal neutrino emission from its surface.

The neutrino signal from SN 1987A (Fig. 7) was observed by the $\bar{\nu}_e p \to n e^+$ reaction in several detectors [75]. The number of events, their energies, and the distribution over several seconds corresponds well to theoretical expectations and thus has been taken as a confirmation of the standard picture that a compact remnant formed which emitted its energy by quasi-thermal neutrino emission. Detailed statistical analyses of the data were performed in [85,86].

The signal does show a number of "anomalies." The average $\bar{\nu}_e$ energies infered from the IMB and Kamiokande observations are quite different [87,88]. The large time gap of 7.3 s between the first 8 and the last 3 Kamiokande events looks

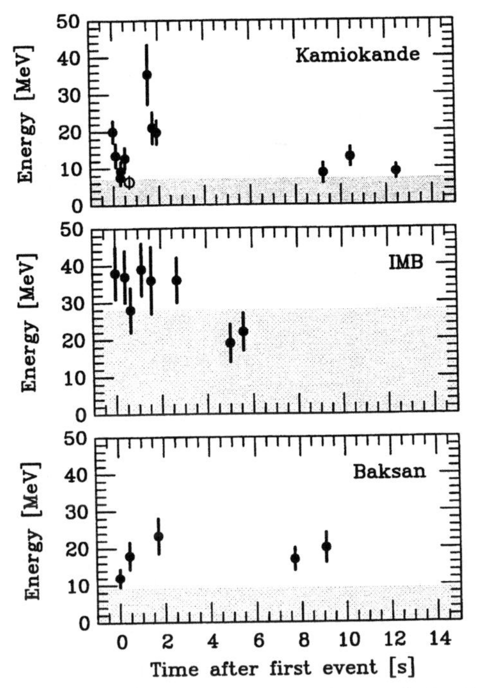

FIGURE 7. SN 1987A neutrino observations at Kamiokande [82], IMB [83] and Baksan [84]. The energies refer to the secondary positrons from the reaction $\bar{\nu}_e p \to n e^+$. In the shaded area the trigger efficiency is less than 30%. The clocks have unknown relative offsets; in each case the first event was shifted to $t = 0$. In Kamiokande, the event marked as an open circle is attributed to background.

worrisome [89]. The distribution of the final-state positrons from the $\bar{\nu}_e p \to n e^+$ capture reaction should be isotropic, but is found to be significantly peaked away from the direction of the SN [52,90,91]. Short of other explanations, these features have been blamed on statistical fluctuations in the sparse data.

B Signal Dispersion

A dispersion of the neutrino burst can be caused by a time-of-flight delay from a nonvanishing neutrino mass [92]. The arrival time from SN 1987A at a distance D would be delayed by

$$\Delta t = 2.57\ \mathrm{s} \left(\frac{D}{50\ \mathrm{kpc}}\right) \left(\frac{10\ \mathrm{MeV}}{E_\nu}\right)^2 \left(\frac{m_\nu}{10\ \mathrm{eV}}\right)^2. \tag{4}$$

As the $\bar{\nu}_e$'s were registered within a few seconds and had energies in the 10 MeV range, m_{ν_e} is limited to less than around 10 eV. Detailed analyses reveal that the pulse duration is consistently explained by the intrinsic SN cooling time and that $m_{\nu_e} \lesssim 20$ eV is implied as something like a 95% CL limit [85,93].

The apparent absence of a time-of-flight dispersion effect of the $\bar{\nu}_e$ burst was also used to constrain a "millicharge" of these particles (they would be deflected in the galactic magnetic field) [1,94], a quantum field theory with a fundamental length scale [95], and deviations from the Lorentzian rule of adding velocities [96]. Limits on new long-range forces acting on the neutrinos [97–101] seem to be invalidated in the most interesting case of a long-range leptonic force by screening from the cosmic background neutrinos [102].

The SN 1987A observations confirm that the visual SN explosion occurs several hours after the core-collapse and thus after the neutrino burst. Again, there is no apparent time-of-flight delay of the relative arrival times between the neutrino burst and the onset of the optical light curve, allowing one to confirm the equality of the relativistic limiting velocity for these particle types to within 2×10^{-9} [103,104]. Moreover, the Shapiro time delay in the gravitational field of the galaxy between neutrinos and photons is equal to within about 4×10^{-3} [105], constraining certain alternative theories of gravity [106,107].

C Energy-Loss Argument

The late events in Kamiokande and IMB reveal that the signal duration was not anomalously short. Very weakly interacting particles would freely stream from the inner core, removing energy which otherwise powers the late-time neutrino signal. Therefore, its observed duration can be taken as evidence against such novel cooling effects. This argument has been advanced to constrain axion-nucleon couplings [108–116], majorons [117–123], supersymmetric particles [124–131], and graviton emission in quantum-gravity theories with higher dimensions [132]. It has also been used to constrain right-handed neutrinos interacting by a Dirac mass [109,133–143], mixed with active neutrinos [144,145], interacting through right-handed currents [109,146–149], a magnetic dipole moment [150–154], or an electric form factor [155,156]. Many of these results will be reviewed in Secs. V and VI in the context of specific particle-physics hypotheses.

Here we illustrate the general argument with axions (Sec. VI) which are produced by nucleon bremsstrahlung $NN \to NNa$ so that the energy-loss rate depends on the axion-nucleon Yukawa coupling g_{aN}. In Fig. 8 we show the expected neutrino-signal duration as a function of g_{aN}. With increasing g_{aN}, corresponding to an increasing energy-loss rate, the signal duration drops sharply. For a sufficiently large g_{aN}, however, axions no longer escape freely; they are trapped and thermally emitted from the "axion sphere" at unit optical depth. Beyond some coupling strength axions are less important than neutrinos and cannot be excluded.

However, particles which are on the "strong interaction" side of this argument need not be allowed. They could be important for the energy-transfer during the infall phase and they could produce events in the neutrino detectors. For example, "strongly coupled" axions in a large range of g_{aN} are actually excluded because they would have produced too many events by their absorption on ^{16}O [157].

Likewise, particles on the free-streaming side can cause excessive events in the neutrino detectors. For example, right-handed neutrinos escaping from the inner core could become "visible" by decaying into left-handed states [158] or by spin-precessing in the galactic magnetic field if they have a dipole moment.

Returning to the general argument, one can estimate a limit on the energy-loss rate on the free-streaming side by the simple criterion that the new channel should be less effective than the standard neutrino losses, corresponding to [24]

$$\epsilon_x \lesssim 10^{19}\ \text{erg g}^{-1}\ \text{s}^{-1} \quad \text{at} \quad \rho = 3\times10^{14}\ \text{g cm}^{-3}, \quad T = 30\ \text{MeV}. \tag{5}$$

The density is the core average, the temperature an average during the first few seconds. Some authors find higher temperatures, but for a conservative limit one should probably stick to a value at the lower end of the plausible range. At these conditions the nucleons are partially degenerate while the electrons are highly degenerate. Several detailed numerical studies reveal that this simple criterion corresponds to approximately halving the neutrino signal duration [24].

A simple analytic treatment is far more difficult on the trapping side; see [110] for an example in the context of axions.

The SN 1987A energy-loss argument tends to be most powerful at constraining new particle interactions with nucleons. Therefore, one typically needs to calculate the interaction rate with a hot and dense nuclear medium which is dominated by many-body effects. Besides the sparse data, the theoretical treatment of the emission rate is the most problematic aspect of this entire method.

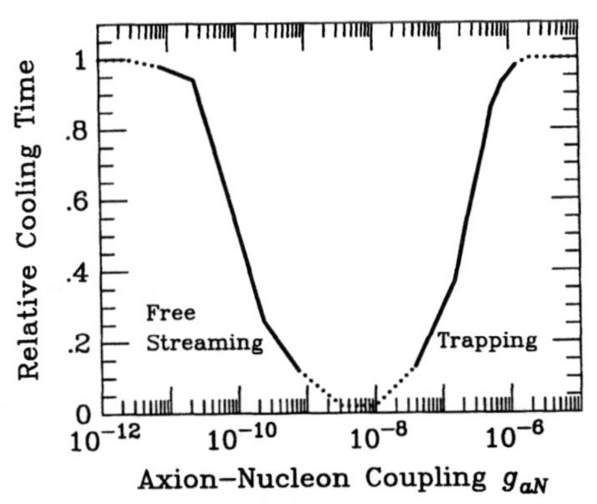

FIGURE 8. Relative duration of SN neutrino cooling as a function of the axion-nucleon coupling. Freely streaming axions are emitted from the entire core volume, trapped ones from the "axion sphere." The solid line follows from the numerical calculations [113,114]; the dotted line is an arbitrary continuation.

D Neutrino Spectra and Neutrino Oscillations

Neutrino oscillations can have several interesting ramifications in the context of SN physics because the temporal and spectral characteristics of the emission process depend on the neutrino flavor [79–81,171]. The simplest case is that of the "prompt ν_e burst" which represents the deleptonization of the outer core layers at about 100 ms after bounce when the shock wave breaks through the edge of the collapsed iron core. This "deleptonization burst" propagates through the mantle and envelope of the progenitor star so that resonant oscillations take place for a large range of mixing parameters between ν_e and some other flavor, notably for most of those values where the MSW effect operates in the Sun [159–168]. In a water Cherenkov detector one can see this burst by ν_e-e-scattering which is forward peaked, but one would have expected only a fraction of an event from SN 1987A. The first event in Kamiokande may be attributed to this signal, but this interpretation is statistically insignificant.

The next few 100 ms is when the shock wave stalls at a few 100 km above the core and needs rejuvenating. The efficiency of neutrino heating can increase by resonant flavor oscillations which swap the ν_e flux with, say, the ν_τ one. Therefore, what passes through the shock wave as a ν_e was born as a ν_τ at the proto neutron star surface. It has on average higher energies and thus is more effective at transfering energy. In Fig. 9 the shaded range of mixing parameters is where supernovae are helped to explode, assuming a "normal" neutrino mass spectrum with $m_{\nu_e} < m_{\nu_\tau}$ [169]. Below the shaded region the resonant oscillations take place beyond the shock wave and thus do not affect the explosion.

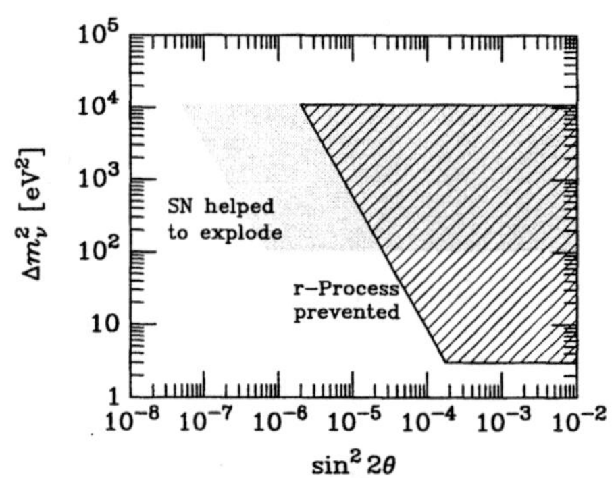

FIGURE 9. Mass difference and mixing between ν_e and ν_μ or ν_τ where a spectral swap would occur to help explode supernovae, schematically after [169], and where it would prevent r-process nucleosynthesis, schematically after [183–185].

The logic of this scenario depends on deviations from strictly thermal neutrino emission at some blackbody "neutrino sphere." The neutrino cross sections are very energy dependent and different for different flavors so that the concept of a neutrino sphere is rather crude—the spectra are neither thermal nor equal for the different flavors [170,171]. The dominant opacity source for ν_e is the process $\nu_e+n \to p+e^-$, for $\bar\nu_e$ it is $\bar\nu_e+p \to n+e^+$, while for $\nu_{\mu,\tau}$ and $\bar\nu_{\mu,\tau}$ it is neutral-current scattering on nucleons. Therefore, unit optical depth is at the largest radius (and lowest medium temperature) for ν_e, and deepest (highest temperature) for $\nu_{\mu,\tau}$ and $\bar\nu_{\mu,\tau}$. In typical calculations one finds a hierarchy $\langle E_{\nu_e}\rangle : \langle E_{\bar\nu_e}\rangle : \langle E_{\rm others}\rangle \approx \frac{2}{3} : 1 : \frac{5}{3}$ with $\langle E_{\bar\nu_e}\rangle = 14$–$17$ MeV [81]. The SN 1987A observations imply a somewhat lower range of $\langle E_{\bar\nu_e}\rangle \approx 7$–$14$ MeV [85,88].

A few seconds after core bounce the shock wave has long taken off, leaving behind a relatively dilute "hot bubble" above the neutron-star surface. This region is one suspected site for the r-process heavy-element synthesis, which requires a neutron-rich environment [176–182]. The neutron-to-proton ratio, which is governed by the beta reactions $\nu_e + n \to p + e^-$ and $\bar\nu_e + p \to n + e^+$, is shifted to a neutron-rich phase if $\langle E_{\nu_e}\rangle < \langle E_{\bar\nu_e}\rangle$ as for standard neutrino spectra. Resonant oscillations can swap the ν_e flux with another one, inverting this hierarchy. In the hatched range of mixing parameters shown in Fig. 9 the r-process would be disturbed [183–186]. On the other hand, $\nu_e \to \nu_s$ oscillations into a sterile neutrino could actually help the r-process by removing some of the neutron-stealing ν_e's [187,189].

If the mixing angle between ν_e and some other flavor is large, the $\bar\nu_e$ flux from a SN contains a significant fraction of oscillated states that were born as $\bar\nu_\mu$ or $\bar\nu_\tau$ and thus should have higher average energies. The measured SN 1987A event energies are already somewhat low, a problem so strongly exacerbated by oscillations that a large mixing angle solution of the solar neutrino deficit poses a problem [88,93,188]. This conclusion, however, depends on the standard predictions for the average neutrino energies which may not hold up to closer scrutiny [115,173–175].

V LIMITS ON NEUTRINO PROPERTIES

A Masses and Mixing

Astrophysics and cosmology play a fundamental role for neutrino physics as the properties of stars and the universe at large provide some of the most restrictive limits on nonstandard properties of these elusive particles. Therefore, it behoves us to summarize what the astrophysical arguments introduced in the previous sections teach us about neutrinos.

Unfortunately, stars do not tell us very much about neutrino masses, the holy grail of neutrino physics. The current discourse [190–192] centers on the interpretation of the solar [3] and atmospheric [193] neutrino anomalies and the LSND experiment [194,195] which all provide overwhelming evidence for neutrino oscillations. Solar neutrinos imply a Δm_ν^2 of about 10^{-5} eV^2 (MSW solutions) or 10^{-10} eV^2 (vacuum oscillations), atmospheric neutrinos 10^{-3}–10^{-2} eV^2, and the LSND experiment 0.3–8 eV^2. Taken together, these results require a fourth flavor, a sterile neutrino, which is perhaps the most spectacular implication of these experiments, but of course also the most suspicious one.

Core-collapse supernovae appear to be the only case in stellar astrophysics, apart from the solar neutrino flux, where neutrino oscillations can be important. However, Fig. 9 reveals that the experimentally favored mass differences negate a role of neutrino oscillations for the explosion mechanism or r-process nucleosynthesis, except perhaps when sterile neutrinos exist [187,189]. Oscillations affect the interpretation of the SN 1987A signal [88,93,188] and that of a future galactic SN [196–198].

Oscillation experiments reveal only mass *differences* so that one still needs to worry about the absolute neutrino mass scale. The absence of anomalous SN 1987A signal dispersion (Sec. IV B) gives us a limit [85,93]

$$m_{\nu_e} \lesssim 20 \text{ eV}, \tag{6}$$

somewhat weaker than current laboratory bounds. A high-statistics observation of a galactic SN by a detector like Superkamiokande could improve this to something like 3 eV by using the fast rise-time of the neutrino burst as a measure of dispersion effects [199]. If the neutrino mass differences are indeed very small, this limit carries over to the other flavors. One can derive an independent mass limit on ν_μ and ν_τ in the range of a few 10 eV if one identifies a neutral-current signature in a water Cherenkov detector [200–204], or if one has an additional measurement in a future neutral-current detector [205,206].

The SN 1987A energy-loss argument (Sec. IV C) provides a limit on a neutrino Dirac mass of [24,109,133–136,140]

$$m_\nu(\text{Dirac}) \lesssim 30 \text{ keV}. \tag{7}$$

It is based on the idea that trapped Dirac neutrinos produce their sterile component with a probability of about $(m_\nu/2E_\nu)^2$ in collisions and thus feed energy into an invisible channel. This result was of interest in the discourse on Simpson's 17 keV neutrino which is now only of historical interest [207].

B Dipole and Transition Moments

1 Electromagnetic Form Factors

Neutrino electromagnetic interactions would provide for a great variety of astrophysical implications. In a vacuum, the most general neutrino interaction with the electromagnetic field is [208,209]

$$\mathcal{L}_{\text{int}} = -F_1\bar{\psi}\gamma_\mu\psi A^\mu - G_1\bar{\psi}\gamma_\mu\gamma_5\psi\partial_\mu F^{\mu\nu} - \tfrac{1}{2}\bar{\psi}\sigma_{\mu\nu}(F_2 + G_2\gamma_5)\psi F^{\mu\nu}, \tag{8}$$

where ψ is the neutrino field, A^μ the electromagnetic vector potential, and $F^{\mu\nu}$ the field-strength tensor. The form factors are functions of Q^2 with Q the energy-momentum transfer. In the $Q^2 \to 0$ limit F_1 is a charge, G_1 an anapole moment, F_2 a magnetic, and G_2 an electric dipole moment.

If neutrinos are electrically strictly neutral, $F_1(0) = 0$, they still have a charge radius, usually defined as $\langle r^2 \rangle = 6\partial F_1(Q^2)/e\partial Q^2|_{Q^2=0}$. This form factor provides for a contact interaction, not for a long-range force, and as such a correction to processes with Z^0 exchange [210–214]. As astrophysics provides no precision test for the effective strength of neutral-current interactions, this form factor is best probed in laboratory experiments [215]. Likewise, the anapole interaction vanishes in the $Q^2 \to 0$ limit and thus represents a correction to the standard neutral-current interaction, with no apparent astrophysical consequences.

The most interesting possibility are magnetic and electric dipole and transition moments. If the standard model is extended to include neutrino Dirac masses, the magnetic dipole moment is $\mu_\nu = 3.20 \times 10^{-19}\, \mu_{\rm B}\, m_\nu/{\rm eV}$ where $\mu_{\rm B} = e/2m_e$ is the Bohr magneton [208,209]. An electric dipole moment ϵ_ν violates CP, and both are forbidden for Majorana neutrinos. Including flavor mixing implies electric and magnetic transition moments for both Dirac and Majorana neutrinos, but they are even smaller due to GIM cancellation. These values are far too small to be of any experimental or astrophysical interest. Significant neutrino electromagnetic form factors require a more radical extension of the standard model.

2 Plasmon Decay in Stars

Dipole or transition moments allow for several interesting processes (Fig. 10). For the purpose of deriving limits, the most important case is $\gamma \to \nu\bar\nu$ which is kinematically possible in a plasma because the photon acquires a dispersion relation which roughly amounts to an effective mass. Even without anomalous couplings, the plasmon decay proceeds because the charged particles of the medium induce an effective neutrino-photon interaction. Put another way, even standard neutrinos have nonvanishing electromagnetic form factors in a medium [216,217]. The standard plasma process [218–220] dominates the neutrino production in white dwarfs or the cores of globular-cluster red giants.

The plasma process was first used in [221] to constrain neutrino electromagnetic couplings. Numerical implementations of the nonstandard rates in stellar-evolution calculations are [26,32,34,70]. The helium-ignition argument in globular clusters (Sec. III A 2), equivalent to Eq. (2), implies a limit [24,25,33,34]

$$\mu_\nu \lesssim 3 \times 10^{-12}\, \mu_{\rm B}, \tag{9}$$

applicable to magnetic and electric dipole and transition moments for Dirac and Majorana neutrinos. Of course, the final-state neutrinos must be lighter than the photon plasma mass which is around 10 keV for the relevant conditions.

The corresponding laboratory limits are much weaker [239]. The most restrictive bound is $\mu_{\nu_e} < 1.8\times10^{-10}\, \mu_{\rm B}$ at 90% CL from a measurement of the $\bar\nu_e$-e-scattering cross section involving reactor sources. A significant improvement should become possible with the MUNU experiment [222], but it is unlikely that the globular-cluster limit can be reached anytime soon.

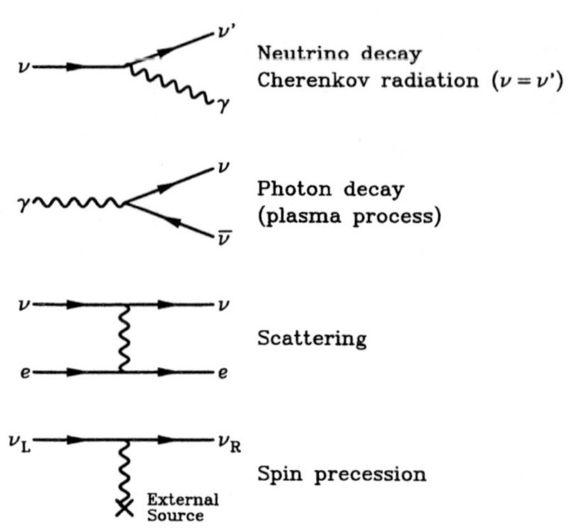

FIGURE 10. Processes with neutrino electromagnetic dipole or transition moments.

3 Radiative Decay

A neutrino mass eigenstate ν_i may decay to another one ν_j by the emission of a photon, where the only contributing form factors are the magnetic and electric transition moments. The inverse radiative lifetime is found to be [208,209]

$$\tau_\gamma^{-1} = \frac{|\mu_{ij}|^2 + |\epsilon_{ij}|^2}{8\pi}\left(\frac{m_i^2 - m_j^2}{m_i}\right)^3 = 5.308\ \mathrm{s}^{-1}\left(\frac{\mu_{\rm eff}}{\mu_{\rm B}}\right)^2\left(\frac{m_i^2 - m_j^2}{m_i^2}\right)^3\left(\frac{m_i}{\mathrm{eV}}\right)^3, \quad (10)$$

where μ_{ij} and ϵ_{ij} are the transition moments while $|\mu_{\rm eff}|^2 \equiv |\mu_{ij}|^2 + |\epsilon_{ij}|^2$. Radiative neutrino decays have been constrained from the absence of decay photons of reactor $\bar\nu_e$ fluxes [223], the solar ν_e flux [224,225], and the SN 1987A neutrino burst [226–230]. For $m_\nu \equiv m_i \gg m_j$ these limits can be expressed as

$$\frac{\mu_{\rm eff}}{\mu_{\rm B}} \lesssim \begin{cases} 0.9\times10^{-1}\ (\mathrm{eV}/m_\nu)^2 & \text{Reactor } (\bar\nu_e), \\ 0.5\times10^{-5}\ (\mathrm{eV}/m_\nu)^2 & \text{Sun } (\nu_e), \\ 1.5\times10^{-8}\ (\mathrm{eV}/m_\nu)^2 & \text{SN 1987A (all flavors)}, \\ 1.0\times10^{-11}(\mathrm{eV}/m_\nu)^{9/4} & \text{Cosmic background (all flavors)}. \end{cases} \quad (11)$$

In this form the SN 1987A limit applies for $m_\nu \lesssim 40$ eV. The decay of cosmic background neutrinos would contribute to the diffuse photon backgrounds, excluding the shaded areas in Fig. 11. They are approximately delineated by the dashed line, corresponding to the bottom line in Eq. (11). More restrictive limits obtain for certain masses above 3 eV from the absence of emission features from several galaxy clusters [237,238,240].

For low-mass neutrinos the m_ν^3 phase-space factor in Eq. (10) is so punishing that the globular-cluster limit is the most restrictive one for m_ν below a few eV, i.e. in

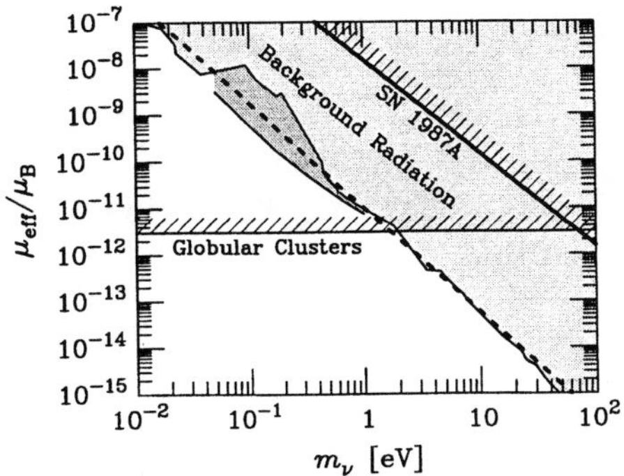

FIGURE 11. Astrophysical limits on neutrino dipole moments. The light-shaded background-radiation limits are from [234], the dark-shaded ones from [235,236], the dashed line is the approximation formula in Eq. (11), bottom line.

the mass range which today appears favored from neutrino oscillation experiments. Turning this around, the globular-cluster limit implies that radiative decays of low-mass neutrinos do not seem to have observable consequences.

For masses above something like 30 eV one must invoke fast invisible decays in order to avoid a conflict with the cosmological mass limit. In this case radiative decay limits involve the total lifetime as another parameter; the SN 1987A limits have been interpreted in this sense in [24,229,231,233].

4 *Spin-Flip Scattering*

The magnetic or electric dipole interaction couples neutrino fields of opposite chirality. In the relativistic limit this implies that a neutrino flips its helicity in an "electromagnetic collision," in the Dirac case producing the sterile component. The active states are trapped in a SN core so that spin-flip collisions open an energy-loss channel in the form of sterile states. Conversely, the SN 1987A energy-loss argument (Sec. IV C) allows one to derive a limit [151,154],

$$\mu_\nu(\text{Dirac}) \lesssim 3 \times 10^{-12}\,\mu_\text{B}, \qquad (12)$$

for both electric and magnetic dipole and transition moments. It is the same as the globular-cluster limit Eq. (9), which however includes the Majorana case.

Spin-flip collisions would also populate the sterile Dirac components in the early universe and thus increase the effective number of thermally excited neutrino degrees of freedom. Full thermal equilibrium attains for $\mu_\nu(\text{Dirac}) \gtrsim 60 \times 10^{-12}\,\mu_\text{B}$ [31,241]. In view of the SN 1987A and globular-cluster limits this result assures us that big-bang nucleosynthesis remains undisturbed.

VI AXIONS AND OTHER PSEUDOSCALARS

A Interaction Structure

New spontaneously broken global symmetries imply the existence of Nambu-Goldstone bosons which are massless and as such the most natural case, besides neutrinos, for using stars as particle-physics laboratories. Massless scalars would lead to new long-range forces so that we may presently focus on pseudoscalars. The most prominent example are axions which were proposed more than twenty years ago as a solution to the strong CP problem [242–245]; for reviews see [246,247] and for the latest developments the proceedings of a topical conference [248]. We use axions as a generic example—it will be obvious how to extend the following results and discussions to other cases.

Actually, axions are only "pseudo Nambu-Goldstone bosons" in that the spontaneously broken chiral Peccei-Quinn symmetry $U_{\rm PQ}(1)$ is also explicitly broken, providing these particles with a small mass

$$m_a = 0.60\ \text{eV}\ \frac{10^7\ \text{GeV}}{f_a}. \tag{13}$$

Here, f_a is the Peccei-Quinn scale, an energy scale which is related to the vacuum expectation value of the field which breaks $U_{\rm PQ}(1)$. The properties of Nambu-Goldstone bosons are always related to such a scale which is the main quantity to be constrained by astrophysical arguments, while Eq. (13) is specific to axions and allows one to express limits on f_a in terms of m_a.

In order to calculate the axionic energy-loss rate from stellar plasmas one needs to specify the interaction with the medium constituents. The interaction with a fermion j (mass m_j) is generically

$$\mathcal{L}_{\rm int} = \frac{C_j}{2f_a}\,\bar\Psi_j\gamma^\mu\gamma_5\Psi_j\partial_\mu a \quad \text{or} \quad -i\,\frac{C_j m_j}{f_a}\,\bar\Psi_j\gamma_5\Psi_j a, \tag{14}$$

where Ψ_j is the fermion and a the axion field, C_j is a model-dependent coefficient of order unity, $g_{aj} \equiv C_j m_j/f_a$ plays the role of a Yukawa coupling, and $\alpha_{aj} \equiv g_{aj}^2/4\pi$ that of an "axionic fine structure constant." The derivative form of the interaction is more fundamental in that it is invariant under $a \to a + a_0$ and thus respects the Nambu-Goldstone nature of these particles. The pseudoscalar form is usually equivalent, but one has to be careful when calculating processes where two Nambu-Goldstone bosons are attached to one fermion line, for example an axion and a pion attached to a nucleon [109,249–252].

The dimensionless couplings C_i depend on the detailed implementation of the Peccei-Quinn mechanism. Limiting our discussion to "invisible axion models" where f_a is much larger than the scale of electroweak symmetry breaking, one generically distinguishes between models of the DFSZ type (Dine, Fischler, Srednicki [253], Zhitnitskiĭ [254]) and of the KSVZ type (Kim [255], Shifman, Vainshtein,

Zakharov [256]). The latter provides no tree-level couplings to the standard quarks or leptons, yet axions couple to nucleons by their generic mixing with the neutral pion. The latest analysis gives numerically [116]

$$C_p = -0.34, \qquad C_n = 0.01 \tag{15}$$

with a statistical uncertainty of about ± 0.04 and an estimated systematic uncertainty of roughly the same magnitude. The tree-level couplings to standard quarks and leptons in the DFSZ model depend on an angle β which measures the ratio of vacuum expectation values of two Higgs fields. One finds [116]

$$C_e = \tfrac{1}{3}\cos^2\beta, \quad C_p = -0.07 - 0.46\cos^2\beta, \quad C_n = -0.15 + 0.38\cos^2\beta, \tag{16}$$

with similar uncertainties as in the KSVZ case.

The CP-conserving interaction between photons and pseudoscalars is commonly expressed in terms of an inverse energy scale $g_{a\gamma}$ according to

$$\mathcal{L}_{\text{int}} = -\tfrac{1}{4} g_{a\gamma} F_{\mu\nu} \tilde{F}^{\mu\nu} a = g_{a\gamma} \mathbf{E} \cdot \mathbf{B}\, a, \tag{17}$$

where F is the electromagnetic field-strength tensor and $\tilde{F}$ its dual. For axions

$$g_{a\gamma} = -\frac{\alpha}{2\pi f_a} C_\gamma, \qquad C_\gamma = \frac{E}{N} - 1.92 \pm 0.08, \tag{18}$$

where E/N is the ratio of the electromagnetic over color anomaly and as such a model-dependent ratio of small integers. In the DFSZ model or grand unified models one has $E/N = 8/3$ so that $C_\gamma \approx 0.75$, but one can also construct models with $E/N = 2$, significantly reducing the axion-photon coupling [257]. The value of C_γ in a great variety of cases was reviewed in [258,259].

B Limits on the Interaction Strength

1 *Photons*

The interaction with fermions or photons allows for numerous reactions which can produce axions in stars and thus allow one to derive limits on their coupling strength. Beginning with photons, pseudoscalars interact according to Eq. (17) which allows for the decay $a \to 2\gamma$. In stellar plasmas it also allows for the Primakoff conversion $\gamma \leftrightarrow a$ in the electric fields of electrons and nuclei [35]—see Fig. 1. For low-mass pseudoscalars the emission rate was calculated for various degrees of electron degeneracy in [45,71,260], superseding an earlier calculation where screening effects had been ignored [40].

The helioseismological constraint on solar energy losses then leads to Eq. (1) as a bound on $g_{a\gamma}$. It is shown in Fig. 12 ("Sun") in the context of other bounds;

similar plots are found in [239,24,261–263]. For axions the relationship between $g_{a\gamma}$ and m_a is indicated by the heavy solid line, assuming $E/N = 8/3$.

One may also search directly for solar axions. One method ("helioscope") is to direct a dipole magnet toward the Sun, allowing solar axions to mutate into x-rays by the inverse Primakoff process [265,266]. A first pilot experiment was not sensitive enough [267], but the exposure time was significantly increased in a new experiment in Tokyo where a dipole magnet was gimballed like a telescope so that it could follow the Sun [268,269]. The resulting limit $g_{a\gamma} \lesssim 6 \times 10^{-10}\ \mathrm{GeV}^{-1}$ is more restrictive than Eq. (1). Another helioscope project was begun in Novosibirsk several years ago [271], but its current status has not been reported for some time. An intruiging project (SATAN) at CERN would use a decommissioned LHC test magnet that could be mounted on a turning platform to achieve reasonable periods of alignment with the Sun [272]. This setup could begin to compete with the globular-cluster limit of Eq. (19).

The axion-photon transition in a macroscopic magnetic field is analogous to neutrino oscillations and thus depends on the particle masses [270]. For a large mass difference the transition is suppressed by the momentum mismatch of particles with equal energies. Therefore, the Tokyo limit applies only for $m_a \lesssim 0.03$ eV. In a next step one will fill the helioscope with a pressurized gas, giving the photon a dispersive mass to prevent the momentum mismatch.

An alternative method is "Bragg diffraction," where one uses the strong electric field of a crystal lattice which has large Fourier components for the required momentum transfer [273–275]. One has used Ge detectors which were originally built

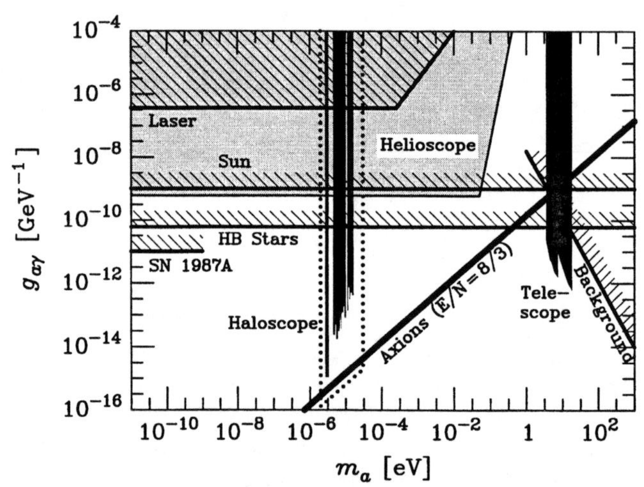

FIGURE 12. Limits to the axion-photon coupling $g_{a\gamma}$ as defined in Eq. (17). They apply to any pseudoscalar except for the "haloscope" search which assumes that these particles are the galactic dark matter; the dotted region marks the projected sensitivity range of the ongoing dark-matter axion searches. For higher masses than shown here the pertinent limits are reviewed in [262].

to search for neutrinoless double-beta decay and for WIMP dark matter; the crystal serves simultaneously as a Primakoff "transition agent" and as an x-ray detector. A first limit of the SOLAX Experiment [276] of $g_{a\gamma} \lesssim 27 \times 10^{-10}$ GeV^{-1} is not yet compatible with Eq. (1) and thus not self-consistent. In the future one may reach this limit, but prospects to go much further appear dim [277].

The Primakoff conversion of stellar axions can also proceed in the magnetic fields of Sun spots or in the galactic magnetic field so that one might expect anomalous x- or γ-ray fluxes from the Sun [278], the red supergiant Betelgeuse [279], or SN 1987A [280,281]. The latter case yields $g_{a\gamma} \lesssim 0.1 \times 10^{-10}$ GeV^{-1} for nearly massless pseudoscalars with $m_a \lesssim 10^{-9}$ eV. A similar limit obtains from the isotropy of the cosmic x-ray background which would be modified by the conversion to axions in the galactic magnetic field [282]. Axion-photon conversion in the magnetic fields of stars, the galaxy, or the early universe were also studied in [270,283–290], but no additional limits emerged.

The existence of massless pseudoscalars would cause a photon birefringence effect in pulsar magnetospheres, leading to a differential time delay between photons of opposite helicity and thus to $g_{a\gamma} \lesssim 0.5 \times 10^{-10}$ GeV^{-1} [291].

A laser beam in a laboratory magnetic field would also be subject to vacuum birefringence [292], adding to the QED Cotton-Mouton effect. First pilot experiments [263,293] did not reach the QED level. Two vastly improved current projects should get there [294,295], but they will stay far away from the "axion line" in Fig. 12. With a laser beam in a strong magnet one can also search for the Primakoff axion production and subsequent back-conversion, but a pilot experiment naturally did not have the requisite sensitivity [296]. The exclusion range of current laser experiments is schematically indicated in Fig. 12.

The most important limit on the photon coupling of pseudoscalars derives from the helium-burning lifetime of HB stars in globular clusters, i.e. from Eq. (3),

$$g_{a\gamma} \lesssim 0.6 \times 10^{-10}\ \mathrm{GeV}^{-1}. \tag{19}$$

For $m_a \gtrsim 10$ keV this limit quickly degrades as the emission is suppressed when the particle mass exceeds the stellar temperature. For a fixed temperature, the Primakoff energy-loss rate decreases with increasing density so that Eq. (2) implies a less restrictive constraint. Equation (19) was first stated in [24], superseding the slightly less restrictive but often-quoted "red-giant bound" of [14]—see the discussion after Eq. (3). With Eq. (18) one finds for axions

$$m_a C_\gamma \lesssim 0.3\ \mathrm{eV} \quad \text{and} \quad f_a/C_\gamma \gtrsim 2 \times 10^7\ \mathrm{GeV}. \tag{20}$$

In the DFSZ model and grand unified models we have $E/N = 8/3$ or $C_\gamma = 0.75$ so that $m_a \lesssim 0.4$ eV and $f_a \gtrsim 1.5 \times 10^7$ GeV (Fig. 13). It is conceivable, however, that $E/N = 2$ and thus C_γ very small, significantly degrading these bounds.

On the basis of their two-photon coupling alone, pseudoscalars can reach thermal equilibrium in the early universe. Their subsequent $a \to 2\gamma$ decays would contribute to the cosmic photon backgrounds [262,264], excluding a non-trivial

m_a-$g_{a\gamma}$-range (Fig. 12). Some of the pseudoscalars would end up in galaxies and clusters of galaxies. Their decay would produce an optical line feature that was not found [240,297,298], leading to the "telescope" limits in Fig. 12. For axions they exclude an approximate mass range 4–14 eV even for a small C_γ.

Axions with a mass in the μeV (10^{-6} eV) range could be the dark matter of the universe (Sec. VI C). The Primakoff conversion in a microwave cavity placed in a strong magnetic field ("haloscope") allows one to search for galactic dark-matter axions [265]. Two pilot experiments [299,300] and first results from a full-scale search [301] already exclude a range of coupling strength shown in Fig. 12. The new generation of full-scale experiments [248,301–303] should cover the dotted area in Fig. 12, perhaps leading to the discovery of axion dark matter.

2 Electrons

Pseudoscalars which couple to electrons are produced by the Compton process $\gamma + e^- \to e^- + a$ [36,41,43,45,46,49] and by the electron bremsstrahlung process $e^- + (A,Z) \to (A,Z) + e^- + a$ [42,45,47,67,68,71]. A standard solar model yields an axion luminosity of [45] $L_a = \alpha_{ae}\, 6.0 \times 10^{21}\, L_\odot$ where α_{ae} is the axion electron "fine-structure constant" as defined after Eq. (14). The helioseismological constraint $L_a \lesssim 0.1\, L_\odot$ of Sec. II B implies $\alpha_{ae} \lesssim 2 \times 10^{-23}$. White-dwarf cooling gives [24,66] $\alpha_{ae} \lesssim 1.0 \times 10^{-26}$, while the most restrictive limit is from the delay of helium ignition in low-mass red-giants [49] in the spirit of Eq. (2)

$$\alpha_{ae} \lesssim 0.5 \times 10^{-26} \quad \text{or} \quad g_{ae} \lesssim 2.5 \times 10^{-13}. \tag{21}$$

For $m_a \gtrsim T \approx 10$ keV this limit quickly degrades because the emission from a thermal plasma is suppressed. With Eq. (14) one finds for axions

$$m_a C_e \lesssim 0.003 \text{ eV} \quad \text{and} \quad f_a/C_e \gtrsim 2 \times 10^9 \text{ GeV}. \tag{22}$$

In KSVZ-type models $C_e = 0$ at tree level so that no interesting limit obtains. In the DFSZ model $m_a \cos^2\beta \lesssim 0.01$ eV and $f_a/\cos^2\beta \gtrsim 0.7 \times 10^9$ GeV. Since $\cos^2\beta$ can be very small, there is no generic limit on m_a.

3 Nucleons

The axion-nucleon coupling strength is primarily constrained by the SN 1987A energy-loss argument [108–116]. The main problem is to estimate the axion emission rate reliably. In the early papers it was based on a somewhat naive calculation of the bremsstrahlung process $NN \to NNa$, using quasi-free nucleons which interact perturbatively by a one-pion exchange potential. Assuming an equal axion coupling g_{aN} to protons and neutrons this treatment leads to the g_{aN}-dependent shortening of the SN 1987A neutrino burst of Fig. 8. However, in a dense medium the bremsstrahlung process likely saturates, reducing the naive emission rate by as much as an order of magnitude [115]. With this correction, and assuming that the neutrino burst was not shortened by more than half, one reads from Fig. 8 an excluded range

$$3 \times 10^{-10} \lesssim g_{aN} \lesssim 3 \times 10^{-7}. \tag{23}$$

With Eq. (14) this implies an exclusion range

$$0.002 \text{ eV} \lesssim m_a C_N \lesssim 2 \text{ eV} \quad \text{and} \quad 3 \times 10^6 \text{ GeV} \lesssim f_a/C_N \lesssim 3 \times 10^9 \text{ GeV}. \tag{24}$$

For KSVZ axions the coupling to neutrons disappears while $C_p \approx -0.34$. With a proton fraction of about 0.3 one estimates an effective $C_N \approx 0.2$ so that [24,115]

$$0.01 \text{ eV} \lesssim m_a \lesssim 10 \text{ eV} \quad \text{and} \quad 0.6 \times 10^6 \text{ GeV} \lesssim f_a \lesssim 0.6 \times 10^9 \text{ GeV} \tag{25}$$

is excluded.

In a detailed numerical study the values for C_n and C_p appropriate for the KSVZ model and for the DFSZ model with different choices of $\cos^2\beta$ were implemented [116]. For KSVZ axions one finds a limit $m_a \lesssim 0.008$ eV, while it varies between about 0.004 and 0.012 eV for DFSZ axions, depending on $\cos^2\beta$. In view of the large overall uncertainties it is probably good enough to remember $m_a \lesssim 0.01$ eV as a generic limit (Fig. 13).

Axions on the "strong interaction side" of the exclusion range Eq. (23) would have produced excess counts in the neutrino detectors by their absorption on oxygen if $1 \times 10^{-6} \lesssim g_{aN} \lesssim 1 \times 10^{-3}$ [157]. For KSVZ axions this crudely translates into 20 eV $\lesssim m_a \lesssim$ 20 keV as an exclusion range (Fig. 13).

4 Hadronic Axion Window

This limit as well as the "trapping side" of the energy-loss argument have not been studied in as much detail because the relevant m_a range is already excluded by the globular-cluster argument (Fig. 13) which, however, depends on the axion-photon interaction which would nearly vanish in models with $E/N = 2$. In this case a narrow gap of allowed axion masses in the neighborhood of 10 eV may exist between the two SN arguments ("hadronic axion window").

In this region one can derive interesting limits from globular-cluster stars where axions can be emitted by nuclear processes, causing a metallicity-dependent modification of the core mass at helium ignition [48]. It is intruiging that in this window axions could play a cosmological role as a hot dark matter component [304]. Usually, of course, axions are a *cold* dark matter candidate. Moreover, in this window it may be possible to detect a 14.4 keV monochromatic solar axion line which is produced by transitions between the first excited and ground state of ^{57}Fe. In the laboratory one can then search for axion absorption which would give rise to x-rays as ^{57}Fe de-excites [305]. A recent pilot experiment did not have enough sensitivity to find axions [306], but a vastly improved detector is now in preparation in Tokyo (private communication by S. Moriyama and M. Minowa).

C Cosmological Limits

The astrophysical axion mass limits are particularly interesting when juxtaposed with the cosmological ones which we thus briefly review. For $f_a \gtrsim 10^8$ GeV cosmic axions never reach thermal equilibrium in the early universe. They are produced by a nonthermal mechanism which is intimately intertwined with their Nambu-Goldstone nature and which implies that their contribution to the cosmic density is proportional to $f_a^{1.175}$ and thus to $m_a^{-1.175}$. The requirement not to "overclose" the universe with axions thus leads to a *lower* mass limit.

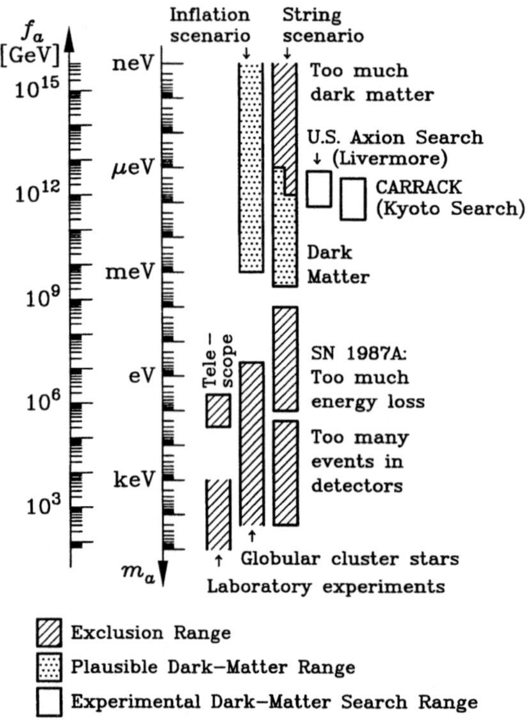

FIGURE 13. Astrophysical and cosmological exclusion regions (hatched) for the axion mass m_a, or equivalently the Peccei-Quinn scale f_a. The globular-cluster limit depends on the axion-photon coupling; it was assumed that $E/N = 8/3$ as in GUT models or the DFSZ model. The SN 1987A limits depend on the axion-nucleon couplings; the shown case corresponds to the KSVZ model and approximately to the DFSZ model. The dotted "inclusion regions" indicate where axions could plausibly be the cosmic dark matter. Most of the allowed range in the inflation scenario requires fine-tuned initial conditions. In the string scenario the plausible dark-matter range is somewhat controversial as indicated by the step in the low-mass end of the "inclusion bar." Also shown is the projected sensitivity range of the search experiments for galactic dark-matter axions.

There are two generic cosmological scenarios. If inflation occurred after the Peccei-Quinn symmetry breaking or if $T_{\rm reheat} < f_a$, the initial axion field takes on $a_{\rm i} = f_a\Theta_{\rm i}$ throughout the universe, where $0 \leq \Theta_{\rm i} < \pi$ is the initial "misalignment" of the QCD Theta parameter [307–310]. If $\Theta_i \sim 1$ one obtains a critical density in axions for $m_a \sim 1$ μeV, but since Θ_i is unknown there is no strict cosmological limit on m_a. However, the possibility to fine-tune $\Theta_{\rm i}$ is limited by inflation-induced quantum fluctuations which lead to temperature fluctuations of the cosmic microwave background [311–314]. In a broad class of inflationary models one thus finds an upper limit to m_a where axions could be the dark matter. According to the most recent discussion [314] it is about 10^{-3} eV (Fig. 13).

If inflation did not occur at all or if it occurred before the Peccei-Quinn symmetry breaking with $T_{\rm reheat} > f_a$, cosmic axion strings form by the Kibble mechanism [315,316]. Their motion is damped primarily by axion emission rather than gravitational waves. After axions acquire a mass at the QCD phase transition they quickly become nonrelativistic and thus form a cold dark matter component. Unknown initial conditions no longer enter, but details of the string mechanism are sufficiently complicated to prevent an exact prediction of the axion density. On the basis of Battye and Shellard's treatment [317,318] and assuming that axions are the cold dark matter of the universe one finds a plausible mass range of m_a = 6–2500 μeV [239]. Sikivie et al. [319–321] predict somewhat fewer axions, allowing for somewhat smaller masses if axions are the dark matter.

Either way, the ongoing full-scale search experiments for galactic dark matter axions (Sec. VI B 1 and Fig. 12) in Livermore (U.S. Axion Search [301]) and in Kyoto (CARRACK [302,303]) aim at a cosmologically well-motivated range of axion masses (Fig. 13).

VII CONCLUSION

Stellar-evolution theory together with astronomical observations, the SN 1987A neutrino burst, and certain x- and γ-ray observations provide a number of well-developed arguments to constrain the properties of low-mass particles. The most successful examples are globular-cluster stars where the "energy-loss argument" was condensed into the simple criteria of Eqs. (2) and (3) and SN 1987A where it was summarized by Eq. (5). New particle-physics conjectures must first pass these and other simple astrophysical standard tests before being taken too seriously.

A showcase example for the interplay between astrophysical limits with laboratory experiments and cosmological arguments is provided by the axion hypothesis. The laboratory and astrophysical limits push the Peccei-Quinn scale to such high values that it appears almost inevitable that axions, if they exist at all, play an important role as a cold dark matter component. This makes the direct search for galactic axion dark matter a well-motivated effort. Other important standard limits are on neutrino electromagnetic form factors—laboratory experiments will have a difficult time catching up.

Most of the theoretical background relevant to this field could not be touched upon in this brief overview. The physics of weakly coupled particles in stars is a nice playing ground for "particle physics in media" which involves field theory at finite temperature and density (FTD), many-body effects, particle dispersion and reactions in magnetic fields and media, oscillations of trapped neutrinos, and so forth. It is naturally in the context of SN theory where such issues are of particular interest, but even the plasmon decay $\gamma \to \nu\bar{\nu}$ in normal stars or the MSW effect in the Sun are interesting cases. Particle physics in media and its astrophysical and cosmological applications is a fascinating topic in its own right.

Much more information of particle-physics interest may be written in the sky than has been deciphered as yet. Other objects or phenomena should be considered, perhaps other kinds of conventional stars, perhaps more exotic phenomena such as γ-ray bursts. The particle-physics lessons to be learned from them are left to be discussed in a future review!

REFERENCES

1. Bahcall J., *Neutrino Astrophysics*, Cambridge: Cambridge University Press, 1989.
2. Castellani V. *et al.*, *Phys. Rept.* **281**, 309 (1997).
3. Bahcall J.N., Krastev P.I., and Smirnov A.Yu., *Phys. Rev. D* **58**, 096016 (1998).
4. Glendenning N.K., *Compact Stars*, New York: Springer, 1997.
5. Weber F., *Pulsars as Astrophysical Laboratories for Nuclear and Particle Physics*, Bristol: IOP Publishing, 1999.
6. Alcock C., and Olinto A.V., *Annu. Rev. Nucl. Part. Sci.* **38**, 161 (1988).
7. Madsen J., and Haensel P. (editors), *Strange Quark Matter in Physics and Astrophysics*, *Nucl. Phys. B (Proc. Suppl.)* vol. 24B (1992).
8. Vassiliadis G. *et al.* (editors), *Proc. Int. Symp. Strangeness and Quark Matter, Sept. 1–5, 1994, Crete, Greece*, Singapore: World Scientific, 1995.
9. Kolb E.W., and Turner M.S., *The Early Universe*, Reading, Mass.: Addison-Wesley, 1990.
10. Freese K., and Krasteva E., astro-ph/9804148.
11. Jungman G., Kamionkowski M., and Griest K., *Phys. Rept.* **267**, 195 (1996).
12. Frieman J.A., Dimopoulos S., and Turner M.S., *Phys. Rev. D* **36**, 2201 (1987).
13. Schlattl H., Weiss A., and Raffelt G., *Astrop. Phys.*, in press (1999).
14. Raffelt G., and Dearborn D., *Phys. Rev. D* **36**, 2211 (1987).
15. Christensen-Dalsgaard J., Thompson D.O., and Gough D.O., *Astrophys. J.* **378**, 413 (1991).
16. Degl'Innocenti S., Dziembowski W.A., Fiorentini G., and Ricci B., *Astrop. Phys.* **7**, 77 (1997).
17. Basu S. *et al.*, *Mon. Not. R. Astr. Soc.* **292**, 243 (1997).
18. Carlson E.D., and Salati P., *Phys. Lett. B* **218**, 79 (1989).
19. Raffelt G., and Starkman G., *Phys. Rev. D* **40**, 942 (1989).
20. Buonanno R. *et al.*, *Mem. Soc. Astron. Ital.* **57**, 391 (1986).
21. Renzini A., and Fusi Pecci F., *Annu. Rev. Astron. Astrophys.* **26**, 199 (1988).

22. Clayton D.D., *Principles of Stellar Evolution and Nucleosynthesis*, Chicago: University of Chicago Press, 1968.
23. Kippenhahn R., and Weigert A., *Stellar Structure and Evolution*, Berlin: Springer, 1990.
24. Raffelt G.G., *Stars as Laboratories for Fundamental Physics*, Chicago: University of Chicago Press, 1996.
25. Raffelt G., *Astrophys. J.* **365**, 559 (1990).
26. Castellani M., and Degl'Innocenti S., *Astrophys. J.* **402**, 574 (1993).
27. Catelan M., de Freitas Pacheco J.A., and Horvath J.E., *Astrophys. J.* **461**, 231 (1996).
28. Buzzoni A. *et al.*, *Astron. Astrophys.* **128**, 94 (1983).
29. Raffelt G., and Dearborn D., *Phys. Rev. D* **37**, 549 (1988).
30. Sutherland P., *et al.*, *Phys. Rev. D* **13**, 2700 (1976).
31. Fukugita M., and Yazaki S., *Phys. Rev. D* **36**, 3817 (1987).
32. Raffelt G., Dearborn D., and Silk J., *Astrophys. J.* **336**, 64 (1989).
33. Raffelt G.G., *Phys. Rev. Lett.* **64**, 2856 (1990).
34. Raffelt G., and Weiss A., *Astron. Astrophys.* **264**, 536 (1992).
35. Dicus D.A. *et al.*, *Phys. Rev. D* **18**, 1829 (1978).
36. Mikaelian K.O., *Phys. Rev. D* **18**, 3605 (1978).
37. Dicus D.A. *et al.*, *Phys. Rev. D* **22**, 839 (1980).
38. Georgi H., Glashow S.L., and Nussinov S., *Nucl. Phys. B* **193**, 297 (1981).
39. Barroso A., and Branco G.C., *Phys. Lett. B* **116**, 247 (1982).
40. Fukugita M., Watamura S., and Yoshimura M., *Phys. Rev. Lett.* **48**, 1522 (1982).
41. Fukugita M., Watamura S., and Yoshimura M., *Phys. Rev. D* **26**, 1840 (1982).
42. Krauss L.M., Moody J.E., and Wilczek F., *Phys. Lett. B* **144**, 391 (1984).
43. Brodsky S.J. *et al.*, *Phys. Rev. Lett.* **56**, 1763 (1986).
44. Pantziris A., and Kang K., *Phys. Rev. D* **33**, 3509 (1986).
45. Raffelt G., *Phys. Rev. D* **33**, 897 (1986).
46. Chanda R., Nieves J.F., and Pal P.B., *Phys. Rev. D* **37**, 2714 (1988).
47. Raffelt G., *Phys. Rev. D* **41**, 1324 (1990).
48. Haxton W.C., and Lee K.Y., *Phys. Rev. Lett.* **66**, 2557 (1991).
49. Raffelt G., and Weiss A., *Phys. Rev. D* **51**, 1495 (1995).
50. Hoffmann S., *Phys. Lett. B* **193**, 117 (1987).
51. Raffelt G., *Phys. Rev. D* **38**, 3811 (1988).
52. van der Velde J.C., *Phys. Rev. D* **39**, 1492 (1989).
53. Grifols J.A., and Massó E., *Phys. Lett. B* **173**, 237 (1986).
54. Grifols J.A., Massó E., and Peris S., *Mod. Phys. Lett. A* **4**, 311 (1989).
55. Bouquet A., and Vayonakis C.E., *Phys. Lett. B* **116**, 219 (1982).
56. Fukugita M., and Sakai N., *Phys. Lett. B* **114**, 23 (1982).
57. Anand J.D. *et al.*, *Phys. Rev. D* **29**, 1270 (1984).
58. Sweigart A.V., and Gross P.G., *Astrophys. J. Suppl.* **36**, 405 (1978).
59. Shapiro S.L., and Teukolsky S.A., *Black Holes, White Dwarfs, and Neutron Stars*, New York: John Wiley, 1983.
60. Fleming T.A., Liebert J., and Green R.F., *Astrophys. J.* **308**, 176 (1986).
61. Liebert J., Dahn C.C., and Monet D.G., *Astrophys. J.* **332**, 891 (1988).

62. Koester D., and Schönberner D., *Astron. Astrophys.* **154**, 125 (1986).
63. Mestel L., *Mon. Not. R. Astr. Soc.* **112**, 583 (1952).
64. Kepler S.O. *et al.*, *Astrophys. J.* **378**, L45 (1991).
65. Kepler S.O. *et al.*, *Baltic Astron.* **4**, 221 (1995).
66. Raffelt G., *Phys. Lett. B* **166**, 402 (1986).
67. Nakagawa M., Kohyama Y., and Itoh N., *Astrophys. J.* **322**, 291 (1987).
68. Nakagawa M. *et al.*, *Astrophys. J.* **326**, 241 (1988).
69. Wang J., *Mod. Phys. Lett. A* **7**, 1497 (1992).
70. Blinnikov S.I., and Dunina-Barkovskaya N.V., *Mon. Not. R. Astr. Soc.* **266**, 289 (1994).
71. Altherr T., Petitgirard E., and del Río Gaztelurrutia T., *Astrop. Phys.* **2**, 175 (1994).
72. Isern J., Hernanz M., and García-Berro E., *Astrophys. J.* **392**, L23 (1992).
73. Schramm D.N., and Truran J.W., *Phys. Rept.* **189**, 89 (1990).
74. Raffelt G.G., *Mod. Phys. Lett. A* **5**, 2581 (1990).
75. Koshiba M., *Phys. Rept.* **220**, 229 (1992).
76. Brown G.E., Bethe H.A., and Baym G., *Nucl. Phys. A* **375**, 481 (1982).
77. Bethe H.A., *Rev. Mod. Phys.* **62**, 801 (1990).
78. Petschek A.G. (editor), *Supernovae*, New York: Springer, 1990.
79. Cooperstein J., *Phys. Rept.* **163**, 95 (1988).
80. Burrows A., *Annu. Rev. Nucl. Part. Sci.* **40**, 181 (1990).
81. Janka H.T., in: *Proc. Vulcano Workshop 1992: Frontier Objects in Astrophysics and Particle Physics*, ed. by F. Giovannelli and G. Mannocchi, *Conf. Proc. Soc. Ital. Fis.* Vol. 40 (1993).
82. Hirata K.S. *et al.*, *Phys. Rev. D* **38**, 448 (1988).
83. Bratton C.B. *et al.*, *Phys. Rev. D* **37**, 3361 (1988).
84. Alexeyev E.N. *et al.*, *Pis'ma Zh. Eksp. Teor. Fiz.* **45**, 461 (1987) [*JETP Lett.* **45**, 589 (1987)].
85. Loredo T.J., and Lamb D.Q., in: *Proc. Fourteenth Texas Symposium on Relativistic Astrophysics*, ed. by E. J. Fenyves, *Ann. N.Y. Acad. Sci.* **571**, 601 (1989).
86. Loredo T.J. *From Laplace to Supernova SN 1987A: Bayesian Inference in Astrophysics*, Ph.D. Thesis, University of Chicago (1995).
87. Janka H.T., and Hillebrandt W., *Astron. Astrophys.* **224**, 49 (1989).
88. Jegerlehner B., Neubig F., and Raffelt G., *Phys. Rev. D* **54**, 1194 (1996).
89. Lattimer J.M., and Yahil A., *Astrophys. J.* **340**, 426 (1989).
90. LoSecco J.M., *Phys. Rev. D* **39**, 1013 (1989).
91. Kiełczewska D., *Phys. Rev. D* **41**, 2967 (1990).
92. Zatsepin G.I., *Pis'ma Zh. Eksp. Teor. Fiz.* **8**, 333 (1968) [*JETP Lett.* **8**, 205 (1968)].
93. Kernan P.J., and Krauss L.M., *Nucl. Phys. B* **437**, 243 (1995).
94. Barbiellini G., and Cocconi G., *Nature* **329**, 21 (1987).
95. Fujiwara K., *Phys. Rev. D* **39**, 1764 (1989).
96. Atzmon E., and Nussinov S., *Phys. Lett. B* **328**, 103 (1994).
97. Pakvasa S., Simmons W.A., and Weiler T.J., *Phys. Rev. D* **39**, 1761 (1989).
98. Grifols J.A., Massó E., and Peris S., *Phys. Lett. B* **207**, 493 (1988).
99. Grifols J.A., Massó E., and Peris S., *Astrop. Phys.* **2**, 161 (1994).

100. Fiorentini G., and Mezzorani G., *Phys. Lett. B* **221**, 353 (1989).
101. Malaney R.A., Starkman G.D., and Tremaine S., *Phys. Rev. D* **51**, 324 (1995).
102. Dolgov A.D., and Raffelt G.G., *Phys. Rev. D* **52**, 2581 (1995).
103. Longo M.J., *Phys. Rev. D* **36**, 3276 (1987).
104. Stodolsky L., *Phys. Lett. B* **201**, 353 (1988).
105. Krauss L.M., and Tremaine S., *Phys. Rev. Lett.*, **60**, 176 (1988).
106. Coley A.A., and Tremaine S., *Phys. Rev. D* **38**, 2927 (1988).
107. Almeida L.D., Matsas G.E.A., and Natale A.A., *Phys. Rev. D* **39**, 677 (1989).
108. Ellis J., and Olive K.A., *Phys. Lett. B* **193**, 525 (1987).
109. Raffelt G., and Seckel D., *Phys. Rev. Lett.* **60**, 1793 (1988).
110. Turner M.S., *Phys. Rev. Lett.* **60**, 1797 (1988).
111. Mayle R. *et al.*, *Phys. Lett. B* **203**, 188 (1988).
112. Mayle R. *et al.*, *Phys. Lett. B* **219**, 515 (1989).
113. Burrows A., Turner M.S., and Brinkmann R.P., *Phys. Rev. D* **39**, 1020 (1989).
114. Burrows A., Ressell T., and Turner M.S., *Phys. Rev. D* **42**, 3297 (1990).
115. Janka H.T., Keil W., Raffelt G., and Seckel D., *Phys. Rev. Lett.* **76**, 2621 (1996).
116. Keil W. *et al.*, *Phys. Rev. D* **56**, 2419 (1997).
117. Grifols J.A., Massó E., and Peris S., *Phys. Lett. B* **215**, 593 (1988).
118. Aharonov Y., Avignone III F.T., and Nussinov S., *Phys. Rev. D* **37**, 1360 (1988).
119. Aharonov Y., Avignone III F.T., and Nussinov S., *Phys. Lett. B* **200**, 122 (1988).
120. Aharonov Y., Avignone III F.T., and Nussinov S., *Phys. Rev. D* **39**, 985 (1989).
121. Choi K., Kim C.W., Kim J., and Lam W.P., *Phys. Rev. D* **37**, 3225 (1988).
122. Choi K., and Santamaria A., *Phys. Rev. D* **42**, 293 (1990).
123. Chang S., and Choi K., *Phys. Rev. D* **49**, 12 (1994).
124. Ellis J. *et al.*, *Phys. Lett. B* **215**, 404 (1988).
125. Lau K., *Phys. Rev. D* **47**, 1087 (1993).
126. Nowakowski M., and Rindani S.D., *Phys. Lett. B* **348**, 115 (1995).
127. Grifols J.A., Massó E., and Peris S., *Phys. Lett. B* **220**, 591 (1989).
128. Grifols J.A., Mohapatra R.N., and Riotto A., *Phys. Lett. B* **400**, 124 (1997).
129. Grifols J.A., Mohapatra R.N., and Riotto A., *Phys. Lett. B* **401**, 283 (1997).
130. Grifols J.A., Massó E., and Toldra R., *Phys. Rev. D* **57**, 614 (1998).
131. Dicus D.A., Mohapatra R.N., and Teplitz V.L., *Phys. Rev. D* **57**, 578 (1998); (E) *ibid.* **57**, 4496 (1998).
132. Arkani-Hamed N., Dimopoulos S., and Dvali G., hep-ph/9807344.
133. Gaemers K.J.F., Gandhi R., and Lattimer J.M., *Phys. Rev. D* **40**, 309 (1989).
134. Grifols J.A., and Massó E., *Phys. Lett. B* **242**, 77 (1990).
135. Gandhi R., and Burrows A., *Phys. Lett. B* **246**, 149 (1990); (E) *ibid.* **261**, 519 (1991).
136. Mayle R. *et al.*, *Phys. Lett. B* **317**, 119 (1993).
137. Maalampi J., and Peltoniemi J.T., *Phys. Lett. B* **269**, 357 (1991).
138. Turner M.S., *Phys. Rev. D* **45**, 1066 (1992).
139. Pantaleone J., *Phys. Rev. D* **46**, 510 (1992).
140. Burrows A., Gandhi R., and Turner M.S., *Phys. Rev. Lett.* **68**, 3834 (1992).
141. Goyal A., and Dutta S., *Phys. Rev. D* **49**, 3910 (1994).
142. Babu K.S., Mohapatra R.N., and Rothstein I.Z., *Phys. Rev. D* **45**, 5 (1992).

143. Babu K.S., Mohapatra R.N., and Rothstein I.Z., *Phys. Rev. D* **45** 3312 (1992).
144. Kainulainen K., Maalampi J., and Peltoniemi J.T., *Nucl. Phys. B* **358**, 435 (1991).
145. Raffelt G., and Sigl G., *Astrop. Phys.* **1**, 165 (1993).
146. Barbieri R., and Mohapatra R.N., *Phys. Rev. D* **39**, 1229 (1989).
147. Grifols J.A., and Massó E., *Nucl. Phys. B* **331**, 244 (1990).
148. Grifols J.A., Massó E., and Rizzo T.G., *Phys. Rev. D* **42**, 3293 (1990).
149. Rizzo T.G., *Phys. Rev. D* **44**, 202 (1991).
150. Lattimer J.M., and Cooperstein J., *Phys. Rev. Lett.* **61**, 23 (1988).
151. Barbieri R., and Mohapatra R.N., *Phys. Rev. Lett.* **61**, 27 (1988).
152. Goyal A., and Dutta S., *Phys. Rev. D* **49**, 5593 (1994).
153. Goyal A., Dutta S., and Choudhury S.R., *Phys. Lett. B* **346**, 312 (1995).
154. Ayala A., D'Olivo J.C., and Torres M., hep-ph/9804230.
155. Mohapatra R.N., and Rothstein I.Z., *Phys. Lett. B* **247**, 593 (1990).
156. Grifols J.A., and Massó E., *Phys. Rev. D* **40**, 3819 (1989).
157. Engel J., Seckel D., and Hayes A.C., *Phys. Rev. Lett.* **65**, 960 (1990).
158. Dodelson S., Frieman J.A., and Turner M.S., *Phys. Rev. Lett.* **68**, 2572 (1992).
159. Mikheyev S.P., and Smirnov A.Yu., *Zh. Eksp. Teor. Fiz.* **91**, 7 (1986) [*Sov. Phys. JETP* **64**, 4 (1986)].
160. Arafune J. *et al.*, *Phys. Rev. Lett.* **59**, 1864 (1987).
161. Arafune J. *et al.*, *Phys. Lett. B* **194**, 477 (1987).
162. Lagage P.O. *et al.*, *Phys. Lett. B* **193**, 127 (1987).
163. Minakata H. *et al.*, *Mod. Phys. Lett. A* **2**, 827 (1987).
164. Nötzold D., *Phys. Lett. B* **196**, 315 (1987).
165. Walker T.P., and Schramm D.N., *Phys. Lett. B* **195**, 331 (1987).
166. Kuo T.K., and Pantaleone J., *Phys. Rev. D* **37**, 298 (1988).
167. Minakata H., and Nunokawa H., *Phys. Rev. D* **38**, 3605 (1988).
168. Rosen S.P., *Phys. Rev. D* **37**, 1682 (1988).
169. Fuller G.M. *et al.*, *Astrophys. J.* **389**, 517 (1992).
170. Janka H.T., and Hillebrandt W., *Astron. Astrophys. Suppl.* **78**, 375 (1989).
171. Janka H.T., *Astrop. Phys.* **3**, 377 (1995).
172. Hardy S., Janka H.T., and Raffelt G., *Work in progress* (1999).
173. Suzuki H., *Num. Astrophys. Japan* **2**, 267 (1991).
174. Suzuki H., in: *Frontiers of Neutrino Astrophysics, Proc. of the International Symposium on Neutrino Astrophysics, 19–22 Oct. 1992, Takayama/Kamioka, Japan*, ed. by Y. Suzuki and K. Nakamura, Tokyo: Universal Academy Press, 1993.
175. Hannestad S., and Raffelt G., *Astrophys. J.* **507**, 339 (1998).
176. Woosley S.E., and Hoffmann R.D., *Astrophys. J.* **395**, 202 (1992).
177. Meyer B.S. *et al.*, *Astrophys. J.* **399**, 656 (1992).
178. Witti J., Janka H.T., and Takahashi K., *Astron. Astrophys.* **286**, 841 (1994).
179. Takahashi K., Witti J., and Janka H.T., *Astron. Astrophys.* **286**, 857 (1994).
180. Meyer B.S., *Annu. Rev. Astron. Astrophys.* **32**, 153 (1994).
181. Meyer B.S., *Astrophys. J.* **449**, L55 (1995).
182. Meyer B.S., McLaughlin G.C., and Fuller G.M., *Phys. Rev. C* **58**, 3696 (1998).
183. Qian Y.Z. *et al.*, *Phys. Rev. Lett.* **71**, 1965 (1993).
184. Qian Y.Z., and Fuller G.M., *Phys. Rev. D* **51**, 1479 (1995).

185. Sigl G., *Phys. Rev. D* **51**, 4035 (1995).
186. Pantaleone J., *Phys. Lett. B* **342**, 250 (1995).
187. Caldwell D.O., Fuller G., and Qian Y.Z., Quoted after D.O. Caldwell, astro-ph/9812026.
188. Smirnov A.Yu., Spergel D.N., and Bahcall J.N., *Phys. Rev. D* **49**, 1389 (1994).
189. Nunokawa H., Peltoniemi J.T., Rossi A., and Valle J.W.F., *Phys. Rev. D* **56**, 1704 (1997).
190. Valle J.W.F., in: *Proc. New Trends in Neutrino Physics, Tegernsee, Ringberg Castle, Germany, 24–29 May 1998*, to be published; hep-ph/9809234.
191. Kayser B., in: *Proc. 29th International Conference on High-Energy Physics (ICHEP 98), Vancouver, Canada, 23–29 July 1998*, to be published; hep-ph/9810513.
192. Smirnov A.Yu., in: *Proc. 5th International WEIN Symposium: A Conference on Physics Beyond the Standard Model (WEIN 98), Santa Fe, New Mexico, 14–21 June 1998*, to be published; hep-ph/9901208.
193. Fukuda Y. *et al.* (Superkamiokande Collaboration), *Phys. Rev. Lett.* **81**, 1562 (1998).
194. Athanassopoulos C. *et al.*, *Phys. Rev. Lett.* **77**, 3082 (1996).
195. Athanassopoulos C. *et al.*, *Phys. Rev. Lett.* **81**, 1774 (1998).
196. Qian Y.Z., and Fuller G.M., *Phys. Rev. D* **49**, 1762 (1994).
197. Choubey S., Majumdar D., and Kar K., hep-ph/9809424.
198. Fuller G.M., Haxton W.C., and McLaughlin G.C., astro-ph/9809164.
199. Totani T., *Phys. Rev. Lett.* **80**, 2039 (1998).
200. Seckel D., Steigman G., and Walker T., *Nucl. Phys. B* **366**, 233 (1991).
201. Krauss L.M. *et al.*, *Nucl. Phys. B* **380**, 507 (1992).
202. Fiorentini G., and Acerbi C., *Astrop. Phys.* **7**, 245 (1997).
203. Beacom J.F., and Vogel P., *Phys. Rev. D* **58**, 093012 (1998).
204. Beacom J.F., and Vogel P., *Phys. Rev. D* **58**, 053010 (1998).
205. Cline D. *et al.*, *Phys. Rev. D* **50**, 720 (1994).
206. Smith P.F., *Astrop. Phys.* **8**, 27 (1997).
207. Morrison D.R.O., *Nature* **366**, 29 (1993).
208. Mohapatra R.N., and Pal P., *Massive Neutrinos in Physics and Astrophysics*, Singapore: World Scientific, 1991.
209. Winter K. (editor), *Neutrino Physics*, Cambridge: Cambridge University Press, 1991.
210. Lucio J.L., Rosado A., and Zepeda A., *Phys. Rev. D* **31**, 1091 (1985).
211. Auriemma G., Srivastava Y., and Widom A., *Phys. Lett. B* **195**, 254 (1987).
212. Degrassi G., Sirlin A., and Marciano W.J., *Phys. Rev. D* **39**, 287 (1989).
213. Musolf M.J., and Holstein B.R., *Phys. Rev. D* **43**, 2956 (1991).
214. Góngora A., and Stuart R.G., *Z. Phys. C* **55**, 101 (1992).
215. Salati P., *Astrop. Phys.* **2**, 269 (1994).
216. D'Olivo J.C., Nieves J.F., and Pal P.B., *Phys. Rev. D* **40**, 3679 (1989).
217. Altherr T., and Salati P., *Nucl. Phys. B* **421**, 662 (1994).
218. Adams J.B., Ruderman M.A., and Woo C.H., *Phys. Rev.* bf 129, 1383 (1963).
219. Zaidi M.H., *Nuovo Cim.* **40**, 502 (1965).

220. Haft M., Raffelt G., and Weiss A., *Astrophys. J.* **425**, 222 (1994); (E) *ibid.* **438**, 1017 (1995).
221. Bernstein J., Ruderman M.A., and Feinberg G., *Phys. Rev.* **132**, 1227 (1963).
222. Broggini C. *et al.* (MUNU Collaboration), *Nucl. Phys. B (Proc.Suppl)* **70**, 188 (1999).
223. Oberauer L., von Feilitzsch F., and Mössbauer R.L., *Phys. Lett. B* **198**, 113 (1987).
224. Cowsik R., *Phys. Rev. Lett.* **39**, 784 (1977).
225. Raffelt G., *Phys. Rev. D* **31**, 3002 (1985).
226. Chupp E.L., Vestrand W.T., and Reppin C., *Phys. Rev. Lett.* **62**, 505 (1989).
227. Oberauer L. *et al.*, *Astrop. Phys.* **1**, 377 (1993).
228. von Feilitzsch F., and Oberauer L., *Phys. Lett. B* **200**, 580 (1988).
229. Kolb E.W., and Turner M.S., *Phys. Rev. Lett.* **62**, 509 (1989).
230. Bludman S.A., *Phys. Rev. D* **45**, 4720 (1992).
231. Jaffe A.H., and Turner M.S., *Phys. Rev. D* **55**, 7951 (1997).
232. Miller R.S., *A Search for Radiative Neutrino Decay and its Potential Contribution to the Cosmic Diffuse Gamma-Ray Flux*, Ph.D. Thesis, Univ. New Hampshire (1995).
233. Miller R.S., Ryan J.M., and Svoboda R.C., *Astron. Astrophys. Suppl. Ser.* **120**, 635 (1996).
234. Ressell M.T., and Turner M.S., *Comments Astrophys.* **14**, 323 (1990).
235. Biller S.D. *et al.*, *Phys. Rev. Lett.* **80**, 2992 (1998).
236. Raffelt G.G., *Phys. Rev. Lett.* **81**, 4020 (1998).
237. Henry R.C., and Feldmann P.D., *Phys. Rev. Lett.* **47**, 618 (1981).
238. Davidsen A.F. *et al.*, *Nature* **351**, 128 (1991).
239. Caso C. *et al.*, *Eur. Phys. J. C* **3**, 1 (1998).
240. Bershady M.A., Ressel M.T., and Turner M.S., *Phys. Rev. Lett.* **66**, 1398 (1991).
241. Elmfors P., Enqvist K., Raffelt G., and Sigl G., *Nucl. Phys. B* **503**, 3 (1997).
242. Peccei R.D., and Quinn H.R., *Phys. Rev. Lett.* **38**, 1440 (1977).
243. Peccei R.D., and Quinn H.R., *Phys. Rev. D* **16**, 1791 (1977).
244. Weinberg S., *Phys. Rev. Lett.* **40**, 223 (1978).
245. Wilczek F., *Phys. Rev. Lett.* **40**, 279 (1978).
246. Kim J.E., *Phys. Rept.* **150**, 1 (1987).
247. Cheng H.Y., *Phys. Rept.* **158**, 1 (1988).
248. Sikivie P., *Proc. Axion Workshop, Univ. of Florida, Gainesville, Florida, USA, 13–15 March 1998*, to be published in *Nucl. Phys. B (Proc. Suppl.)* (1999).
249. Carena M., and Peccei R.D., *Phys. Rev. D* **40**, 652 (1989).
250. Choi K., Kang K., and Kim J.E., *Phys. Rev. Lett.* **62**, 849 (1989).
251. Turner M.S., Kang H.S., and Steigman G., *Phys. Rev. D* **40**, 299 (1989).
252. Iwamoto N., *Phys. Rev. D* **39**, 2120 (1989).
253. Dine M., Fischler W., and Srednicki M., *Phys. Lett. B* **104**, 199 (1981).
254. Zhitnitskiĭ A.P., *Yad. Fiz.* **31**, 497 (1980) [*Sov. J. Nucl. Phys.* **31**, 260 (1980)].
255. Kim J.E., *Phys. Rev. Lett.* **43**, 103 (1979).
256. Shifman M.A., Vainshtein A.I., and Zakharov V.I., *Nucl. Phys. B* **166**, 493 (1980).
257. Kaplan D.B., *Nucl. Phys. B* **260**, 215 (1985).
258. Cheng S.L., Geng C.Q., and Ni W.T., *Phys. Rev. D* **52**, 3132 (1995).

259. Kim J.E., *Phys. Rev. D* **58**, 055006 (1998).
260. Raffelt G., *Phys. Rev. D* **37**, 1356 (1988).
261. Massó E., and Toldrà R., *Phys. Rev. D* **52**, 1755 (1995).
262. Massó E., and Toldrà R., *Phys. Rev. D* **55**, 7967 (1997).
263. Cameron R. *et al.*, *Phys. Rev. D* **47**, 3707 (1993).
264. Mori F., *Mod. Phys. Lett. A* **11**, 715 (1996).
265. Sikivie P., *Phys. Rev. Lett.* **51**, 1415 (1983); (E) *ibid.* **52**, 695 (1984).
266. van Bibber K., Morris D., McIntyre P., and Raffelt G., *Phys. Rev. D* **39**, 2089 (1989).
267. Lazarus D.M. *et al.*, *Phys. Rev. Lett.* **69**, 2333 (1992).
268. Moriyama S. *et al.*, *Phys. Lett. B* **434**, 147 (1998).
269. Moriyama S., *Direct Search for Solar Axions by Using Strong Magnetic Field and X-Ray Detectors*, Ph.D. Thesis (in English), University of Tokyo (1998).
270. Raffelt G., and Stodolsky L., *Phys. Rev. D* **37**, 1237 (1988).
271. Vorob'ev P.V., and Kolokolov I.V., astro-ph/9501042 (unpublished).
272. Zioutas K. *et al.*, astro-ph/9801176, submitted to *Nucl. Instrum. Meth. A.*
273. Buchmüller W., and Hoogeveen F., *Phys. Lett. B* **237**, 278 (1990).
274. Paschos E.A., and Zioutas K., *Phys. Lett. B* **323**, 367 (1994).
275. Creswick R.J. *et al.*, *Phys. Lett. B* **427**, 235 (1998).
276. SOLAX Collaboration (Avignone III F.T. *et al.*), *Phys. Rev. Lett.* **81**, 5068 (1998).
277. Cebrián S. *et al.*, *Astrop. Phys.*, in press (1999).
278. Carlson E.D., and Tseng L.S., *Phys. Lett. B* **365**, 193 (1996).
279. Carlson E.D., *Phys. Lett. B* **344**, 245 (1995).
280. Brockway J.W., Carlson E.D., and Raffelt G.G., *Phys. Lett. B* **383**, 439 (1996).
281. Gifols J.A., Massó E., and Toldrà R., *Phys. Rev. Lett.* **77**, 2372 (1996).
282. Krasnikov S.V., *Phys. Rev. Lett.* **76**, 2633 (1996).
283. Morris D.E., *Phys. Rev. D* **34**, 843 (1986).
284. Yoshimura M., *Phys. Rev. D* **37**, 2039 (1988).
285. Yanagida T., and Yoshimura M., *Phys. Lett. B* **202**, 301 (1988).
286. Gnedin Y.N., and Krasnikov S.V., *Zh. Eksp. Teor. Fiz.* **102**, 1729 (1992) [*Sov. Phys. JETP* **75**, 933 (1992)].
287. Gnedin Y.N., and Krasnikov S.V., *Astron. Lett.* **20**, 72 (1994).
288. Gnedin Y.N., *Comments Astrophys.* **18**, 257 (1996).
289. Gnedin Y.N., *Astrophys. Space Sci.* **249**, 125 (1997).
290. Carlson E.D., and Garretson W.D., *Phys. Lett. B* **336**, 431 (1994).
291. Mohanty S., and Nayak S.N., *Phys. Rev. Lett.* **70**, 4038 (1993); (E) *ibid.* **71**, 1117 (1993); (E) *ibid.* **76**, 2825 (1996).
292. Maiani L., Petronzio R., and Zavattini E., *Phys. Lett. B* **175**, 359 (1986).
293. Semertzidis Y. *et al.*, *Phys. Rev. Lett.* **64**, 2988 (1990).
294. Bakalov D. *et al.*, *Quantum Semiclass. Opt.* **10**, 239 (1998).
295. Lee S. *et al.*, *Fermilab Proposal E-877* (1995).
296. Ruoso G. *et al.*, *Z. Phys. C* **56**, 505 (1992).
297. Ressell M.T., *Phys. Rev. D* **44**, 3001 (1991).
298. Overduin J.M., and Wesson P.S., *Astrophys. J.* **414**, 449 (1993).
299. Wuensch W.U. *et al.*, *Phys. Rev. D* **40**, 3153 (1989).

300. Hagmann C. *et al.*, *Phys. Rev. D* **42**, 1297 (1990).
301. Hagmann C. *et al.*, *Phys. Rev. Lett.* **80**, 2043 (1998).
302. Ogawa I., Matsuki S., and Yamamoto K., *Phys. Rev. D* **53**, R1740 (1996).
303. Yamamoto K., and Matsuki S., in: *Proc. 2nd International Workshop on the Identification of Dark Matter (IDM 98), Buxton, England, 7–11 Sept. 1998*, to be published; hep-ph/9811487.
304. Moroi T., and Murayama H., *Phys. Lett. B* **440**, 69 (1998).
305. Moriyama S., *Phys. Rev. Lett.* **75**, 3222 (1995).
306. Krčmar M. *et al.*, *Phys. Lett. B* **442**, 38 (1998).
307. Preskill J., Wise M., and Wilczek F., *Phys. Lett. B* **120**, 127 (1983).
308. Abbott L., and Sikivie P., *Phys. Lett. B* **120**, 133 (1983).
309. Dine M., and Fischler W., *Phys. Lett. B* **120**, 137 (1983).
310. Turner M.S., *Phys. Rev. D* **33**, 889 (1986).
311. Lyth D.H., *Phys. Lett. B* **236**, 408 (1990).
312. Turner M.S., and Wilczek F., *Phys. Rev. Lett.* **66**, 5 (1991).
313. Linde A., *Phys. Lett. B* **259**, 38 (1991).
314. Shellard E.P.S., and Battye R.A., astro-ph/9802216.
315. Davis R.L., *Phys. Lett. B* **180**, 225 (1986).
316. Davis R.L., and Shellard E.P.S., *Nucl. Phys. B* **324**, 167 (1989).
317. Battye R.A., and Shellard E.P.S., *Nucl. Phys. B* **423**, 260 (1994).
318. Battye R.A., and Shellard E.P.S., *Phys. Rev. Lett.* **73**, 2954 (1994); (E) *ibid.* **76**, 2203 (1996).
319. Harari D., and Sikivie P., *Phys. Lett. B* **195**, 361 (1987).
320. Hagmann C., and Sikivie P., *Nucl. Phys. B* **363**, 247 (1991).
321. Chang S., Hagmann C., and Sikivie P., *Phys. Rev. D* **59**, 023505 (1999).

Physics at the LHC

David Rousseau[†]

CERN, EP division, CH-1211 Geneva 23, Switzerland
[†] *Now at CPPM, 163 Avenue de Luminy, Case 907, 13288 Marseille Cedex 09, France*

Abstract. A sample of the physics potential of the Large Hadron Collider is given. The ATLAS and CMS detectors are described with the constraints of the difficult LHC environment. Potential discovery of the Standard Model and MSSM Higgs bosons is demonstrated. The reach of these experiments and of the B-physics dedicated experiment LHCb for CP violation, B_s^0 mixing and rare B decays is also discussed.

INTRODUCTION

The LHC will be a p-p collider running at 14 TeV center-of-mass energy, with a luminosity of 10^{34} $cm^{-2}s^{-1}$. It will be commissioned in 2005 at CERN, Geneva, Switzerland.

The motivation for such a machine stems from the desire of a complete coverage of the Standard Model. The Standard Model has been tested with great precision mainly at LEP e^+e^- collider ($\sqrt{s}$= 90 GeV to 190 GeV) and at Tevatron p-$\bar{p}$ collider ($\sqrt{s}$=1.4 TeV) [1]. With the recent discovery of the top quark [2], it is rather complete. The Higgs boson is now the only (but essential) missing piece. Finding the Higgs boson from the expected LEPII reach ($\sim$100 GeV/c^2) up to the theoretical bound ($\sim$1 TeV/c^2) is a major goal in particle physics.

If the Higgs boson is not found in this range, the Standard Model needs a new electroweak symmetry breaking mechanism. Even if it is found, it still leaves many questions unanswered in the Standard Model, such as (but not limited to): Why is the fermion mass hierarchy as it is, spanning the MeV to 200 GeV range ? Why are there so many free parameters (19, not counting neutrino possible masses and mixing [3])? What is the real source of CP violation ? A very wide range of models try to answer these questions such as (but not limited to, again): various flavour of supersymmetry (Minimal, Gauge mediated, R parity conserving or violating), composite models, technicolor,... These models often predict a deviation of the couplings of the Standard Model particles, and new particles below 1 TeV/c^2. It is hence needed to measure with even more precision than now Standard Model parameters (triple gauge boson couplings, W and top mass, CP violation in B decays) and to look for new massive particles.

CP490, *Particles and Fields: Eighth Mexican School,* edited by J. C. D'Olivo et al.

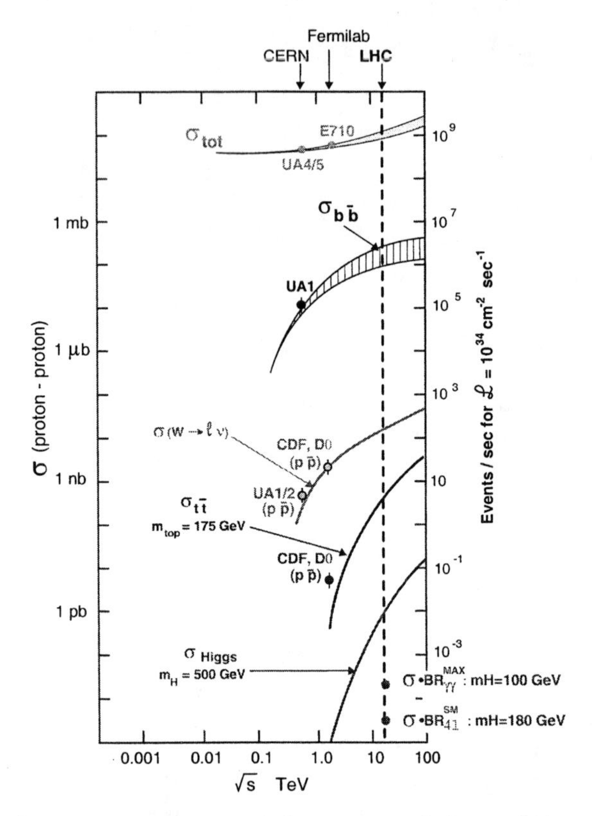

FIGURE 1. Cross-section of some typical processes at hadron collider as a function of $\sqrt{s}$.

The reason why a high energy proton proton collider is the right machine for such studies is that in these collisions the cross-section for heavy particle production is high, and it increases by several orders of magnitude for objects with masses in the range 100 GeV/c^2-1 TeV/c^{2}when $\sqrt{s}$ increase from 1.4 TeV to 14 TeV (see Fig. 1 [4]). Also the cross-section is a smoothly increasing function of energy, so that no energy scan is necessary to find new particles. Finally, proton-proton collisions are used instead of proton-antiproton collisions, as in past and present colliders, to avoid the very inefficient and luminosity limiting antiproton production. At these energies, only for some very specific valence-valence dominated process (like Z' production) the proton-proton cross-section is significatively less than the proton-antiproton one.

Large number of events will be obtained each year: 10^5 Higgs bosons for $M_{\rm H}{\sim}500$ GeV/c^2 (Higgs boson discovery and couplings), 10^6 Z' for $M_{\rm Z'}{\sim}1$ TeV/c^2 (new heavy gauge bosons), 10^7 top quark pairs compared to a total of 10^5 at Tevatron(top mass, couplings and rare decays), 10^8 W$\to\ell\nu$ boson decays (at low luminosity), compared to a total of 10^3 at LEPII, 10^6 at Tevatron (precise W mass, triple boson couplings), 10^{12} $b\bar{b}$ pair (at low luminosity) compared to a total of 10^8 in near future B-factories CLEO III, BaBar and Belle. However, absolute rates is not a fair comparison w.r.t e^+e^- colliders because: (i) trigger and QCD background is much more difficult (ii) the high total cross-section meaning a high number of interactions poses severe constraints on the detectors (see Section I B 1) (iii) the measured final state 4-momentum is very different from the incoming particles 4-momentum, because the proton remnants are lost down the beam-pipe (only in the transverse plane can momentum balance be assumed) (iv) absolute normalization (luminosity measurement and cross-sections computing) is less precise. It will be demonstrated that these difficulties notwithstanding, the LHC physics program is unique in its reach and diversity.

These lectures are organised as follows: the first section describes the LHC accelerator and the two general purpose experiments, ATLAS and CMS. The second section details the search for Standard Model and Minimal Supersymmetric Model Higgs Bosons. The last section describes the dedicated B-physics experiment LHCb, with a sample of CP-violation measurement and other B physics subjects. Other Standard Model physics (top physics, W mass, gauge bosons coupling...) and the search for new particles and supersymmetry are not described here (see [5] for a review). The 4th experiment, ALICE, and its heavy ion physics program is also not discussed [6].

I THE LHC COLLIDER AND THE ATLAS AND CMS EXPERIMENTS

A The LHC machine

The LHC machine will be housed in the 27km circumference tunnel currently housing the LEP e^+e^- collider. For this given circumference, the limiting factor on the beam energy is the 8.3 Tesla magnetic field reached by the superconducting dipoles. Each dipole provides two beam tubes with opposite sign magnetic field. The injection will be provided by the same refurbished accelerator chain that is providing injection to LEP. To provide the design luminosity, protons are grouped in bunches of $\sim 10^{11}$ protons colliding in the middle of each experiment every 25 ns. The bunches are squeezed and provide a luminous region of 15 μm transverse radius and 5.5 cm half length (with gaussian distribution). The machine will run at 10^{33} cm^{-2}s^{-1} ("low luminosity") during the first three years, accumulating 10^4 pb^{-1} per year. This will allow delicate measurements (W and top mass for example) and channels with a difficult trigger to be studied (most of the B physics program). The luminosity will then be increased up to the design "high luminosity", 10^{34} cm^{-2}s^{-1}.

B Overview of the LHC detectors

ATLAS [7] and CMS [8] are the two general purpose detectors. Their design is now frozen and construction has begun. Each experiment involves about 1700 physicists. LHCb [9] (described in Section III B 3) is a smaller experiment dedicated to B physics, which is still in the designing stage. It involves some 400 physicists.

1 Pile-up

The high luminosity required and the high total inelastic cross-section σ=70mb have the consequence that for each bunch-crossing (every 25 ns), in average 23 inelastic interactions occurs. The simultaneous occurrence of these "minimum bias events" which will be superimposed to any interesting event is the so called "pile-up" effect. The main consequences are described below.

First, two useful variables must be defined. Since the primary interaction between partons is not in its center-of-mass, particles are boosted along the beam direction. The p_T relative to the beam direction, and the pseudo-rapidity $\eta = -\log\tan(\frac{\theta}{2})$ are natural variables instead of momentum and polar angle θ.

Inclusive particle production in minimum bias events is uniform in ϕ the azimuthal angle, uniform in η up to $\eta \simeq 4$ (see Fig. 2) and peaked at very low p_T. The density of particles per unit of pseudo-rapidity in a minimum bias event is 6 charged

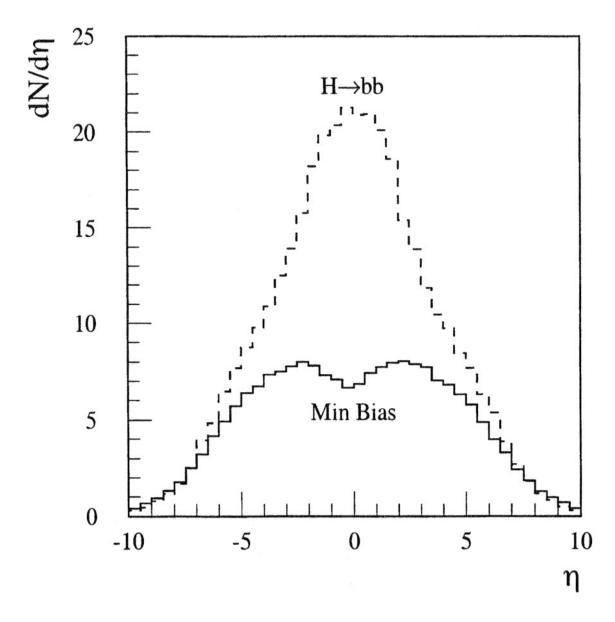

FIGURE 2. Number of track per unit pseudo-rapidity as a function of pseudo-rapidity in a minimum-bias event (solid) compared to a H→ $b\bar{b}$ event (dashed) for M_H=100 GeV.

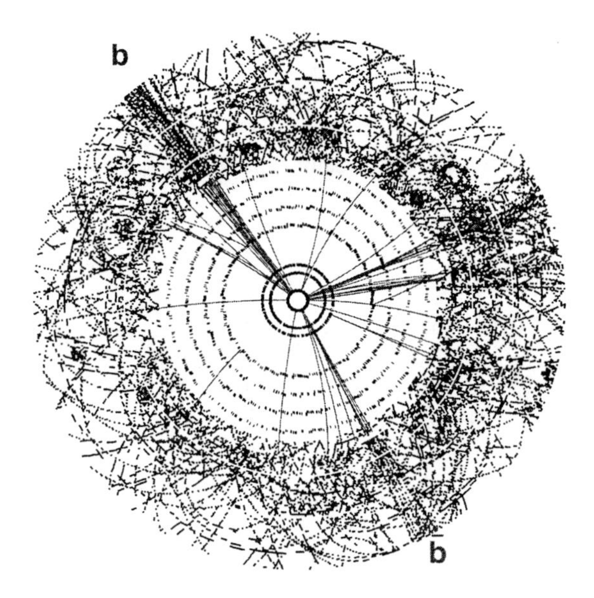

FIGURE 3. H→b$\bar{b}$ decay event at high luminosity in ATLAS tracker. All hits in the pixel, strip and straw detector and the reconstructed tracks with p_T more than 1 GeV/c are shown.

tracks (mainly pions) with $<p_T>\sim 0.6$ GeV/c, and 6 neutrals (mainly photons from π^0 decay) with $<p_T>\sim 0.3$ GeV/c. With a detector coverage of $-2.5 < \eta < 2.5$, about 750 charged tracks cross the detector every 25 ns. The calorimeters see about 1500 particles per bunch crossing for a total 700 GeV transverse energy. Fig. 2 compares the particle density in one minimum bias event to the one in a typical H→ b$\bar{b}$ decay. The H→ b$\bar{b}$ events are more central but clearly at high luminosity with 23 minimum bias events superimposed, only 10% of the particles come from the interesting events. However, the p_T of H→ b$\bar{b}$ event particles is much harder and especially these particles will be grouped in a small number of jets, when particles from pile-up will be uniformly distributed. Inside a jet, pile-up particles contribute to only 20% of the multiplicity, as it can be seen on the event display on Fig. 3.

The probability that two superimposed events mimic a rare interesting one (for example two independent Z events mimicking H→ZZ decay) is always negligible. In practice, pile-up can be considered as a noise rather than real additional events. It adds random hits in the tracking detectors and random energy in the calorimeters. It is a source of severe constraints on the design of all parts of the detector, including trigger and data acquisition, but it does not affect severely the performance of the detector. However some precision measurement (W and top mass) can only be done at low luminosity, as some trigger difficult channels like B-physics.

2 *Radiation damage*

Radiation damage at the LHC is mainly due to particles from pp interactions, and not from beam halo and beam losses as in present machines, so that it can be calculated with some accuracy. Two different processes occur. Ionizing particles deposit energy in materials (2 MeV g^{-1} cm^{-2} for a Minimum Ionizing Particle). Damages are proportional to the density of particles, which decrease like the square of the distance to the beam. Detectors nearest to the beam-pipe are the most affected. Neutrons are created in hadronic showers in the calorimeters. Low momentum neutrons (0.1-20 MeV/c) random walk through the whole detector. They cause crystal defects to semi-conductors, tracking Silicon and Gallium-Arsenide sensors and on-detector electronics. Radiation hard technology is mandatory and has been a major successful R&D effort in the past years.

3 *Basic detector requirements*

The first requirement is that the detectors must survive the ~10 years operation in the harsh environment described in previous section, or be easily replaceable. It should have the largest acceptance in pseudo-rapidity. It should be able to tag rare events by identifying e, μ and τ leptons, photons and b quark jets. It should be able to measure particles 3-momenta and charged particle impact parameters, jets energy and direction and missing transverse energy (E_T^{miss}). It should be granular to minimize pile-up effect, and accurate in time for the same reason, to avoid mixing consecutive bunch crossings.

The final detector design is a compromise between these requirements and feasibility and costs. A few channels have been used as benchmarks, namely (i) H$\rightarrow \gamma\gamma$, and H$\rightarrow$ ZZ$^{(*)}\rightarrow e^+e^-e^+e^-$ for electromagnetic calorimetry and tracking, (ii) H$\rightarrow$ b$\bar{\text{b}}$ for tracking (iii) $H \rightarrow WW \rightarrow qql\nu$ for calorimetry (jets)(iv) A$\rightarrow \tau\tau$, H$\rightarrow$ ZZ$^{(*)}\rightarrow \ell\ell\nu\nu$ for for hadronic calorimetry (E_T^{miss}) (v) H$\rightarrow$ ZZ$^{(*)}\rightarrow\mu^+\mu^-\mu^+\mu^-$ for muons. Many other different channels have been looked at as early as possible in the detector design, to spot all possible shortcomings.

C ATLAS and CMS detectors

The layout of the two detectors is broadly similar with, for the CMS detector shown on Fig. 4: the inner tracker, the electromagnetic calorimeter, the hadronic calorimeter, the superconducting solenoid (in ATLAS, the calorimeters are outside the solenoid) and the muon detector.

1 *Tracking*

Tracking is the reconstruction of charged tracks, with emphasis on momentum and impact parameter. A high magnetic field provided by a superconducting

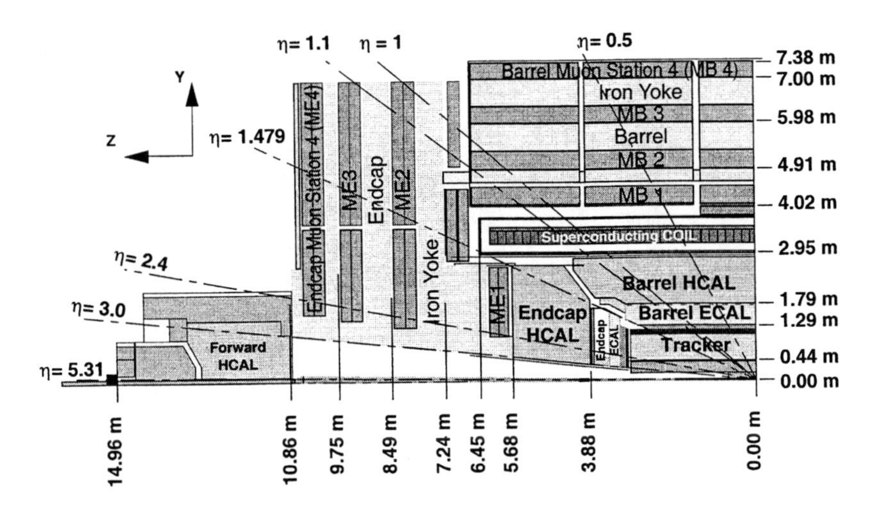

FIGURE 4. Longitudinal cut view of the CMS detector. The detector is symmetric w.r.t the y axis.

solenoid is used in ATLAS (2 Teslas) and CMS (4 Teslas): it is needed to have good momentum resolution in a small volume. Both experiments make extensive use of semi-conductor detectors which are fast (compared to drift chambers), precise ($\sim$10-20μm) and radiation hard.

The inner part of tracking detector suffer from high radiation level and high density of tracks. Both experiments have chosen to use pixel detectors which directly give 3D points and are radiation hard. The enormous number of channels ($\sim 10^8$) necessitates complex on-detector electronic.

The middle part of the tracking detectors is covered by less expensive strip detectors. Strip detectors give two 2D measurements for each track, which can be confused if the density of tracks is too high.

The outer part of the tracking detectors needs large areas, so that cheaper technologies must be used there. CMS uses Micro-Strips Gas Chambers (MSGC): the small drift distance allows a fast signal. ATLAS uses a Transition Radiation Tracker with 2mm diameters straws with maximum drift-time 38ns; the modest precision of 180 μm is more than compensated by the high number of points ($\sim$36) on a given tracks. Transition Radiation detection allows additional electron identification. In both ATLAS and CMS, the tracker covers acceptance up to $|\eta| = 2.5$.

The material budget inside the tracker volume should be kept to a minimum. A realistic detector description includes active detectors, support structures, on-detector electronics, high-voltage and readout cables and connectors. Material causes degradation of the performance of the detector (both tracking and calorimetry) from a variety of effect: multiple scattering, electron Bremsstrahlung, photon conversion, hadronic interactions. The radiation length of the ATLAS and CMS

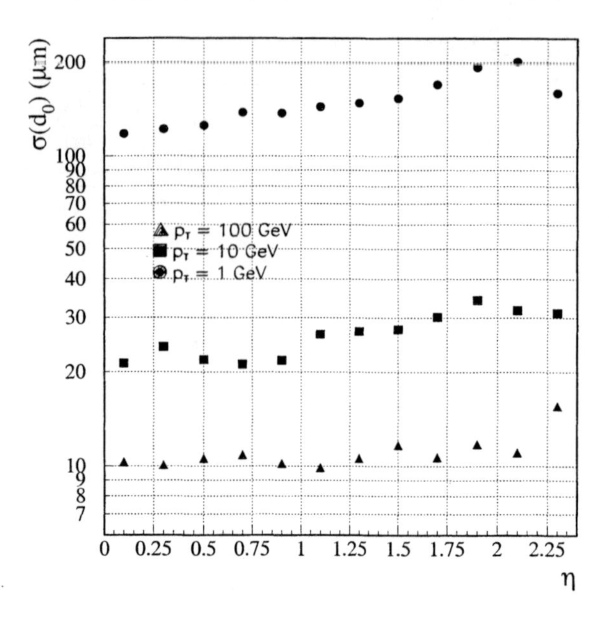

FIGURE 5. Resolution on the transverse impact parameter in the CMS tracker.

tracker is $\sim$ 30% at $\eta = 0$ and peaks at 60% in the barrel end-cap transition region. Fig. 5 [10] shows the degradation at low momentum of the impact parameter resolution of the CMS tracker.

2 Calorimetry

The electromagnetic and hadronic calorimetry measure jets energy and direction and missing transverse p_{T}. The electromagnetic calorimetry has the additional role of identifying and measuring electrons and photons.

Hadron calorimetry should have an acceptance up to $|\eta| = 5$ to provide good $E_{\mathrm{T}}^{\mathrm{miss}}$ measurement and forward jet tagging. Fig. 2 shows that an important fraction of the particles goes to very high η. Losing these particles degrades the transverse energy resolution as illustrated in section II F 2. Hermiticity is also very important to prevent energetic jets going into cracks to cause fake $E_{\mathrm{T}}^{\mathrm{miss}}$. Quarks in heavy particle decay appear as jets in the calorimeter; a good granularity is necessary to separate nearby jets as in high p_{T} W decays of heavy Higgses. Atlas uses a Tile calorimeter while CMS uses a lead sampling calorimeter.

Electromagnetic calorimetry has the additional granularity requirements to distinguish the narrow shower caused by electron and photon from the broader one caused by jets or even π^0's, in an acceptance $|\eta| < 2.5$ (matching tracker acceptance). ATLAS use a liquid argon accordion calorimeter while CMS uses crystals. ATLAS calorimeter is segmented in depth for better electron or photon/jet separation. Handling the ~700 GeV transverse energy per bunch crossing due to pile-up is done with a special electronic shaping of the calorimeters signal, which give a weight unity to the triggering bunch crossing and smaller negative weights to a tens of consecutive bunch crossing, so that the integral of pile-up energies is null for any luminosity.

3 Muons

Muon identification and measurement is needed from the lowest p_T (~6 GeV/c, for B physics and b-jet tagging) to the highest values (~1 TeV/c for heavy gauge bosons physics). Tracking in magnetic field outside the inner tracking chamber is mandatory to fulfill these requirements. CMS has chosen to instrument the return flux of its 4 Tesla solenoid in the hadron calorimeter, leading to a compact design. Atlas has chosen to track the muon in additional air toroids. In both cases, the best performance is obtained from combining the track segment in the muon detector with the track segment in the tracking detector.

4 Photons and electrons

Photons and electrons are identified from the electromagnetic shower shape, and lack of energy in hadronic calorimeter. Typical performance is a jet rejection of ~ 3000 for 80% efficiency.

For photon, an additional rejection of a factor 3 against electromagnetic jets can be obtained using the fine pseudo-rapidity segmentation of the calorimeter, capable of resolving $\pi^0 \rightarrow \gamma\gamma$decay.

Additional handles for electrons is the presence of a track with momentum matching the cluster energy and (ATLAS only) transition radiation identification. Typical performance is a rejection higher than 10^5 for 80% efficiency.

5 Trigger

The role of the trigger is to reduce the initial 40MHz rate to 100Hz readout, which will still produce 100MBytes of data per second, amounting to 10^6 GBytes per year. The bulk of cross-section being jets of moderate p_T, the trigger is done on signatures: electron, muon, τ lepton, photon, high E_T^{miss}, high p_T jets.

All these signatures have a rapidly falling p_T spectrum. The output rates are tuned by changing the p_T thresholds, requiring isolation or two or more signatures to open the window to lower p_T. A delicate balance of triggers allows the coverage of all the physics channels.

II STANDARD MODEL HIGGS BOSON SEARCH

The Higgs boson, if it exists, lies in the range delimited by direct search at LEPII on the low side , and theoretical arguments on the high side. Direct search in the channel $e^+e^- \rightarrow HZ$ at LEPII can set limit close to the kinematical limit $\sqrt{s} - M_Z$. Current limit obtained with $\sqrt{s}$=183 GeV in 1997 is M_H >90 GeV/c^2 at 95% Confidence Limit [11]. LEPII running at $\sqrt{s}$=200 GeV in year 2000 will probably set the mass limit to ~110 GeV/c^2. On the high mass range, a Higgs boson with mass in excess of 1 TeV is theoretically excluded because WW scattering would violate unitarity.

Overall fit of electroweak parameters allows a constraint on the Higgs boson mass through radiative corrections. A low Higgs boson mass is favoured with $M_H = 76^{+85}_{-47}$ GeV/c^2, or M_H<262 GeV/c^2 at 95% Confidence Level [12,1]. The ATLAS and CMS detectors have been designed to allow Higgs boson discovery from 80 GeV/c^2 up to 1 TeV/c^2, covering the whole allowed range with a good overlap with LEPII limit (which is necessary in case of non-standard Higgs couplings).

The different diagrams for Higgs boson production are shown on Fig. 6, and the corresponding cross-sections on Fig. 7. The main diagram is gluon fusion (A). WW/ZZ fusion (B) is non-negligible, especially at high mass. Due to the large mass of the W/Z, the recoiling quarks are visible in the detector forward region ($2 < |\eta| < 5$), which provide an additional useful signature. The associated production diagrams (C1 and C2) are ~50 times smaller at low mass, but are useful

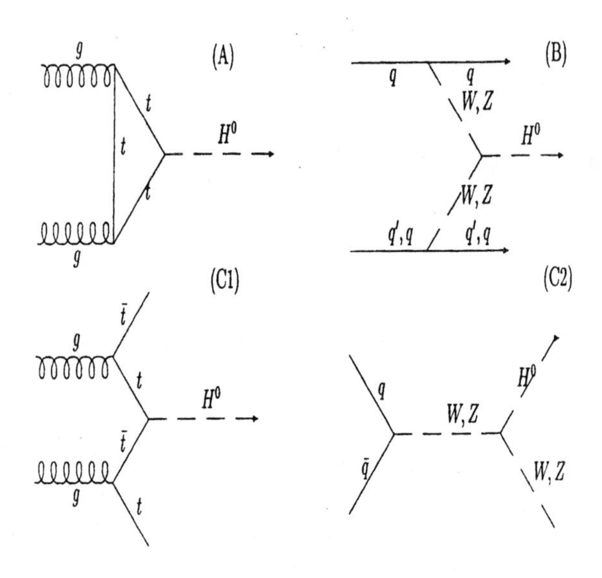

FIGURE 6. Main diagrams for Higgs boson production at the LHC.

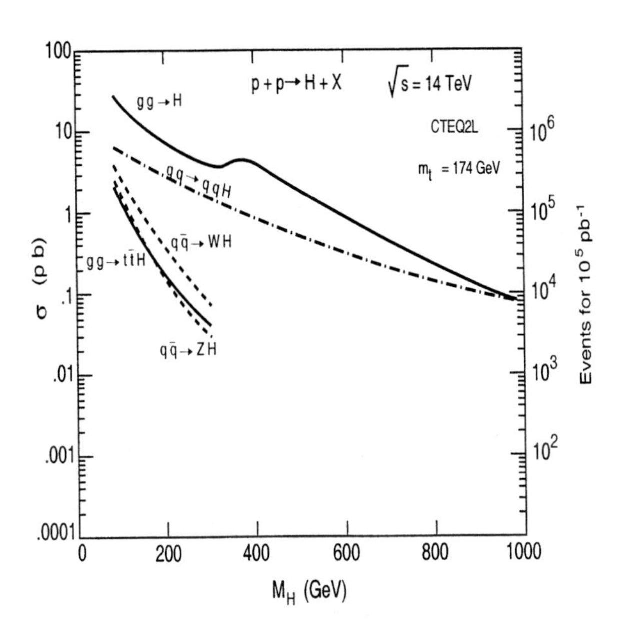

FIGURE 7. Cross-sections of the different Higgs boson production mechanisms as a function of the Higgs boson mass.

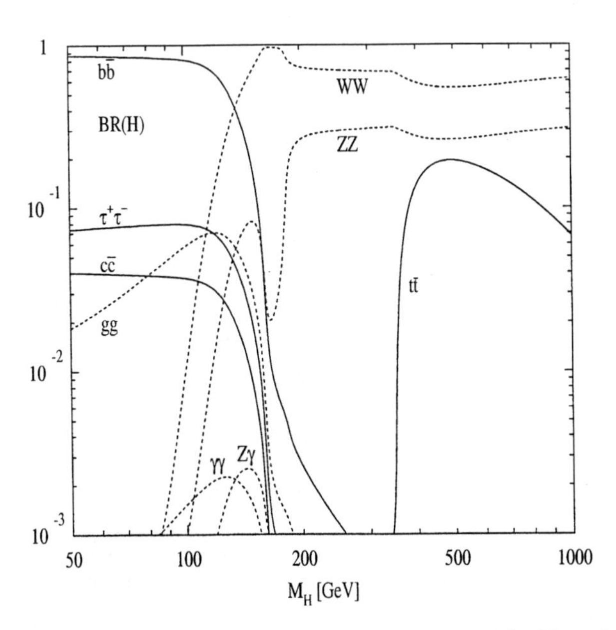

FIGURE 8. Higgs boson branching ratios as a function of the Higgs boson mass.

also for the additional signatures they provide. The calculations of the cross-section usually do not take into account higher-order corrections (so-called K factors) which would enlarge signal and backgrounds cross-section by a factor 1.4 to 1.8.

The Standard Model Higgs boson branching ratios are shown on Fig. 8 [13]. Two different regimes are seen. For Higgs boson mass well below $2M_W$, the Higgs boson decay predominantly in the heaviest fermions, mainly $b\bar{b}$ (90%). The small ($\sim$0.2% branching ratio) decay into two photons is in fact the main discovery channel, in the absence of other signatures. The Higgs boson width is small. Above $2M_W$ the decay in two gauge bosons is favoured: 2/3 in WW, 1/3 in ZZ. Higgs search is performed in the cleaner ZZ mode. For $2M_W<M_H<2M_Z$, the W's in H$\rightarrow$WW are on shell while one Z in H$\rightarrow$ZZ is still offshell. This causes a dip in the H$\rightarrow$ZZ branching ratio curve which will remain visible on the final significance plot (Fig. 10). The Higgs width become comparable to the experimental width at $M_H \simeq 200\text{GeV/c}^2$, significatively larger above.

The Higgs boson search strategy is a trade-off between rate and background. Jets final state have higher rate and backgrounds, compared to lepton and photon, and may require additional signatures (associated production). The Z boson main decay modes are (i) e^+e^- and $\mu^+\mu^-$ (3.4% each) the cleanest signatures (ii) $\tau^+\tau^-$(3.4%), a background to rare A$\rightarrow\tau^+\tau^-$(iii) $\nu\nu$(20%) (iv) $q\bar{q}$ (70%) among which $b\bar{b}$ is 15%, background to rarer H$\rightarrow$ $b\bar{b}$. The W boson main decay modes are (i) $e\nu$ and $\mu\nu$ (11% each) the cleanest signature, but only the p_T of the neutrino can be evaluated and only if it is the only one in the event (ii) $\tau\nu$ (11%) (iii) $q\bar{q}$ jets (67%). The inclusive W and Z cross-sections are typically 1000 times larger than that of a light Higgs boson.

The result of the trade-off between visible cross-section and backgrounds is the following channels. In the low mass range (M_H<120 GeV/c^2) H$\rightarrow$ $\gamma\gamma$ in direct production and H$\rightarrow$ $\gamma\gamma$ and H$\rightarrow$ $b\bar{b}$ in associated production. In the intermediate mass region (120 GeV/c^2 $<M_H<$ $2M_Z$), H$\rightarrow$ZZ* $\rightarrow\ell^+\ell^-\ell^+\ell^-$. In the high mass region, ($M_H>2M_Z$) H$\rightarrow$ZZ$\rightarrow\ell^+\ell^-\ell^+\ell^-$, H$\rightarrow$ZZ$\rightarrow\ell^+\ell^-\nu\nu$, H$\rightarrow$ZZ$\rightarrow\ell^+\ell^-q\bar{q}$ and H$\rightarrow$WW$\rightarrow$ $\ell\nu q\bar{q}$. These channels are now described in details, concentrating on the ATLAS experiment (CMS reach is similar).

A Low mass range: H$\rightarrow$ $\gamma\gamma$

This channel is the most demanding for the detectors. The large jet background require photon identification with jet rejection of order 10^4. The photon background with the highest cross-section are photons from quark Bremsstrahlung. These photons being emitted inside jets, they are easily removed with isolation requirement. The main remaining background is the inclusive pair production of isolated photons, which is indistinguishable from the signal. It is thus important to have the best mass resolution to be able to observe the peak on the background. Part of the mass resolution comes from the intrinsic energy resolution of the electromagnetic calorimeter. Photons that convert in the material upstream the calorimeter have

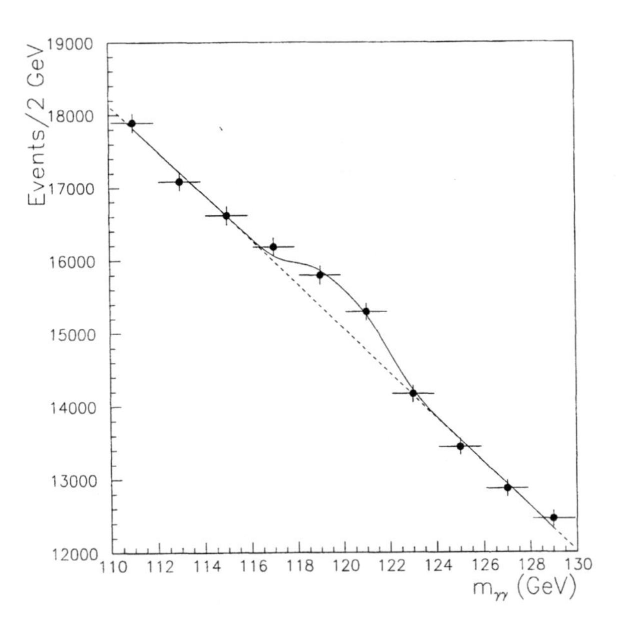

FIGURE 9. $\gamma\gamma$ mass plot with 10^5 pb^{-1} for $M_{\rm H}$=120 GeV/c^2. Note the broken vertical scale. There are about 1200 signal events in the peak.

a poorer resolution, especially if the conversion happens at small radius, because the electron-positron pair opens up in the magnetic field. It is thus necessary to identify these conversions using the outside part of the tracker and use a larger cluster size and different energy correction. The large uncertainty (~6cm) on the primary vertex along the beam yields a large uncertainty on the angle between the photons, contributing also to the mass resolution. Some angle information can be obtained from the calorimeter itself, when measurements at different depths are available (CMS end-cap and ATLAS). The most precise information is obtained if associated tracks coming from the underlying event can be unambiguously reconstructed. This turns out to be difficult at high luminosity, with in average 23 minimum bias vertices on top of the main one.

The photon-jet and jet-jet background can be reduced to 10% of the irreducible backgrounds. A very special situation occurs for $M_{\rm H}\simeq M_{\rm Z}$ (which is already excluded by LEPII direct search, but with Standard Model couplings only). The cross-section for Z$\rightarrow$e$^+$e$^-$ is 25000 times higher than that of H$\rightarrow \gamma\gamma$, meaning that the probability to mistake an electron for a photon should be less than 0.2%, so that pseudo Z$\rightarrow$"γ" "γ" does not peak below the H$\rightarrow \gamma\gamma$. This probability is of order of the probability for an electron to lose 95% of its energy in the tracker. A special algorithm able to follow an electron after hard Bremsstrahlung has been successfully developed in Atlas reaching this goal, within a small photon efficiency loss.

An example of a mass plot obtained with this channel is shown on Fig. 9 [7]. It's only in a limited range around M_{H}=120 GeV that the significance of this channel is greater than 6, which means it should be supplemented by other channels.

The significance quoted here and later on is the number of signal events expected divided by square root of the number of background events in a 1.4σ mass window. The significance gives (in standard deviation) the probability for the background to fake the signal. When the number of expected signal and background events is small, Poisson statistics should be used with the same meaning. A 5σ significance is believed to be necessary for claiming a discovery. The significance scales like the square root of the luminosity and significance for different channels for a given mass may be added in quadrature. Other less quantifiable aspects like the uncertainty on the level and shape of the backgrounds are not included in $\mathcal{S}$ but should not be forgotten.

The cross-sections for associated Higgs boson production are ~50 times smaller than direct production. However, a high p_{T} lepton in W, Z or top decays allows a very significant reduction of the continuum background. The main remaining background is Z$\gamma\gamma$ final state, where at least one of the photon comes from FSR of the Z. The significance is better than that of direct production at 100 GeV/c^2 mass and below and only slightly lower at higher mass. With 10^5 pb^{-1}, ~14 signal events are expected on a background of ~4. This very different regime allows an independent confirmation of the possible signal, and give information on Higgs couplings.

B Low mass range: H→ b$\overline{\mathrm{b}}$

The Higgs →b$\overline{\mathrm{b}}$ decay mode is the main one (branching ratio ~80%) for a low mass Higgs. However, b jets final state is much more difficult than the $\gamma\gamma$ final state, w.r.t trigger and tagging. Also H→ b$\overline{\mathrm{b}}$ mass resolution is ~15 GeV/c^2 compared to 1 GeV/c^2 for H→ $\gamma\gamma$, because hadronic calorimetry resolution is intrinsically worse than electromagnetic calorimetry, and b jets are especially difficult because of the presence of a neutrino from semi-leptonic decay in 20% of the b-jets. Finally, inclusive high p_{T} b$\overline{\mathrm{b}}$ production is much larger than $\gamma\gamma$; in particular. For these reasons, Higgs boson search in H→ b$\overline{\mathrm{b}}$ is only doable in associated production, using the additional signatures for trigger and background rejection.

Tagging of b-jet against c-jets, light quark jets and gluon jets relies on three facts: B hadron have long lifetime (implying large impact parameter, 400μm in average), are heavy (implying large track multiplicity, 5 in average), and decay semi-leptonically in 20% of the cases. b-jet tagging relies on computing a tagging variable from tracks impact parameter (rejecting fake large impact parameter caused by resolution tails, secondaries, strange particle decays) combined with soft-lepton identification inside jets (rejecting non-b lepton like photon conversion and Dalitz decay for electrons, K/π in flight decay for muons). Typical performance at 60% efficiency is a rejection of 80 against uds-jet, 40 against gluon jet and 10 against c-jet. The c-jet rejection is poor because charm hadrons do have some lifetime, mass and semi-leptonic decays. Gluon jet rejection is lower than for quark jets, because of possible g$\rightarrow$c$\bar{c}$ or b$\bar{b}$ fragmentation.

The main background for the channel WH, with W$\rightarrow$ $\ell\nu$ and H$\rightarrow$ b$\bar{b}$ is the W+jets channel, the rejection of which depends crucially of b-jet tagging. Another background is t$\bar{t}$ production where the final state of t$\bar{t}$ production is two W's and two b's, against one W and two b's for the signal; events with additional jets or leptons are hence rejected. Finally the WZ$\rightarrow$ $\ell\nu$b$\bar{b}$ process is irreducible but has a somewhat lower cross-section; good understanding of the b-jet tagging performance is necessary to properly subtract this background for Higgs mass below 100 GeV/c^2. The significance of this channel is above 5σ for $M_{\rm H}$<100 GeV/c^2, so that it complements nicely the H$\rightarrow$ $\gamma\gamma$ channel, so that the low mass range is well covered (see Fig. 10).

C Intermediate mass: H$\rightarrow$ZZ$^{(*)}$ $\rightarrow\ell^+\ell^-\ell^+\ell^-$

This decay mode requires two high $p_{\rm T}$ leptons consistent with the Z mass and two lower $p_{\rm T}$ leptons with significant mass. The irreducible background from continuum ZZ$^*/\gamma^*$ $\rightarrow\ell^+\ell^-\ell^+\ell^-$ is small thanks to the good Higgs boson mass resolution of order 1.5 GeV/c^2. The reducible backgrounds are $\ell^+\ell^-$b($\rightarrow$ ℓ^-)$\bar{\rm b}$($\rightarrow$ ℓ^+) final states (from t$\bar{t}$ or Z$\rightarrow$ b$\bar{b}$), where the b-jets decay semi-leptonically. Direct leptons from Z decays can be distinguished from leptons from b-jets because they are isolated and have impact parameter consistent with zero. Typical numbers of events expected after 3 years at low luminosity is 20 signal events on a few background events.

The significance of this channel is high, with a dip for Higgs boson mass around $2M_{\rm W}$, due to the dip in the branching ratio curve (see Fig. 8). The channel H$\rightarrow$WW$\ell\nu\ell\nu$ (not described here) may be used in this limited range.

Above $2M_{\rm Z}$ the Higgs boson decay in two on-shell Z's with a 1/3 branching ratio. The analysis is very similar to the previous channel with one more Z mass constraint, allowing more background rejection. The continuum background is lower, but the Higgs boson width gets larger than the detector resolution. The significance is excellent. The observability is limited to Higgs boson mass less than 700 GeV/c^2, because the width gets very large and the signal rate decreases faster than the continuum background rate. More copious channel needs to be searched for to cover the high mass range.

D High mass range

The cross-section for the channel H$\rightarrow$ZZ$\rightarrow\ell^+\ell^-\nu\nu$ is three times higher than the previous one. The signature is high missing energy ($E_{\rm T}^{\rm miss}$>100 GeV), with the missing $p_{\rm T}$ vector approximately back-to-back to Z$\rightarrow\ell^+\ell^-$. The irreducible background is continuum ZZ production. The reducible background is Z+jets production, where the jets can fake missing energy through b semi-leptonic decay. The hermiticity of the detector is crucial in this channel. A good understanding of the tail of the missing energy at high luminosity is also needed. A double forward jet tag can be used to enhance VV fusion (which is 30% of the Higgs cross-section) against continuum ZZ production, especially at high mass.

The cross-section for the channel H→WW→$\ell\nu q\bar{q}$ is 150 times higher than the four leptons decay mode. The signature is a high p_T lepton and missing p_T consistent with the W mass, two close jets ($\Delta R<0.2$) consistent with the W mass. Not much additional central energy should be present, in contrast to the main background from $t\bar{t}$ decay, where two additional b jets are present. Requiring additional forward jets is mandatory to reduce the background further. The signal obtained is a clear broad peak (significance $\sim 10\sigma$ for Higgs mass up to 1 TeV/c^2). A small problem is that the background has the same reconstructed mass distribution as the signal. However, changing the lepton and jet thresholds change the shape of the background while leaving the signal unchanged, allowing an important cross-check.

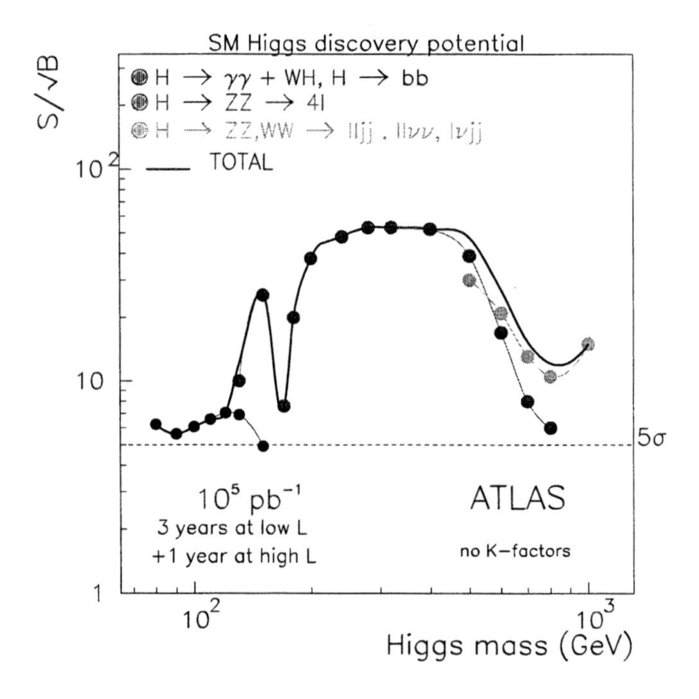

FIGURE 10. Observability of the Standard Model Higgs boson in ATLAS.

E Summary of Higgs searches

Taking simultaneously into account the channels described, the overall significance of the Higgs search from the Atlas experiment is always greater than 5σ (see Fig. 10 [14]). Several channels are available in all cases (considering the $H\rightarrow ZZ^{(*)}$ channel involves electrons and/or muons), allowing independent confirmation. CMS reach is similar, allowing also cross-checks between experiments.

Once the Higgs boson is found, a measurement of its properties is a mandatory test of the Standard Model. After three years at low luminosity and three years at high luminosity, the mass (a free parameter of the Standard Model) can be measured with a 0.1% precision up to 400 GeV/c^2, degrading to 1% at 700 GeV/c^2. The width can be measured with 10% precision for mass above 300 GeV/c^2, otherwise it is smaller than the experimental resolution. The spin of the Higgs boson can be checked by analyzing the angular distribution $H\rightarrow \gamma\gamma$ decay at low mass, and the azimuthal Z decay correlation in $H\rightarrow ZZ^{(*)}$. When more than one channel is available, ratio of branching ratios can be compared to Standard Model prediction. At low mass, the ratio of direct production (through a $t\bar{t}$ loop) to associated production would be sensitive to new unexpected heavy particles in the loop.

F Minimal Supersymmetric Model Higgs bosons search

In the Minimal Supersymmetric Model, the Higgs sector contains two charged ($H^{\pm}$) and three neutral (h, H, A) physical states. All masses and couplings can be expressed as a function of two parameters, which are usually chosen to be: M_A, the mass of the CP-odd boson, and $\tan\beta$ the ratio of the vacuum expectation value of the Higgs doublets. Any search for a given Higgs boson in a given channel can be translated in an exclusion plot in the (M_A,$\tan\beta$) plane. Radiative corrections (top quark, SuSy particles) are important in the masses and couplings calculation [15]. For example, the simple tree level relation $M_h < M_Z \cos(2\beta)$ was long thought to be in favour of a discovery of h at LEP. When radiative corrections are taken into account the limit on M_h rise to 115 GeV/c^2 which is reached for small $\tan\beta$.

The strategy is to cover the full (M_A,$\tan\beta$) plane with possibly independent channels, to be redundant and to be able to distinguish MSSM scenarios. The method is first to translate the Standard Model searches into contours in the (M_A,$\tan\beta$), second to look for specific MSSM channels such as: $A/H\rightarrow\tau^+\tau^-$, $t\rightarrow bH^+ (H^+ \rightarrow \tau\nu)$, $H\rightarrow hh\rightarrow b\bar{b}\gamma\gamma$, $A\rightarrow Zh\rightarrow \ell^+\ell^- b\bar{b}$, $A/H\rightarrow t\bar{t}$... The final exclusion plot (Fig. 12) shows that LEPII will set limits M_A>100 GeV/c^2 and $\tan\beta > 3$. Channels covering the parameter space excluded by LEPII should still be considered to provide a good overlap with LEPII and to provision non MSSM couplings. However, for the brevity of this review, only some of the channels non covered by LEPII will be described. Also, this study is restricted to the case when all other SuSy particles are heavy (masses $\sim$1 TeV/c^2) and with no mixing in the third generation.

1 *Standard Model channels*

The h→γγ (direct and associated production) and h→b$\bar{b}$ channels cover already 90% of the parameter space after one year at high luminosity. More precisely an upper limit of ~120 GeV/c^2 can be placed on M_A (see Fig. 12).

2 *A,H→$\tau^+\tau^-$*

The decay A,H→$\tau^+\tau^-$ is strongly enhanced in a large region of the parameter space. A and H are degenerated in mass if M_A>150 GeV/c^2. One τ lepton decaying to electron or muon provides the trigger. The second τ is identified as a very narrow jet in the calorimeter associated to only one charged track, which allow a rejection of 100 against jets for 50% τ efficiency. The missing p_T is the sum of the p_T of the two ν_τ. Making the additional hypothesis that the direction of each neutrino is always collinear to each visible τ decay allows the reconstruction of the mass of the A/H, provided the two τ's are not back-to-back. The resolution on the mass is 15 GeV/c^2 at 90 GeV/c^2, but can be improved at the expense of efficiency by cutting on the angle between the τ's, especially at low mass to distinguish A decay from Z→$\tau^+\tau^-$ (see Fig. 11 [16]). The mass resolution would be twice worse if the pseudo-rapidity coverage of the calorimetry was limited to 3 instead of 5. This

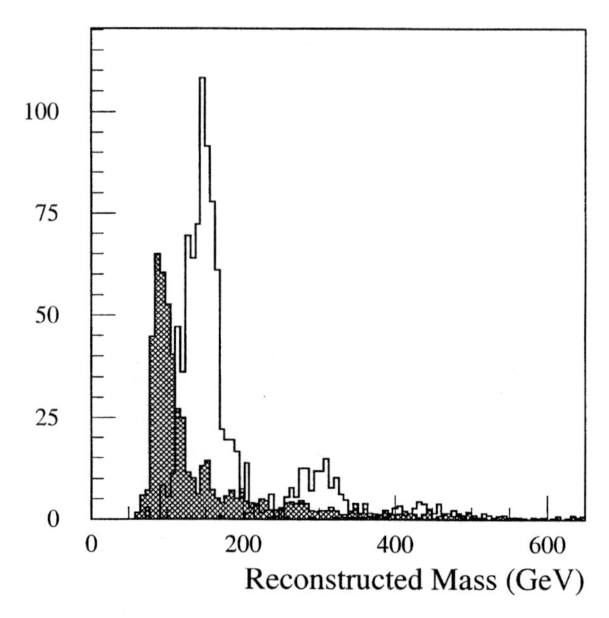

FIGURE 11. A→$\tau^+\tau^-$reconstruction for M_A=150, 300 and 450 GeV/c^2 (white). Backgrounds (black) with the Z→$\tau^+\tau^-$ peak.

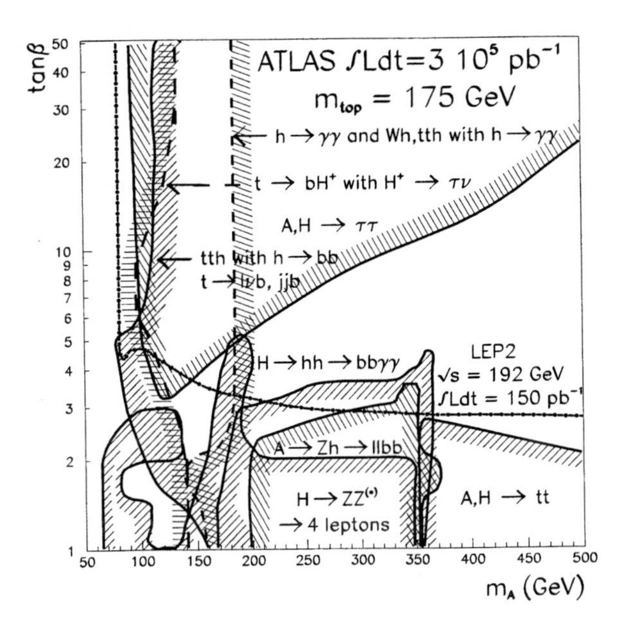

FIGURE 12. Observability of the MSSM Higgs bosons, after 3 years at high luminosity.

channel covers a large fraction of the parameter space, already at low luminosity (Fig. 12).

3 $t \rightarrow H^+(\tau\nu)b$

If the charged H^+ mass is less than the top quark mass, the branching ratio for $t \rightarrow H^+b$ is expected to be large with a minimum for $\tan\beta \sim 7.5$. With the very large $t\bar{t}$ production at the LHC this decay is promising. $t\bar{t}$ events are selected with one tagging $t \rightarrow W(\ell\nu)b$, requiring a high p_T lepton and two b-jets. The number events with a τ is counted and compared to the main background $t \rightarrow W(\tau\nu)b$. The H^+ can be observed for M_A up to $\sim$130 GeV/c^2(Fig. 12).

4 Summary of MSSM Higgs search

All MSSM Higgs searches are summarized on the somewhat crowded Fig. 12 (from [17], with $h \rightarrow b\bar{b}$ added). The parameter space is entirely covered with more than one channel. MSSM Higgs bosons can be distinguished from SM Higgs boson almost everywhere, except at large M_A and moderate $\tan\beta$, where $h \rightarrow \gamma\gamma$ and $h \rightarrow b\bar{b}$ are like the corresponding Standard Model H channels. Allowing for lighter SuSy particles change some contour but not the overall picture. Also in this case new channels with Higgs boson decay to SuSy particles can be searched for.

III B PHYSICS AT THE LHC

The origin and nature of CP violation in the Standard Model is not really understood, even if it can be accommodated with a phase in the CKM matrix. Numerous models try to describe CP violation as the shadow of a more fundamental theory: for example, superweak models [18] predicts the absence of direct CP violation in kaon decay ($\epsilon'/\epsilon = 0$) and of any CP violation in B decays, others [19] predict well defined value of the CKM parameters. The interest of B physics for a precise test of the Standard Model description of CP violation is twofold: first, CP violation is expected to be large in this system (of order 1), second, the B mass being much larger than $\Lambda_{\rm QCD}$ decay amplitudes can be more accurately related to CKM matrix parameter than in the case of the K system. However, the branching ratio of the decay modes of interest being of order 10^{-5}, sufficient statistics have not yet been accumulated to have evidence for CP violation in the B system. At the LHC approximately 10^{12} $\rm b\overline{b}$ pair will be produced per year and per experiment during the first 3 years of running at low luminosity (10^{33} $\rm cm^{-2}s^{-1}$). This unprecedented statistics will allow numerous precise measurements to be done. The ATLAS and CMS experiments will be joined in this physics by LHCb, an experiment dedicated to B physics.

A The Unitarity Triangle

Many measurements in the B system can be summarized in the so-called "Unitarity Triangle", which is now introduced. A lot of other measurements (form factors, excited states, branching ratios ...) do not appear directly as constraints on the Unitarity Triangle but are valuable in showing the validity of the framework in which B decays are calculated (perturbative QCD, Heavy Quark Effective, Theory,...). They are not described here.

1 Introduction

The mass eigenstates of the quarks are not weak eigenstates. The mixing is described by the Cabibbo-Kobayashi-Maskawa matrix [20], conventionally written as follows:

$$\rm V_{CKM} = \begin{pmatrix} V_{ud} & V_{us} & V_{ub} \\ V_{cd} & V_{cs} & V_{cb} \\ V_{td} & V_{ts} & V_{tb} \end{pmatrix}$$

This unitary matrix can be described as a function of four parameters, one of which being the phase which makes it complex. This phase is the source of CP violation in the Standard Model. Wolfenstein [21] has proposed an expansion of the matrix in terms of $\lambda \simeq 0.22$ the Cabibbo Angle:

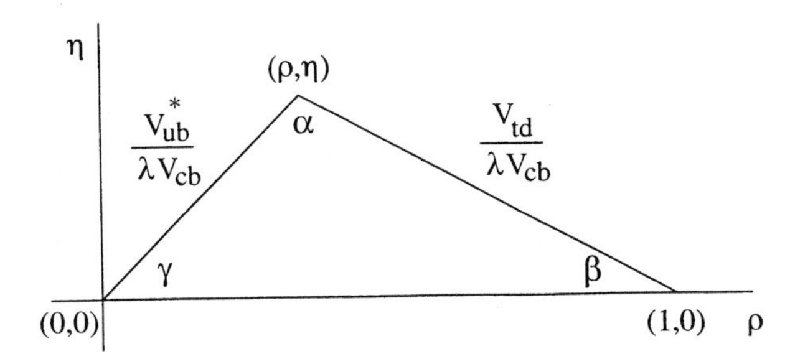

FIGURE 13. The Unitarity Triangle.

$$V_{CKM} \simeq \begin{pmatrix} 1-\frac{\lambda^2}{2} & \lambda & A\lambda^3(\rho - i\eta) \\ -\lambda & 1-\frac{\lambda^2}{2} & A\lambda^2 \\ A\lambda^3(1-\rho-i\eta) & -A\lambda^2 & 1 \end{pmatrix} + \mathcal{O}(\lambda^4).$$

Of all the unitarity relations, the one implying the first and third columns is particularly interesting because it involves three terms of the same order λ^3:

$$V_{\rm td}V^*_{\rm tb} + V_{\rm cd}V^*_{\rm cb} + V_{\rm ud}V^*_{\rm ub} = 0$$

Applying Wolfenstein approximation and normalizing by $V_{\rm cb}$, this relation describes the Unitarity Triangle in the complex plane (Fig. 13) (the triangle obtained from the first and third line is slightly different at $O(\lambda^4)$, which should be taken into account given the precision reached by the LHC experiments). The area of the triangle needs to be non-zero ($\eta \neq 0$) to have CP violation. The name of the game is to estimate from a variety of measurements (the two non-unit lengths and the three angles, each of them obtained from different processes) the position of the apex of the triangle. The ultimate goal being twofold: (i) estimate with high precision the four free parameters of the Standard Model hiding in the CKM matrix to test models predicting this value (ii) test the consistency of all the measurements to possibly have evidence for new physics: the more constraints the more information on which measurement departs from the Standard Model. In 4th generation models, the CKM matrix would not be unitary; in SuSy with three generations the CKM matrix is still unitary and inconsistencies in the measurements would come from neglecting SuSy particles in decay diagrams ; in models with spontaneous CP violation, the CKM matrix is real, so that the angles and sides of the Unitarity Triangle would not match.

2 Constraints on the Unitarity triangle

Current constraints on the Unitarity Triangle are now briefly described [22–24]. $\lambda = V_{\rm us} = \sin\theta_c$ is known from K and Λ semi-leptonic decays with precision $\sim 0.8\%$.

V_{cb}=Aλ^2 is measured from inclusive and exclusive semi-leptonic B decays with precision 5%. $|V_{ub}/V_{cb}|$ is measured from charmless semi-leptonic B decays with precision of $\sim$ 25%, dominated by theoretical uncertainties (heavy to light transition). Measurement of Δm_d in $B_d^0 \leftrightarrow \overline{B}_d^0$ oscillation allows to reach V_{td} with large ($\sim$ 25%) theoretical uncertainties. A measurement of Δm_s in $B_s^0 \leftrightarrow \overline{B}_s^0$ oscillations depends of V_{ts}=V_{cb} with similar theoretical uncertainties, which are reduced to less than 10% in the ratio $\Delta m_d/\Delta m_s$(only limits on Δm_s are available now). CP violation in the Kaon system (ϵ_K) brings another constraint, which is the only one where CP violation is actually measured.

An overall fit of all these constraints has been performed by various authors [25,23,24], with broadly consistent results. The outcome of these fits is the following (see Fig. 14 [23]): (i) CP violation in the Kaon system is in good agreement with parameters measured in the B system (ii) measurements in the B system alone are still consistent with a flat Unitarity Triangle (iii) these fits give non-trivial values for yet unmeasured parameters. For example [23] quotes $\sin 2\beta = 0.68 \pm 0.1$ (which means that CP violation in the $B_d^0 \rightarrow J/\psi K_S^0$ channel is indeed large), $6 < \Delta m_s < 21$ ps^{-1} at 95% confidence level (which means Δm_s is just beyond the reach of current experiments which is 13.8 ps^{-1} at 95% Confidence Level [26]), other angles are also estimated: $\sin 2\alpha = -0.1 \pm 0.40$ and $\gamma = (64 \pm 12)°$.

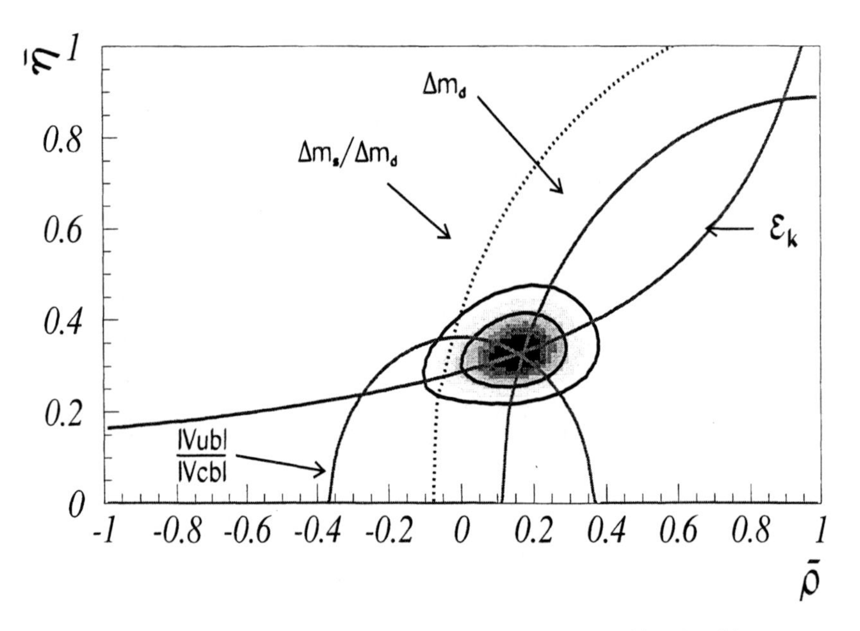

FIGURE 14. Fit of the Unitarity Triangle apex, with the 68% and 95% contours.

3 Future improvements of the Unitarity Triangle

Improvement of the experimental knowledge of the Unitarity Triangle relies on better experiments and, most important, better general theoretical understanding of B decays. Existing measurement can still be improved at existing facilities (LEP,CLEO) and will be remade at near-future B factories: V_{cb}, V_{ub}. Also improvements in Lattice QCD [27] will allow a better extraction of the CKM parameters from existing measurements. New measurements (notably ϵ'/ϵ) in Kaon physics are planned. Important new measurements in B physics (among others, see Ref [28] for a review) are (i) $\mathrm{B_s^0}$ mixing (Δm_{s}) (ii) CP violation in $\mathrm{B_d^0}\rightarrow \mathrm{J}/\psi \mathrm{K_S^0}$ (angle β) (iii) CP violation in $\mathrm{B_d^0}\rightarrow\pi^+\pi^-$(angle α but with theoretical uncertainties due to penguins) (iv) Rare muonic or semi-muonic decays $\mathrm{B_d^0}/\mathrm{B_s^0}\rightarrow\mu^+\mu^-(\mathrm{X})$ (related to V_{td} and V_{ts}). First evidence for CP violation will probably be seen at Tevatron or Babar or Belle. But precise measurements will only be done at the LHC: the 68% Confidence Level contour on Fig. 14 will be as small as one dot of the dotted line.

4 B decay examples

Two typical B decay diagrams are shown on Fig. 15 [9]. The penguin diagram lead to the same final state as the tree one, with which it will interfere. Penguin diagrams may involve different heavy particle loop, in particular unexpected ones, which is a reason why new physics may be discovered first in B decays. On the

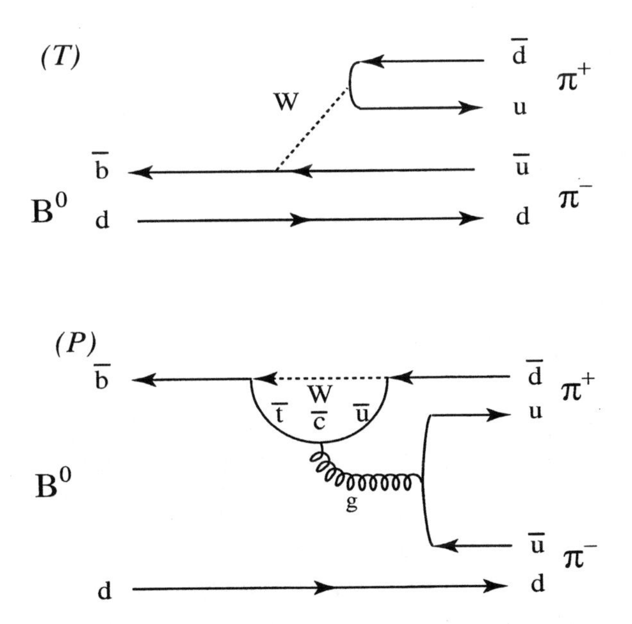

FIGURE 15. The tree diagram (T) and a penguin diagram (P) generating $\mathrm{B_d^0}\rightarrow\pi^+\pi^-$.

other hand, penguin diagrams can be considered as a way for the Standard Model to produce Flavour Changing Neutral Currents (FCNC, forbidden at tree level) such as b→d or b→s transitions, which could hide genuine FCNC due to new physics.

B Experimental Overview

Maximum CP violation arise when the decay $B_d^0 \to f$ interfere with the decay $B_d^0 \to \overline{B}_d^0 \to f$, where f is a CP eigen-state. Taking as example the decay $B_d^0 \to J/\psi K_S^0$ the time-dependent CP asymmetry has the following form:

$$A(t) = \frac{\Gamma(B_d^0 \to J/\psi K_S^0) - \Gamma(\overline{B}_d^0 \to J/\psi K_S^0)}{\Gamma(B_d^0 \to J/\psi K_S^0) + \Gamma(\overline{B}_d^0 \to J/\psi K_S^0)} = \sin 2\beta \sin(\Delta m_d t),$$

where the label B_d^0 or $\overline{B}_d^0$ indicates the initial state flavour.

Any algorithm for tagging the initial state is characterized by its efficiency $\epsilon = (N^+ + N^-)/N_S$ and its mistag fraction $W = N^-/(N^+ + N^-)$, where N^+ (N^-) is the number of correctly (incorrectly) tagged signal events. The tagging dilution is $D_{tag} = 1 - 2W$. Finally, the tagging quality factor is $Q = \epsilon D_{tag}^2$, which is the equivalent loss in luminosity compared to a perfect tag with $D = \epsilon = 1$.

The time-integration of the CP asymmetry is an additional source of dilution: $D_{int} = (\sin(xt_0/\tau) + x\cos(xt_0/\tau))/(1 + x^2)$, where t_0 is the minimum accepted proper-time and x the usual mixing parameter. For B_d^0 decays, D_{int} is of order 0.6. If a fit to the time-dependent asymmetry is performed, D_{int} is replaced by D_{fit} which is one for a perfect time resolution and no threshold on the proper-time. Time-dependent fit is mandatory to observe CP asymmetry in B_s^0 decays because of the rapid $B_s^0 \leftrightarrow \overline{B}_s^0$ oscillation.

Finally, the observed asymmetry is (neglecting the production and detection asymmetry expected to be less than 0.01):

$$A_{obs} = D_{tag} D_{bkg} D_{int} \sin 2\beta,$$

and the sensitivity,

$$\delta(\sin 2\beta) = \frac{\sqrt{1 - A_{obs}^2}}{D_{tag} D_{int} \sqrt{D_{bkg} N_S}}.$$

The precision of a CP violation measurement in B's depend on the statistics collected (N_S), which is a combination of high cross-section, high luminosity and high trigger efficiency. It is necessary to identify cleanly (D_{bkg}) the final states, which means identifying leptons and K_S^0 (for $B_d^0 \to J/\psi K_S^0$ channel), hadrons (π/K/p separation is important for the $B_d^0 \to \pi^+\pi^-$, and other hadronic decay modes). Good impact parameter resolution allows very high combinatorial background rejection through vertex consistency. Background is also reduced through the mass resolution depending mainly from momentum resolution. Impact parameter and momentum resolution allows proper-time resolution which is the limiting factor to B_s^0 mixing sensitivity and CP violation in B_s^0 decay channels (D_{int}). Tagging the flavour of the initial state (D_{tag}) is always possible once the experiment has been optimized to the above mentioned tasks (see section III B 4): it relies more on ingenuity than on detector performance. The current and future experimental scene is now reviewed.

1 B physics experiments

Existing and future B physics experiments (not necessarily dedicated to B physics) are of four different types. (i) e^+e^- colliders at the $\Upsilon(4S)$ resonance (CLEO, 5 10^6 BB pairs accumulated until 1997, new run with upgraded detector starting in 2000), which decay into $B_d^0\overline{B}_d^0$ or $B_u^+B_u^-$ meson pairs nearly at rest. No additional particles are produced in B events: constraint to the beam energy allows excellent mass resolution. The two B decay products are mixed. The detectors are blind to very low momentum particles in the lab frame, which is almost the B rest frame, which is a source of systematics. No B proper-time information is available, which prevents the measurement of CP asymmetry in this particular case where the two B's evolve coherently.(ii) e^+e^- colliders at the Z pole (LEP experiments up to 1995). B's are produced back to back with a large boost ($<\gamma>\sim 6$), so that B decay products do not mix, proper-time can be measured accurately and inclusive flavour tagging is possible. Also B_s^0 and Λ_b are produced. Statistics is smaller, 800.000 per experiment. (iii) Asymmetric e^+e^- collider (BaBar [29], Belle [30], data-taking starting in year 2000 with the goal of 5.10^7 BB pairs collected) at the $\Upsilon(4S)$ resonance cumulate the advantages of the two previous colliders (except B_s^0's and Λ_b's). (iv) Hadronic colliders have a much larger $b\overline{b}$ rate. But the experimental environment is much less clean, in particular for the trigger.

2 B physics at ATLAS and CMS

About 10^{12} $b\overline{b}$ pair per year are produced at low luminosity. The cross-section, extrapolated from Tevatron energy, is uncertain within a factor of two. The average relativistic boost is moderate ($<\gamma>\sim 3$). The trigger at first level relies on a B semi-leptonic decay to an electron or muon, which requires as low a threshold as possible (for example, the trigger efficiency is decreased by a factor of two if the threshold increases from 6 GeV/c p_T to 8 GeV/c). Exclusive decay-modes of interest are then selected online. In contrast to existing experiments, it will not be possible to test new theoretical ideas on new decay modes once data taking has started. ATLAS and CMS performance are similar with some difference. CMS can set a lower threshold on lepton trigger and has better momentum resolution at small p_T, thanks to its high magnetic field. ATLAS straw tracker allows electron identification down to p_T=1GeV/c, and good K_S^0 reconstruction.

For B physics it is sufficient to measure the decay products of the two B's produced in an event. The first B is the decay of interest, while the second B allows the tagging of the initial flavour, and helps triggering. B's produced in pp interactions are correlated in pseudo-rapidity, so that a detector covering only a limited range of pseudo-rapidity is enough. LHCb is a detector dedicated to B physics with a "fixed target"-like geometry covering rapidity +2 to +5 (see Fig. 16 [9]) (A similar experiment BTeV [31] has recently been proposed on the Tevatron, but with lower $b\bar{b}$ rate).

The advantages of such a geometry are numerous:(i) it is possible to implement a Cerenkov detector (RICH) with only the radiator being on the trajectory of particles, while photon detectors are on the side, allowing excellent particle identification. (ii) the detector support structure and services (high-voltage and readout cables) can be put outside the acceptance, thus limiting the material budget, allowing a good tracking resolution even at small momentum. (iii) Working at large pseudo-rapidities, moderate p_T from B physics correspond to large momentum, which reduces again multiple-scattering, helps calorimeter resolution and proper-time resolution for a given decay-length resolution. (iv) there is enough room to make a magnetic spectrometer with a large lever-arm for excellent momentum resolution (v) it is easy to access parts of the detector (vi) the detector covering a

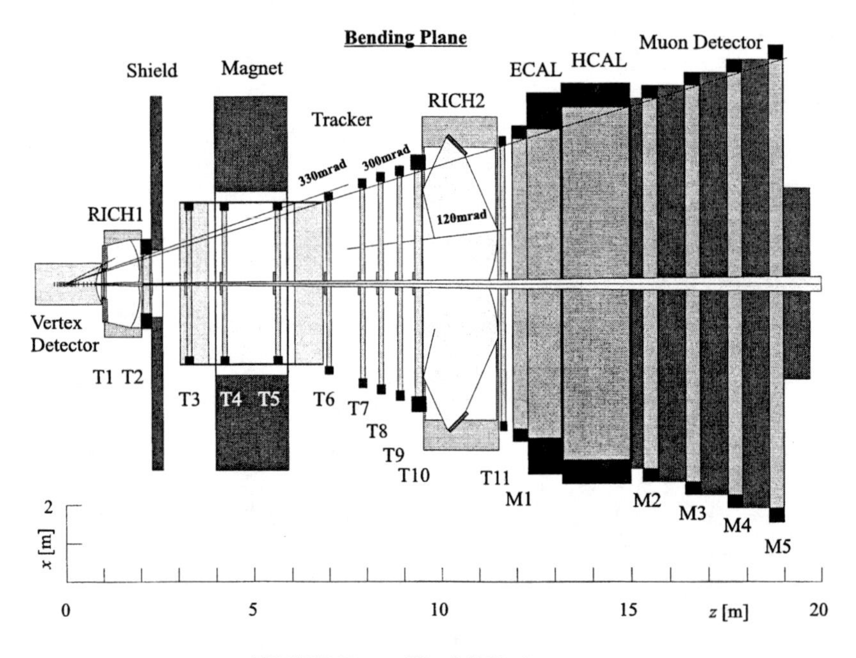

FIGURE 16. The LHCb detector.

relatively small area is relatively cheap. LHCb will run with a luminosity trimmed down to 2 $10^{32}\mathrm{cm}^{-2}\mathrm{s}^{-1}$ to reduce pile-up difficulties. The reduced luminosity will be compensated by an efficient inclusive trigger (using leptons, high p_{T} hadrons, vertex). Finally, the experiment will run for approximately 10 years for B physics, which will allow lengthy interaction with theorists on experimental results which could require trigger modification.

4 Initial state flavour tagging

The lepton tagging uses the correlation between the flavour of the $\mathrm{B_d^0}$ and the charge of the lepton from the decay of the second B hadron in the event. The mistag fraction of order 20% is due partly to the possible mixing of the second B hadron, partly to the cascade $\mathrm{b} \to \mathrm{c} \to \ell$ decays. The efficiency is 100% for the sample where the trigger lepton is used for tagging, but only ~5% for the self-triggering J/ψ where an additional lepton is searched for.

When no lepton is available, more inclusive methods are needed. Their high mistag fraction compared to the lepton tag is compensated by a much higher efficiency [32]. Kaon tagging (only possible at LHCb because good K/π separation is needed) uses the correlation between the flavour of the $\mathrm{B_d^0}$ and the sign of the charged kaon from the decay chain b→c→s of the other B. Mistags come from fragmentation kaon (limited by impact parameter requirement) $\mathrm{B_d^0}$ mixing, $\mathrm{B_s^0}$'s (at least two kaon in the final state), additional $\mathrm{s\bar{s}}$ popping in the decay chain.

The jet on the same side of the $\mathrm{B_d^0}$ decay can also be used in two ways. The momentum weighted charge of the jet retains some correlation to the charge of the primary b quark. The B-π algorithm uses the correlation between the flavour of the B meson and the charge of the adjacent hadron in the fragmentation chain, which is close in phase space to the B. Both algorithms have similar performance and are highly correlated. The jet containing the other B may also be used, provided it can be found unambiguously.

C A sample of possible measurements

The LHC reach in terms of B physics is now illustrated through four different channels, among many currently under study.

1 $\sin 2\beta$: the $\mathrm{J}/\psi\mathrm{K_S^0}$ channel

The decay $\mathrm{B_d^0} \to \mathrm{J}/\psi\mathrm{K_S^0}$ is generally considered to be the gold-plated channel for showing evidence for CP violation in the B system. It is theoretically clean, with penguin diagrams contributing by less than 1%. Experimentally, the decay has also a clean signature with the leptonic decay of the J/ψ.

For all experiments two samples are available: (i) self-triggered J/ψ are more abundant because they do not rely on a particular decay on the other side (ii) J/ψ triggered by the other B decay are less abundant but flavour tagging is easier (in the case of ATLAS and CMS, the triggering lepton can be used for tagging with 100% efficiency).

K^0_S are reconstructed from pairs of tracks forming a detached vertex reconstructed in the tracker. The J/ψ and the identified K^0_S [33] are then combined to form a B candidate, with good vertex probability (see Ref. [32] for details). The clean B^0_d mass plot is shown in Fig. 17 [34]. The dominant combinatorial background is the combination of a true J/ψ from B decay, with a K^0_S from fragmentation, for which extrapolation back onto the beam-line is sometimes not precise enough.

The statistical sensitivity on $\sin 2\beta$ for ATLAS, CMS, and LHCb for one year of running is of order 1%. Production and detection asymmetry, tagging performance will be measured on control channels such as $B^0_d \rightarrow J/\psi K^{*0}$ and $B^+_u \rightarrow J/\psi K^+$ where no asymmetry is expected. Systematical errors and theoretical errors from residual penguins contribution will be of same order.

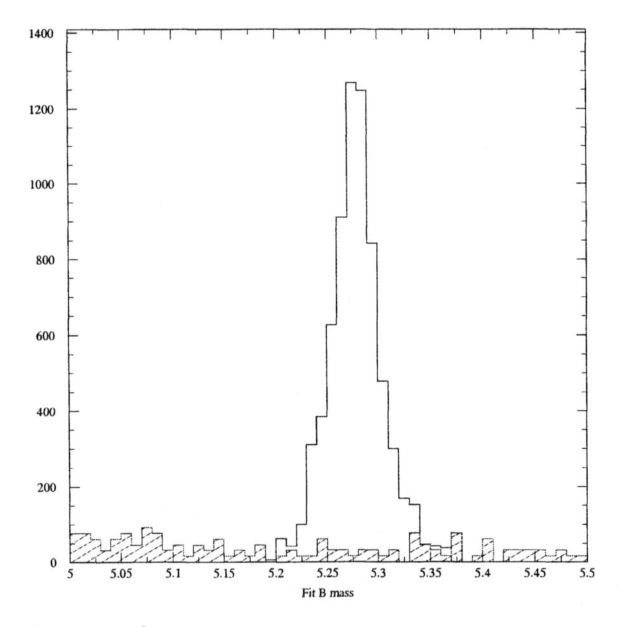

FIGURE 17. ATLAS $J/\psi K^0_S$ mass spectrum, with signal (white) and backgrounds (hatched).

2 $\sin 2\alpha$: *the* $\pi^+\pi^-$ *channel*

The decay asymmetry in the $B_d^0 \to \pi^+\pi^-$ channel is related to the angle α:

$$A(t) = \frac{\Gamma(B_d^0 \to \pi^+\pi^-) - \Gamma(\overline{B}_d^0 \to \pi^+\pi^-)}{\Gamma(B_d^0 \to \pi^+\pi^-) + \Gamma(\overline{B}_d^0 \to \pi^+\pi^-)} = a\cos(\Delta m_d t) + b\sin(\Delta m_d t)$$

with a (b) the direct (indirect) CP violation component,

$$a = -2\frac{A_P}{A_T}\sin\delta\sin\alpha$$

$$b = -\sin 2\alpha - 2\frac{A_P}{A_T}\cos\delta\cos 2\alpha\sin\alpha,$$

where A_P/A_T is the ratio of the penguin and tree amplitude, δ is the penguin and tree amplitude phase difference. It is assumed that $\alpha = \pi - (\beta + \gamma)$. Since there are three unknowns and only two observables (a and b), and since δ is not believed to be calculable, a theoretical estimation of A_P/A_T is needed. Much theoretical work is currently going on on this subject, to which measurements from the near-future B-factory will be valuable inputs. To set the scale, Fleischer and Mannel [35] estimate that, with the help of the measurement of the $B_u^+ \to \pi^+ K^0$ and $B_u^+ \to \pi^+\pi^0$ branching ratios, the theoretical uncertainty on the penguin contribution would yield an uncertainty on α of less than 3°. A small value of $\sin 2\alpha$ (as favoured today by Unitarity Triangle fit) could mean no observable CP asymmetry in this channel, but a tight bound on $\sin 2\alpha$ would in any case be a valuable constraint.

Events are reconstructed by combining opposite sign particles well separated from the beam-line, with a mass close to the B_d^0 mass. The first difficulty of this channel is the absence of constraint which causes a huge combinatorial background, of order 10^6 larger than the signal before vertexing. Combinatorial particles can be fought by stringent vertex requirements. The remaining background from the random crossing of unrelated track is reduced by fighting all sources of displaced tracks: poor quality tracks, secondaries, tracks from two unrelated B's or D's. Since only "crazy" tracks remain, the use of full simulation and reconstruction is mandatory to measure accurately the tracking resolution, including tails.

The second difficulty is the important physics background from other two-body B decays, which are poorly known today. The list of background branching ratios and yields is shown in Table 1. The $B_d^0 \to K^+\pi^-$ and $\pi^+\pi^-$ branching ratios are recent, still imprecise first measurements from CLEO [36]. The other branching ratios have been inferred from the first two with simple SU(3) symmetry, except the Λ_b decays which have been set to their experimental upper limits [37].

For the ATLAS and CMS experiments which do not have $p/K/\pi$ separation the most abundant background is the $B_d^0 \to K^+\pi^-$ decay, also because it peaks close to the signal (Fig. 18, [34]). In contrast, the LHCb signal is very clean (Fig. 19, [9]), demonstrating the strength of such a dedicated experiment.

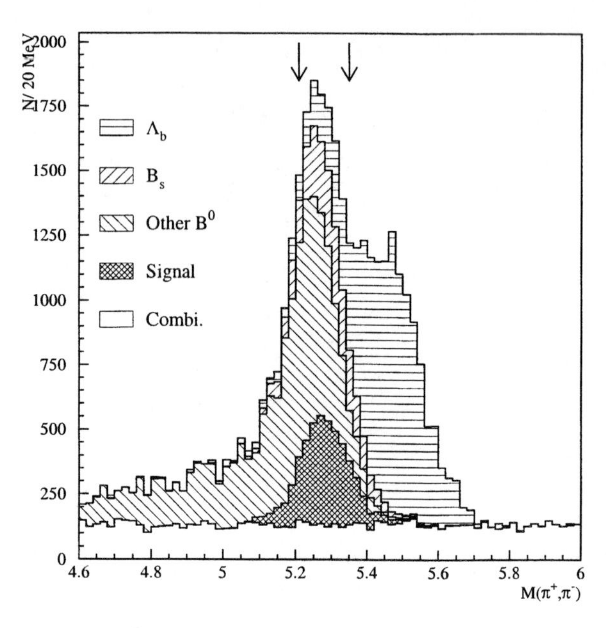

FIGURE 18. ATLAS $\pi^+\pi^-$ mass spectrum. The arrows indicate the signal window.

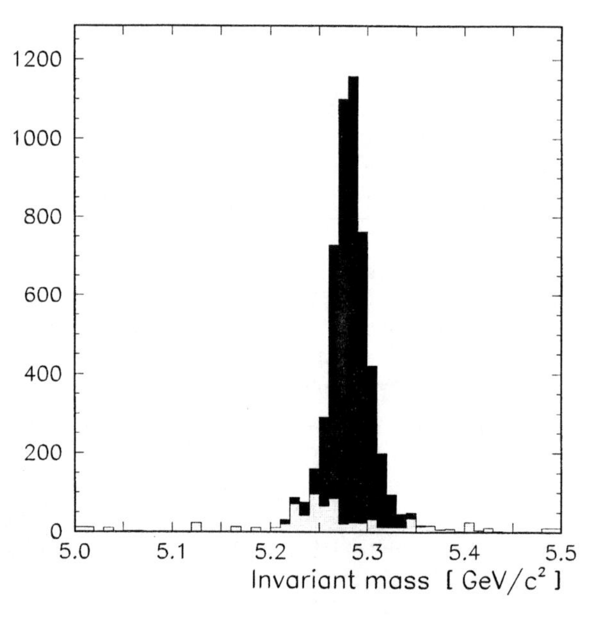

FIGURE 19. LHCb $\pi^+\pi^-$ mass spectrum with particle identification. Signal (black). Backgrounds, except combinatorial (grey).

TABLE 1. Branching ratios and yields in the $B_d^0 \to \pi^+\pi^-$ channel, for ATLAS with a $\pm 1\sigma$ mass cut.

Channel	BR's	Yield
$B_d^0 \to \pi^+\pi^-$	0.7×10^{-5}	2500
$B_d^0 \to K^+\pi^-$	1.5×10^{-5}	4600
$B \to \rho\pi^-$	0.7×10^{-5}	70
$B \to \pi^+\pi^-\pi$	5.0×10^{-5}	<20
$B_s^0 \to \pi^+K^-$	0.7×10^{-5}	600
$B_s^0 \to K^+K^-$	1.5×10^{-5}	1300
$\Lambda_b \to p\pi^-$	8.0×10^{-5}	600
$\Lambda_b \to pK^-$	8.0×10^{-5}	1000
Comb.	-	900
Total bkg	-	9000

By year 2005, near-future B factories will have measured precisely the branching ratios of the two-body decay channels. A 5% relative uncertainty on these branching ratios, will give a *relative* uncertainty on the asymmetry of less than 5%. More problematic for ATLAS and CMS is the possibility that these backgrounds exhibit some CP asymmetry themselves (in fact the asymmetry in $B_d^0 \to K^+\pi^-$ could help constraint the difficult angle γ [38]). A more sophisticated fitting method will have to be used, taking into account on a event by event basis the invariant mass and the lifetime. The Λ_b decays (which does not oscillate) and the B_s^0 decays (with rapid oscillation period ~ 0.5 ps) will then be distinguished from B_d^0 decays (with slow oscillation period ~ 14 ps). ATLAS and CMS will measure some combination of the CP asymmetries in these decays.

The statistical uncertainty on $\sin 2\alpha$ in the absence of penguin decays is 0.12 for ATLAS, 0.06 for LHCb. Allowing for penguin decays with $A_P/A_T = 0.2 \pm 0.02$ degrades the sensitivity by a factor 1.2 to 3, depending of the real values of α and δ. Systematic uncertainties arising from asymmetry in the detection efficiency and from the lepton tagging are expected to be of order 0.01, as in the $B_d^0 \to J/\psi K_S^0$ channel.

3 B_s^0 *mixing*

The interest of B_s^0 mixing (Δm_s) is twofold (i) as explained in Section III A 2 the ratio V_{ts}/V_{td} is related to the ratio $\Delta m_s/\Delta m_d$ with a 10% theoretical uncertainty (ii) as B_s^0 oscillation proceeds through a box diagram involving the top quark, new particles in the box could make it faster. B_s^0 mixing measurement relies on the observation of the time oscillations:

$$\frac{N_{unmix} - N_{mix}}{N_{unmix} + N_{mix}} = \cos(\Delta m_s t),$$

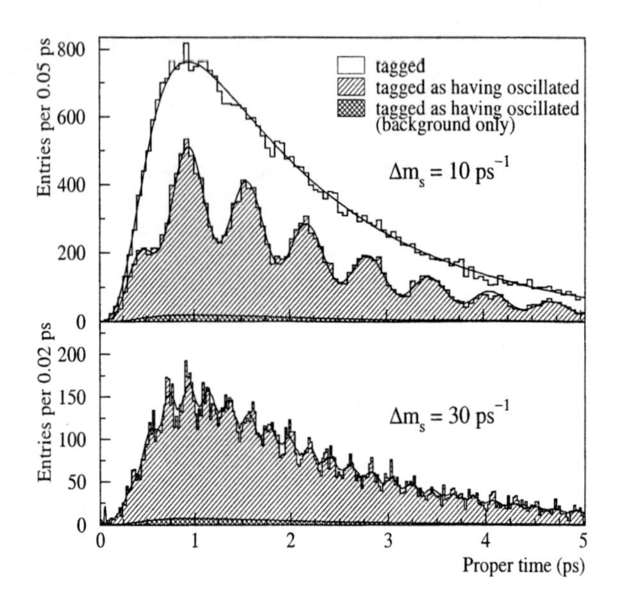

FIGURE 20. Proper-time distribution of $B_s^0 \to D_s^+ \pi^-$ events generated with two different values of Δm_s in LHCb. The curves display the result of the oscillation fit.

where N_{unmix} is the number of B_s^0 ($\overline{B}_s^0$) decaying as $B_s^0(\overline{B}_s^0)$, N_{mix} is the number of B_s^0 ($\overline{B}_s^0$) decaying as $\overline{B}_s^0(B_s^0)$, t is the proper-time. Measuring Δm_s requires reconstruction of a clean final state allowing the tagging of flavor at decay time (for example $B_s^0 \to D_s^+ \pi^-$), and initial state tagging (as in Section III B 4). The statistical significance $\mathcal{S}$ of a B_s^0 oscillation signal can approximately be written [39]:

$$\mathcal{S} = \sqrt{\frac{N}{2}} f_S (1 - 2W) e^{-\frac{(\Delta m_s \sigma_t)^2}{2}}$$

where N is the total number of tagged event, f_S is the fraction of signal, W the mistag fraction and σ_t the proper-time resolution, which is the key ingredient that sets the Δm_s value beyond which oscillations cannot be resolved anymore. Since $t = (lM)/(Pc)$, proper-time resolution means primarily decay-length resolution (good impact parameter resolution) and also good momentum resolution and high average momentum. These factors favour LHCb which proper-time resolution is 0.04 ps compared to 0.07 ps for ATLAS and CMS.

As a consequence ATLAS and CMS will be able to observe B_s^0 oscillations for Δm_s up to 25 ps^{-1} (covering Standard Model expectation), while LHCb reaches up to 48 ps^{-1}. Fig. 20 shows the observed oscillations in LHCb experiment [9]. Once oscillations are observed Δm_s can be measured with an accuracy better than 1 per mille.

4 *Rare decays*

The semi-muonic decays $B_d^0 \rightarrow \mu^+\mu^-\rho^0$, $B_d^0 \rightarrow \mu^+\mu^-K^{*0}$, are pure penguins decays, involving the matrix elements V_{td} and V_{ts}. In fact:

$$\frac{N(B_d^0 \rightarrow \rho^0\mu^+\mu^-)}{N(B_d^0 \rightarrow K^{*0}\mu^+\mu^-)} = k_d \left(\frac{V_{td}}{V_{ts}}\right)^2,$$

where k_d is a SU3 breaking term, known with 10% theoretical uncertainty. The corresponding branching ratios are of order 10^{-6} [28]. The equivalent decays to e^+e^- with similar branching ratios are not considered because of lower trigger efficiency.

The trigger and reconstruction of these events proceeds as for usual $B \rightarrow J/\psi(\rightarrow \mu^+\mu^-)X$ decays, except that precisely the J/ψ and $\psi(2S)$ regions should be avoided. After three years at low luminosity the ratio V_{td}/V_{ts} will be measured with 6% statistical error (for ATLAS or CMS) and 5% theoretical uncertainty [33,40]. This measurement is as precise as the value of the ratio obtained from $\Delta m_s/\Delta m_d$ but with different theoretical uncertainties. Also new physics (for example charged Higgs exchange [41]) would modify the Standard Model prediction for the mixing and these decays in a different way.

The purely muonic decays $B_d^0 \rightarrow \mu^+\mu^-$and $B_s^0 \rightarrow \mu^+\mu^-$ have expected branching ratios of order 10^{-10} and 3.10^{-9} respectively. These annihilation penguins are helicity suppressed by a factor m_ℓ^2/m_B^2. The corresponding $B_d^0 \rightarrow e^+e^-$ decay is quasi-null while the more abundant $B_d^0 \rightarrow \tau^+\tau^-$ decay suffers from the lack of a clear τ signature. Being very small in the Standard Model, these decays are rather sensitive to new physics.

The decay involving two muons with reconstructed mass consistent to the B_d^0 mass is a very clear signature for both trigger and reconstruction, even at high luminosity, which favour ATLAS and CMS w.r.t LHCb. After three years at low luminosity a first measurement with 30% precision can be obtained for $B_s^0 \rightarrow \mu^+\mu^-$, improving to 10% after three years at high luminosity [33]. Limits of a few 10^{-10} on the $B_d^0 \rightarrow \mu^+\mu^-$decay can be set.

IV CONCLUSION

The Standard Model Higgs boson, if it exists, and if it is not discovered elsewhere before, will be discovered at the LHC. The full mass range is covered by several channels in the two ATLAS and CMS experiments. The MSSM Higgs bosons also, if they exist, will be discovered at the LHC, and be distinguished from the Standard Model one in most regions of (M_A,tan β) plane.

CP violation will be very accurately measured by all three LHC experiments (ATLAS, CMS and LHCb), in particular in the $J/\psi K_S^0$ channel related to $\sin 2\beta$. The measurement of $\sin 2\alpha$ in the $B_d^0 \rightarrow \pi^+\pi^-$channel will be less precise because of penguin contamination, even more for ATLAS and CMS compared to LHCb in the absence of K/π separation. B_s^0 mixing can be measured at ATLAS and CMS if it is consistent with current Unitarity Triangle fit, by LHCb even if twice larger. Rare muonic and semi-muonic decays provide another handle to measure the raito V_{td}/V_{ts}, and are sensitive to new physics. The area of the allowed region of the apex of the Unitarity Triangle will decrease by more than two orders of magnitude, so that it may reveal new phenomenom beyond the Standard Model.

It is the strength of the LHC that high p_T and low p_T physics meet to test the Standard Model to unprecedented precision, and to have the potential to go far beyond.

ACKNOWLEDGEMENTS

I would like to thanks the organizers for inviting me to the VIIIth Mexican School on Particles and Fields in beautiful Oaxaca, where zapotecan hieroglyphs and string theory so nicely met, or was it the opposite ?

REFERENCES

1. Hollik, W., these proceedings.
2. Abe, F., *et al.* (CDF Collaboration), Phys. Rev. Lett.**74** 2626 (1995); Abachi, S., *et al.* (D0 collaboration), Phys. Rev. Lett. **74** 2632 (1995).
3. Peccei, R., these proceedings.
4. Pauss, F., and Dittmar, M. ETHZ-IPP PR-98-09, in Proceedings of ASI Summer School, Ste-Croix, 1998.
5. Dittmar, M., ETHZ-IPP PR-98-08, Lectures given at Summer School on Hidden Symmetries and Higgs Phenomena, Zuoz, 1998.
6. ALICE Technical Proposal CERN/LHCC/95-71 (1995).
7. ATLAS Technical Proposal, CERN/LHCC/94-43 (1994); ATLAS Detector and Physics Performance TDR, in preparation.
8. CMS Technical Proposal, CERN/LHCC 94-38 (1994).
9. LHCb Technical Proposal, CERN/LHCC 98-4 (1998).
10. CMS Collaboration, The Tracker Project Technical Design Report, CERN/LHCC 98-6 (1998).
11. McNamara, P., to appear in Proceedings of the International Conference on High Energy Physics, Vancouver 1998.
12. The LEP Collaboration CERN-EP/99-15.
13. Djouadi, A. Kalinowski, J., and Spira, M., Comput. Phys. Commun. **108**, 56 (1998).
14. ATLAS Collaboration, Calorimeter Performance Technical Design Report, CERN/LHCC 96-40.
15. M. Carena *et al.*, Phys. Lett. **B355** 209 (1995).
16. D. Cavalli *et al.*, ATLAS Communication, ATL-COM-PHYS-99-010.
17. Richter-Was, E., *et al.*, Int. J. Mod. Phys. **A 13** 137 (1998).
18. Georgi, H., and Glashow, S.L., HUTP-98-A048, Jul. 1998, hep-ph/9807399.

19. Rodriguez-Jauregui, E., these proceedings;
Mondragon, A., Rodriguez-Jauregui, E., IFUNAM-FT98-12, (1999), submitted to Phys. Rev. **D**.
20. Kobayashi, M., and Maskawa, T., Prog. Theor. Phys. **49**, 652 (1973).
21. Wolfenstein, L., Phys. Rev. Lett. **51** 1945. (1983)
22. Caso, C. *et al.*(Particle Data Group), Eur. Phys. J. **C3** 1 (1998).
23. Paganini, P., *et al.* Phys. Scripta **58** 556 (1998),
Parodi, F., *et al.* LAL-98-49, hep-ph/9802289.
24. Mele, S., CERN-EP-98-133, Aug. 1998, submitted to Phys. Lett. **B**.
25. Ali, A., and London, D., Z. Phys. **C65** 431 (1995).
26. Parodi, F., to appear in Proceedings of the International Conference on High Energy Physics, Vancouver 1998.
27. Gupta, R., these proceedings.
28. Buras, A., and Fleischer, R., TUM-HEP-275-97, Apr 1997, In *Buras, A.J. (ed.), Lindner, M. (ed.): Heavy flavours II 65-238. World Scientific (1997).
29. The BaBar Physics Book: Physics at an Asymmetric B Factory, BaBar Collaboration (Harrison, P.F., (ed.) et al.), SLAC-R-0504, Oct 1998.
30. Belle Technical Design Report, KEK-Report 95-1, April 1995.
31. Butler, J. for the BTeV collaboration, to appear in Proceedings of the International Conference on Hyperons, Charm and Beauty Hadrons, Genova, Italy, 1998.
32. Tartarelli, G.F., Nucl. Instr. and Meth. **A 408** (1998) 110.
33. Sherwood, P. for the ATLAS collaboration, to appear in Proceedings of the International Conference on Hyperons, Charm and Beauty Hadrons, Genova, Italy, 1998.
34. Rousseau, D. for the ATLAS collaboration, to appear in Proceedings of the International Conference on Hyperons, Charm and Beauty Hadrons, Genova, Italy, 1998.
35. Fleischer, R. and Mannel, T., Phys. Lett. **B397** 269 (1997).
36. Godang, R., *et al.* (CLEO Collaboration), Phys. Rev. Lett. **80** 3456 (1998).
37. Buskulic, D., *et al.* (ALEPH Collaboration), Phys. Lett. **B384** 471 (1996).
38. Fleischer, R., CERN-TH/98-60, to appear in Eur. Phys. J. **C**.
39. Moser, H.-G., and Roussarie, A. , Nucl. Instr. and Methods, **A 384** 491 (1997).
40. Stefanescu, J., for the CMS collaboration, to appear in Proceedings of the International Conference on Hyperons, Charm and Beauty Hadrons, Genova, Italy, 1998
41. Hewett, J.L., Wells, J.D., Phys. Rev. **D55** 5549 (1997).

Notes on black holes and three dimensional gravity

Máximo Bañados

Departamento de Física Teórica, Universidad de Zaragoza, Ciudad Universitaria 50009, Zaragoza, Spain.

Abstract. In these notes we review some relevant results on 2+1 quantum gravity. These include the Chern-Simons formulation and its affine Kac-Moody symmetry, the asymptotic algebra of Brown and Henneaux, and the statistical mechanics description of 2+1 black holes. A brief introduction to the classical and semiclassical aspects of black holes is also included. The level of the notes is basic assuming only some knowledge on Statistical Mechanics, General Relativity and Yang-Mills theory.

I INTRODUCTION

During the last three years we have witnessed a rapid progress in the string theory description of general relativity. Successfull computations of black hole entropy for extremal and near extremal solutions [1,2] have made it clear that the string theory degrees of freedom describes the expected semiclassical behaviour of general relativity. This is in sharp constrast with the more standard approach to quantum gravity either based on the path integral approach or the Wheeler-de Witt equation which has provided little information about the fundamental degrees of freedom giving rise to the Bekenstein-Hawking entropy. In the Loop representation approach to quantum gravity, a computation of the black hole entropy has been proposed [3,4]. However, in this formulation it is still obscure how to introduce dynamics, and only the kinematics of spin networks is under control.

In this contribution we shall consider neither string theory nor loop quantum gravity. Instead, we work in the very simple setting of three-dimensional quantum gravity whose Lagrangian describes a well-defined quantum field theory [5,6]. As motivations to study three-dimensional gravity, let us mention the following aspects of it. (i) It is a mathematically simple theory which combines three important branches of physics: General Relativity, Yang-Mills theory (with a Chern-Simons action), and two-dimensional Conformal Field Theory. (ii) The mathematical tools are surprinsingly similar to those used in string theory, with a centrally extended Virasoro algebra [7] as one of its main ingredients. (iii) The space of solutions

CP490, *Particles and Fields: Eighth Mexican School,* edited by J. C. D'Olivo et al.

contains particle solutions [8] and black holes [9], thus making it interesting from the dynamical point of view.

In these notes, we shall mainly be interested in quantum black holes in three dimensions. Our goal is to give, in a somehow self-contained way, a derivation of Strominger's [10] proposal for the statistical mechanical origin of the three-dimensional black hole entropy. We refer the reader to [11–14] for the stringy aspects of Strominger's result. See also [15] for a recent review. We shall concentrate here on the gravitational aspects. For a detailed and complete treatment of three dimensional gravity we refer to the recent book by Carlip [16].

In Sec. II we shall briefly review, at the most basic level, some of the main properties of the Schwarzschild solution, as well as the three-dimensional black hole [9]. In Sec. III we review the Chern-Simons formulation of three-dimensional gravity. Particular empahsis is given to the absence of bulk degrees of freedom, and a quick derivation of the affine Kac-Moody algebra is presented. Finally, in Sec. IV we derive the Brown-Henneaux conformal algebra, and its statistical mechanical [10,45] implications.

II CLASSICAL AND SEMICLASSICAL BLACK HOLES

A The black hole spacetime

The Schwarzschild metric ($r > r_0$),

$$ds^2 = -(1 - 2M/r)dt^2 + (1 - 2M/r)^{-1}dr^2 + r^2 d\Omega^2, \tag{1}$$

is an exact solution of the Einstein vacuum equations

$$G_{\mu\nu} = R_{\mu\nu} - \frac{1}{2}g_{\mu\nu}R = 0, \tag{2}$$

representing the geometry outside a collapsing star of mass M and radius r_0. One of the most surprising predictions of General Relativity, which caused much confusion in the past, is the appearance of a singularity in the metric for the particular value of r:

$$r =: r_+ = 2M \qquad \text{(Event Horizon)}. \tag{3}$$

If the radius of the star r_0 is less that r_+ then the solution (1), which is valid for $r > r_0$, has a singularity at $r = r_+$. Furthermore, in the region $0 < r < r_+$ where the metric is again regular, r is a timelike coordinate while t is spacelike. Finally, at $r = 0$ the curvature blows up making gravitational forces divergent there. This means, in particular, that no observer can reach the singularity without being destroyed. The possibility of making experiments near the singularity is prevented by another fact: any observer that crosses the event horizon $r = r_+$ will never come back, at least not according to the classical Einstein equations. We shall prove this below. Quantum mechanically, particles can tunnel out of the black hole and escape to infinity. This is Hawking's famous discovery of black hole radiation [26]. However, according to Hawking's description there is no correlation between the particles that fall into the black hole with the ones that escape. This point is actually a matter of discussion and there is no agreement yet. We will not have time here to describe in any detail this very interesting work. We refer the interested reader, for example, to [24] for a review with an extensive list of references.

Let us now briefly show how to deal with the $r = r_+$ singularity in (1). This will allow us to see why observers cannot travel back once they have crossed the horizon. We shall also infer the value of Hawking's temperature via a geometrical argument.

The analysis that follows does not depend on the details of the Schwarzschild solution but only on some general properties of black holes. We consider general metrics in d dimensions of the form,

$$ds^2 = -f(r)dt^2 + f^{-1}(r)dr^2 + ds^2_{d-2}, \tag{4}$$

where ds^2_{d-2} represents the metric of a S_{d-2} sphere, or some other compact or non-compact surface. The function $f(r)$ satisfies the following two properties: (i) There exists a value of r denoted as r_+ such that $f(r_+) = 0$; (ii) The derivative of f at r_+ is different from zero,

$$\alpha \equiv \frac{1}{2}\left.\frac{df(r)}{dr}\right|_{r=r_+}, \qquad \alpha \neq 0. \tag{5}$$

Most known (non-extreme!) black holes have a metric of this form, or at least there is a plane on which the metric near $r = r_+$ looks like the first two terms in (4). The extreme black holes do not fall into the above class of metrics because the function $f(r)$ has a second order zero and thus $\alpha = 0$. These black holes play an important role in string theory because they are related to BPS states.

For the Schwarzschild black hole, the function f is given by $f(r) = 1 - 2M/r$ and $f'(r_+) = 1/2M$. This means that $\alpha = 1/4M$ which is indeed different from zero. Other examples are: The Reissner-Nordstrom black hole with $f(r) = 1 - 2M/r + e^2/r^2$ and e is the electric charge. In this case, α is different from zero provided $M \neq e$; The 2+1 black hole (to be studied in detail in the next section) with $f(r) = -M + r^2/l^2$ and l is related to a cosmological constant; The d-dimensional Schwarzschild solution with $f(r) = 1 - 2M/r^{d-3}$; plus all the (non-extreme) stringy black holes [27], as well as other higher dimensional situations [28]. Students are encouraged to compute the value of α for each of these black holes, as we shall see soon, this number is essentially Hawking's temperature for each of these objects.

The metric (4) is singular at the event horizon, just as the Schwarzschild metric is. To cure this singularity we introduce the following new set of coordinates. We change $\{r, t\}$ to $\{u, v\}$ according to,

$$\begin{aligned} u &= g(r)\cosh \alpha t, \\ v &= g(r)\sinh \alpha t, \end{aligned} \tag{6}$$

where the function $g(r)$ is defined by,

$$g(r) = \exp\left(\alpha \int^r \frac{dr'}{f(r')}\right). \tag{7}$$

This change of coordinates has the following properties. The event horizon $r = r_+$ is mapped into the lines $u = \pm v$. The metric in terms of u, v reads,

$$ds^2 = \Omega^2(r)(-dv^2 + du^2) + r^2 d\Omega^2, \tag{8}$$

where the function,

$$\Omega^2(r) = \frac{f(r)}{\alpha^2 g^2(r)}, \tag{9}$$

is regular at $r = r_+$. The regularity of Ω^2 holds provided $f(r)$ has a simple pole at $r = r_+$. It is easy to see using L'Hopital rule that the zero in $f(r)$ is cancelled by $g^2(r)$ provided α is chosen as in (5). Note that the above coordinate change does not depend on the details of the function $f(r)$, provided it has a single pole at $r = r_+$. Of course, our formulae for the conformal factor and change of coordinates reproduce the usual expressions when applied to particular situations like the Schwarzschild black hole (see [25])..

The coordinates u and v are called Kruskal coordinates and their range is $-\infty < u, v < \infty$. These coordinates can be compactified (see [25] for more details on this points) and led to the Penrose diagram shown in Fig. 1[1]. Region **I** is the black hole exterior $r > r_+$ and region **II** its interior ($r < r_+$). It should be clear from the figure that an observer situated in region **II** cannot go back to region **I** because he or she would need to travel faster than light. The fate of any future-directed (timelike) observer is to hit the singularity.

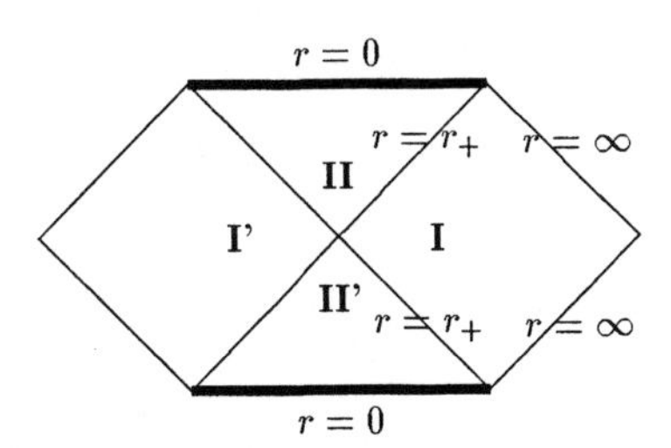

FIGURE 1. Penrose diagram for a Schwarzschild black hole.

---

1) The Penrose diagram shown here assumes asymptotic flatness. This means that the function $f(r)$ satisfies $f(r \to \infty) \to 1$. This is not the case for asymptotically anti-de Sitter black holes for which $f \to r^2$. The Penrose diagram in that case can still be drawn and differs only in the asymptotic structure, not the properties surrounding the horizon. See second Ref. in [9].

B Semiclassical black holes. The Gibbons-Hawking approximation

The combination of Euclidean field theory together with the coordinate change (6) suggest in a very direct way that black holes should have a non-zero temperature. The Euclidean formalism (sometimes called the Euclidean sector) is obtained by setting $\tau = it$, and the metric (1) becomes Euclidean. Consider again the change of coordinates (6) in the Euclidean formalism. The hyperbolic functions will be replaced by their trigonometrical versions and it is clear that the Euclidean time variable needs to be an angle $0 \leq \alpha\tau < 2\pi$. In the Euclidean sector, and near the horizon, the change (6) is nothing but the relation between polar and cartesian coodinates in $\Re^2$. In fact, the topology of the Euclidean Schwarzschild black hole is $\Re^2 \times S_2$ where the origin of $\Re^2$ is situated at the horizon $r = r_+$. The Euclidean sector does not see the inner region of the black hole $r < r_+$.

Following the usual practice of Euclidean field theory we define the inverse temperature as the Euclidean time period ($\hbar = 1$),

$$\beta = \frac{2\pi}{\alpha}. \tag{10}$$

So far this is only a mathematical trick with no real physics meaning. However, it turns out that the temperature $T = 1/\beta$ defined in (10) coincides exactly with Hawking's evaporation temperature. For the Schwarzschild black hole, we recall that $\alpha = 1/4M$, this yields the famous Hawking result,

$$T_H = \frac{1}{\beta_H} = \frac{1}{8\pi M}. \tag{11}$$

Now, integrating the first law $dM = TdS$ we find the Bekenstein-Hawking formula for black hole entropy,

$$S = \frac{A}{4}, \tag{12}$$

where $A = 4\pi r_+^2$ is the area of the event horizon ($r_+ = 2M$).

For our purposes, this "derivation" of the black hole temperature and entropy has an important meaning: geometry knows that black holes radiates. In other words, the very deep origin of Hawking's process is not contained only on the matter fields surrounding a black hole but rather on the gravitational (perhaps string) degrees of freedom. This point of view is further supported by the Gibbons-Hawking calculation of the Schwarzschild black hole partition function which we now describe.

Let us briefly review here the results presented in [17] in the simplest case of a non-rotating black hole. Our main tool will be again the analogy between Euclidean field theory and statistical mechanics.

Consider the functional integral,

$$Z[h] = \int Dg\, e^{-I[g,h]}, \tag{13}$$

where,

$$I[g,h] = -\frac{1}{16\pi G}\left(\int_M \sqrt{g}R + 2\int_{\partial M}\sqrt{h}K\right). \tag{14}$$

is the Euclidean gravitational action appropriated to fix the metric at the boundary. Here h_{ij} is the 3-metric induced on ∂M. The boundary term is added to the action to ensure that I has an extremum when h is fixed. Dg denotes the sum, modulo diffeomorphisms, over all metrics with h_{ij} fixed. As it is well known, the formula (13) is purely formal and cannot be given a precise mathematical meaning. This is because gravity is not renormalizable and the perturbation expansion for (13) is not well-defined. To make things worst, the action (14) is not bounded from below/above, not even in the Euclidean formulation. It is possible to find sequences of Euclidean manifolds M_i for which the value of the action (14) goes to minus/plus infinity [18].

Although (13) cannot be computed in general, its saddle point approximation around some classical solutions gives interesting results. Incidentally, we mention here that the evaluation of the action I on classical solutions has become crucial in the recently discovered adS/CFT correspondence [19–21]. The first example of an evaluation of (13) was performed by Gibbons and Hawking [17] who considered the Euclidean Schwarzschild black hole (1) with mass M. The mass M and Euclidean period β are related by (11) in order to avoid singularities (sources) in the Euclidean metric. The value [2] of Z in the saddle point (1) is

$$Z[\beta] \sim e^{-\beta^2/16\pi}. \tag{15}$$

This result is quite remarkable. The thermodynamical formula for the average energy $M = -\partial \log Z/\partial\beta$ reproduces (11) and confirms that β is the inverse temperature of the black hole. In the same way, the average entropy $S = \ln Z - \beta\partial_\beta \ln Z$ reproduces (12). This result confirms once again that the black hole thermal properties are present in a pure quantum theory of gravity, and not only in the interaction of a classical background metric with quantized fields.

Hawking's discovery of black hole evaporation is one of the most important results in the theory of general relativity and quantum mechanics. We refer the reader, for example, to the classic books by Birrell and Davis [29] and Wald [30] for a detailed discussion on quantum black holes, and in particular, quantum field theory on curved spacetimes. These books were written before black holes became important in string theory. See [31] for a review on the string theory approach to black holes.

We shall now depart from the Schwarzschild four dimensional black hole and go down to three dimensions where a black hole solution exists [9] having many of the features of the Schwarzschild solution, but it is far simpler mathematically.

[2)] Actually, the value of I diverges and needs to be regularized. See [17] for details and [33,34] for an evaluation of I using Hamiltonian methods on which the regularization is automatic.

Consider the action for three-dimensional gravity with a negative cosmological constant $\Lambda = -2/l^2$,

$$I = \frac{1}{16\pi G} \int \sqrt{-g} \left(R + \frac{2}{l^2} \right) dx^3. \tag{16}$$

In three dimensions it is convenient to keep the fundamental constants because, due to the cosmological constant, there are two fundamental length parameters: Plank's length $l_p = \hbar G$ and the cosmological radius l.

The equations of motion derived from this action are solved by the (non-rotating) three-dimensional black hole [9]

$$ds^2 = -\left(-8MG + \frac{r^2}{l^2} \right) dt^2 + \left(-8MG + \frac{r^2}{l^2} \right)^{-1} dr^2 + r^2 d\varphi^2. \tag{17}$$

Angular momentum as well as electric charge can be added easily, see [9].

As for the Schwarzschild metric, we can go to the Euclidean sector and discover that the time coordinate is periodic. The associated temperature is,

$$T_3 = \frac{\sqrt{M}}{2\pi l^2}, \tag{18}$$

and the entropy is again given by (12), but now $A = 2\pi r_+$ is the perimeter length of the horizon.

A word of caution is necessary here. Contrary to the Schwarzschild case, the metric (17) is not asymptotically flat. This means that the Euclidean period cannot be defined as the proper length of the time coordinate at infinity. Note that the limit $r \to \infty$ of the Euclidean Schwarzschild metric yields a well defined metric at the "boundary" (infinity is not really a boundary) with the topology $S_1 \times S_2$. S_1 corresponds to the periodic time coordinate, while S_2 to the angular sphere. The limit $r \to \infty$ of (17) is not well defined. At infinity, one can only define a conformal class of metrics . This is a three-dimensional example of the adS/CFT correspondence [19–21] first studied in [7]. A more rigorous definition for the temperature can be given by noticing that the topology of (17) in the Euclidean sector is a solid torus. The temperature is related to the complex structure of the torus [37,38] by,

$$\tau = \frac{\beta}{2\pi} \left(\Omega + \frac{i}{l} \right), \tag{19}$$

where β is the period of the Euclidean time coordinate, and Ω is the angular velocity. In the non-rotating case, $\Omega = 0$.

As before, one can write down the three-dimensional partition function in the saddle point approximation provided by the solution (17). This yields [9] (see also [52] for a Lagrangian approach),

$$Z_3 \sim e^{\pi^2 l^2/(2G\beta)}, \tag{20}$$

and it is direct to check that the thermodynamical formulae for the average energy and entropy is consistent with (18).

Our main motivation to study three dimensional quantum gravity is to try to give a precise meaning to the formula (13) in three dimensions. In other words, we hope that in three dimensions (13) could be well-defined mathematically, and provide the semiclassical limit (20). If this is true, then one should be able to extract from the exact formula for Z which are the degrees of freedom giving rise to the black hole entropy.

The main mathematical device that we shall use is the Chern-Simons formulation of three dimensional gravity [5,6]. This formulation makes manifest the fact that three dimensional gravity does not have any bulk degrees of freedom and it is renormalizable. Still, this does not mean that the problem is trivial because the relevant group, see below, is $SL(2,C)$ which is not compact. The evaluation of the partition function for the black hole problem using the Chern-Simons formulation was initiated by Carlip [44]. Some clarification on the boundary conditions and the role of ensembles can be found in [37]. A string theory approach can be found in [38]. Further developments on the modular properties of the partition function in three dimensions have recently appeared in [39].

III 2+1 GRAVITY AS A CHERN-SIMONS THEORY.

A First order form of the Euclidean action

We start with the (Palatini) Euclidean action for three dimensional gravity with a negative cosmological constant $\Lambda = -2/l^2$,

$$I[g_{\mu\nu}, \Gamma^{\rho}_{\mu\nu}] = \frac{1}{16\pi G}\int \sqrt{g}\left(g^{\mu\nu}R_{\mu\nu}(\Gamma) + \frac{2}{l^2}\right) \tag{21}$$

The discovery of Achúcarro and Townsend is that in three dimensions one can replace the metric by two Yang-Mills fields such that both the structure of the action and equations of motion simplifies enormously.

This is achieved in various steps. First, we use the Palatini formalism. The idea of this formalism is to note that the Ricci tensor $R_{\mu\nu}$ depends on the metric only through the Christoffel symbol $\Gamma^{\rho}_{\mu\nu}$. Then, it follows that if one treats $g_{\mu\nu}$ and $\Gamma^{\rho}_{\mu\nu}$ as independent variables in the action (21), the equations of motion yield the expected relation $g_{\mu\nu;\lambda} = 0$ between the metric and connection. Next, we make a change of coordinates from the coordinate basis ∂_μ to orthonormal coordinates on which the metric is flat. The matrix that makes this change is called the triad and is defined by the formula,

$$g_{\mu\nu} = e^a_{\ \mu}\,\eta_{ab}\,e^b_{\ \nu} \tag{22}$$

Clearly, $e^a_{\ \mu}$ is defined only up to a (local) Lorentz rotation because if Λ is an element of the Lorentz group then, by definition, $\Lambda\eta\Lambda^{-1} = \eta$. Equation (22) is nothing but the transformation of a tensor under a change of coordinates described by the matrix e^a_μ. We also need to transform the Christoffel symbol which is not a tensor but we know its transformation law under e^a_μ,

$$\Gamma^\sigma_{\mu\nu} = e^\sigma_a \omega^a{}_{b\nu} e^b_\mu + e^\sigma_a e^a_{\mu,\nu} \tag{23}$$

where w^{ab}_μ, known for historical reasons as the spin connection, is the new 'Christoffel symbol' in the new coordinates. Eq. (23) is often written in the literature as $e^a_{\mu;\nu} = 0$ where the semicolon denoted full covariant derivative, or as $D_\mu e^a_\nu = \Gamma^\rho_{\nu\mu} e^a_\rho$ where D_μ denotes covariant derivative in the spin connection. These formulae are, of course, equivalent to (23). Note that in (23) we have only transform two indices. The reason is that the Christoffel symbol is a 1-form connection for the group $GL(4,\Re)$, $\Gamma^\mu{}_\nu = \Gamma^\mu{}_{\nu\rho} dx^\rho$. The next object we would like to write in the new coordinates is the curvature tensor. The curvature tensor is a tensorial 2-form, for that reason we only transform two of its four indices as,

$$R^{\lambda\sigma}{}_{\mu\nu} = e^\lambda_a e^\sigma_b R^{ab}{}_{\mu\nu} \tag{24}$$

where $R^{ab} = dw^{ab} + w^a{}_c {\scriptstyle\wedge} w^{cb}$. With formulae (22), (23) and (24) at hand we can prove the identity

$$\int \epsilon_{abc} R^{ab}{\scriptstyle\wedge} e^c = \int \sqrt{g} R \tag{25}$$

The relevant steps are (we go from left to right),

$$\begin{aligned}\int \epsilon_{abc} R^{ab}{\scriptstyle\wedge} e^c &= \int \epsilon^{\mu\nu\lambda} \epsilon_{abc} \left(\frac{1}{2} R^{ab}{}_{\mu\nu}\right) e^c_\lambda \\ &= \frac{1}{2} \int \epsilon^{\mu\nu\lambda} \epsilon_{abc} R^{\alpha\beta}{}_{\mu\nu} e^a_\alpha e^b_\beta e^c_\lambda \\ &= \frac{1}{2} \int \epsilon^{\mu\nu\lambda} \epsilon_{\alpha\beta\lambda}\, e\, R^{\alpha\beta}{}_{\mu\nu}.\end{aligned} \tag{26}$$

In the second line we have used (24), and in the third line $\epsilon_{abc} e^a_\alpha e^b_\beta e^c_\lambda = e \epsilon_{\alpha\beta\lambda}$ with e equal to the determinant of e^a_μ, and $\epsilon^{\mu\nu\lambda}\epsilon_{\alpha\beta\lambda} = \delta^{[\mu\nu]}_{[\alpha\beta]}$ (recall that we are working in the Euclidean formalism). It should be clear that the last line in (26) is equal to the right hand side of (25).

Collecting all formulae together we arrive at the new action for three-dimensional gravity,

$$I[e^a, w^{ab}] = \frac{1}{16\pi G} \int \epsilon_{abc} \left(R^{ab} + \frac{1}{3l^2} e^a {\scriptstyle\wedge} e^b \right) {\scriptstyle\wedge} e^c. \tag{27}$$

The action $I[e,w]$ is equal to the action $I[g,\Gamma]$ shown in (21). Besides notation issues there is a conceptual consequence. The action (27) is perfectly well defined even is the metric is degenerate. In this sense, the triad formulation provides a generalization for the Einstein-Hilbert action.

The last step before we can write the Chern-Simons action is to define the new spin connection[3] w^a and curvature R^a,

---

3) With this definition of w^a the torsion becomes $T^a = de^a + \epsilon^a{}_{bc} w^b \wedge e^c$.

$$w^a = -(1/2)\epsilon^a{}_{bc}w^{bc}, \qquad R^a = -(1/2)\epsilon^a{}_{bc}R^{bc} \tag{28}$$

with $R^a = dw^a + \frac{1}{2}\epsilon^a{}_{bc}w^b{\wedge}w^c$.

We are now ready to make the connection with Chern-Simons theory. Let x be a complex number and let A^a and $\bar{A}^a$ to fields related to e and w by,

$$A^a = w^a + xe^a, \qquad \bar{A}^a = w^a - xe^a. \tag{29}$$

The relation between Chern-Simons theory and three dimensional general relativity follows from the equality:

$$\begin{aligned} 2e_aR^a + \frac{x^2}{3}\epsilon_{abc}e^ae^be^c = & \frac{1}{2x}(A_adA^a + \frac{1}{3}\epsilon_{abc}A^aA^bA^c) \\ & -\frac{1}{2x}(\bar{A}_ad\bar{A}^a + \frac{1}{3}\epsilon_{abc}\bar{A}^a\bar{A}^b\bar{A}^c) + dB. \end{aligned} \tag{30}$$

This relation is true regardless the signature of spacetime or sign of the cosmological constant. Just plug (29) into the right hand side of (30) and obtain the left hand side. (dB is a total derivative term.)

Depending on the signature of spacetime and cosmological constant x need to be complex or real. We shall be interested here in the Euclidean gravity with a negative cosmological constant. In the case, x is purely imaginary.

B Chern-Simons action

¿From the equality (30) with $x = i/l$ it follows that the Einstein-Hilbert action (21) for Euclidean three dimensional gravity can be written in the form,

$$I[g,\Gamma] = iI[A] - iI[\bar{A}], \tag{31}$$

where $I[A]$ is the Chern-Simons action,

$$I[A] = \frac{k}{4\pi}\int \mathrm{Tr}(AdA + \frac{2}{3}A^3), \tag{32}$$

at level[4],

$$k = -\frac{l}{4G}. \tag{33}$$

In (31) we have defined

$$A = A^aJ_a, \quad \bar{A} = \bar{A}^aJ_a. \tag{34}$$

[4]) The sign of k depends on the identity $\sqrt{g} = \pm e$ where e is the determinant of the triad. This sign determines the relative orientation of the coordinate and orthonormal basis. We have chosen here the plus sign which means that we work with $e > 0$.

where the $SU(2)$ generators are given by,

$$J_1 = \frac{i}{2}\begin{pmatrix} 0 & 1 \\ 1 & 0 \end{pmatrix}, \quad J_2 = \frac{1}{2}\begin{pmatrix} 0 & -1 \\ 1 & 0 \end{pmatrix}, \quad {}_3 = \frac{i}{2}\begin{pmatrix} 1 & 0 \\ 0 & -1 \end{pmatrix}, \tag{35}$$

and satisfy $[J_a, J_b] = \epsilon_{ab}{}^c J_c$, $\mathrm{Tr}(J_a J_b) = -(1/2)\delta_{ab}$. Note that $\bar{A}$ is not the complex conjugate of A.

Let F^a and $\bar{F}^a$ the Yang-Mills curvatures associated to A^a and $\bar{A}^a$. From the point of view of the equations of motion, the relation between Chern-Simons theory and general relativity is contained in the fact that the Chern-Simons equations,

$$F^a = 0, \qquad \bar{F}^a = 0, \tag{36}$$

are equivalent to the three-dimensional Einstein equations. Thus, studying the space of solutions of (36) we are studying general relativity.

The 1-form A^a is a $SL(2, C)$ Yang-Mills gauge field because in (29) x is imaginary. For Minkowskian gravity $x = 1/l$ is real and the relevant group is $SO(2,1) \times SO(2,1)$.

C Chern-Simons dynamics. Kac-Moody symmetry

Once we have proved the equality between the Chern-Simons and gravitational actions we can forget about metrics and work with Yang-Mills fields which are much more tractable. We should keep in mind however that the Chern-Simons action is a generalization to general relativity in the sense that it can accept degenerate metrics.

The classical dynamics of Chern-Simons theory is simple to analyse. First, we note that the Chern-Simons action is already in Hamiltonian form. In the 2+1 decomposition of the gauge field $A^a = A^a_0 dt + A^a_i dx^i$, the Chern-Simons action reads,

$$I[A_i, A_0] = \frac{k}{8\pi}\int dt \int_\Sigma \epsilon^{ij}\delta_{ab}(A^a_i \dot{A}^b_i - A^a_0 F^b_{ij}). \tag{37}$$

The coordinates x^i are local coordinates on the spatial surface denoted by Σ. This action has $2N$ dynamical fields A^a_i ($a = 1, ..., N$; $i = 1, 2$) and N Lagrange multipliers A^a_0. The dynamical fields satisfy the basic equal-time Poisson bracket algebra,

$$\{A^a_i(x), A^b_j(y)\} = \frac{4\pi}{k}\epsilon_{ij}\delta^{ab}\delta^2(x, y). \tag{38}$$

The equation of motion with respect to A_0 leads to the constraint equation,

$$\frac{k}{8\pi}\epsilon^{ij} F^a_{ij} \approx 0. \tag{39}$$

which (after properly taken into account boundary condition and boundary terms if the spatial surface has a boundary) generates the gauge transformations $\delta A_i^a = D_i\lambda^a$ in the Poisson bracket (38).

Because the equations of motion of Chern-Simons theory are $F = 0$ we know that there are no local degrees of freedom in this theory. It is instructive however to count the number of degrees of freedom per point using the Dirac formalism. We have $2N$ dynamical variables subject to N constrains. These constraints are of first class and generate the N local gauge transformations. Thus the total number of local degrees of freedom is indeed zero. This does not mean that the action is trivial. There are an infinite number of degrees of freedom associated to the breakdown of gauge invariance at the boundary, plus a finite number associated to holonomies along non-contractible loops. Here we shall not consider the holonomies. We refer the reader to [6,22].

The boundary degrees of freedom in Chern-Simons theory can be understood in many different ways. Their existence was first indicated in [49], and Carlip [43,44] first pointed out that they may be responsible for the three-dimensional black hole entropy.

These degrees of freedom are somehow a matter of interpretation rather than a specific calculation. The point is that, at the boundary, is incorrect to identify configurations that differ by a gauge transformation. As discussed in [43] this follows from boundary terms arising in the transformation of the Chern-Simons action under gauge transformations. Alternatively, following [32], one can see that at the boundary the transformations $\delta A_i^a = D_i\lambda^a$ are not generated by constraints and therefore they do not represent proper gauge transformations. In summary, the symmetry is still there but its interpretation is different.

This point can be exhibited in the following calculation. This analysis is taken from [50] and [54], with minor modifications. To simplify the notation, let us use differential form notation in the spatial manifold $A = A_i dx^i$. We rewrite the action (37) in the form,

$$I[A, A_0] = \frac{k}{4\pi}\int dt \int_\Sigma (A \wedge \dot{A} - A_0 F), \tag{40}$$

where the symbol $\int$ includes the trace Tr. The constraint $F = 0$ implies,

$$A = g^{-1}dg, \tag{41}$$

from where we derive two useful identities,

$$\delta A = D(g^{-1}\delta g), \qquad \dot{A} = D(g^{-1}\dot{g}). \tag{42}$$

D represents the covariant derivative in the flat connection A given in (41): $D = d + [A, \ ]$. Our goal is to compute the commutator of two solutions of the form (41).

Consider a non-canonical Lagrangian of the form $L = l_a(z)\dot{z}^a$ whose variation reads

$$\delta L = \delta z^a \sigma_{ab} \dot{z}^b, \qquad \sigma_{ab}(z) = \partial_a l_b - \partial_b l_a. \tag{43}$$

If σ is non-degenerate, the Poisson bracket of z^a with itself is given by

$$\{z^a, z^b\} = J^{ab}(z) \tag{44}$$

where J is the inverse of σ, $J^{ab}\sigma_{bc} = \delta^a_c$. The Jacobi identity for J follows from the Bianchi identity for σ. If $L = p\dot{q}$ the above construction yields $[q, p] = 1$, as expected. Following [54], we shall use this method to compute the Poisson bracket between solutions of the form (41).

The idea is to replace the solution (41) in the action (40) and compute its variation on the surface (41). Since after replacing (41) in (40) only the kinetic term survives the variation of I reads,

$$\begin{aligned}
\delta I &= -\frac{k}{2\pi}\int \dot{A}{\scriptstyle\wedge}\delta A, \\
&= -\frac{k}{2\pi}\int_\Sigma D(g^{-1}\dot{g}){\scriptstyle\wedge} D(g^{-1}\delta g), \\
&= \frac{k}{2\pi}\int_{\partial\Sigma} D_\varphi(g^{-1}\dot{g})\, g^{-1}\delta g.
\end{aligned} \tag{45}$$

The last equality follows from $D{\scriptstyle\wedge}D = F = 0$. We thus find that the variation of I on the surface (41) depends only on the boundary values of g. This is of course the well known fact that the variation of the WZW action can be written as a local functional of the boundary. It also means that the only non-trivial degrees of freedom arise at the boundary, and they are the values of g at the boundary. Using (42), the variation of I can be written as,

$$\delta I = \frac{k}{2\pi}\int_{\partial\Sigma} \dot{A}_\varphi \frac{1}{D_\varphi}\delta A_\varphi \tag{46}$$

where $1/D_\varphi$ is the inverse of the operator $D_\varphi = \partial_\varphi + [A_\varphi, \;]$ which we assume exists (we exclude functions satisfying $D_\varphi f = 0$). Comparing this variation with (43) and (44), we find the Poisson bracket of A_φ with itself,

$$\{A_\varphi, A_\varphi\} = \frac{2\pi}{k} D_\varphi \tag{47}$$

where the derivative term in D_φ should be understood as the derivative of a Dirac delta function. Finally, we make a Fourier expansion,

$$A(\varphi) = \frac{2}{k}\sum_n T^a_n e^{in\varphi} \tag{48}$$

and obtain the quantum commutator ($\hbar = 1$)

$$[T^a_n, T^b_m] = i\epsilon^{ab}{}_c T^c_{n+m} + n\frac{k}{2}\delta^{ab}\delta_{n+m} \tag{49}$$

Some comments are in order here.

(i) It is clear that the equations of motion do not force A_φ to be zero. Actually, in the sector with chiral boundary conditions A_φ is arbitrary. On the other hand, A_φ generates "gauge" transformations acting on itself. Indeed, let $Q(\eta) = (k/2\pi) \int \eta_a A^a_\varphi$, it follows directly from (47) that,

$$\delta A^a_\varphi = [A^a_\varphi, Q] = D_\varphi \eta^a. \tag{50}$$

However, here the interpretation is quite different because (50) is not generated by the "Gauss law" constraint $F = 0$. Instead, it is generated by A_φ which is different from zero. The symmetry (50) is a global -not gauge- symmetry. This means that configurations which differ by a transformation of the form (50) are physically distinct. This is the origin of boundary degrees of freedom in Chern-Simons theory.

(ii) We have only computed the bracket between the values of the gauge field A, not the group element g. This will be enough for our purposes but we remark that the problem of computing the bracket of $g(x)$ with itself leads to interesting constructions which involve quantum groups. Another remarkable application of Chern-Simons theory which we will not consider here is knot theory [49].

(iii) The algebra (49) is known as affine, or Kac-Moody, $SU(2)$ algebra. This algebra is a non-Abelian generalization of the usual Heisenberg algebra $[a_n, a_m] = n\delta_{n+m}$. Note that the last term in (49) is precisely the algebra of three oscillators. The first term couples them and, for example, alter the number of degrees of freedom (degeneracy). Unitary representations for (49) are well understood (see, for example, [47,58,48]) and they exist provided k is an integer.

(iv) Finally, an exercise for interested students: derive (49) starting from (38) by fixing the gauge $A_r = 0$, solving the constraints $F = 0$ and constructing the Dirac bracket. Note that the constraint $F = 0$ is a differential equation which on a manifold with boundary will necessarily lead to an integration function. Identify this function with A_φ above.

IV THE BROWN-HENNEAUX CONFORMAL SYMMETRY

As a final point, we briefly mention one the main application of the affine algebra (49) to three-dimensional gravity. The content of this section follows the original papers [7,41] for the derivation of the conformal algebra, [55–57] for the $SU(2)_k \to$ Virasoro reduction, and [10,23,45] for the statistical interpretation of the conformal algebra. See [35,40,59] for other aspects and further developments.

Let us show how does the Brown-Henneaux conformal algebra for anti-de Sitter spacetimes is derived from (49). We follow [41]. (See [36] for an alternative derivation based on a twisted Sugawara construction, and [46] for a supersymmetric generalization.)

It was pointed out in [41] that the full affine algebra does not represent the dynamics of anti-de Sitter spacetimes. Indeed, computing the metric associated to the boundary conditions invariant under (49), one discoveres that they match the boundary conditions found in [7] only if one imposes the additional restrictions [41]

$$T^3_n = 0, \qquad T^+_n = \delta^0_n, \tag{51}$$

on the affine generators ($T^{\pm} = T^1 \pm iT^2$).

These reductions conditions were first studied in [55] in the context of two-dimensional gravity. It was shown in that reference that the residual algebra leads to a Virasoro algebra with a central charge $c = -6k$. Starting from (49) this result can be proved as follows. We regard (51) as a set of second class constraints to be imposed in the algebra (49). We then construct the Dirac[5] bracket $[\ ,\ ]^*$ which is invariant under (51): $[T^3_n, X]^* = [T^+_n, X]^* = 0$ for all X. The only remaining component T^-_n can be renamed as $L_n = (1/k)T^-_n$ and it follows that, in the Dirac bracket, L_n satisfies the Virasoro algebra

$$[L_n, L_m]^* = (n-m)L_{n+m} + \frac{c}{12}n(n^2-1)\delta_{n+m} \tag{52}$$

with central charge $c = -6k$ (see [40] for an explicit calculation). From the value of k given in (33) we find,

$$c = \frac{3l}{2G}, \tag{53}$$

which is the correct Brown-Henneaux central charge [7].

The $c = 3l/2G$ Virasoro algebra was discovered in 1986 [7]. However only recently [10] it was pointed out that it plays a central role in the understanding of quantum three-dimensional black holes. The idea is the following.

The first input is that the zero modes L_0 and $\bar{L}_0$ of the Virasoro algebra are related to the mass and spin of anti-de Sitter spacetime as [7],

$$Ml = L_0 + \bar{L}_0 - \frac{c}{12}, \qquad J = L_0 - \bar{L}_0. \tag{54}$$

For the black hole [9], these two parameters are related to the inner and outer horizon via,

---

5) For those not familiar with the Dirac bracket formalism, see [53] for a complete treatment. The idea is to find the Poisson bracket acting on a system with constraints. For example, a free particle in three dimensions with a canonical kinetic term $\int p_i\dot{q}^i$ has the standard Poisson bracket structure. Suppose we decide to restrict the movement of the particle according to $q^3 = 0, p_3 = 0$. The new Poisson bracket is the same as before with the only modification that the coordinates q^3 and p_3 are removed. There are cases, however, in which the constraints are complicated functions of the canonical variables and one can not remove the right coordinates just by inspection. Let us consider a system with variables z^a and a Poisson bracket $[z^a, z^b] = J^{ab}$. Now, we impose the restrictions $\phi_\alpha(z) = 0$ such that $\det C_{\alpha\beta} \neq 0$ where $C_{\alpha\beta} = [\phi_\alpha, \phi_\beta]$. The Dirac bracket $[\ ,\ ]^* = [\ ,\ ] - [\ ,\chi_\alpha]C^{\alpha\beta}[\chi_\beta,\ ]$ is antisymmetric, satisfies the Jacobi identity and is invariant under the constraints, $[X, \phi_\alpha]^* = 0$ for all X.

$$Ml = \frac{r_+^2 + r_-^2}{8Gl}, \qquad J = \frac{2r_+ r_-}{8Gl}. \tag{55}$$

The Virasoro algebra (52) represents a symmetry of the theory, just like the angular momentum algebra, $[L_i, L_j] = i\varepsilon_{ijk}L_k$, is the symmetry algebra of a rotational invariant Lagrangian. Suppose that the algebra (52) is the symmetry algebra associated to some conformal field theory which is unitary ($L_0, \bar{L}_0 \geq 0$) and modular invariant. Modular invariance implies that the partition function,

$$Z[\tau] = \mathrm{Tr}\, e^{2\pi i\tau(L_0 - c/24) - 2\pi i\bar{\tau}(\bar{L}_0 - c/24)}, \tag{56}$$

satisfies

$$Z[\tau'] = Z[\tau], \qquad \tau' = \frac{a\tau + b}{c\tau + d}, \tag{57}$$

for any $a, b, c, d \in Z$ and $ad - bc = 1$. The parameter τ is the modular parameter, or complex structure, of the torus on which the CFT is defined. We recall that the partition function (56) has a precise interpretation in the black hole manifold. The Euclidean black hole has the topology of a solid torus whose modular parameter is given in (19) and $L_0 + \bar{L}_0$ is the Hamiltonian of the theory (up to an additive constant that we discuss below). This is actually implicit in (54).

Since $Z[\tau]$ is modular invariant, we can evaluate $Z[-1/\tau]$ in the limit $\mathrm{Im}(\tau) \to 0$. Assuming $L_0, \bar{L}_0 \geq 0$ we obtain,

$$Z[\tau] \sim \left| \exp\left(\frac{2\pi i c}{24\tau}\right)\right|^2. \tag{58}$$

¿From (19), $\tau = i\beta/2\pi l$ (non-rotating case), and (53) we find exactly the semiclassical Gibbons-Hawking limit (20). (Exercise: generalize this to the rotating case.) This is the canonical [9,37,38,52] version of the results obtained in [10]. Note that the limit $\mathrm{Im}(\tau) \to 0$ corresponds to small β and, according to (18), large values of M. This is a characteristic of the three dimensional black hole not shared by the Schwarzschild solution. The temperature in three dimensions decreases with the mass, the specific heat is positive and the canonical ensemble is well-defined.

A microcanonical calculation follows by writing the partition function (56) in the form,

$$Z[\tau] = \sum_{L_0, \bar{L}_0} \rho(L_0, \bar{L}_0) e^{2\pi i\tau(L_0 - c/24) - 2\pi i\bar{\tau}(\bar{L}_0 - c/24)}, \tag{59}$$

where $\rho(L_0, \bar{L}_0)$ is the number of states with eigenvalues $L_0, \bar{L}_0$. Using the approximation (58) in (59) one can extract the number of states $\rho(L_0, \bar{L}_0)$ by a contour integral obtaining,

$$\rho(L_0, \bar{L}_0) \sim e^{2\pi\sqrt{cL_0/6} + 2\pi c\sqrt{c\bar{L}_0/6}}. \tag{60}$$

This is known as Cardy formula. It is amusing to check that using (53), (54) and (55), the associated entropy is exactly equal to the Bekenstein-Hawking value $S = A/4G$ with $A = 2\pi r_+$ [10].

Actually, the above calculation is true provided the black hole mass is large enough: $Ml >> c/12$ (see (54)). The shift $-c/12$ appearing in (54) (which should be written as $-c/24 - c/24$) is the source of a number of issues. For unitary theories, on which the above calculation makes sense, it means that the mass spectrum is $M \geq -c/12$ and thus not only black holes enter in the partition function but also the conical singularities (particle solutions) introduced in [8]. Curiously when writing canonical expressions for the Virasoro generators, either using the Liouville approach [41] or the twisted Sugawara operator [36], one finds $M \geq 0$. This looks fine because the entropy should be associated to black holes spacetimes having horizons and not to the particle solutions. However, if one restricts the spectrum to positive masses, then the saddle point approximation (58) is not valid. In summary, the CFT whose symmetry is generated by (52), and that we assumed existed, does not seem to be related to general relativity.

We shall end here. See [45,11] for discussions on this last point, [59] for a proposal to resolve this problem within general relativity, and [11,13,14] for the string theory side of it.

ACKNOWLEDGMENTS

The author would like to thank the organizers of the VIII Mexican School on Particles and Fields for the kind invitation to deliver these lectures. I would also like to thank S. Carlip, M. Henneaux, M.Ortiz and A.Ritz for many conversations and correspondence which had been very helpful to understand the ideas presented here. Financial support from CICYT (Spain) grant AEN-97-1680, and the Spanish postdoctoral program of Ministerio de Educación y Ciencia is also acknowledged.

REFERENCES

1. A. Strominger and C. Vafa, Phys. Lett. **B379**, 99 (1996)
2. C.G. Callan and J. Maldacena, Nucl.Phys. **B472**, 591-610 (1996).
3. C. Rovelli, Phys. Rev. Lett **77**, 3288 (1996).
4. A. Ashtekar, J. Baez, A. Corichi, K. Krasnov, Phys. Rev. Lett **80**, 904 (1998).
5. A. Achúcarro and P.K. Townsend, Phys. Lett. **B180**, 89 (1986).
6. E. Witten, Nucl. Phys. **B 311**, 46 (1988).
7. J.D. Brown and M. Henneaux, Commun. Math. Phys. **104**, 207 (1986).
8. S. Deser, R. Jackiw and G. 't Hooft, Ann. Phys. **152**, 220 (1984); S. Deser and R. Jackiw, Ann. Phys. (NY), **153**, 405 (1984)
9. M. Bañados, C. Teitelboim and J.Zanelli, Phys. Rev. Lett. **69**, 1849 (1992); M. Bañados, M. Henneaux, C. Teitelboim and J.Zanelli, Phys. Rev. **D48**, 1506 (1993).
10. A. Strominger, High Energy Phys. **02** 009 (1998).

11. E. Martinec, "Conformal field theory, geometry, and entropy", hep-th/9809021
12. M. Henningson, K. Skenderis, J.High Energy Phys 9807, 023 (1998).
13. A. Giveon, D. Kutasov and N. Seiberg, "Comments on string theory on adS(3)", e-Print Archive: hep-th/9806194
14. J. de Boer, H. Ooguri, H. Robins and J. Tannenhauser, "String theory on adS_3, hep-th/9812046
15. K. Skenderis, "Black holes and branes in string theory", hep-th/9901050.
16. S. Carlip, *Quantum Gravity in 2+1 Dimensions*, Cambridge University Press (1998).
17. G. Gibbons and S.W. Hawking, Phys.Rev. **D15**, 2752 (1977)
18. G. Gibbons, Phys.Lett. **A61**, 3 (1977).
19. J. Maldacena, Adv. Theor. Math. Phys. **2**, 231 (1998).
20. S.S. Gubser, I.R. Klebanov, A.M. Polyakov, Phys. Lett. **B428**, 105 (1998)
21. E. Witten, Adv. Theor. Math. Phys. **2** 253 (1998)
22. J.E. Nelson and T. Regge, Nucl.Phys.**B328**, 190 (1989); Phys.Lett. **B272**, 213 (1991); Commun. Math. Phys. **141**, 211 (1991).
23. D. Birmingham, I. Sachs and A. Sen, Phys.Lett. **B424**, 275-280 (1998)
24. D.N. Page, "Black hole information", hep-th/9305040
25. S.W. Hawking and G.F.R. Ellis (1973), *The Large Scale Structure of Space-time*, Cambridge University Press.
26. S.W.Hawking, Nature (London) **248**, 30 (1974).
27. G. Howowitz, "The Dark Side of String Theory: Black Holes and Black Strings", hep-th/9210119.
28. M. Bañados, C. Teitelboim and J.Zanelli, Phys. Rev. **D49**, 975 (1994).
29. N.D. Birrell and P.C.W. Davies, "Quantum Fields in curved space time", Cambridge University Press (1982).
30. R.M. Wald (1984), *General Relativity* University Press, Chicago, USA.
31. G.T. Horowitz, "The origin of black hole entropy in string theory", gr-qc/9604051.
32. T. Regge and C. Teitelboim, Ann. Phys. (N.Y.) **88**, 286 (1974).
33. J.D.Brown, E.A.Martinez and J.W.York, Phy. Rev. Lett. **66**, 2281 (1991)
34. M. Bañados, C. Teitelboim and J.Zanelli, Phys. Rev. Lett. **72**, 957 (1994).
35. P. Navarro and J. Navarro-Salas, Phys. Lett. **B439**, 262 (1998).
36. M. Bañados, Phys. Rev. **D52**, 5816 (1995).
37. M. Bañados, T. Brotz and M. Ortiz, "Boundary dynamics and the statistical mechanics of the 2+1 dimensional black hole", hep-th/9802076, To appear in Nucl.Phys.B
38. J. Maldacena, A. Strominger, "adS(3) black holes and a stringy exclusion principle". e-Print Archive: hep-th/9804085
39. T. Brotz, M. Ortiz and A. Ritz, "On Modular Invariance and 3D Gravitational Instantons", hep-th/9903222.
40. M. Bañados, "Three dimensional quantum geometry and black holes", hep-th/9901148
41. O. Coussaert, M. Henneaux, P. van Driel, Class.Quant.Grav. **12**, 2961 (1995).
42. S. Carlip, C. Teitelboim, Phys.Rev. **D51**, 622 (1995).
43. S. Carlip, Phys. Rev. **D51**, 632 (1995)
44. S. Carlip, Phys. Rev. **D55**, 878, (1997)
45. S. Carlip, Class. Quant. Grav. **15** 3609 (1998)

46. M. Bañados, K. Bautier, O. Coussaert, M. Henneaux and M. Ortiz, Phys. Rev. **D58** 085020 (1998)
47. D. Gepner and E. Witten, Nucl.Phys. **B278**, 493 (1986).
48. P. Di Francesco, P. Mathieu, D. Senechal. "Conformal field theory" New York, USA: Springer (1997) 890 p.
49. E. Witten, Commun. Math. Phys. **121**, 351 (1989).
50. G. Moore and N. Seiberg, Phys.Lett. **B220**, 422 (1989); S. Elitzur, G. Moore, A. Schwimmer and N. Seiberg, Nucl. Phys. **B326**, 108 (1989)
51. P. Goddard, A. Kent and D. Olive, Comm. Math.Phys. **103**, 105, (1986).
52. M. Bañados, F. Méndez Phys. Rev. **D58** 104014 (1998)
53. M. Henneaux and C. Teitelboim, *Quantization of Gauge Systems* (Princeton University Press, Princeton, 1992).
54. E. Witten, Commun. Math. Phys. **92**, 455 (1984).
55. A.M. Polyakov, Int. J. Mod. Phys. **A5** (1990) 833.
56. A. Alekseev and S. Shatashvili, Nucl. Phys. **B323**, 719 (1989).
57. P. Forgács, A. Wipf, J. Balog, L. Fehér and L. O'Raifeartaigh, Phys. Lett. **227 B** (1989) 214.
58. P. Goddard and D. Olive, Int. Journ. Mod. Phys. **A1**, 303, (1986).
59. M. Bañados, "Twisted sectors in three dimensional gravity", hep-th/9903178

Heterotic/Type-II Duality and its Field Theory Avatars

Elias Kiritsis*

* *Theory Division, CERN*
CH-1211, Geneva 23, Switzerland
and
Physics Department, University of Crete
PO Box 2208, 71003 Heraclion, Greece

Abstract. In these lecture notes, I will describe heterotic/type-II duality in six and four dimensions. When supersymmetry is the maximal N=4 it will be shown that the duality reduces in the field theory limit to the Montonen-Olive duality of N=4 Super Yang-Mills theory. We will consider further compactifications of type II theory on Calabi-Yau manifolds. We will understand the physical meaning of geometric conifold singularities and the dynamics of conifold transitions. When the CY manifold is a K3 fibration we will argue that the type-II ground-state is dual to the heterotic theory compactified on K3×T^2. This allows an exact computation of the low effective action. Taking the field theory limit, $\alpha' \to 0$, we will recover the Seiberg-Witten non-perturbative solution of N=2 gauge theory.

INTRODUCTION

In the past four years, important progress has been made towards understanding non-perturbative phenomena in supersymmetric field theories and string theory, starting from the pioneering works [1]- [6]. We do know now that the five different supersymmetric string theories in ten dimensions are connected [3,4] and we take much more seriously solitonic excitations [3]. Moreover, a type of such excitations, D-branes [5], provide the appropriate microscopic degrees of freedom essential for reproducing the Bekenstein entropy of black holes [6].

In the context of field theory, strong-weak coupling duality ideas and supersymmetry were instrumental in understanding the low energy effective action of N=1 [1] and N=2 [2] gauge theories. Moreover, string theory provides a geometric interpretation of the various tools used in field theory approaches.

In these notes we will discuss a small subset of such dualities. The master duality to describe is heterotic/type-II duality in six dimensions. According to it, the strong coupling limit of heterotic string theory compactified on T^4 is equivalent to the weak coupling limit of the type-IIA string compactified on K3 and vice versa.

CP490, *Particles and Fields: Eighth Mexican School,* edited by J. C. D'Olivo et al.

By compactifying further on a two-torus, we will show that this duality implies in the field theory limit the Montonen-Olive [7] duality of N=4 Yang-Mills theory [8].

We will further consider descendants of this duality once supersymmetry is reduced from N=4 to N=2 in four dimensions. We will find that heterotic string theory on K3×T^2 is dual to type-IIA theory on a Calabi-Yau manifold that is a K3 fibration over S^2. This correspondence will provide a tool to calculate the low energy effective action exactly. Taking the field theory limit we will find the Seiberg-Witten solution.

Also in need of understanding are the conifold singularities of CY manifolds and their implications for the low energy physics of type-IIB string theory compactified on them. We will find that when such singularities appear, they are due to non-perturbative hypermultiplets becoming massless. Moreover, conifold transitions will be explained as new (Higgs) branches of the moduli space that open up when several hypermultiplets become massless.

Basic string theory expositions can be found in [9]- [13]. There are excellent reviews on non-perturbative string theory [14]- [21], D-branes [22]- [25], black holes [26,27] and field theory [28]- [32], and we urge the reader to consult them.

HETEROTIC/TYPE-II DUALITY IN SIX DIMENSIONS

The duality relation that we are going to discuss here is that of the heterotic string compactified to six dimensions on T^4 and the type-IIA string compactified on K3. Both theories have N=2 supersymmetry in six dimensions. Both theories have the same massless spectrum, containing the N=2 supergravity multiplet and twenty vector multiplets.

The six-dimensional tree-level heterotic effective action in the σ-model frame is

$$S_6^{\rm het} = \int d^6x \, \sqrt{-\det G} e^{-\Phi} \Big[R + \partial^\mu \Phi \partial_\mu \Phi - \frac{1}{12} \hat{H}^{\mu\nu\rho} \hat{H}_{\mu\nu\rho} - \tag{1}$$

$$- \frac{1}{4} (\hat{M}^{-1})_{ij} F^i_{\mu\nu} F^{j\mu\nu} + \frac{1}{8} {\rm Tr}(\partial_\mu \hat{M} \partial^\mu \hat{M}^{-1}) \Big] ,$$

where $i = 1, 2, \ldots, 24$ and

$$\hat{H}_{\mu\nu\rho} = \partial_\mu B_{\nu\rho} - \frac{1}{2} \hat{L}_{ij} A^i_\mu F^j_{\nu\rho} + \text{cyclic} . \tag{2}$$

It can be obtained from the ten-dimensional heterotic effective action using the Kaluza-Klein ansatz (see appendix C of [13]). The moduli scalar matrix $\hat{M}$ is

$$\hat{M} = \begin{pmatrix} G^{-1} & G^{-1}C & G^{-1}Y^t \\ C^t G^{-1} & G + C^t G^{-1} C + Y^t Y & C^t G^{-1} Y^t + Y^t \\ Y G^{-1} & Y G^{-1} C + Y & \mathbf{1}_{16} + Y G^{-1} Y^t \end{pmatrix} , \tag{3}$$

where $\mathbf{1}_{16}$ is the sixteen-dimensional unit matrix and

$$C_{\alpha\beta} = B_{\alpha\beta} - \frac{1}{2} Y^I_\alpha Y^I_\beta\,. \tag{4}$$

The moduli are the metric of the internal T^4, $G_{\alpha\beta}$, $\alpha, \beta = 1, 2, 3, 4$, the associated antisymmetric tensor $B_{\alpha\beta}$, as well as the internal components of the ten-dimensional Cartan gauge fields (Higgs scalars) Y^I_α, $I = 1, 2, \cdots, 16$. More details on toroidal reduction can be found in appendix C of [13]. The moduli matrix $\hat{M}$ satisfies

$$\hat{M}^T \hat{L} \hat{M} = \hat{M} \hat{L} \hat{M} = \hat{L} \quad , \quad \hat{M}^{-1} = \hat{L} \hat{M} \hat{L}\,, \tag{5}$$

where $\hat{L}$ is the invariant O(4,20) metric. Thus, $\hat{M} \in$ O(4,20). The low energy effective action (1) is invariant under continuous O(4,20,$\mathbb{R}$) transformations.

Exercise. Show that (1) is invariant under

$$A^i_\mu \to U^{-1}_{ij} A^j_\mu \quad , \quad \hat{M} \to U \hat{M} U^T \tag{6}$$

where $U \in$ O(4,20):

$$U \hat{L} U^T = \hat{L} \tag{7}$$

This continuous symmetry is broken to the discrete T-duality symmetry O(4,20,$\mathbb{Z}$) by the massive states coming from the internal lattice (KK and winding states). This is described in more detail in appendix B of [13]. Here we should note that the lattice momentum and winding integers, $q^i \sim (m_\alpha, n_\alpha, Q^I)$ are the electric charges of the gauge fields A^i_μ. We should remind the reader that, for generic values of the moduli, the gauge group is $U(1)^{24}$. However, for special values of the moduli, non-abelian symmetry appears.

Going to the Einstein frame by $G_{\mu\nu} \to e^{\Phi/2} G_{\mu\nu}$, we obtain

$$S^{\text{het}}_{D=6} = \int d^6x\, \sqrt{-G} \Bigg[R - \frac{1}{4} \partial^\mu \Phi \partial_\mu \Phi - \frac{e^{-\Phi}}{12} \hat{H}^{\mu\nu\rho} \hat{H}_{\mu\nu\rho} - \tag{8}$$

$$- \frac{e^{-\frac{\Phi}{2}}}{4} (\hat{M}^{-1})_{ij} F^i_{\mu\nu} F^{j\mu\nu} + \frac{1}{8} \mathrm{Tr}(\partial_\mu \hat{M} \partial^\mu \hat{M}^{-1}) \Bigg]\,.$$

We will now consider the effective action of type-IIA theory compactified on K3. This theory has again N=(1,1) supersymmetry in six dimensions. The tree-level type-IIA effective action in the σ-model frame is [35]

$$S^{IIA}_{K3} = \int d^6x \sqrt{-\det G_6} e^{-\Phi} \Bigg[R + \nabla^\mu \Phi \nabla_\mu \Phi - \frac{1}{12} H^{\mu\nu\rho} H_{\mu\nu\rho} + \tag{9}$$

$$+\frac{1}{8}Tr(\partial_\mu \hat{M}\partial^\mu \hat{M}^{-1})\Big] - \frac{1}{4}\int d^6x\sqrt{-\det G}(\hat{M}^{-1})_{ij}F^i_{\mu\nu}F^{j\mu\nu} +$$

$$+\frac{1}{16}\int d^6x\epsilon^{\mu\nu\rho\sigma\tau\upsilon}B_{\mu\nu}F^i_{\rho\sigma}\hat{L}_{ij}F^j_{\tau\upsilon} ,$$

where $i, j = 1, 2, \ldots, 24$. Some comments about the form of the effective action are in order here. Although all terms of this effective action come from the sphere (tree level), the factors of the dilaton in front are different from the heterotic one. The terms associated to the supergravity multiplet $(g_{\mu\nu}, B_{\mu\nu}, \Phi)$ have the standard dilaton dependence since they come from the NS-NS sector. The gauge fields, however, are descendants of the R-R forms, and thus have an anomalous dilaton dependence [25]. The CP-odd term in the effective action is a descendant of the CP-odd term of eleven-dimensional supergravity [34], cubic in the three-form. Consequently there is no dilaton dependence. Transforming again to the Einstein frame we obtain

$$S^{IIA}_{D=6} = \int d^6x\sqrt{-G}\Big[R - \frac{1}{4}\partial^\mu\Phi\partial_\mu\Phi - \frac{1}{12}e^{-\Phi}H^{\mu\nu\rho}H_{\mu\nu\rho} - \tag{10}$$

$$-\frac{1}{4}e^{\Phi/2}(\hat{M}^{-1})_{ij}F^i_{\mu\nu}F^{j\mu\nu} + \frac{1}{8}\mathrm{Tr}(\partial_\mu\hat{M}\partial^\mu\hat{M}^{-1})\Big] + \frac{1}{16}\int d^6x\epsilon^{\mu\nu\rho\sigma\tau\varepsilon}B_{\mu\nu}F^i_{\rho\sigma}\hat{L}_{ij}F^j_{\tau\varepsilon} ,$$

where $\hat{L}$ is the O(4,20) invariant metric. Notice the following differences between the two Einstein actions (8), (10): the heterotic $\hat{H}_{\mu\nu\rho}$ contains the Chern-Simons term (2) while the type-IIA one does not. The type-IIA action instead contains a parity-odd term coupling the gauge fields and $B_{\mu\nu}$. Both effective actions have a continuous O(4,20,$\mathbb{R}$) symmetry, which is broken in the string theory to the T-duality group O(4,20,$\mathbb{Z}$).

We will denote by a prime the fields of the type-IIA theory (Einstein frame) and without a prime those of the heterotic theory. Then the heterotic/type-II duality is stated by the following exercise:

Exercise. Derive the equations of motion stemming from the actions (8) and (10). Show that the two sets of equations of motion are equivalent through the following (duality) transformations

$$\Phi' = -\Phi \quad , \quad G'_{\mu\nu} = G_{\mu\nu} \quad , \quad \hat{M}' = \hat{M} \quad , \quad A'^i_\mu = A^i_\mu , \tag{11}$$

$$e^{-\Phi}\hat{H}_{\mu\nu\rho} = \frac{1}{6}\frac{\epsilon_{\mu\nu\rho}{}^{\sigma\tau\varepsilon}}{\sqrt{-G}}H'_{\sigma\tau\varepsilon} , \tag{12}$$

where the data on the right-hand side are evaluated in the type-IIA theory.

Since the string coupling g_s is related to the dilaton as $g_s^2 = \langle e^{\Phi} \rangle$, then the above duality gives

$$g_s^{\text{het}} \leftrightarrow \frac{1}{g_s^{\text{IIA}}} . \tag{13}$$

This is a non-perturbative correspondence. It is the generalization to six dimensions of the Montonen-Olive duality. Here the role of the gauge fields is played by the antisymmetric tensor, whose field strength is dualized in (12). The coupling constant is also inverted. However, here the analogues of electric and magnetic charges are carried by strings rather than point-like particles.[1]

This duality was first suggested in [3] and several qualitative tests that we will review below were discussed in [4].

In order for the above duality to work, we will have to verify a few things. We will start from the heterotic string. There, the fundamental heterotic string is the string that couples electrically to $B_{\mu\nu}$. In order for the duality to be correct, there should be a solitonic string, which should couple magnetically to $B_{\mu\nu}$[2]. Moreover, this solitonic string should be the fundamental type-IIA string on K3. This has been verified explicitly in [35]. Similarly, the heterotic string, as viewed from the perturbative type-IIA string, is a solitonic string. There are, however, dualities that imply the one we are discussing here. Consider the compactification of M-theory on $S^1 \times$K3, which is equivalent to type-IIA compactified on K3. As you have learned in Narain's lectures, all vectors descend from the RR one- and three-forms of the ten-dimensional type-IIA theory, and these descend from the three-form of M-theory to which the membrane and five-brane couple. The membrane wrapped around S^1 would give a string in six dimensions. As in ten dimensions, this is the perturbative type-IIA string. There is another string, however, obtained by wrapping the five-brane around the whole K3. This is the heterotic string [36].

We can look at a more detailed situation. It is well known that, we can have on the heterotic side gauge symmetry enhancement once the moduli take some special values. At such points, extra gauge bosons become massless, and since they are charged under the $U(1)^{24}$ generic gauge symmetry, they generate a non-abelian group. Four out of the 24 U(1) gauge bosons belong to the supergravity multiplet (those that come from the supersymmetric side of the string). These cannot participate in the gauge group enhancement.

If duality is correct, we would like to see a similar phenomenon on the type-IIA side. Here however we have an obstacle: all gauge bosons come from the R-R sector and, as you have learned already, there are no charged states under R-R forms in the perturbative type-II string. So if there are such charged gauge bosons that become massless for a special value of the moduli, then such states are non-perturbative on the type-II side.

---

1) $B_{\mu\nu}$, being a two-form, can be integrated naturally on a two-surface. It thus couples naturally to strings through $\int_{M_2} B$, where M_2 is the world-sheet of the string.

2) In six dimensions the magnetic dual of a string is still a string.

The gauge bosons in six dimensions are descending from the R-R one- and three-forms in ten dimensions. The one-form in ten dimensions is paired with the unique closed zero-form (constant) on K3 to give a vector in six dimensions. The three-form can be expanded into a sum of 22 gauge bosons in six dimensions times the 22 closed but not exact (harmonic) forms on K3. Finally, the three-form in ten dimensions also gives a three-form in six dimensions paired with the zero-form of K3. A three-form in six dimensions is dual to a one-form, so this gives rise to an extra vector that can be associated to the volume form of K3. Thus, the 24 six-dimensional gauge bosons are in a one-to-one correspondence with the even cohomology of K3[3]. The O(4,20) invariant metric $\hat{L}$ is the intersection form of the even cohomology of K3. We can count the number of moduli as follows. We know from mathematical work [37] that the moduli space of Ricci-flat metrics for K3 is the coset $O(3,19)/(O(3)\times O(19))$. There are also moduli associated with the antisymmetric tensor on K3, and these are obviously in a one-to-one correspondence with the 22 harmonic two-forms on K3 (3 of them are self-dual and 19 are anti-self-dual). Finally we have the dilaton. It can be shown that the full moduli space is now $O(4,20)/(O(4)\times O(20))$, the same as in the heterotic case. It takes a more delicate analysis to show that the discrete group $O(4,20,\mathbb{Z})$ is also a symmetry on this moduli space [37].

To come back to our problem, we have to find non-perturbative states charged under the six-dimensional U(1) gauge bosons. In ten dimensions, we know of non-perturbative states charged under the R-R forms: they are the D-branes [5]. In the IIA theory the D0-brane is electrically charged under the one-form, the D2-brane is electrically charged and the D4-brane magnetically charged under the three-form.[4]

Before we continue looking for candidate charged gauge bosons, we will have to ensure a specific property: that these states can become massless at some special points (or subvarieties) of the moduli space. Massless supermultiplets contain many less states than generic massive ones. Thus, massless states can appear only if the massive states are in short (BPS) supermultiplets[5]. This is good news, since BPS multiplets are protected from quantum corrections and they provide a window to strong coupling physics.

The D0-brane is a particle in both ten and six dimensions which, by definition, carries the electric charge of the ten-dimensional vector. Moreover, it is a BPS state preserving half of the supersymmetry. Furthermore, we know from ten dimensions that there are marginal bound states of D0-branes for any charge $m \in Z$ [38]. Moreover, the existence of such bound states is required by the type-IIA/M-theory connection [4]. Thus, these are some of the states we are looking for. The D4-brane can also give rise to BPS particles in six dimensions, when its world-volume

---

[3] A very good review on the geometrical properties of K3 surfaces and their relevance to string theory can be found in [37].

[4] There are also higher branes that are magnetically charged, but we will not need them in six dimensions. They will be, however, useful in four.

[5] The properties of BPS multiplets and their relevance to duality are analysed in appendix E of [13].

is wrapped around the whole of K3. This is a wrapping that preserves half of the original supersymmetry. Thus, a wrapped D4-brane is a BPS state breaking half of the original supersymmetry. Moreover, it was magnetically charged under the three-form in ten dimensions, which means that it is electrically charged under the vector in six dimensions obtained by dualizing the three-form. Multiple wrappings produce states with multiple charge.

So far we have found (non-perturbative) BPS states that are charged under the two six-dimensional gauge bosons dual to the zero- and four-form of K3. The rest are not difficult to find. They correspond to D2-branes wrapped supersymmetrically around non-trivial supersymmetric two-cycles of K3. In fact, a basis of such two-cycles is precisely the second homology of K3. They are in a one-to-one correspondence with the second cohomology of K3. Thus, they provide the left-over 22 charges. Now, the mass of such gauge bosons is proportional to the tension of the appropriate D2-brane times the area of the compact surface it is wrapped around. We can thus see that we can make, by adjusting moduli, one or more two-cycles have vanishing area. The associated charged particles will then become massless and will provide an enhancement of the gauge group. The K3 geometry close to a set of vanishing two-cycles is that of a four-dimensional ALE space and has an ADE classification. If the K3 moduli are tuned appropriately, K3 as an algebraic variety[6] can locally (near the singularity and in some suitable coordinate patch) be written as:

$$W_{K3} \;=\; \epsilon\,[\,\mathcal{W}^{ALE}_{ADE}\,] + \mathcal{O}(\epsilon^2) \;=\; 0\;, \tag{14}$$

where $\epsilon \to 0$. Moreover, $S_{ADE}\,:\;\mathcal{W}^{ALE}_{ADE}(x_i) \;=\; 0$ is the **A**symptotically **L**ocally **E**uclidean space with ADE singularity at the origin; it is a non-compact space obtained by excising a small neighbourhood around the singularity on K3. This kind of space is essentially given by the ADE simple singularities up to adding extra quadratic pieces that do not change the structure of the singularity:

$$\mathcal{W}^{ALE}_{A_{n-1}} = W_{A_{n-1}}(x_1, u) + {x_2}^2 + {x_3}^2 \tag{15}$$

$$\mathcal{W}^{ALE}_{D,E} = W_{D,E}(x_1, x_2, u) + {x_3}^2 \tag{16}$$

where

$$W_{A_{n-1}}(x, u) = x^n - \sum_{l=0}^{n-2} u_{l+2}x^{n-2-l} \tag{17}$$

$$W_{D_n}(x_1, x_2, u) = {x_1}^{n-1} + \tfrac{1}{2}x_1\,{x_2}^2 - \sum_{l=1}^{n-1} u_{2l}\,{x_1}^{n-l-1} - \tilde{u}_n x_2$$

$$W_{E_6}(x_1, x_2, u) = {x_1}^3 + {x_2}^4 - u_2x_1x_2^2 - u_5x_1x_2 - u_6x_2^2 - u_8x_1 - u_9x_2 - u_{12}$$

---

[6] Defined as the vanishing locus of an appropriate polynomial.

$$W_{E_7}(x_1, x_2, u) = x_1{}^3 + x_1 x_2{}^3 - u_2 x_1^2 x_2 - u_6 x_1^2 - u_8 x_1 x_2 -$$
$$-u_{10} x_2^2 - u_{12} x_1 - u_{14} x_2 - u_{18}$$
$$W_{E_8}(x_1, x_2, u) = x_1{}^3 + x_2{}^5 - u_2 x_1 x_2^3 - u_8 x_1 x_2^2 - u_{12} x_2^3 -$$
$$-u_{14} x_1 x_2 - u_{18} x_2^2 - u_{24} x_2 - u_{30} \ ,$$

The parameters u_i are parametrizing the relative way in which several two-cycles shrink to zero. In fact, the ADE singularities can be characterized in a most uniform and natural way if they are written, exactly as above, as ALE space singularities in terms of three variables. By definition, $S_{ADE} = \mathbb{C}^3/[\mathcal{W}_{ADE} = 0]$, but one may also write $S_{ADE} = \mathbb{C}^2/\Gamma$, where $\Gamma = \mathcal{Z}_n, \mathcal{D}_n, \mathcal{T}, \mathcal{O}, \mathcal{I} \subset SL(2, \mathbb{C})$ are the discrete isometry groups of the sphere that are canonically associated with the ADE groups. You can find a more detailed discussion of this in [37].

The above picture can be made more precise by considering the world-volume action for D4-branes which is the (4+1)-dimensional DBI action. The reason is that D0-brane bound-states can be viewed as four-dimensional instanton configurations [39][7] while D2-branes are given by fluxes of $F_{\mu\nu}$ through non-trivial two-dimensional cycles [40].

A quantitative test of this duality has been performed in [41]. At tree level, the heterotic string on T^4 has a gravitational Chern-Simons correction to the field strength of $B_{\mu\nu}$ [9]. This correction is of higher order in derivatives, but it is important for anomaly cancellation. In terms of ω, the spin-connection one-form, the curvature two-form can be written as $R = d\omega + \omega^2$. The gravitational Chern-Simons three-form is

$$CS_{\text{grav}} = \text{Tr}\left[\omega d\omega + \frac{2}{3}\omega^3\right] \quad , \quad d\, CS_{\text{grav}} = \text{Tr}\left[R \wedge R\right] \tag{18}$$

The anomaly related coupling of the heterotic string is obtained by substituting

$$\hat{H} = dB + CS_{\text{grav}} - CS_{\text{gauge}} \tag{19}$$

in (2).

Exercise. Redo the heterotic/type-IIA duality transformation by now keeping track of the extra gravitational Chern-Simons term in (19). Show that it implies the presence of a $B \wedge \text{Tr}[R \wedge R]$ term in the type-IIA string effective action and calculate its coefficient. Such a term should come from a type-IIA one loop diagram.

This term was directly calculated in [41] and found to agree with the duality conjecture.

7) The number of D0-branes is the instanton number.

This type of duality (string/string duality in six dimensions) fits nicely in a wider web of dualities. In two dimensions, a scalar (zero form) is dual to a scalar and this is the standard T-duality of two-dimensional σ-models. In four dimensions, a one-form is dual to a one-form and this is the usual Montonen-Olive SL(2,$\mathbb{Z}$) duality. In six dimensions we have the string/string duality described above. In eight dimensions we have a similar duality that is dealt with in the following exercise.

Exercise. Consider the type-IIA theory compactified on a two-torus with volume T_2 and antisymmetric tensor $B_{\alpha\beta} = T_1\epsilon_{\alpha\beta}$. $T = T_1 + iT_2$ is the Kähler modulus of the torus. Use the standard Kaluza-Klein ansatz and show that $T \in SU(1,1)/U(1)$ with an SL(2,$\mathbb{R}$) invariant action. Also show that the other complex modulus of the torus U combines with the dilaton as well as the two scalars obtained from the internal components of the one-form to an element of $SL(3)/SO(3)$ with an SL(3,$\mathbb{R}$)-invariant kinetic term. It can be shown that the full symmetry of the massless effective action (including the other fields) is SL(3,$\mathbb{R}$)$\times$SL(2,$\mathbb{R}$). This continuous symmetry is broken by massive states. We know already that the T-duality SL(2,$\mathbb{Z}$)$\times$SL(2,$\mathbb{Z}$) is embedded in the symmetry above. Thus, the remaining symmetry compatible with T-duality is SL(3,$\mathbb{Z}$)$\times$SL(2,$\mathbb{Z}$) where the SL(2,$\mathbb{Z}$) acts on T with fractional transformations. This symmetry is known as U-duality [3]. Derive now the terms involving the three-form and show that they are given by (set all two-forms to zero, to avoid complications with Chern-Simmons interactions)

$$S_3 = \frac{1}{2\cdot 4!}\int d^8x\left[\sqrt{-g}\;T_2\;G_{\mu\nu\rho\sigma}G^{\mu\nu\rho\sigma} + \frac{1}{48}T_1\;\epsilon^{\mu\nu\rho\sigma\mu'\nu'\rho'\sigma'}G_{\mu\nu\rho\sigma}G_{\mu'\nu'\rho'\sigma'}\right] \quad (20)$$

where G is the field stregth of the three-form. The three-form couples electrically to the D2-brane and magnetically to a D4-brane with two of its directions wrapped around the two-torus. Doing a $T \to -1/T$ duality, the electric and magnetic branes are interchanged (for the effect of T-duality on branes see [25]). This is therefore the appropriate analogue of electric-magnetic duality for membranes.

Finally in ten dimensions a four-form is dual to a four-form and this is reflected in the SL(2,$\mathbb{Z}$) duality of the type IIB string.

HETEROTIC/TYPE-II DUALITY IN FOUR DIMENSIONS

We will now further compactify both theories on a two-torus down to four dimensions and examine the consequences of the six-dimensional string/string duality. Now there are N=4 supersymmetries in four dimensions. The relevant massless supermultiplets are

• The supergravity multiplet. It contains the metric, six vectors (the graviphotons), two scalars (or one scalar and one antisymmetric tensor), four Majorana fermions and four gravitini.

• The vector multiplet. It contains a vector, four Majorana spinors and six scalars.

Both the type-IIA and heterotic theories, upon further reduction to four dimensions, give N=4 supergravity coupled to 22 abelian vector multiplets.

In both cases we use the standard Kaluza-Klein ansatz described in appendix C of [13]. The four-dimensional dilaton becomes, as usual,

$$\phi = \Phi - \frac{1}{2}\log[\det G_{\alpha\beta}] \,, \tag{21}$$

where $G_{\alpha\beta}$ is the metric of T^2 and $B_{\alpha\beta} = \epsilon_{\alpha\beta}B$ is the antisymmetric tensor. We obtain

$$S^{\text{het}}_{D=4} = \int d^4x\sqrt{-g}e^{-\phi}\left[R + L_B + L_{\text{gauge}} + L_{\text{scalar}}\right] \,, \tag{22}$$

where

$$L_{g+\phi} = R + \partial^\mu\phi\partial_\mu\phi \,, \tag{23}$$

$$L_B = -\frac{1}{12}H^{\mu\nu\rho}H_{\mu\nu\rho} \,, \tag{24}$$

with

$$\begin{aligned} H_{\mu\nu\rho} &= \partial_\mu B_{\nu\rho} - \frac{1}{2}\left[B_{\mu\alpha}F^{A,\alpha}_{\nu\rho} + A^\alpha_\mu F^B_{a,\nu\rho} + \hat{L}_{ij}A^i_\mu F^j_{\nu\rho}\right] + \text{cyclic} \\ &\equiv \partial_\mu B_{\nu\rho} - \frac{1}{2}L_{IJ}A^I_\mu F^J_{\nu\rho} + \text{cyclic} \,. \end{aligned} \tag{25}$$

The matrix

$$L = \begin{pmatrix} 0 & 0 & 1 & 0 & \vec{0} \\ 0 & 0 & 0 & 1 & \vec{0} \\ 1 & 0 & 0 & 0 & \vec{0} \\ 0 & 1 & 0 & 0 & \vec{0} \\ \vec{0} & \vec{0} & \vec{0} & \vec{0} & \hat{L} \end{pmatrix} \tag{26}$$

is the O(6,22) invariant metric. Also

$$C_{\alpha\beta} = \epsilon_{\alpha\beta}B - \frac{1}{2}\hat{L}_{ij}Y^i_\alpha Y^j_\beta \,, \tag{27}$$

so that

$$L_{\text{gauge}} = -\frac{1}{4}\Big\{\left[(\hat{M}^{-1})_{ij} + \hat{L}_{ki}\hat{L}_{lj}Y^k_\alpha G^{\alpha\beta}Y^l_\beta\right] \; F^i_{\mu\nu}F^{j,\mu\nu} + G^{\alpha\beta} \; F^B_{\alpha,\mu\nu}F^{\mu\nu}_{B,\beta} +$$

$$+\left[G_{\alpha\beta}+C_{\gamma\alpha}G^{\gamma\delta}C_{\delta\beta}+Y^i_\alpha(\hat{M}^{-1})_{ij}Y^j_\beta\right]\ F^{A,a}_{\mu\nu}F^{\beta,\mu\nu}_A-2G^{\alpha\gamma}C_{\gamma\beta}\ F^B_{\alpha,\mu\nu}F^{A,\beta,\mu\nu}- \quad (28)$$

$$-2\hat{L}_{ij}Y^i_\alpha G^{\alpha\beta}\ F^j_{\mu\nu}F^{B,\mu\nu}_\beta+2(Y^i_\alpha(\hat{M}^{-1})_{ij}+C_{\gamma\alpha}G^{\gamma\beta}\hat{L}_{ij}Y^i_\beta)\ F^{a,A}_{\mu\nu}F^{j,\mu\nu}\Big\}$$

$$\equiv-\frac{1}{4}(M^{-1})_{IJ}F^I_{\mu\nu}F^{J,\mu\nu}\ ,$$

where the index I takes 28 values. For the scalars

$$L_{\text{scalar}}=\partial_\mu\phi\partial^\mu\phi+\frac{1}{8}\text{Tr}[\partial_\mu\hat{M}\partial^\mu\hat{M}^{-1}]-\frac{1}{2}G^{\alpha\beta}(\hat{M}^{-1})_{ij}\partial_\mu Y^i_\alpha\partial^\mu Y^j_\beta+$$

$$+\frac{1}{4}\partial_\mu G_{\alpha\beta}\partial^\mu G^{\alpha\beta}-\frac{1}{2\text{det}G}\left[\partial_\mu B+\epsilon^{\alpha\beta}\hat{L}_{ij}Y^i_\alpha\partial_\mu Y^j_\beta\right]\left[\partial^\mu B+\epsilon^{\alpha\beta}\hat{L}_{ij}Y^i_\alpha\partial^\mu Y^j_\beta\right] \quad (29)$$

$$=\partial_\mu\phi\partial^\mu\phi+\frac{1}{8}\text{Tr}[\partial_\mu M\partial^\mu M^{-1}]\,.$$

We will now go to the standard axion basis in terms of the usual duality transformation in four dimensions. First we will pass to the Einstein frame by

$$g_{\mu\nu}\to e^{-\phi}g_{\mu\nu}\,, \quad (30)$$

so that the action becomes

$$S^{\text{het,E}}_{D=4}=\int d^4x\sqrt{-g}\ \left[R-\frac{1}{2}\partial^\mu\phi\partial_\mu\phi-\frac{1}{12}e^{-2\phi}\hat{H}^{\mu\nu\rho}\hat{H}_{\mu\nu\rho}- \right. \quad (31)$$

$$\left.-\frac{1}{4}e^{-\phi}(M^{-1})_{IJ}F^I_{\mu\nu}F^{J,\mu\nu}+\frac{1}{8}\text{Tr}(\partial_\mu M\partial^\mu M^{-1})\right]\,.$$

In four dimensions a massless two-index antisymmetric tensor has the same degrees of freedom as a massless (pseudo)-scalar, usually called an axion. The two can be interchanged by a duality transformation. The axion a can be introduced as

$$e^{-2\phi}H_{\mu\nu\rho}=\frac{\epsilon_{\mu\nu\rho}{}^\sigma}{\sqrt{-g}}\partial_\sigma a\,. \quad (32)$$

The transformed equations come from the following action:

$$\tilde{S}^{\text{het}}_{D=4}=\int d^4x\sqrt{-g}\ \left[R-\frac{1}{2}\partial^\mu\phi\partial_\mu\phi-\frac{1}{2}e^{2\phi}\partial^\mu a\partial_\mu a-\frac{1}{4}e^{-\phi}(M^{-1})_{IJ}F^I_{\mu\nu}F^{J,\mu\nu} \right. \quad (33)$$

$$\left.+\frac{1}{4}a\ L_{IJ}F^I_{\mu\nu}\tilde{F}^{J,\mu\nu}+\frac{1}{8}\text{Tr}(\partial_\mu M\partial^\mu M^{-1})\right]\ ,$$

where

$$\tilde{F}^{\mu\nu}=\frac{1}{2}\frac{\epsilon^{\mu\nu\rho\sigma}}{\sqrt{-g}}F_{\rho\sigma}\,. \quad (34)$$

Finally, defining the complex S field

$$S = S_1 + iS_2 = a + i\, e^{-\phi}\,, \tag{35}$$

we obtain

$$\tilde{S}^{\text{het}}_{D=4} = \int d^4x\sqrt{-g}\,\Big[R - \frac{1}{2}\frac{\partial^\mu S\partial_\mu \bar{S}}{S_2^2} - \frac{1}{4}S_2(M^{-1})_{IJ}F^I_{\mu\nu}F^{J,\mu\nu} \tag{36}$$

$$+\frac{1}{4}S_1\, L_{IJ}F^I_{\mu\nu}\tilde{F}^{J,\mu\nu} + \frac{1}{8}\text{Tr}(\partial_\mu M\partial^\mu M^{-1})\Big]\,.$$

Now consider the type-IIA action (9). Going through the same procedure and introducing the axion through

$$e^{-2\phi}H_{\mu\nu\rho} = \frac{\epsilon_{\mu\nu\rho}{}^{\sigma}}{\sqrt{-g}}\left[\partial_\sigma a + \frac{1}{2}\hat{L}_{ij}Y^i_\alpha\delta_\sigma Y^j_\beta\epsilon^{\alpha\beta}\right]\,, \tag{37}$$

we obtain the following four-dimensional action in the Einstein frame

$$\tilde{S}^{IIA}_{D=4} = \int d^4x\sqrt{-g}\left[R + L^{\text{even}}_{\text{gauge}} + L^{\text{odd}}_{\text{gauge}} + L_{\text{scalar}}\right]\,, \tag{38}$$

with

$$L^{\text{even}}_{\text{gauge}} = -\frac{1}{4}\int d^4x\sqrt{-g}\,\Big[e^{-\phi}G^{\alpha\beta}\left(F^B_{\alpha,\mu\nu} - B_{\alpha\gamma}F^{A,\gamma}_{\mu\nu}\right)\left(F^{B,\mu\nu}_\beta - B_{\alpha\delta}F^{\delta,\mu\nu}_A\right) + \tag{39}$$

$$+e^{-\phi}G_{\alpha\beta}F^{A,\alpha}_{\mu\nu}F^{\beta,\mu\nu}_A + \sqrt{\det G_{\alpha\beta}}(\hat{M}^{-1})_{ij}\left(F^i_{\mu\nu} + Y^i_\alpha F^{A,\alpha}_{\mu\nu}\right)\left(F^{j,\mu\nu} + Y^j_\beta F^{\beta,\mu\nu}_A\right)\Big]\,,$$

$$L^{\text{odd}}_{\text{gauge}} = \frac{1}{2}\int d^4x\epsilon^{\mu\nu\rho\sigma}\Big[\frac{1}{4}aF^B_{\alpha,\mu\nu}F^{A,\alpha}_{\rho\sigma} + \frac{1}{2}\epsilon^{\alpha\beta}\hat{L}_{ij}Y^i_\beta F^B_{\alpha,\mu\nu}\left(F^j_{\rho\sigma} + \frac{1}{2}Y^j_\gamma F^{A,\gamma}_{\rho\sigma}\right) \tag{40}$$

$$-\frac{1}{8}\epsilon^{\alpha\beta}\hat{L}_{ij}B_{\alpha\beta}\left(F^i_{\mu\nu} + Y^i_\gamma F^{A,\gamma}_{\mu\nu}\right)\left(F^j_{\rho\sigma} + Y^j_\delta F^{A,\delta}_{\rho\sigma}\right)\Big]$$

,

$$L_{\text{scalar}} = -\frac{1}{2}(\partial\phi)^2 + \frac{1}{4}\partial^\mu G_{\alpha\beta}\partial_\mu G^{\alpha\beta} - \frac{1}{2\det G}\partial_\mu B\partial^\mu B + \frac{1}{8}\text{Tr}[\partial_\mu\hat{M}\partial^\mu\hat{M}^{-1}] + \tag{41}$$

$$-\frac{1}{2}e^{2\phi}\left(\partial_\mu a + \frac{1}{2}\hat{L}_{ij}\epsilon^{\alpha\beta}Y^i_\alpha\partial^\mu Y^j_\beta\right)^2 - \frac{1}{2}e^{\phi}\sqrt{\det G_{\alpha\beta}}(\hat{M}^{-1})_{ij}G^{\alpha\beta}\partial_\mu Y^i_\alpha\partial^\mu Y^j_\beta\,.$$

We will use unprimed fields to refer to the heterotic side and primed ones for the type-II side. We will work out the implications of the six-dimensional duality relations (11) and (12) in four dimensions. From (11), we obtain

$$e^{-\phi} = \sqrt{\det G'_{\alpha\beta}}\quad,\quad e^{-\phi'} = \sqrt{\det G_{\alpha\beta}}\,, \tag{42}$$

$$\frac{G_{\alpha\beta}}{\sqrt{\det G_{\alpha\beta}}} = \frac{G'_{\alpha\beta}}{\sqrt{\det G'_{\alpha\beta}}} \quad , \quad A'^{\alpha}_{\mu} = A^{\alpha}_{\mu} \,, \tag{43}$$

$$g_{\mu\nu} = g'_{\mu\nu} \quad \text{Einstein frame}\,, \tag{44}$$

$$M' = M \quad , \quad A^{i}_{\mu} = A'^{i}_{\mu} \quad , \quad Y^{i}_{\alpha} = Y'^{i}_{\alpha} \,. \tag{45}$$

Finally, the relation (12) implies

$$a = B' \quad , \quad a' = B \tag{46}$$

and

$$\frac{1}{2}\frac{\epsilon_{\mu\nu}{}^{\rho\sigma}}{\sqrt{-g}}\epsilon^{\alpha\beta}F^{B'}_{\beta,\rho\sigma} = e^{-\phi}G^{\alpha\beta}\left[F^{B}_{\beta,\mu\nu} - C_{\beta\gamma}F^{A,\gamma}_{\mu\nu} - \hat{L}_{ij}Y^{i}_{\beta}F^{j}_{\mu\nu}\right] - \frac{1}{2}a\frac{\epsilon_{\mu\nu}{}^{\rho\sigma}}{\sqrt{-g}}F^{A,\alpha}_{\rho\sigma} \,, \tag{47}$$

which is an electric-magnetic duality transformation on the $B_{\alpha,\mu}$ gauge fields (see appendix H of [13]). It is easy to check that this duality maps the scalar heterotic terms to the type-IIA ones and vice versa.

In the following, we will keep for simplicity the four moduli of the two-torus and the 16 Wilson lines Y^{i}_{α} associated with it. In the heterotic case we will define the T, U moduli of the torus and the complex Wilson lines as

$$W^{i} = W^{i}_{1} + iW^{i}_{2} = -Y^{i}_{2} + UY^{i}_{1} \,, \tag{48}$$

$$G_{\alpha\beta} = \frac{T_2 - \frac{\sum_i (W^i_2)^2}{2U_2}}{U_2}\begin{pmatrix} 1 & U_1 \\ U_1 & |U|^2 \end{pmatrix} \quad , \quad B = T_1 - \frac{\sum_i W^i_1 W^i_2}{2U_2} \,. \tag{49}$$

Altogether we have the complex field S$\in$ SU(1,1)/U(1) and the T, U, W^i moduli $\in \frac{O(2,18)}{O(2)\times O(18)}$. Then the relevant scalar kinetic terms can be written as

$$L^{\text{het}}_{\text{scalar}} = -\frac{1}{2}\partial_{z^i}\partial_{\bar{z}^j}K(z_k, \bar{z}_k)\; \partial_\mu z^i \partial^\mu \bar{z}^j \,, \tag{50}$$

where the Kähler potential is

$$K = \log\left[S_2\left(T_2U_2 - \frac{1}{2}\sum_i (W^i_2)^2\right)\right] \,. \tag{51}$$

In the type-IIA case the complex structure is different: (48) remains the same but

$$G_{\alpha\beta} = \frac{T_2}{U_2}\begin{pmatrix} 1 & U_1 \\ U_1 & |U|^2 \end{pmatrix} \quad , \quad B = T_1 \,. \tag{52}$$

Also

$$S = a - \frac{\sum_i W_1^i W_2^i}{2U_2} + i\left(e^{-\phi} - \frac{\sum_i (W_2^i)^2}{2U_2}\right) . \tag{53}$$

Here $T \in$ SU(1,1)/U(1) and $S, U, W^i \in \frac{O(2,18)}{O(2)\times O(18)}$. In this language the duality transformations (42)-(46) become

$$S' = T \quad , \quad T' = S \quad , \quad U = U' \quad , \quad W^i = W'^i . \tag{54}$$

In the type-IIA string, there is an SL(2,$\mathbb{Z}$) T-duality symmetry acting on T by fractional transformations. This is a good symmetry in perturbation theory. We also expect it to be a good symmetry non-perturbatively, since it is a discrete remnant of a gauge symmetry and is not expected to be broken by non-perturbative effects. Then heterotic/type-II duality implies that there is an SL(2,$\mathbb{Z}$) S-symmetry that acts on the coupling constant and the axion. This is a non-perturbative symmetry from the point of view of the heterotic string. It acts as an electric-magnetic duality on all the 28 gauge fields. In the field theory limit it implies an S-duality symmetry for N=4 super-Yang-Mills theory in four dimensions [7,8].

So far we have seen that S, T interchange corresponds to heterotic/type-IIA duality. There is also a perturbative T-duality that corresponds to $T \leftrightarrow U$ interchange and corresponds to interchanging the type-IIA and type-IIB strings. This is a special case of mirror symmetry. Thus, in total we have S, T, U triality between IIA, IIB and heterotic strings in four dimensions with N=4 supersymmetry.

The BPS mass formula for N=4 supersymmetry in four dimensions is known [42]. We parametrize the electric and magnetic charges in terms of the integer-valued 28-vectors $\vec{\alpha}, \vec{\beta}$ and the moduli as follows:

$$\vec{Q}_e = \frac{1}{\sqrt{2}\, S_2} M(\vec{\alpha} + S_1\, \vec{\beta}) \quad , \quad \vec{Q}_m = \frac{1}{\sqrt{2}} L\, \vec{\beta} . \tag{55}$$

This parametrization incorporates automatically the Dirac-Schwinger-Zwanziger-Witten quantization condition for dyons with a θ-angle (proportional to S_1). The BPS mass formula can then be expressed in two equivalent ways

$$M^2_{BPS} = \frac{S_2}{4}\left[Q_e^t \tilde{M}_+ Q_e + Q_m^t \tilde{M}_+ Q_m + 2\sqrt{(Q_e^t \tilde{M}_+ Q_e)(Q_m^t \tilde{M}_+ Q_m) - (Q_e^t \tilde{M}_+ Q_m)^2}\,\right] \tag{56}$$

$$= \frac{1}{4S_2}(\alpha^t + S\beta^t)M_+(\alpha + \bar{S}\beta) + \frac{1}{2}\sqrt{(\alpha^t M_+ \alpha)(\beta^t M_+ \beta) - (\alpha^t M_+ \beta)^2}$$

with $M_+ = M + L$ and $\tilde{M}_+ = L M_+ L$. The square-root factor in the above expressions is proportional to the difference of the two central charges squared: depending on whether this vanishes or not, the representation preserves 1/2 or 1/4 of the supersymmetries, and is thus either short or intermediate. For perturbative BPS states of the heterotic string, $\vec{\beta}=0$. They belong to short multiplets. Their mass reads

$$M^2_{BPS,\text{pert}} = \frac{1}{4S_2}\,\alpha^t M_+ \alpha = \frac{1}{4S_2}\,p_L^2 \ . \tag{57}$$

The factor of S_2 is there because masses are measured in units of M_{Planck}.

The BPS mass formula is manifestly invariant under O(6,22,$\mathbb{Z}$) acting on the charge vectors as

$$\alpha \to \Omega\,\alpha \quad , \quad \beta \to \Omega\,\beta \ . \tag{58}$$

It is also invariant under SL(2,$\mathbb{Z}$)$_S$ acting on the fields as

$$S \to \frac{aS+b}{cS+d} \quad , \quad M \to M \quad , \quad F^i_{\mu\nu} \to (c\,S_1+d)F^i_{\mu\nu} + c\,S_2\,(ML)_{ij}\ {}^*F^j_{\mu\nu} \ , \tag{59}$$

with $ad-bc=1$, and on the vectors as

$$\begin{pmatrix}\vec{\alpha}\\ \vec{\beta}\end{pmatrix} \to \begin{pmatrix}a & -b\\ -c & d\end{pmatrix}\begin{pmatrix}\vec{\alpha}\\ \vec{\beta}\end{pmatrix} \ . \tag{60}$$

It can be checked that (56) is SL(2,$\mathbb{Z}$)$_S$-invariant.

We will analyse the BPS mass formula in the subspace of the full (6,22) moduli space described previously. In other words, we will keep the 4 real moduli of a two-torus as well as the 16 Wilson-line moduli Y^i_α, $i=9,2,\cdots,24$, $\alpha=1,2$, associated with the two-torus. This is a subspace of the full moduli space and it is the coset space O(2,18)/(O(2)×O(18)).

The perturbative part of (57) (electric charges only) becomes the well-known O(2,18)-invariant mass formula

$$M^2_{\text{per}} = \frac{1}{4S_2\left(T_2U_2 - \sum_i \frac{\text{Im}W_i^2}{2}\right)}\left|-m_1U+m_2+Tn_1+(TU-\sum_i\frac{W_i^2}{2})n_2+W_iq^i\right|^2 . \tag{61}$$

The quantity under the square root in (56) is a perfect square,

$$\sqrt{(\alpha\cdot M_+\cdot\alpha)(\beta\cdot M_+\cdot\beta)-(\alpha\cdot M_+\cdot\beta)^2} = |\alpha\cdot F\cdot\beta| \ . \tag{62}$$

Then (we assume for the moment that the number in the absolute value is positive),

$$M_+ + i\,F = e^K\,R \ , \tag{63}$$

where R is the following complex matrix

$$R = \begin{pmatrix} R_{11} & R_{12} \\ R_{21} & R_{22} \end{pmatrix} \quad ; \tag{64}$$

R_{11} is a 4×4 matrix given by

$$R_{11} = \begin{pmatrix} |U|^2 & -U & -U\bar{T} & U\left(\sum_i \frac{\bar{W}_i^2}{2} - \bar{T}\bar{U}\right) \\ -\bar{U} & 1 & \bar{T} & \bar{T}\bar{U} - \sum_i \frac{\bar{W}_i^2}{2} \\ -T\bar{U} & T & |T|^2 & T\left(\bar{T}\bar{U} - \sum_i \frac{\bar{W}_i^2}{2}\right) \\ \bar{U}\left(\frac{\sum_i W_i^2}{2} - TU\right) & TU - \sum_i \frac{W_i^2}{2} & \bar{T}\left(TU - \sum_i \frac{W_i^2}{2}\right) & |TU - \sum_i \frac{W_i^2}{2}|^2 \end{pmatrix} \tag{65}$$

R_{12} is a 4×24 matrix:

$$R_{12} = \begin{pmatrix} \cdots & -U\bar{W}_i & \cdots \\ \cdots & \bar{W}_i & \cdots \\ \cdots & T\bar{W}_i & \cdots \\ \cdots & \left(TU - \frac{1}{2}\sum_i W_i^2\right)\bar{W}_i & \cdots \end{pmatrix} \tag{66}$$

while

$$R_{21} = \overline{R_{12}^t} \tag{67}$$

and finally R_{22} is a 24×24 matrix

$$R_{22,ij} = W_i \bar{W}_j \ . \tag{68}$$

The matrix R satisfies $R^t = \bar{R}$, which implies (as it should) that M_+ is symmetric and F is antisymmetric.

Then the BPS formula can be cast in the form (when $\alpha \cdot F \cdot \beta$ is positive)

$$M^2_{BPS,+} = \frac{(\alpha + S\beta) \cdot R \cdot (a + \bar{S}\beta)}{4\, S_2 \left(T_2 U_2 - \frac{1}{2}\sum_i \mathrm{Im} W_i^2\right)} \ , \tag{69}$$

and by direct computation we finally obtain

$$M^2_{BPS,+} = \frac{1}{4S_2\left(T_2U_2 - \sum_i \frac{\mathrm{Im}W_i^2}{2}\right)} \Bigg| -m_1 U + m_2 + Tn_1 + \left(TU - \sum_i \frac{W_i^2}{2}\right) n_2 +$$

$$+ W_i q^i + S[-\tilde{m}_1 U + \tilde{m}_2 + T\tilde{n}_1 + \left(TU - \sum_i \frac{W_i^2}{2}\right)\tilde{n}_2 + \tilde{q}^i W_i] \Bigg|^2 \tag{70}$$

We will now derive how heterotic/type-II duality acts on the 28 electric and 28 magnetic charges. Label the electric charges by a vector $(m_1, m_2, n_1, n_2, q^i)$, where

m_i are the momenta of the two torus, n_i are the respective winding numbers, and q^i are the rest of the 24 charges. For the magnetic charges we write the vector $(\tilde{m}_1, \tilde{m}_2, \tilde{n}_1, \tilde{n}_2, \tilde{q}^i)$. Because of (47) we have the following duality map:

$$\begin{pmatrix} m_1 \\ m_2 \\ n_1 \\ n_2 \\ q^i \end{pmatrix} \to \begin{pmatrix} m_1 \\ m_2 \\ \tilde{n}_2 \\ -\tilde{n}_1 \\ q^i \end{pmatrix} , \quad \begin{pmatrix} \tilde{m}_1 \\ \tilde{m}_2 \\ \tilde{n}_1 \\ \tilde{n}_2 \\ \tilde{q}^i \end{pmatrix} \to \begin{pmatrix} \tilde{m}_1 \\ \tilde{m}_2 \\ -n_2 \\ n_1 \\ \tilde{q}^i \end{pmatrix} . \tag{71}$$

One can compute the spectrum of BPS multiplets both short and intermediate. In the heterotic string, BPS states are ground-states on the supersymmetric side. They break half of the supersymmetry, and their mass formula is the O(6,22)-invariant formula (57) coming from the torus. Their multiplicities can be calculated using helicity-supertrace formulae (see appendices E and G of [13]) and they are generated by the expansion coefficients of $1/\eta^{24}$. There are no perturbative BPS states that break 1/4 of the supersymmetry. On the other hand, in the type-II side the situation is different. First, perturbatively, the only charges present are those associated with momenta and windings of the two torus. The reason is that the other gauge fields come from the R-R sector and, as such, have no perturbative charged states. The states that are ground-states both on the left and the right are BPS states that break half of the original supersymmetry. Their multiplicity at any mass level is the same as at the massless level. There are states that are ground-states only on the left or the right. Such states break 1/4 of the supersymmetry. However, their associated helicity supertrace vanishes. This implies that such states can be paired into long N=4 representations and as such, are not protected from non-renormalization theorems.

There is a direct quantitative test of this duality [43,44]. It involves computing the quantum corrections to the Riemann squared term $R_{\mu\nu\rho\sigma}R^{\mu\nu\rho\sigma}$ (or the CP-odd term $\mathrm{Tr}[R \wedge R]$ in the low-energy effective action. In the type-IIB theory, this threshold comes entirely from one loop. One can argue [44] that there are no further perturbative and non-perturbative corrections. In heterotic string theory, there is a contribution at tree level but all other perturbative corrections vanish. There can be, however, non-perturbative corrections coming from the heterotic 5-brane wrapping around T^6.

By a direct one-loop computation the full one-loop type-II result is [43]

$$\begin{aligned} \Delta_{\text{gr}}^{IIA}(T) &= -36\log(T_2|\eta(T)|^4) = \\ &= 12\pi T_2 - 36\log(T_2) + 72\sum_{N=1}^{\infty}\left(\sum_{p|N}\frac{1}{p}\right)\left[e^{2\pi i N T} + e^{-2\pi i N \bar{T}}\right] = \\ &= 12\pi T_2 - 36\log(T_2) + \mathcal{O}(e^{-4\pi T_2}) \end{aligned} \tag{72}$$

where we have expanded also for large volume T_2. Under heterotic/type-II duality $S \leftrightarrow T$ and it predicts that the respective heterotic threshold is given by

$$\Delta_{\rm gr}^{\rm het}(S) = -36\log(S_2|\eta(S)|^4) =$$

$$= 12\pi S_2 - 36\log(S_2) + 72\sum_{N=1}^{\infty}\left(\sum_{p|N}\frac{1}{p}\right)\left[e^{2\pi i NS} + e^{-2\pi i N\bar{S}}\right] \ . \tag{73}$$

Remember that $S_2 = 1/g_4^2$, the four-dimensional string coupling, so that the above result contains a tree-level term proportional to S_2, which agrees with the tree-level heterotic answer. This is guaranteed by the successful matching of the $B\wedge R\wedge R$ term in six dimensions discussed at the end of last section. We can further see that it does not contain any other perturbative contributions. The peculiar logarithmic term, it is due to the infrared divergence of the physical threshold. Finally, the exponential terms correspond to instanton corrections. As we will argue below these are the contributions expected from NS5-brane instantons.

The heterotic NS5-brane is a BPS 5-brane that breaks half of the supersymmetry. The most important terms of its world-volume action are of the Nambu-Goto type,

$$S_{5-{\rm brane}} = T_5\int d^6\xi\ e^{-\Phi}\sqrt{\det\hat{G}} + iT_5\int d^6\xi\ \hat{D}_{012345} + \cdots \tag{74}$$

The induced fields are defined as

$$\hat{G}_{ab} = G_{\mu\nu}\frac{\partial x^\mu}{\partial\xi^a}\frac{\partial x^\nu}{\partial\xi^b} \tag{75}$$

and similarly for the six-form $D_{\mu_1\cdots,\mu_6}$ that is the dual of $B_{\mu\nu}$ in ten dimensions. x^μ are coordinates in ten-dimensional spacetime whereas ξ^a are the coordinates of the six-dimensional world-volume. The dots in (74) stand for interactions that are not relevant for our analysis. The tension T_5 can be obtained by saturating the Nepometchie-Teitelboim quantization condition [45] (the analogue of the Dirac quantization condition for branes) which in ten dimensions reads

$$T_p T_{6-p} = \frac{2\pi n}{2\kappa_{10}^2} \tag{76}$$

The electric dual of the heterotic NS5-brane is the perturbative heterotic string with tension $T_1 = 1/(2\pi\alpha')$. Using $2\kappa_{10}^2 = (2\pi)^7\alpha'^4$ and (76) for n=1, we obtain the NS5-brane tension

$$T_5 = \frac{1}{(2\pi)^5\alpha'^3} \tag{77}$$

Remember though that the full tension is $e^{-\Phi}T_5 = T_5/g_s^2$ where g_s is the ten-dimensional heterotic string coupling.

Exercise. Use the definition of the six-form D,

$$dB =^* dD \quad \rightarrow \quad (\partial_{\mu_1} D_{\mu_2 \cdots \mu_7} + \text{cyclic}) = \frac{1}{3!} \frac{\epsilon_{\mu_1 \cdots \mu_7}{}^{\mu_8 \mu_9 \mu_{10}}}{\det\, G} (\partial_{\mu_8} B_{\mu_9 \mu_{10}} + \text{cyclic}) \quad (78)$$

and the definition of the four-dimensional axion (32) to show that when the ten-dimensional $B_{\mu\nu}$ has only four-dimensional (transverse) dependence then

$$\frac{1}{6!} \epsilon^{\mu_1 \cdots \mu_6} D_{\mu_1 \cdots \mu_6} = a \quad (79)$$

which implies that for constant axion

$$\int d^6\xi\ \hat{D}_{012345} = (2\pi)^6 \alpha'^3\ a \quad (80)$$

An instanton that can contribute to the (non-perturbative) renormalization of the Tr R^2 coupling must be a configuration that preserves half of the spacetime supersymmetry. It must also be a Euclidean solution and it must have a finite classical action. The only regular solution of the heterotic string preserving half of the supersymmetry is the NS5-brane, [46]. So we must consider the NS5-brane rotated to Euclidean space. Moreover, in order for it to have a finite action, its (Euclidean) world-volume must wrap (supersymmetrically) around a six-dimensional compact manifold. In our case this manifold is the compactification manifold T^6. We will now calculate $e^{-S_{Class}}$ and show that it has the form expected from duality in (73).

Once the world-volume of the Euclidean NS5-brane wraps N times around T^6 (of volume $(2\pi)^6 \alpha'^3 V_6$) we obtain

$$T_5 \int d^6\xi\ e^{-\Phi} \sqrt{\det \hat{G}} = 2\pi N \frac{V_6}{g_s^2} = 2\pi N \frac{1}{g_4^2} = 2\pi N\ S_2 \quad (81)$$

where in the last equality we have introduced the four-dimensional (dimensionless) heterotic string coupling as the imaginary part of the complex S field in (35). Putting this together with the CP-odd part we obtain

$$e^{-S_{class}} = e^{2\pi i N S} \quad (82)$$

which is holomorphic in S. For an anti-NS5-brane instanton the imaginary part flips sign and we obtain

$$e^{-S_{class}} = e^{-2\pi i N \bar{S}} \quad (83)$$

These instanton correction factors have exactly the form predicted by duality in (73). A direct instanton calculation of the determinant proportional to $\sum_{p|N} \frac{1}{p}$ is still lacking.

There are further N=4 D=4 type-II ground-states that are dual to heterotic string ground-states. Some of them are not left-right symmetric so that they do not have a direct geometrical interpretation. Moreover, the Montonen-Olive duality relevant in the heterotic side corresponds to proper subgroups of SL(2,$\mathbb{Z}$)$_S$. Similar tests for these extended dualities (corresponding to calculating the R^2 and other thresholds) have been carried out in [44,47].

CONIFOLD SINGULARITIES AND CONIFOLD TRANSITIONS

In the previous sections we have provided substantial evidence that type-IIA theory compactified on K3 is related by weak-strong coupling duality to the heterotic string compactified on T^4. Both theories have N=2 supersymmetry in six dimensions. By compactifying further on T^2 we obtain a dual pair with N=4 space-time supersymmetry. One might wonder whether duality maps can be extended to ground-states with less supersymmetry in four dimensions. In these lectures we will discuss the case of N=2 supersymmetry, which is well understood. We will first discuss the geometric phenomenon of conifold singularities and conifold transitions and we will develop the physical picture lying behind it. In the next section we will analyse N=2 ground-states that in the field theory limit $\alpha' \to 0$, will reproduce the Seiberg-Witten field theory results.

In an N=2 supersymmetric theory in four dimensions, the relevant massless supersymmetric multiplets are:

• `Supergravity multiplet`. It contains the graviton, a vector (the graviphoton) and two gravitini.

• `Vector multiplet`. It contains a massless vector, a complex scalar and two real fermions.

• `Hypermultiplet`. It contains four scalars and two real fermions.

The most general N=2 supergravity is specified by two types of data. The first is the prepotential, a holomorphic function of the scalars belonging to the abelian vector multiplets. The scalars corresponding to the Cartan of the gauge group are moduli (they have flat potential) and their expectation values are unrestricted by the dynamics. They break the gauge group generically to the Cartan subgroup. They parametrize the vector moduli space $\mathcal{M}_V$. The geometry of $\mathcal{M}_V$ is constrained by supersymmetry to be what is known as "special geometry". All relevant low energy couplings between vector multiplets, as well as the supergravity multiplet, can be written in terms of the prepotential. Explicit expressions can be found in [19] and in more detail in [48]. We will summarize some of them here for convenience.

Picking the gauge group to be abelian is without loss of generality since any non-abelian gauge group can be broken to the maximal abelian subgroup by giving expectation values to the scalar partners of the abelian gauge bosons. Denote the graviphoton by A^0_μ, the rest of the gauge bosons by A^i_μ, $i = 1, 2, \ldots, N_V$, and the scalar partners of A^i_μ as $T^i, \bar{T}^i$. Although the graviphoton does not have a scalar partner, it is convenient to introduce one. The theory has a scaling symmetry, which allows us to set this scalar equal to 1. We will introduce the complex coordinates Z^I, $I = 0, 1, 2, \ldots, N_V$, which will parametrize the vector moduli space $\mathcal{M}_V$. We will denote by $F(Z^I)$, the holomorphic prepotential. It must be a homogeneous function of the coordinates of degree 2: $Z^I\ F_I = 2$, where $F_I = \frac{\partial F}{\partial Z^I}$. For example, the Kähler potential is

$$K = -\log\left[i(\bar{Z}^I\ F_I - Z^I \bar{F}_I)\right], \tag{84}$$

which determines the metric $G_{I\bar{J}} = \partial_I \partial_{\bar{J}} K$ of the kinetic terms of the scalars. We can fix the scaling freedom by setting $Z^0 = 1$, and then $Z^i = T^i$ are the physical moduli. The Kähler potential becomes

$$K = -\log\left[2\left(f(T^i) + \bar{f}(\bar{T}^i)\right) - (T^i - \bar{T}^i)(f_i - \bar{f}_i)\right], \tag{85}$$

where $f(T^i) = -iF(Z^0 = 1, Z^i = T^i)$ and $f_i = \partial f / \partial T^i$. The Kähler metric $G_{i\bar{j}}$ has the following property

$$R_{i\bar{j}k\bar{l}} = G_{i\bar{j}} G_{k\bar{l}} + G_{i\bar{l}} G_{k\bar{j}} - e^{-2K} W_{ikm} G^{m\bar{m}} \bar{W}_{\bar{m}\bar{j}\bar{l}}, \tag{86}$$

where $W_{ijk} = \partial_i \partial_j \partial_k f$. Since there is no potential, the only part of the bosonic action left to be specified is the kinetic terms for the vectors:

$$\mathcal{L}^{\text{vectors}} = -\frac{1}{4}\Xi_{IJ} F^I_{\mu\nu} F^{J,\mu\nu} - \frac{\theta_{IJ}}{4} F^I_{\mu\nu} \tilde{F}^{J,\mu\nu}, \tag{87}$$

where

$$\Xi_{IJ} = \frac{i}{4}[N_{IJ} - \bar{N}_{IJ}] \quad , \quad \theta_{IJ} = \frac{1}{4}[N_{IJ} + \bar{N}_{IJ}], \tag{88}$$

$$N_{IJ} = \bar{F}_{IJ} + 2i \frac{\text{Im}\ F_{IK}\ \text{Im}\ F_{JL} Z^K Z^L}{\text{Im}\ F_{MN} Z^M Z^N}. \tag{89}$$

The N=2 BPS states have masses of the form

$$M^2_{BPS} = \frac{|e_I\ Z^I + q^I\ F_I|^2}{\text{Im}(Z^I \bar{F}_I)}, \tag{90}$$

where e_I, q^I are the electric and magnetic charges of the state.

There is also another part of the theory to be determined and this is the interactions of hypermultiplets. Supersymmetry constrains the space of hypermultiplets to be a quaternionic space. Moreover, when hypermultiplets obtain expectation values, they break (part of) the gauge group; some of them become massive and decouple from the massless spectrum. The quaternionic manifold loses some dimensions through what is known in mathematics as a "hyper-kähler quotient", which represents nothing else than the N=2 Higgs effect.

It is a non-trivial requirement of supersymmetry that the vector and hyper moduli spaces do not mix. Thus, the total moduli space is a direct product $\mathcal{M}_V \times \mathcal{M}_H$. The only couplings allowed between vector multiplets and hypermultiplets are the minimal couplings of charged hypermultiplets.

In the type-II string, a ground state with N=2 supersymmetry in four dimensions can be constructed by compactifying type-IIA or type-IIB theory on a CY three-fold. In such a compactification one of the supersymmetries is coming from the left-moving sector and the other from the right-moving sector. We will denote it by (1,1). We will now derive the massless spectrum of such compactifications. An important ingredient is the number of various harmonic forms of a CY three-fold. There is a single zero-form and no one-forms. There are h_{11} (1,1)-forms and no (2,0)- or (0,2)-forms. A characteristic of CY manifolds is that there are unique (3,0) and (0,3) forms Ω and $\bar{\Omega}$. Ω is used to define the period integrals of the manifold. There are also h_{21} (2,1) and (1,2) forms. The rest of the forms are given by Poincaré duality.

Let us first describe the massless spectrum of type-IIA theory compactified on a CY manifold to four non-compact dimensions. In the NS-NS sector, the ten-dimensional metric gives rise to a four-dimensional metric and $(h_{11} + 2h_{12} + 2)$ scalars, the antisymmetric tensor to a four-dimensional axion as well as h_{11} scalars, while the dilaton gives an extra scalar. In the R-R sector, the three-form gives h_{11} vectors and $(2h_{12} + 2)$ scalars, while the vector gives a vector in four dimensions. In total, apart from the supergravity multiplet, we have $N_V = h_{11}$ vector multiplets and $N_H = h_{12} + 1$ hypermultiplets. An important observation is that the dilaton belongs to a hypermultiplet. Since the scalars of the vector multiplets are associated with the (1,1) forms, the classical vector moduli space is the same as the moduli space of complexified Kähler structures, $K + iB$. Moreover, we have seen that N=2 supersymmetry forbids neutral couplings between vector multiplets and hypermultiplets. Since the dilaton (string coupling) is in a hypermultiplet, this means that the tree-level M_V is exact! Notice that all vectors come from the R-R sector and have, thus, no perturbative charged states much like in cases we had seen before. On the other hand the hypermultiplets are $h_{21} + 1$ in number. One contains the dilaton, while the others come from the metric and antisymmetric tensor. So here the classical hypermultiplet moduli space is a product of the moduli space of complex structures and the flat R^4 representing the dilaton hypermultiplet. This space is affected by quantum corrections both perturbative and non-perturbative.

Let us now look at type-IIB theory compactified on a CY manifold. The NS-NS sector obviously remains similar. However, the content of the R-R sector is now different. The ten-dimensional axion gives back an axion while the two-index antisymmetric tensor gives $h_{11} + 1$ scalars, the last one coming from dualizing the four-dimensional antisymmetric tensor. The self-dual four-form gives h_{11} scalars and $h_{21} + 1$ vectors. The last one comes from the unique (3,0) form of a CY. In total we have h_{12} vector multiplets and $h_{11} + 1$ hypermultiplets. Thus, in type-IIB compactifications the vector moduli space $\mathcal{M}_V$ is related to the space of complex structures of the CY manifold, while the hypermultiplet moduli space, to complexified Kähler structures. As in the type-IIA case the dilaton is part of a hypermultiplet.

Type-IIB theory compactified on a CY manifold is related to a type-IIA compactification by mirror symmetry. Mirror symmetry interchanges h_{11} and h_{12} and interchanges type-IIA and type-IIB. Thus, type-IIA compactified on a CY manifold is equivalent to type-IIB compactified on its mirror. Mirror symmetry thus interchanges the moduli space of Kähler and complex structures.

We will describe here the moduli space of complex structures and its singularities. We will consider a basis for three-cycles of a CY manifold. The basis is $2(1+h_{12})$-dimensional with elements A_I, B^I satisfying

$$A_I \cap B^J = -B^J \cap A_I = \delta_I{}^J \quad , \quad A_I \cap A_J = B^I \cap B^J = 0 \ , \tag{91}$$

where $\cap$ stands for oriented intersection. This basis is the three-dimensional analogue of the a and b cycles of a two-torus. It is unique up to Sp($1+h_{12}$,Z) transformations:

$$\begin{pmatrix} A_I \\ B^I \end{pmatrix} \rightarrow g \begin{pmatrix} A_I \\ B^I \end{pmatrix} \quad , \quad g \begin{pmatrix} 0 & \mathbf{I} \\ -\mathbf{I} & 0 \end{pmatrix} g^T = \begin{pmatrix} 0 & \mathbf{I} \\ -\mathbf{I} & 0 \end{pmatrix} \ . \tag{92}$$

A choice of complex structure is determined by the periods of the unique holomorphic (3,0) form Ω:

$$F_I = \int_{A_I} \Omega \quad , \quad Z^I = \int_{B^I} \Omega \ . \tag{93}$$

The $Z^I, I = 1, 2, \cdots, h_{12}+1$ are projective coordinates in the moduli space of complex structures. The reason for projectivity is that multiplying Ω by a constant does not change the complex structure. In an appropriate coordinate system, $F_I \sim \frac{\partial F(Z)}{\partial Z^I}$, where $F(Z)$ is holomorphic and will play the role of the prepotential for a type-IIB compactification.

At this point we can describe a physical picture [49], which lies behind what are known as conifold transitions of CY manifolds. A CY manifold can develop a singularity in its moduli space when, by varying its complex moduli, a collection of non-trivial three-cycles shrink to zero size. In certain cases the singular manifold can be further desingularized by blowing up the singularity by replacing the three-cycles by certain two-cycles. This is a topology-changing construction, since it changes the Hodge numbers h_{12} and h_{11}. We would like to consider the interpretation of simple conifold singularities as well as conifold transitions in the context of the type-IIB string compactified on a CY manifold down to four dimensions.

We will consider for concreteness the type-IIB string compactified on a CY manifold, a single three-cycle of which is shrinking to zero. This is the simplest kind of conifold singularity.

Let us pick a basis for three-cycles such that the shrinking cycle is B^1. Then, (93) implies that the conifold singularity is at $Z^1 = 0$. This is a space of complex codimension one and we can thus encircle it with an S^1 and consider the monodromy around it. It turns out that

$$\begin{pmatrix} Z^1 \\ F_1 \end{pmatrix} \to \begin{pmatrix} Z^1 \\ F_1 + Z^1 \end{pmatrix} , \tag{94}$$

which implies that near the singularity,

$$F_1 \sim \frac{1}{2\pi i} Z^1 \log Z^1 + \text{regular} , \tag{95}$$

while the other coordinates have trivial monodromy. If we use the relation between the prepotential and the Kähler metric (84) we obtain

$$G_{1\bar{1}} \sim \log |Z^1|^2 , \tag{96}$$

which in turn implies that the singularity is at a finite distance in the moduli space, and that the curvature is singular there.

The effective action for the vector multiplets is thus singular at the conifold singularity. Moreover we have argued above that the vector moduli space does not receive any quantum corrections. What does this singularity mean?

The logarithmic divergence is reminiscent of the effect of integrating out a state with mass $\sim |Z^1|^2$. In particular, integrating out a single charged hypermultiplet gives precisely rise to such a singularity in the vector moduli space. This led Strominger to propose [49] that the appearance of this singularity indicates that a (non-perturbative) hypermultiplet has become massless at the conifold point. The effective theory will become smooth there if the light hypermultiplet is included in the effective action ("integrated in"). This phenomenon is already familiar from the Seiberg-Witten solution to N=2 SU(2) gauge theory. Here, however, it is due to singularities of the compactification manifold.

Is there such a non-perturbative state with mass $\sim |Z^1|^2$? The answer is yes: the D3 brane of type-IIB can wrap around a non-trivial three-cycle and give a massive BPS hypermultiplet, whose mass is proportional to the volume of the three-cycle. Shrinking this cycle to zero volume makes this hypermultiplet effectively massless.

One would expect that integrating out a non-perturbative state would produce corrections to the effective action that have a non-trivial coupling dependence. As we have seen in the situation at hand this is not the case. The reason is the peculiar (in)dependence of RR gauge couplings of the dilaton.

Let us now consider the general case of conifold transitions. We will have a number k of three-cycles γ^i, $i = 1, 2 \cdots, k$, which go to zero size simultaneously. In order to determine the monodromy, we need the effect of going around the γ^i cycle on a generic three-cycle δ:

$$\delta \to \delta + \langle \delta | \gamma^i \rangle \, \gamma^i \tag{97}$$

where $\langle \ | \ \rangle$ denotes oriented intersection. The singularity is at $Z^1 = Z^2 = \cdots = Z^k = 0$ with $Z^i = \int_{\gamma^i} \Omega$. There may be homology relations between the degenerating cycles γ^i. In that case k is larger than the number N of independent degenerating cycles.

In such a case, there is a mathematical construction known as a conifold transition whereby, at the singularity, we can replace the singular three-cycles by (k-N) singular two-cycles and then blow them up to non-zero size, obtaining a smooth CY manifold with different Betti numbers. What is the interpretation of this construction in terms of the low-energy physics of the associated IIB compactification?

Let us expand the vanishing cycles in the basis of independent cycles

$$\gamma^i = n_a^i B^a \;\;,\;\; i = 1, 2 \cdots, k \;\;,\;\; a = 1, 2 \cdots, N \;\;, \tag{98}$$

where $k \geq N$. This implies that there are relations between γ^i and Z^i when $k > N$. Moreover, the cycles B^a participating in (98) are in one-to-one correspondence with the U(1) gauge fields of the vector multiplets A_μ^a. D3-branes wrapped supersymmetrically around the γ^i cycle give BPS hypermultiplets with mass $M^i = |n_a^i Z^a|$, where n_a^i is the charge of the state with respect to the A_μ^a gauge boson. Thus, we obtain k hypermultiplets that become massless at the conifold point. As argued in the simple conifold singularity, in order to smooth out the effective field theory we must include the hypermultiplets H^i in the low-energy effective field theory.

When we have massless hypermultiplets in an N=2 field theory, then they generically have a potential. Let us denote the four scalars of the hypermultiplet H^i by two complex scalars H_α^i, $\alpha = 1, 2$. Then the potential has the form

$$V_H = \sum_{\alpha,\beta=1}^{2} \sum_{a,b=1}^{N} M^{ab} D_a^{\alpha\beta} D_b^{\alpha\beta} \;\;, \tag{99}$$

where the D-terms are given by

$$D_a^{\alpha\beta} = \sum_{i=1}^{k} n_a^i \left(\bar{H}_\alpha^i H_\beta^i + \bar{H}_\beta^i H_\alpha^i \right) \tag{100}$$

and M^{ab} is a positive-definite matrix. This potential has flat directions given by $D_a^{\alpha\beta} = 0$. They provide $3N$ constraints on 4k real variables. We must also fix N gauge transformations so we are left in total with a 4(k-N) dimensional space of solutions. In this generic case, all N gauge bosons are Higgsed and disappear from the low-energy spectrum. They "eat up" 4N scalars (N=2 super Higgs effect) and we are left with k-N massless hypermultiplets (moduli). Their scalars are the parameters of the generic vanishing of D-terms. Thus, k-N hypermultiplets remain massless and must be retained in the low-energy spectrum.

This agrees precisely with the geometrical picture. The new manifold obtained by the conifold transitions has

$$\tilde{h}_{12} = h_{12} - N \;\;,\;\; \tilde{h}_{11} = h_{11} + k - N \;\;, \tag{101}$$

in agreement with the fact that k three-cycles are lost, whereas k-N two-cycles are blown up. Thus, conifold transitions between topologically distinct CY manifolds correspond to Higgs transitions in the effective N=2 supergravity of the IIB string.

There are non-perturbative transitions that have no geometrical meaning, unlike the conifold transition. There, a compactification manifold degenerates but what lies on the other side of the transition is a left-right asymmetric string ground-state without geometrical interpretation [50].

A further question concerns conifold sigularities in type-IIA compactifications. There, the moduli of the complex structures that control the vanishing of the area of three-cycles belong to the hypermoduli space. Thus, at a conifold singularity, the vector moduli space is regular but the tree level hypermoduli space is singular. Moreover, the type-IIA string has no D3-brane which can give rise to massless particles in the conifold limit.

Here, however, we need to recall that in type-II strings the hypermultiplet moduli space has non-trivial quantum corrections. The reason is that the dilaton is in a hypermultiplet. Moreover, due to the quaternionic constraint and the fact that there are no RR charged states in perturbation theory, no perturbative corrections are expected for the hypermoduli space. On the other hand, instanton corrections are expected [51] due to the IIA Euclidean D2-brane that can have its three-dimensional worldvolume wrapped around the vanishing three-cycle. Moreover, in the conifold limit, such corrections are expected to be important, and they will smooth out the singularity. This was explicitly verified in [52].

N=2 HETEROTIC/TYPE-II DUALITY AND SEIBERG-WITTEN GEOMETRY

In this section we will investigate avatars of heterotic/type-II duality for N=2 ground-states. We will show in particular how to recover the Seiberg-Witten solution in field theory by using heterotic/type-II duality. We closely follow the review of W. Lerche [29] that students are encouraged to read for a more in-depth treatment.

In the case of the heterotic string, in order to obtain a ground-state in four dimensions with N=2 supersymmetry, we will have to compactify on a manifold with SU(2) holonomy and choose an appropriate $E_8 \times E_8$ (or SO(32)) bundle on it. The simplest class of such ground-states are heterotic compactifications on K3$\times T^2$ with an embedding of the spin connection into the gauge connection. This breaks half of the original supersymmetry. In the $E_8 \times E_8$ case we obtain (for generic torus moduli) a gauge group $E_7 \times E_8 \times U(1)^4$, 10 hypermultiplets in the **56** of E_7 and 65 neutral ones.

Four dimensional heterotic N=2 ground-states are characterized by the number of vector multiplets N_V and hyper multiplets N_H. At a generic point of their moduli space, the gauge group is $U(1)^{N_V+1}$. The extra vector is the graviphoton that belongs to the supergravity multiplet. There is a vector multiplet in heterotic ground states that is special: it is the one that contains the dilaton ϕ as one of its two scalars. The other scalar is the axion a, obtained by dualizing the antisymmetric tensor in four dimensions. The vector of the dilaton multiplet as well as the graviphoton come from the supersymmetric side of the heterotic string while all other vectors come from the non-supersymmetric side. They cannot participate perturbatively in enhancement of gauge symmetry. An important observation here is that, unlike the type-II case, the dilaton belongs to a vector multiplet.

Now we come to some important observations that are related to non-renormalization theorems. We have seen that in heterotic N=2 ground-states the heterotic dilaton (which is also the heterotic loop-counting parameter) belongs to a vector multiplet. Thus, supersymmetry forbids its mixing in the hyper-moduli space. This has an important corollary: the structure of the heterotic hyper-moduli space $\mathcal{M}_H$ cannot have quantum corrections (that necessarily will introduce a dilaton dependence). Thus, the tree-level heterotic $\mathcal{M}_H$ is non-perturbatively exact. On the other hand, the heterotic vector moduli space receives quantum corrections. In perturbation theory, they come from one-loop only (since the running of the coupling is related by supersymmetry to a U(1) anomaly that has only one-loop contributions). There can be, however, also non-perturbative corrections.

Let us now consider a dual type-IIA ground-state. Such a ground-state must have (at least) the same number of massless vector and hyper multiplets. Once such a ground-state exists we can use it to obtain the exact quantum vector moduli space $\mathcal{M}_V$. The reason is as follows. Here the dilaton multiplet (that organizes perturbative type-IIA corrections) is a hypermultiplet, and consequently cannot mix into $\mathcal{M}_V$. Thus, the tree-level prepotential that determines the geometry of $\mathcal{M}_V$ gives the exact answer. To calculate the geometry of $\mathcal{M}_V$ one has to do a calculation in the CY σ-model on the sphere. This is a very difficult calculation, since it is known that the geometry of the moduli space of complexified Kähler structures has both perturbative and non-perturbative (world-sheet instantons) α'-corrections. However, mirror symmetry comes to the rescue. By mirror symmetry, it is equivalent to calculate the moduli space of complex structures of the IIB string compactified on the mirror CY manifold. In this case, the geometry of the moduli space has only tree-level σ-model contributions and can be calculated by geometry alone. As is well known [53], this boils down to evaluating period integrals or solving Picard-Fuchs equations, much like what needs to be done to obtain the Seiberg-Witten solution for rigid N=2 gauge theories.

By such a calculation, one can thus obtain the exact prepotential, including both perturbative and non-perturbative corrections from the heterotic point of view. At the end of the exercise we will manage to have the exact structure of the full moduli space by doing tree-level calculations only. This is summarized in Fig. 1 taken from [29].

What is interesting in the above procedure is the following. The non-trivial corrections to the structure of $\mathcal{M}_V$ are due, in the heterotic theory, to space-time instantons. This is mapped by heterotic/type-II duality to effects due to world-sheet instantons on the type-II side.

This program has been implemented in several examples, and we will discuss it in a bit more detail.

The CY manifolds that are dual to N=2 heterotic ground-states have a very

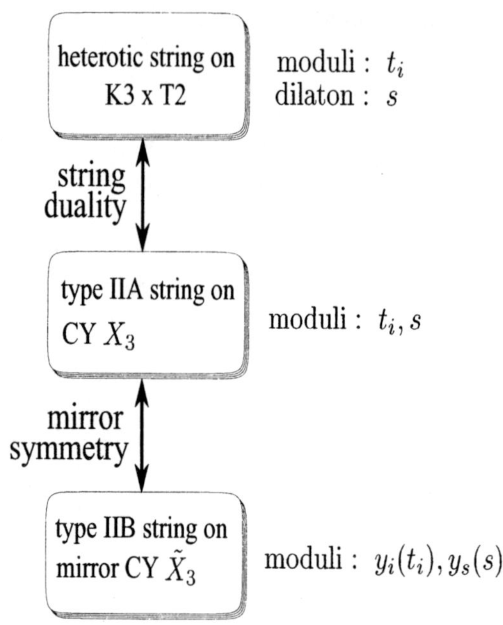

FIGURE 1. A sketch of the procedure to exactly compute the heterotic prepotential.

special structure: namely they must be K3 fibrations. Before we explain what this means, we first point out why such three-folds are what we are looking for. If and only if a three-fold is a K3 fibration, the effective prepotential (in the large volume limit) has the asymptotic form:

$$\mathcal{F}(s,t) = \frac{1}{2} s\, Q_{ij} t^i t^j + \frac{1}{6} C_{ijk} t^i t^j t^k + \dots . \tag{102}$$

Here, Q and C directly reflect the classical intersection properties of two-cycles, and s, t^i are Kähler moduli of the CY, where s is singled out in that it couples only linearly. This means that s can naturally be identified (to leading order) with the semi-classical dilaton of the heterotic string, since the dilaton couples exactly in this way, as can be inferred from a tree-level calculation on the heterotic side.

Here we will study the first and simplest example of such a duality due to Kachru and Vafa [54]. Other examples were discussed, and some analysed in detail in [55,56]. In the heterotic case, we consider first a compactification on T^2 which produces two U(1) vectors from the left-moving sector. By choosing T=U one of the U(1) factors is enhanced to an SU(2). We subsequently compactify on K3 to four dimensions and we embed the 24 instantons in the following way (10,10,4) $\rightarrow$ (E_8, E_8,SU(2)). This freezes T=U and breaks the gauge group to $E_7 \times E_7 \times$ U(1). We also have 3 hypermultiplets in the **56** of each E_7 as well as 59 neutral ones. We can Higgs $E_7 \times E_7$ completely by giving an expectation value to the **56**'s. We are thus left with a U(1) vector multiplet, associated to the T modulus of the torus, the dilaton vector multiplet associated to the complex field $S = a + ie^{-\phi}$ as well as 129 neutral hypermultiplets. Thus, $N_V = 2, N_H = 129$. In heterotic perturbation theory, when $T = i$ the U(1) gauge group is enhanced to SU(2). If there is a type-IIA dual of this heterotic compactification, it must be a CY manifold, which is a K3 fibration with $h_{11} = 2$, $h_{12} = 128$. Such CY manifolds are rare and can be easily identified. Its mirror would provide a IIB dual and it is given by a degree-12 hypersurface in $WP^4_{1,1,2,2,6}$, with defining polynomial

$$W = z_1^{12} + z_2^{12} + z_3^6 + z_4^6 + z_5^2 - 12\psi\ z_1 z_2 z_3 z_4 z_5 - 2\phi\ z_1^6 z_2^6\ . \tag{103}$$

In the above ψ, ϕ parametrize the space of complex structures, which is also the exact vector moduli space in the IIB compactification. They should be mapped to the heterotic moduli T, S under the duality map. For later use we will define

$$x = -\frac{1}{864}\frac{\phi}{\psi^6}\ \ ,\ \ y = \frac{1}{\phi^2}\ . \tag{104}$$

We can exhibit the K3-fibration structure by the change of variables

$$z_1 = \sqrt{x_0}\zeta^{1/12}\ \ ,\ \ z_2 = \sqrt{x_0}\zeta^{-1/12}\phi^{-1/6} \tag{105}$$

so that

$$W = x_0^6\left(\zeta + \frac{y}{\zeta} - 2\right) + z_3^6 + z_4^6 + z_5^2 - 12\frac{\psi}{\phi^{1/6}} x_0 z_3 z_4 z_5\ . \tag{106}$$

If we consider ζ as a parameter, then we have the defining polynomial of a K3 surface:

$$W_{\mathrm{K3}} = x_0^6 + z_3^6 + z_4^6 + z_5^2 + \alpha\ x_0 z_3 z_4 z_5\ ; \tag{107}$$

ζ is parametrizing a sphere, CP^1, whose volume is $1/y$. Thus, W describes a fibre bundle with base CP^1 and fibre K3. The large-volume limit of the base corresponds to $y \to 0$. In that limit the theory becomes six-dimensional and we would expect contact with the heterotic dual.

Let us now study the singularities (conifold locus) of the complex structure moduli space. They are given by the discriminant of the polynomial:

$$(1-x)^2 - x^2 y = 0\ . \tag{108}$$

At large volume of the base, $y = 0$, and we have there a unique singularity $x = 1$. This should correspond to gauge-group enhancement $U(1) \to SU(2)$ in the heterotic side when $T \to i$. Once $y \neq 1$, the singularity splits in two, $x = 1/(1 \pm \sqrt{y})$. This is what we would expect from the field theoretic Seiberg-Witten solution. There, the classical singularity due to the enhancement of the gauge symmetry $U(1) \to$

$SU(2)$ splits beyond perturbation theory to two singularities where monopoles and dyons become massless. There is, however, no gauge-symmetry enhancement non-perturbatively. From the mirror map we can make the identification of heterotic and IIB moduli

$$x = \frac{1728}{j(T)} \quad , \quad y = e^{-S} \ . \tag{109}$$

We have $x = 1$ at $T = i$ and $S = \infty$ at $y = 0$. An interesting test of the correspondence is obtained by comparing with heterotic perturbation theory. By solving the classical Picard-Fuchs equations, the exact prepotential $f(x, y)$ can be obtained. Using the map (109) and expanding for $S \to \infty$, we obtain the heterotic perturbative contributions. From a direct computation in heterotic perturbation theory [57] we obtain

$$f_{\text{pert}} = ST^2 + f^{1-\text{loop}}(S,T) \ , \tag{110}$$

with

$$f_{TTT}^{1-\text{loop}} = \frac{j'(T)^3}{E_4(T)j(T)(j(T)-j(i))^2} \quad , \quad f_{TTS}^{1-\text{loop}} = \frac{j'(T)^2}{E_4(T)j(T)(j(T)-j(i))} \tag{111}$$

while $f_{TSS}^{1-\text{loop}} = f_{SSS}^{1-\text{loop}} = 0$. An expansion of the type-II prepotential, using the map (109), gives the precise perturbative heterotic prepotential as specified above. This is a non-trivial test of consistency for the advocated duality.

We now would like to study the field-theory limit of this duality. To do this we must decouple gravity by taking a suitable $\alpha' \to 0$ limit. We will define the field theory moduli space coordinate u (Higgs field squared) as

$$x - 1 = \alpha' u + \mathcal{O}(\alpha'^2) \ . \tag{112}$$

We must also arrange that the dynamical scale $\Lambda \ll M_P \sim 1/\sqrt{\alpha'}$. This can be done by writing

$$y = e^{-S} = \alpha'^2 \Lambda^4 e^{-\hat{S}} \ , \tag{113}$$

where $\hat{S}$ should correspond to the SU(2) gauge field theory coupling. Using (112) and (113), we can find the singularities from (108) to be

$$u = \pm \Lambda^2 e^{-\hat{S}/2} \ , \tag{114}$$

which would correspond to the two points in the moduli space where monopoles and dyons become massless.

The exact prepotential has an expansion

$$f(T,S) = ST^2 + \sum_{n=0}^{\infty} f_n(T) e^{-nS} \ . \tag{115}$$

We will have to define also an analogue of the Higgs expectation value a as

$$\tilde{T} = \frac{T-i}{T+i} = \sqrt{\alpha'}a + \mathcal{O}(\alpha'^2) \ . \tag{116}$$

The modular transformation $T \to -1/T$ translates to the Weyl transformation $a \to -a$, as it should. We can now rewrite (115) as

$$f = S\tilde{T}^2 + \sum_{n=0}^{\infty} g_n(\tilde{T})e^{-nS} + \text{subleading} \tag{117}$$

$$= \alpha'\hat{S}a^2 + \sum_{n=0}^{\infty} g_n(\sqrt{\alpha'}a)\alpha'^{2n}\Lambda^{4n}e^{-n\hat{S}} + \text{subleading}$$

To recover the Seiberg-Witten solution we must have

$$g_n(\sqrt{\alpha'}a) = c_n(\sqrt{\alpha'}a)^{2-4n} \ , \tag{118}$$

while g_0 should have the appropriate logarithmic divergence. It turns out [56] that if $F_{SW}(a, \Lambda^4)$ is the prepotential of the Seibeg-Witten solution, then the one obtained from type-II string theory, in the above limit is given by

$$F(a, \hat{S}) = \alpha' F_{SW}(a, \Lambda^4 e^{-\hat{S}}) + \mathcal{O}(\alpha'^2) \ . \tag{119}$$

Thus, by taking the field-theory limit in this ground-state, we recover the Seiberg-Witten result.

With this procedure we can recover the field theory non-perturbative solutions for more general gauge groups. The K3 fibration structure turns out to be crucial. To see this, let us assume, for simplicity, that the (mirror) CY, $\tilde{X}_3$, can be represented by some polynomial

$$\tilde{X}_3 : \quad W_{\tilde{X}_3}(x_1, x_2, x_3, x_4, x_5; y_i, y_s) \ = \ 0 \tag{120}$$

in weighted projective space $WP^{2d}_{1,1,2k_1,2k_2,2k_3}$, with overall degree $2d \equiv 2(1 + k_1 + k_2 + k_3)$. Above, y_i denote the moduli, and, in particular, y_s denotes the special distinguished modulus that is related to the heterotic dilaton, $y_s \sim e^{-s}$. The statement that $\tilde{X}_3$ is a K3 fibration of this particular type means that it can be written as

$$\begin{aligned} W_{\tilde{X}_3}(x_j; y_i, y_s) =& \frac{1}{2d}\Big({x_1}^{2d} + {x_2}^{2d} + \frac{2}{\sqrt{y_s}}(x_1 x_2)^d\Big) \\ &+ \widehat{W}\Big(\frac{x_1 x_2}{{y_s}^{1/d}}, x_k; y_i\Big) \ . \end{aligned} \tag{121}$$

Upon the variable substitution

$$\begin{aligned} x_1 &= \sqrt{x_0}\,\zeta^{1/2d} \\ x_2 &= \sqrt{x_0}\,\zeta^{-1/2d}{y_s}^{1/2d} \ , \end{aligned} \tag{122}$$

this gives

$$W_{\tilde{X}_3}(x_j;\zeta,y_i) \;=\; \frac{1}{2d}\Big(\zeta+\frac{y_s}{\zeta}+2\Big)x_0^d+\widehat{W}(x_0,x_k;y_i).$$

In other words, if we now alternatively view ζ as a modulus, and not as a coordinate, then $W_{\tilde{X}_3}(\zeta,x_2,x_3,x_4,x_5;y_i,y_s)\equiv W_{K3}(x_2,x_3,x_4,x_5;\zeta+\frac{y_s}{\zeta},y_i)=0$ describes a K3 – this is precisely what is meant by fibration (of course, if we continue to view ζ as a coordinate, then this equation still describes the Calabi-Yau three-fold). More precisely, ζ is the coordinate of the base CP^1, and $y_s\to 0$ corresponds to the large-base limit – obviously, the fibration looks in this limit locally trivial, and one then expects the theory to be dominated by the "classical" physics of the K3. This is the "adiabatic limit", in which the K3 fibres vary only slowly over the base and where one can apply the original heterotic/type-II duality fibre-wise.

The left-over piece $\widehat{W}$ is precisely such that $\frac{1}{d}x_0^d+\widehat{W}(x_0,x_k;y_i)=0$ describes a K3 in canonical parametrization in WP^d_{1,k_1,k_2,k_3}. Assuming that the K3 is singular of type ADE in some region of the moduli space, we can expand it around the critical point and thereby replace it locally by the ALE normal form (15) of the singularity, $\frac{1}{d}x_0^d+\widehat{W}\sim\epsilon\,\mathcal{W}^{ALE}_{ADE}(x_i,u_k)$. Going to the patch $x_0=1$ and rescaling[8] $y_s=\epsilon^2\Lambda^{2h}$ and $\zeta=\epsilon\,z$, we then obtain the following fibration of the ALE space:

$$W_{\tilde{X}_3}(x_j,z;u_k) \;=\epsilon\Big(z+\frac{\Lambda^{2h}}{z}+2\mathcal{W}^{ALE}_{ADE}(x_j,u_k)\Big)+\mathcal{O}(\epsilon^2)=0. \tag{123}$$

This is not totally surprising: since the CY was a K3 fibration, considering a region in moduli space where the K3 can be approximated by an ALE space simply produces locally a corresponding fibration of the ALE space.

Now, focusing on $G=SU(n)\sim A_{n-1}$ and remembering the definition of the ALE space (15), we see that (123) is exactly the same as the fibred form of the Seiberg-Witten curve [58], apart from the extra quadratic pieces in x_2,x_3 (see the lectures by C. Gomez). Since quadratic pieces do not change the local singularity type, this means that the local geometry of the three-fold in the Seiberg-Witten regime of the moduli space is indeed equivalent to the one of the Seiberg-Witten curve. However, just because of these extra quadratic pieces, the Seiberg-Witten curve itself is, strictly speaking, *not* geometrically embedded in the three-fold, though this distinction is not very important.

One may in fact explicitly integrate out the quadratic pieces in (123), and verify that the *holo*morphic three-form Ω of the three-fold then collapses precisely to the *mero*morphic one-form λ_{SW} that is associated with the Seiberg-Witten curve:

$$\Omega=\frac{d\zeta}{\zeta}\wedge\Big[\frac{dx_1\wedge dx_2}{\frac{\partial W}{\partial x_3}}\Big]\;\stackrel{\epsilon\to 0}{\longrightarrow}\;x_1\frac{dz}{z}\;\equiv\;\lambda_{SW}. \tag{124}$$

This implies that the periods $a_i,a_{D,i}$ are indeed among the periods of the three-fold:

8) This fixes $\epsilon=(\alpha')^{h/2}$ where h is the dual Coxeter number of the ADE group labelling the singularity.

$$(X^I; F_J) \equiv \int_{(\Gamma_{\alpha_I};\Gamma_{\beta^J})} \Omega \overset{\epsilon \to 0}{\supset} \int_{(\alpha_i;\beta^j)} \lambda_{SW} \equiv (a_i; a_{D,j}), \tag{125}$$

and thus that the string effective action, $\mathcal{F} \equiv \frac{1}{2} X^I F_I$, contains the Seiberg-Witten effective action.

We should note here that we have assumed that the mirror $\tilde{X}_3$ is a K3 fibration. However, our starting point was really that the original three-fold X_3 is a K3 fibration, since a priori it is the type-IIA strings on X_3 that are dual to the heterotic string. But in general, if X_3 is a K3 fibration, the mirror $\tilde{X}_3$ is not necessarily a K3 fibration as well. However, our arguments above are nevertheless correct, because all that counts is that *locally*, near the relevant singularity, the mirror is a fibration of an ALE space. One can indeed show, using "local" mirror symmetry, that whenever we have an asymptotically free gauge theory on the type-IIA side, the mirror $\tilde{X}_3$ locally has the required form.

Moreover, note that these ideas carry over to gauge theories with extra matter, although we will be very brief here. In the F-theoretical formulation of gauge-symmetry enhancement [59], there is a systematic way to construct N=2 Yang-Mills theory on the type-IIA side, for almost any matter content. Through local mirror symmetry, this maps over to the type-IIB side and directly produces the relevant Seiberg-Witten curves, which generically exhibit extra matter fields. Similar to (123), one still has ADE singularities fibred over some base S^2, but in general the dependence of z will be more complicated.

Summarizing, we have seen how heterotic/type-II duality extends to N=2 four-dimensional vacua. We have obtained a physical, effective field theory understanding of conifold singularities and transitions. The duality, moreover, explains Seiberg-Witten dynamics in the field-theory limit.

OUTLOOK

In these lectures I have tried to analyse an important part of the duality structure of string theory, namely that of heterotic/type-II duality. Although this duality might not be the most central, conceptually, of string dualities, it is the one that provided us with many independent consistency checks, as well as with a concrete contact with non-perturbative field-theory solutions.

We have analysed heterotic string theory compactified to six dimensions on T^4, as well as type-IIA compactified on K3. We have given several arguments indicating that each one is the strong coupling dual of the other. Moreover the heterotic string is the magnetic analogue of the type-IIA string in six dimensions and vice versa. Most of the BPS states required for the duality to be consistent are supplied by D-branes wrapped supersymmetrically around cycles of the K3 surface.

Upon compactification on an extra T^2, the two theories are still related by duality. Here, however, this duality implies that the S-field of heterotic theory, whose imaginary part is the string coupling constant while its real part is the axion (θ-angle), is interchanged by the Kähler modulus T of the type-IIA two-torus. The perturbative SL(2,$\mathbb{Z}$) duality symmetry acting on T translates through string-string duality, to the non-perturbative Olive-Montonen SL(2,$\mathbb{Z}$)$_S$ duality of N=4 supergravity coupled to Yang-Mills. This map provides a geometrical interpretation of the Olive-Montonen duality.

We have considered further N=2 type-II ground-states in four dimensions, obtained by compactifying the IIB string on a CY manifold. At a point where the manifold has a conifold singularity, the effective field theory of the massless modes (at tree level) becomes singular. Moreover, the tree-level result is exact since, in such type-II ground-states, the dilaton is in a hypermultiplet and thus forbidden by N=2 supersymmetry to mix with the vector multiplets. We have analysed Strominger's interpretation of conifold singularities as being due to non-perturbative hypermultiplets becoming massless there. Moreover, in such a case, new branches of the moduli space open up, by Higgsing some of the vector multiplets. This is represented by the conifold transitions, where some three-cycles shrink to zero and then blown up as two-cycles.

Finally we have established IIA,B duals for heterotic four-dimensional ground-states (compactified on K3$\times T^2$) with N=2 supersymmetry. The associated CY manifolds must be K3 fibrations. We analysed some examples and showed that they reproduce in the field theory limit, the Seiberg-Witten solution.

This is not the end of the story. Many more consistency checks need be done, even for dualities that are thought to be "well established", since puzzles keep popping up once a more detailed look is taken. More dualities await discovery, both in field theory and in string theory. Useful information for the physics of the real world should be taken out from this wealth of non-perturbative behaviour. Let us not forget that two of the most important theoretical problems in the last twenty years are still unsolved: the cosmological-constant problem and supersymmetry breaking. New ideas and, maybe, some form of non-perturbative physics might be necessary for their solution.

Last, but not least, string theory seems to have many more excitations than originally thought. They are responsible for unifying all different-looking perturbative string theories. What is the more natural and comprehensive definition of this single theory?

ACKNOWLEDGMENTS

I would like to thank the organizers of the 8th Mexican School of Particles and Fields, in Oaxaca for their warm hospitality and excellent organization.

REFERENCES

1. N. Seiberg, Phys. Rev. **D49** (1994) 6857, hep-th/9402044.
2. N. Seiberg and E. Witten, Nucl. Phys. **B426** (1994) 19, hep-th/9407087; *ibid.* **B431** (1994) 484, hep-th/9408099.

3. C. Hull and P. Townsend, Nucl. Phys. **B438** (1995) 109, hep-th/9410167.
4. E. Witten, Nucl. Phys. **B443** (1995) 85, hep-th/9503124; hep-th/9507121.
5. J. Polchinski, Phys. Rev. Lett. **75** (1995) 4724, hep-th/9510017.
6. A. Strominger and C. Vafa, Phys. Lett. **B379** (1996) 99, hep-th/9601029.
7. C. Montonen and D. Olive, Phys. Lett. **B72** (1977) 117.
8. A. Sen, Int. J. Mod. Phys. **A9** (1994) 3707; hep-th/9402002; Phys. Lett. **B329** (1994) 217; hep-th/9402032.
9. M. Green, J. Schwarz and E. Witten, *Superstring Theory, Vols I and II*, Cambridge University Press, 1987.
10. D. Lüst and S. Theisen, *Lectures in String Theory*, Lecture Notes in Physics, 346, Springer Verlag, 1989.
11. L. Alvarez-Gaumé and M. Vazquez-Mozo, *Topics in String Theory and Quantum Gravity*, in the 1992 Les Houches School, Session LVII, eds. B. Julia and J. Zinn-Justin, Elsevier Science Publishers, 1995.
12. J. Polchinski, "*What is String Theory?*", hep-th/9411028.
13. E. Kiritsis, "*Introduction to String Theory*", hep-th/9709062, Leuven University Press, 1998.
14. M. Duff, R. Khuri and J. Lu, "*String Solitons*", Phys. Rept. **259** (1995) 213, hep-th/9412184.
15. J. Polchinski, Rev. Mod. Phys. **68** (1996) 1245, hep-th/9607050.
16. W. Lerche, "*Recents Developments in String Theory*", hep-th/9710246.
17. S. Förste and J. Louis, "*Duality in String Theory*", hep-th/9612192.
18. C. Vafa, "*Lectures on Strings and Dualities*", hep-th/9702201.
19. E. Kiritsis, "*Introduction to Non-Perturbative String Theory*", hep-th/9708130.
20. B. de Wit and J. Louis, "*Supersymmetry and Dualities in Various Dimensions*", hep-th/9801132.
21. A. Sen, "*An Introduction to Non-Perturbative String Theory*", hep-th/9802051.
22. J. Polchinski, S. Chaudhuri and C. Johnson, "*Notes on D-branes*", hep-th/9602052.
23. J. Polchinski, "*TASI lectures on D-branes*", hep-th/9611050.
24. W. Taylor, "*Lectures on D-branes, Gauge Theory and M(atrices)*", hep-th/9801182.
25. C. Bachas, "*Lectures on D-branes*", hep-th/9806199.
26. J. Maldacena, "*Black Holes in String Theory.*", hep-th/9806199.
27. A. Peet, "*The Bekenstein Formula and String Theory (N-brane Theory)*", hep-th/9712253.
28. K. Intriligator and N. Seiberg, "*Lectures on Supersymmetric Gauge Theories and Electric-Magnetic Duality*", Nucl. Phys. [Proc. Suppl.] **45BC** (1996) 1; hep-th/9509066.
29. W. Lerche, "*Introduction to Seiberg-Witten Theory and its Stringy Origin*", Nucl. Phys. [Proc. Suppl.] **55B** (1997) 83, hep-th/9611190.
30. L. Alvarez-Gaumé and S.F. Hassan, "*Introduction to S-Duality in N=2 Supersymmetric Gauge Theory*", hep-th/9701069;
L. Alvarez-Gaumé and F. Zamora, "*Duality in Quantum Field Theory (and String Theory)*", hep-th/9709180.
31. M. Peskin, "*Duality in Supersymmetric Yang Mills Theory*", hep-th/9702094.
32. M. Shifman, "*Non-Perturbative Dynamics in Supersymmetric Gauge Theories*", hep-

th/9704114.
33. P. Di Vecchia, "*Duality in N=2,4 Supersymmetric Gauge Theories*", hep-th/9803026.
34. E. Cremmer, B. Julia and J. Scherk, Phys. Lett. **B76** (1978) 409.
35. A. Sen, Nucl. Phys. **B450** (1995) 103; hep-th/9504027.
36. S. Cherkis and J.H. Schwarz, Phys. Lett. **B403** (1997) 225; hep-th/9703062.
37. P. Aspinwall, hep-th/9611137.
38. S. Sethi and M. Stern, Comm. Math. Phys. **194** (1998) 675; hep-th/9705046.
39. M. Douglas, hep-th/9512077.
40. E. Witten, Nucl. Phys. **B460** (1996) 335; hep-th/9510135.
41. C. Vafa and E. Witten, Nucl. Phys. **B447** (1995) 261, hep-th/9505053.
42. M. Čvetic and D. Youm, Phys. Rev. **D53** (1996) 584;hep-th/9507090
M. Duff, J.T. Liu and J. Rahmfeld, Nucl. Phys. **B459** (1996) 125; hep-th/9508094.
43. J. Harvey and G. Moore, Phys. Rev. **D57** (1998) 2323, hep-th/9610237.
44. A. Gregori, E. Kiritsis, C. Kounnas, N. A. Obers, P. M. Petropoulos and B. Pioline, Nucl. Phys. **B510** (1998) 423; hep-th/9708062.
45. R. Nepomechie, Phys. Rev. **D31** (1985) 1921;
C. Teitelboim, Phys. Lett. **B167** (1986) 69.
46. C. Callan, J. Harvey and A. Strominger, Nucl. Phys. **B359** (1991) 611; Nucl. Phys. **B367** (1991) 60.
47. A. Gregori, C. Kounnas and P. M. Petropoulos, hep-th/9808024.
48. L. Andrianopoli, M. Bertolini, A. Ceresole, R. D'Auria, S. Ferrara, P. Fré, and T. Magri, J. Geom. Phys. **23** (1997) 111, hep-th/9605032.
49. A. Strominger, Nucl. Phys. **B451** (1995) 96, hep-th/9504090;
B. R. Greene, D. R. Morrison and A. Strominger, Nucl. Phys. **B451** (1995) 109, hep-th/9504145.
50. E. Kiritsis and C. Kounnas, Nucl. Phys. **B503** (1997) 117, hep-th/9703059.
51. K. Becker, M. Becker and A. Strominger, Nucl. Phys. **B456** (1995) 130; hep-th/9507158;
E. Witten, Nucl. Phys. **B474** (1996) 343; hep-th/9604030.
52. H. Ooguri and C. Vafa, Phys. Rev. Lett. **77** (1996) 3296; hep-th/9608079.
53. P. Candelas, X. de la Ossa, P. Green and L. Parkes, Nucl. Phys. **B359** (1991) 21;
S. Hosono, A. Klemm and S. Theisen, hep-th/9403096.
54. S. Kachru and C. Vafa, Nucl. Phys. **B450** (1995) 69, hep-th/9505105.
55. S. Ferrara, J. Harvey, A. Strominger and C. Vafa, Phys. Lett. **B361** (1995) 59, hep-th/9505162;
A. Klemm, W. Lerche and P. Mayr, Phys. Lett. **B357** (1996) 313, hep-th/9506112;
S. Kachru, A. Klemm, W. Lerche, P. Mayr and C. Vafa, Nucl. Phys. **B459** (1996) 537, hep-th/9508155;
G. Aldazabal, A. Font, L. Ibañez and F. Quevedo, Nucl. Phys. **B461** (1996) 85, hep-th/9510093, Phys. Lett. **B380** (1996) 33, hep-th/9602097.
56. A. Klemm, W. Lerche, P. Mayr, C. Vafa and N. Warner, Nucl. Phys. **B477** (1996) 746, hep-th/9604034;
W. Lerche and N. P. Warner, Phys. Lett. **B423** (1998) 79, hep-th/9608183;
W. Lerche, P. Mayr and N. P. Warner, Nucl. Phys. **B499** (1997) 125, hep-th/9612085.

57. B. de Wit, V. Kaplunovsky, J. Louis and D. Luest, Nucl. Phys. **B451** (1995) 53; hep-th/9504006;
I. Antoniadis, S. Ferrara, E. Gava, K. S. Narain and T. R. Taylor, Nucl. Phys. **B447** (1995) 35; hep-th/9504034.
58. A. Klemm, W. Lerche, S. Yankielowicz and S. Theisen, Phys. Lett. **B344** (1995) 169, hep-th/9411048;
P. C. Argyres and A. E. Faraggi, Phys. Rev. Lett. **74** (1995) 3931, hep-th/9411057;
A. Klemm, W. Lerche and S. Theisen, Int. J. Mod. Phys. **A11** (1996) 1929, hep-th/9505150.
59. C. Vafa, Nucl. Phys. **B469** (1996) 403, hep-th/9602022;
D. Morrison and C. Vafa, Nucl. Phys. **B473** (1996) 74, hep-th/9602114; Nucl. Phys. **B476** (1996) 437, hep-th/9603161;
M. Bershadsky, A. Johansen, T. Pantev, V. Sadov and C. Vafa, hep-th/9612052.
60. J. Maharana and J. Schwarz, Nucl. Phys. **B390** (1993) 3.

Duality symmetries in string theory

Carmen A. Núñez

Instituto de Astronomía y Física del Espacio - CONICET
C.C. 67 - Suc. 28, 1428 Buenos Aires, Argentina

Abstract. The search for a unified theory of quantum gravity and gauge interactions leads naturally to string theory. This field of research has received a revival of interest after the discovery of duality symmetries in recent years. We present a self contained account of some non-perturbative aspects of string theory which have been recently understood. The spectrum and interactions of the five consistent superstring theories in ten dimensions are recollected and the fundamental principles underlying this initial stage in the construction of the theory are briefly reviewed. We next discuss some evidences that these apparently different superstrings are just different aspects of one unique theory. The key to this development is given by the non-perturbative duality symmetries which have modified and improved our understanding of string dynamics in many ways. In particular, by relating the fundamental objects of one theory to solitons of another theory, they have unraveled the presence of extended objects in the theory which stand on an equal footing with strings. We introduce these higher dimensional objects, named $D-branes$, and discuss applications of $D-brane$ physics.

INTRODUCTION

The electroweak and strong interactions are successfully described by a quantum field theory: the $SU(3) \times SU(2) \times U(1)$ Standard Model of particle physics. This model successfully accounts for all known experiments in particle accelerators with extraordinary accuracy. Adding General Relativity, which describes the large scale structure of the universe, this picture of nature can explain the behavior of matter and energy from the subnuclear to the cosmic domain with remarkable precission.

However, this representation of the fundamental interactions cannot be the final theory. On the one hand, the Standard Model is too arbitrary: there are more than 20 arbitrary parameters in the lagrangian and it is not clear why a certain pattern of gauge fields and multiplets appears and not another; it is also unnatural: the very massive particles mix with much lighter ones, thus destroying the hierarchy. On the other hand, many of the interesting solutions to General Relativity predict singularities where the theory breaks down. Furthermore at energies of the order of the Planck mass ($M_P \sim 10^{19} GeV$), where the quantum effects of gravity become relevant, Einstein's theory of General Relativity is non-renormalizable.

CP490, *Particles and Fields: Eighth Mexican School,* edited by J. C. D'Olivo et al.

Historically theoretical physics has developed by successive unifications, *i.e.* by describing different aspects of nature as different manifestations of a decreasing set of principles. Along this path several routes have been followed to overcome the problems posed by the Standard Model and by the unification of Quantum Mechanics and General Relativity. Some of the unifying proposals have been

* Grand Unification: The idea is to unify the three gauge interactions into one, for example $SU(5)$, $O(10)$, etc. This higher gauge symmetry allows to successfully predict some of the free parameters in the Standard Model.

* Higher spacetime dimensionality: The first attempt to unify gravity and electromagnetism, the Kaluza Klein mechanism, postulated five spacetime dimensions. The extra dimension is assumed to be highly curved and the world appears to be four-dimensional. The possibility of having more dimensions allows a single higher dimensional field to give rise to many four dimensional fields.

* Supersymmetry: It provides a link between bosons and fermions which notably reduces the divergence and naturalness problems.

These possibilities motivated an enourmous amount of work. They successfully deal with some of the problems mentioned above. However they do not lead to a consistent quantum theory of gravity. Let us look more closely into the problem of the divergences in quantum gravity. The resolution of a similar problem in the Fermi theory of the weak interaction has played a crucial role in the formulation of the Weinberg-Salam theory, so it is reasonable to assume that it might also help with gravity.

Treating gravity on an equal footing with other field theories one can try to construct the quantum theory of General Relativity by applying the corresponding Feynman rules. Similarly as in QED, divergences arise from loop integrals over large virtual momenta. However there is a major difference between quantum gravity and QED. The number of divergences in the latter theory is finite: all the infinities can be absorbed into a renormalization of the mass, the charge and the wave functions. In quantizing General Relativity, the number of these infinities appears to be limitless. This is because Einstein's lagrangian is quadratic in the first derivatives of the field variables which gives rise to the quadratic momentum dependence of the graviton vertex functions. The divergences grow worse with the order of the diagram. This reflects the dimensionality of Newton's constant G, since in natural units (*i.e.* $\hbar = c = 1$) it has dimensions of length squared. Therefore a dimensionless amplitude of order G^n diverges like $G^n p^{2n}$.

Presently there is only one known way to spread out the gravitational interaction and cut off the divergences: string theory. The basic principle behind constructing a quantum theory of relativistic one dimensional objects is quite simple. There are open strings, with endpoints, and closed strings, which are circles from a topological viewpoint (see figure (1)). All the elementary particles turn out to be different vibrational modes of the string. The propagation of a string from a spacetime configuration A to a spacetime configuration B is depicted in Fig. (2). During this motion the string sweeps out a two dimensional surface known as the string *world sheet* (see figure 2).

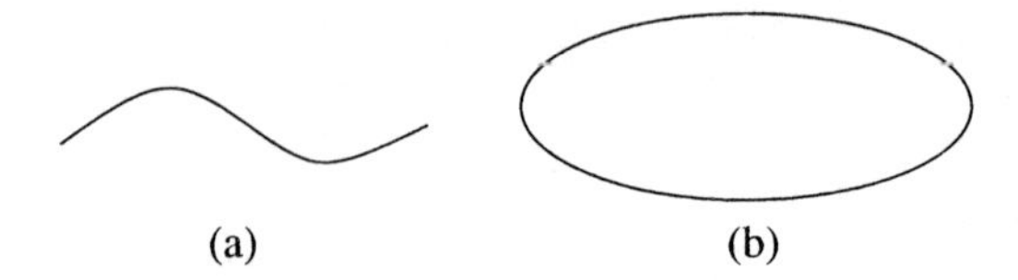

FIGURE 1. a) Open string. b) Closed string

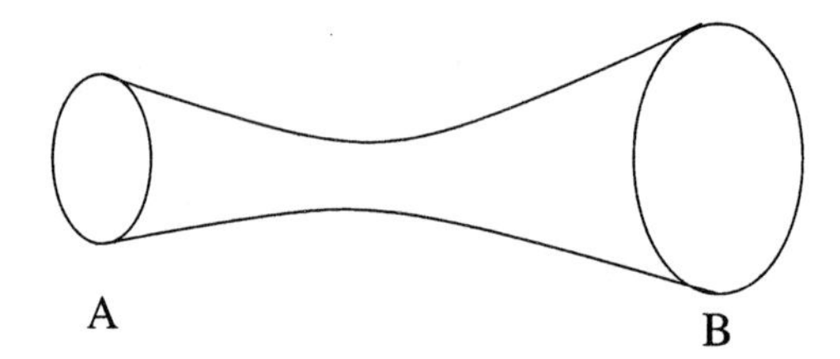

FIGURE 2. A closed string world sheet.

This theory has remarkable properties. It contains

** *Gravity*: the spectrum of every consistent string theory has a massless spin 2 state whose interactions reduce at low energy to General Relativity.

** *A consistent theory of quantum gravity*: this is in contrast to all known quantum field theories of gravity. There is a heuristic argument to understand why string theory is free of short distance divergences. In point particle theories the ultraviolet divergences occur when the interaction vertices coincide, *i.e.* when a=b=c=d in figure (3.a). In the corresponding string diagram (3.b) each particle world line becomes a cylinder and the interactions no longer occur at points.

** *Supersymmetry*: consistent string theories require spacetime supersymmetry.

** *Grand unification*: string theories give rise to gauge groups large enough to include the Standard Model. Some of them lead to the same gauge groups and fermion representations that arise in the unification of the Standard Model.

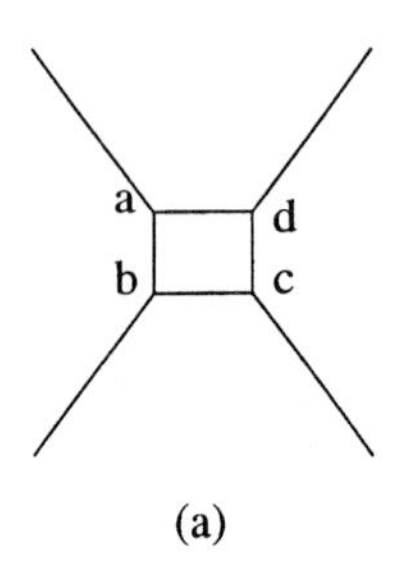

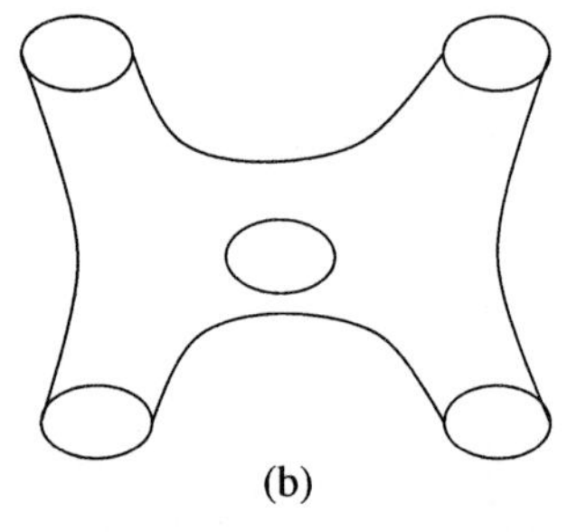

FIGURE 3. (a) One loop Feynman diagram with interaction points a, b, c, d; (b)the corresponding closed string diagram

** *Chiral gauge couplings*: String theory naturally allows chiral gauge couplings unlike previous unified theories which required parity symmetric gauge couplings (the gauge interactions in nature are parity asymmetric).

** *No free parameters*: string theory has no adjustable constants. Moreover there is no freedom analogous to the choice of gauge group representations in field theory.

** *Uniqueness*: There are only five consistent superstring theories at weak coupling. This should be contrasted with the enourmous amount of possible field theories of point particles. Moreover there is growing evidence that when non perturbative effects are taken into account these theories can be understood as different regions of the moduli space of one unique theory, named M theory.

This is a remarkable list arising from the simple supposition of one dimensional objects. However there are still unsolved questions and much remains to be done. The main problem is that while there is a unique theory, it has an enormous number of classical solutions. Upon quantization, each of these is a possible ground state (vacuum) and the physics is different in each.

Until recently our understanding of string theory was limited to perturbation theory. It has now become possible to obtain results at strong coupling and this has led to modify some of the principles previously considered to be fundamental. Spacetime, required to be ten dimensional in perturbation theory, reveals an eleventh dimension. Higher dimensional objects, named $D-branes$, were discovered to be on an equal footing with strings and it is no longer obvious that one dimensional objects should be the starting point in the formulation of the theory. Besides playing a crucial role in the development of non perturbative string theory, $D-branes$ provide a model of black holes that has led to important progress in the understanding of black hole thermodynamics and the microscopic structure of quantum gravity. More surprisingly $D-branes$ have led to a new duality conjecture between string theory on given backgrounds and certain quantum field theories. This novel duality allows a two way map between quantum gravity and gauge field theory.

This set of lectures is organized as follows. In Section I we briefly review string perturbation theory. We shortly outline the basic principles of weakly coupled string theory. In Section II we discuss duality symmetries and their consequences, including a short introduction to M-theory. Finally we present the conclusions and summarize some more recent developments.

I STRING PERTURBATION THEORY

In this section we briefly review the basic principles followed in the formulation of a quantum theory of one dimensional objects. For a complete treatment of string perturbation theory see [1,2].

The string is a one dimensional object which can be open, with endpoints, or closed. The coordinate along the string is conventionally denoted by σ and it runs from 0 to π. A timelike coordinate τ is useful to describe the motion of the string. As it moves in spacetime the string sweeps out a two dimensional surface, the world sheet (see figure (2)), which is mapped in spacetime by the functions $X^\mu(\sigma,\tau), \mu = 0, ..., D-1$. The action for the string is taken to be the area of the world sheet times the string tension, namely

$$S = \frac{1}{2\pi\alpha'}\int d\sigma d\tau \sqrt{h} h^{\alpha\beta}\partial_\alpha X^\mu \partial_\beta X^\nu \eta_{\mu\nu} \tag{1}$$

where $2\pi\alpha'$ is the inverse string tension, $h_{\alpha\beta}$ is the two dimensional metric of the world sheet, $\alpha, \beta = 0, 1$ refer to τ, σ respectively and $\eta_{\mu\nu}$ denotes Minkowski spacetime metric. This action defines a $(1+1)$-dimensional bosonic quantum field theory, but unlike conventional QFT the spatial direction has finite extent. The fields X^μ transform as vectors under spacetime Poincaré transformations and as scalars under reparametrizations of the world sheet.

The action (1) is invariant under Weyl scaling of the two dimensional metric $h_{\alpha\beta} \to e^\phi h_{\alpha\beta}$, under reparametrizations of the world sheet $\sigma^\alpha \to {\sigma^\alpha}'(\sigma^\alpha)$ and under spacetime Poincaré transformations. Weyl invariance distinguishes strings from other extended objects. Moreover higher dimensional generalizations of (1) lead to non renormalizable $(n+1)$-dimensional QFT.

The symmetries of the theory can be used to fix the two dimensional metric. In the conformal gauge $h_{\alpha\beta} = \eta_{\alpha\beta}$, the Euler Lagrange equation is simply the two dimensional wave equation. It must be supplemented with the constraint equation

$$\frac{\delta S}{\delta h^{\alpha\beta}} = 0 \tag{2}$$

which is proportional to the energy momentum tensor $T_{\alpha\beta}$ There are two possible boundary conditions corresponding to closed strings and open strings. The appropriate boundary condition for closed strings is periodicity of the fields, $X^\mu(\tau,\sigma) = X^\mu(\tau,\sigma+\pi)$. In the case of the open string, the variation of the action contains a surface term

$$-\frac{1}{2\pi\alpha'}\int d\tau X'_\mu \delta X^\mu|_{\sigma=0}^{\sigma=\pi} \tag{3}$$

There are two possibilities for the vanishing of this term, namely $X'_\mu = 0$ or $\delta X^\mu = 0$ at $\sigma = 0, \pi$. These are named Neumann and Dirichlet respectively.

The general solution to the massless wave equation can be written as a sum of right moving modes $X^\mu_R(\tau - \sigma)$ and left moving modes $X^\mu_L(\tau+\sigma)$. The periodicity requirement of closed strings determines

$$X^\mu_R = \frac{x^\mu}{2} - i\sqrt{\frac{\alpha'}{2}}\alpha^\mu_0(\tau-\sigma) + i\sqrt{\frac{\alpha'}{2}}\sum_{n\neq 0}\frac{\alpha^\mu_n}{n}e^{-2in(\tau-\sigma)} \tag{4}$$

$$X^\mu_L = \frac{x^\mu}{2} - i\sqrt{\frac{\alpha'}{2}}\tilde{\alpha}^\mu_0(\tau+\sigma) + i\sqrt{\frac{\alpha'}{2}}\sum_{n\neq 0}\frac{\tilde{\alpha}^\mu_n}{n}e^{-2in(\tau+\sigma)} \tag{5}$$

where $\alpha^\mu_n, \tilde{\alpha}^\mu_n$ are Fourier components which will be interpreted as oscillator coordinates. x^μ and $k^\mu = (2alpha)^{-1/2}(\alpha^\mu_0 + \tilde{\alpha}^\mu_0)$ are the center of mass position and momentum of the string respectively. X_R^μ and X_L^μ are required to be real functions, thus x^μ and k^μ are real and α_{-n}^μ is the adjoint of α_n^μ, *i.e.* $\alpha_{-n}^\mu = (\alpha_n^\mu)^\dagger$, $\tilde{\alpha}_{-n}^\mu = (\tilde{\alpha}_n^\mu)^\dagger$.

For the open string the Neumann boundary conditions combine the left and right moving modes into standing waves, namely

$$X^\mu(\sigma,\tau) = x^\mu + 2\alpha' k^\mu \tau + i\sqrt{2}\alpha' \sum_{n\neq 0} \frac{\alpha_n^\mu}{n} e^{-in\tau} cos n\sigma \qquad (6)$$

Historically Dirichlet boundary conditions were not considered until 1989 [6] because they break Poincaré invariance. However it has recently been realized that they play an important role as we will discuss below.

We now turn to the quantization of the bosonic string theory. Several procedures can be followed. In the path integral approach the prescription is to sum the exponential of the action (1) over all possible two dimensional metrics and spacetime fields, *i.e.* formally

$$\int DhDX^\mu e^S \qquad (7)$$

Here we will follow the most traditional approach where the Fourier oscillators are promoted to quantum operators satisfying the following commutation relations

$$[x^\mu, k^\nu] = i\eta^{\mu\nu} \qquad (8)$$

$$[\alpha_m^\mu, \alpha_n^\nu] = [\tilde{\alpha}_m^\mu, \tilde{\alpha}_n^\nu] = m\delta_{m+n}\eta^{\mu\nu}; \quad [\alpha_m^\mu, \tilde{\alpha}_n^\nu] = 0 \qquad (9)$$

These allow to naturally interpret the α_m^μ as harmonic oscillator raising and lowering operators for negative and positive m respectively. The ground state can be defined to be annihilated by the $\alpha_m^\mu, m > 0$,

$$\alpha_m^\mu |0; k >= 0, \quad m > 0 \qquad (10)$$

and similarly for $\tilde{\alpha}_m^\mu$. The center of mass momentum k^μ is necessary to specify the state.

The Fock space is built up by applying the raising operators $\alpha_m^{\mu\dagger}$ to the ground state. It is not positive definite since

$$[\alpha_m^0, \alpha_m^{0\dagger}] = -1 \qquad (11)$$

and therefore a state of the form $\alpha_m^{0\dagger}|0; k >$ has negative norm. The physical states of the theory cannot include these 'ghosts'. Indeed, we still have to impose the constraints, which in the classical theory were shown to correspond to the vanishing of the energy momentum tensor. The corresponding statement in the quantum theory is the requirement that the Fourier modes of the energy momentum tensor annihilate the physical states, namely

$$L_m|\phi>=0,\ m>0 \quad ; \quad L_0|\phi>=a|\phi> \tag{12}$$

where

$$L_m = \frac{1}{2}\sum_{-\infty}^{\infty} \alpha_{m-n} \cdot \alpha_n \tag{13}$$

In the closed string theory we have to consider also the operators $\tilde{L}_m$.

Notice that when $m = 0$ there are normal ordering ambiguities in L_0 and consequently a normal ordering constant a has to be introduced in (12). It can be shown that a ghost free spectrum is only possible for $a = 1$ and spacetime dimension $D = 26$.

The Fourier modes of the energy momentum tensor (13) satisfy the Virasoro algebra,

$$[L_m, L_n] = (m-n)L_{m+n} + \frac{c}{12}m(m^2-1)\delta_{m+n} \tag{14}$$

where c is the central charge.

The mass of a string state can be determined from the second of eqs. (12) by recalling that conventionally $\alpha_0^\mu \propto k^\mu$, thus

$$\alpha' M^2 = -1 + \sum_{n=1}^{\infty} \alpha_{-n} \cdot \alpha_n \tag{15}$$

for the open string and

$$\alpha' M^2 = -4 + 4\sum_{n=1}^{\infty} \alpha_{-n} \cdot \alpha_n = -4 + 4\sum_{n=1}^{\infty} \tilde{\alpha}_{-n} \cdot \tilde{\alpha}_n \tag{16}$$

for the closed string. This shows that the ground state is a tachyon in both the open and closed string theory.

The angular momentum operators

$$J^{\mu\nu} = x^\mu k^\nu - x^\nu k^\mu - i\sum_{n=1}^{\infty} \frac{1}{n}(\alpha_{-n}^\mu \alpha_n^\nu - \alpha_{-n}^\nu \alpha_n^\mu) \tag{17}$$

commute with the Virasoro generators L_m. Therefore the physical state conditions are invariant under Lorentz transformations and the physical states are guaranteed to form Lorentz multiplets.

Let us recall the spectrum of the open bosonic string. We describe it iin the so called light-cone gauge which has the advantage of giving only physical states. The ground state $|0;k>$ is a tachyon with $\alpha' M^2 = -1$; the first excited state $\alpha_{-1}^i|0;k>$ is a massless vector with 24 independent transverse polarizations ($i = 1, ..., 24$). At higher mass levels there is an infinite tower of massive states with $\alpha' M^2 = N - 1$, where $N = 2, 3, ..., \infty$. Gauge groups can be introduced in the open string theory since the endpoints are special points in this theory. It is possible to assume that the open string carries charges at the endpoints and one can introduce a $U(n)$ symmetry group that acts on these charges.

The spectrum of the closed string is given by tensor products of open string states with themselves subject to the condition $(L_0 - \tilde{L}_0)|\phi >= 0$. The ground state is again a tachyon with $\alpha' M^2 = -4$. The next level contains a set of massless states of the form $\alpha^i_{-1}\tilde{\alpha}^j_{-1}|0; k>$ with $SO(24)$ quantum numbers. It can be decomposed into a symmetric and traceless part which transforms under $SO(24)$ as a spin two particle: this is the graviton; a trace term which is a scalar, called the dilaton; and an antisymmetric part transforming as an antisymmetric second rank tensor. There is also an infinite tower of massive states with $\alpha' M^2 = 4(N-1), N = 2, 3, ...\infty$.

The tachyons present in the spectrum of both bosonic string theories can be eliminated by introducing fermions in a spacetime supersymmetric way. As we will see in the next subsection this also reduces the spacetime dimensionality from 26 to 10 dimensions.

A Superstring theory

There are two formulations of superstring theory: the Ramond-Neveu-Schwarz (RNS) [3] and the Green Schwarz (GS) [4] formalisms. The former allows the use of powerful conformal field theory methods of computation [7] but spacetime supersymmetry has to be imposed by hand; the GS description makes spacetime supersymmetry manifest but quantization becomes difficult. Here we briefly review the RNS formulation which relates the spacetime coordinates $X^\mu(\sigma, \tau)$ to fermionic partners $\psi^\mu(\sigma, \tau)$ in a world sheet supersymmetric way. The generalization of the bosonic string action (1) is

$$S = \frac{1}{4\pi} \int d^2z \{\partial X^\mu \overline{\partial} X_\mu - \psi^\mu \overline{\partial} \psi_\mu - \tilde{\psi}^\mu \partial \tilde{\psi}_\mu\} \tag{18}$$

As is conventional we have set $\alpha' = 2$. We have introduced complex world sheet coordinates $z = \sigma + i\tau, \overline{z} = \sigma - i\tau$ and taken an euclidean two dimensional metric. In this notation ∂ denotes $\partial/\partial z$ and $\overline{\partial} = \partial/\partial\overline{z}$. The fields $\psi(\tilde{\psi})$ are respectively holomorphic and antiholomorphic Majorana fermions in two dimensions. They can be taken to be periodic or antiperiodic as z rotates by 2π about the origin, $\psi^\mu(e^{2\pi i}z) = \pm\psi^\mu(z)$. Similarly there are two possible periodicities for $\tilde{\psi}(\overline{z})$. Consequently there are half integer and integer modings for the fields respectively,

$$\psi^\mu(z) = \sum_n \psi^\mu_n \, z^{-n-1/2}; \quad \tilde{\psi}^\mu(\overline{z}) = \sum_n \tilde{\psi}^\mu_n \, \overline{z}^{-n-1/2} \tag{19}$$

In the open superstring theory the surface term arising in the variation of the action requires $\psi^\mu = \pm\tilde{\psi}^\mu$ at each endpoint. Fixing conventionally the plus sign at $\sigma = 0$, the two possible choices at $\sigma = \pi$ determine the Ramond $(\psi^\mu(\pi, \tau) = +\tilde{\psi}^\mu(\pi, \tau))$ or Neveu Schwarz $(\psi^\mu(\pi, \tau) = -\tilde{\psi}^\mu(\pi, \tau))$ sectors of the theory. In the closed superstring theory one can take independently periodic or antiperiodic boundary conditions for ψ and $\tilde{\psi}$, thus defining four sectors referred to as R-R, R-NS, NS-R and NS-NS.

In the supercovariant gauge, the dynamics of X^μ and ψ^μ are given by the Klein Gordon and Dirac equations respectively. These have to be supplemented with the constraint equations, namely the vanishing of the energy momentum tensor and of the supersymmetry generator. To quantize the superstring theory one can follow the steps discussed above for the bosonic string theory. The bosonic fields are exactly as above. The fermionic modes satisfy the anticommutation relations,

$$\{\psi_m^\mu, \psi_n^\nu\} = \delta_{m+n}\eta^{\mu\nu} \tag{20}$$

Notice that the zero modes verify a Dirac gamma matrix algebra.

The Fock space is now constructed from the ground state $|0;k>$ which satisfies

$$\alpha_m^\mu|0;k>=\psi_m^\mu|0;k>=0, \quad m>0, \tag{21}$$

by applying creation operators $\alpha_{-m}^\mu, \psi_{-m}^\mu$. Similarly as above, the constraints have to be imposed on the physical states, namely

$$G_r|\phi>=L_m|\phi>=0, \quad r,m>0 \tag{22}$$

where

$$G_r = \sum_n \alpha_{-n}\cdot\psi_{r+n} \tag{23}$$

is the supersymmetry generator and

$$L_m = \frac{1}{2}\sum_{n=-\infty}^{\infty} :\alpha_{-n}\cdot\alpha_{m+n}: + \frac{1}{2}\sum_{r=-\infty}^{\infty}(r+\frac{1}{2}m):\psi_{-r}\cdot\psi_{m+r}: \tag{24}$$

is the Virasoro generator. Here r is half integer in the NS sector and integer in the R sector. The normal ordering ambiguity for the zero mode of the energy momentum tensor is fixed as

$$L_0|\phi>=\frac{1}{2}|\phi> \text{ in NS }; \qquad L_0|\phi>=0 \text{ in R} \tag{25}$$

and similarly for the antiholomorphic operators. In this case ten spacetime dimensions are necessary to decouple the ghosts.

Let us discuss the spectrum of the open superstring. The mass shell condition receives now contributions also from the fermionic modes. The ground state is a tachyon in the NS sector, $\alpha' M^2 = -\frac{1}{2}|0;k>$; the first excited state is a massless $SO(8)$ vector $\psi_{-1/2}^i|0>$; and then there is an infinite tower of massive states. In the R sector the ground state is a massless Majorana Weyl spinor of $SO(8)$. The massive states organize into representations of $SO(9)$. The spectrum of the closed superstring theory is a product of two copies of the open superstring spectrum.

The superstring theory discussed so far is an inconsistent quantum theory even in ten spacetime dimensions. Gliozzi, Scherk and Olive [5] proposed a truncation of the spectrum, known as GSO projection, which keeps states with even number of fermions. This not only gets rid of the NS tachyon but also produces a spacetime supersymmetric theory, leaving an equal number bosons and fermions at every mass level.

There are five known fully consistent superstring theories in 10 dimensions, namely type IIA, IIB, I, heterotic with gauge group $E_8 \times E_8$ and $SO(32)$ string theories. The type II and heterotic theories are theories of closed strings only whereas the type I theory contains also open strings. Let us briefly review the massless spectrum of these theories in the light-cone gauge.

Type II theories: In this case the GSO projection keeps only states with even number of holomorphic and even number of antiholomorphic fermions. Here there is an ambiguity since the ground state in each of the four sectors can be assigned either an even or odd fermion number. Consistency leaves only two possibilities known as type IIB and type IIA theories. They are defined by assigning the same and opposite fermion number to the left and right moving R ground states respectively. States from the R sector are in the spinor representation of $SO(9,1)$ whereas those from the NS sector are in the tensor representation. Since the product of two spinor representations gives a tensor representation, the states from the NS-NS and R-R sectors are bosonic, and those from the NS-R and R-NS are fermionic. Since type IIA and IIB theories differ only in their R sector, the bosonic states in the NS-NS sector are identical in the two theories. They are a symmetric rank two tensor $G_{\mu\nu}$, an antisymmetric rank two tensor $B_{\mu\nu}$ and a scalar field Φ named the dilaton. The massless states in the RR sector of the IIA theory are a vector A_μ and a rank three antisymmetric tensor $A_{\mu\nu\rho}$ whereas in the type IIB theory they are a scalar χ, a rank two antisymmetric tensor $A_{\mu\nu}$ and a rank four antisymmetric tensor $A_{\mu\nu\rho\sigma}$ with self dual field strength.

The supersymmetry algebra of type IIB theory is known as the chiral $N = 2$ superalgebra and that of type IIA is the non-chiral $N = 2$ superalgebra. Both of them have 32 supersymmetry generators.

Type I theory: Here the world sheet theory is that of the type IIB superstring with two differences. *i*) Type IIB has a symmetry that exchanges the left and right sectors in the world sheet, known as world sheet parity transformation. Type I theory keeps only states in the spectrum which are invariant under this world sheet symmetry. *ii*) Type I theory also includes open string states.

The spectrum of massless bosonic states consists of a symmetric rank two tensor $G_{\mu\nu}$ and a scalar dilaton from the NS-NS closed string sector, an antisymmetric rank two tensor from the R-R closed string sector and 496 gauge fields in the adjoint representation of $SO(32)$ from the open string sector. This spectrum is invariant under the chiral $N = 1$ supersymmetry algebra with 16 generators.

Heterotic theories: The world sheet theory consists of the scalar fields X^{μ} and 8 right moving and 32 left moving Majorana Weyl fermions. In the right moving sector we have as before NS states transforming in the tensor representation and R states transforming in the spinor representation of $SO(9,1)$. However now, unlike in the type II case, the boundary conditions on the left moving fermions do not affect the Lorentz transformation properties of the state. Therefore bosonic states come from states with NS boundary conditions on the right moving fermions and fermionic states come from states with R boundary conditions on the right moving fermions.

There are two possible boundary conditions on the left moving fermions which give rise to two fully consistent string theories. They are:

$SO(32)$ **heterotic string**: Here all the left moving fermions have the same boundary conditions, either all of them periodic or antiperiodic. The GSO projection keeps only states with even number of left moving fermions. The massless bosonic states are a symmetric rank two tensor $G_{\mu\nu}$, an antisymmetric rank two tensor $B_{\mu\nu}$, a scalar field Φ and a set of 496 gauge fields in the adjoint representation of $SO(32)$.

$E_8 \times E_8$ **heterotic string**: In this case the 32 left moving fermions are divided into two groups of 16 each which give rise to four possible boundary conditions. The GSO projection keeps only the states with even number of left moving fermions from both groups. The massless bosonic states are $G_{\mu\nu}$, $B_{\mu\nu}$, Φ and 496 gauge fields filling up the adjoint representation of $E_8 \times E_8$.

The spectrum of both heterotic theories is invariant under the chiral $N = 1$ algebra with 16 supersymmetry generators.

B Interactions

So far we have been discussing free strings. When interactions are included, the theory becomes more interesting. Unlike in QFT where one can choose the interaction vertices, in string theory once we specify the theory there is only one possible interaction vertex. The basic interaction can be viewed as a process in which a single string breaks into two or else two strings join to give a single one. Consider for example a scattering process involving four external strings (figure 4). The prescription for computing the scattering amplitude is to sum over all possible string world sheet bounded by the four strings, weighting with the exponential of the action. If we imagine the time axis running from left to right then the diagram (4.a) represents two strings joining into one string and then splitting into two strings: the analog of a tree diagram in field theory. A more complicated surface is shown in figure (4.b) It represents two strings joining into one, which splits into two, join into one which splits into two again. This is the analog of a one loop graph in field theory. The world sheet of an n-loop closed string diagram is a sphere with n handles which is known as a Riemann surface of genus n.

The relative normalization between the two diagrams in Fig. (4) introduces an arbitrary parameter in string theory, the string coupling constant g_s. However

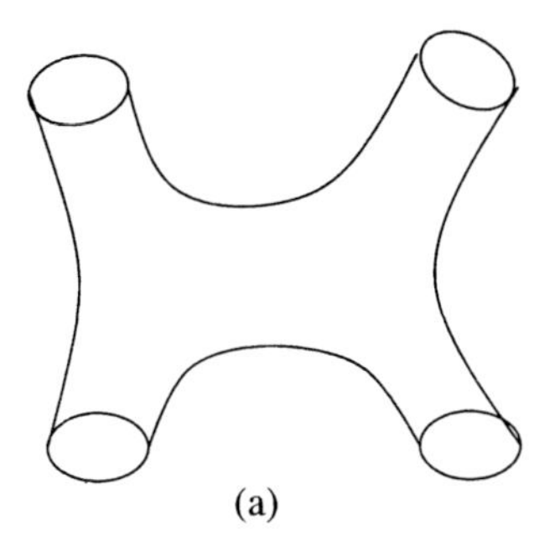

(a)

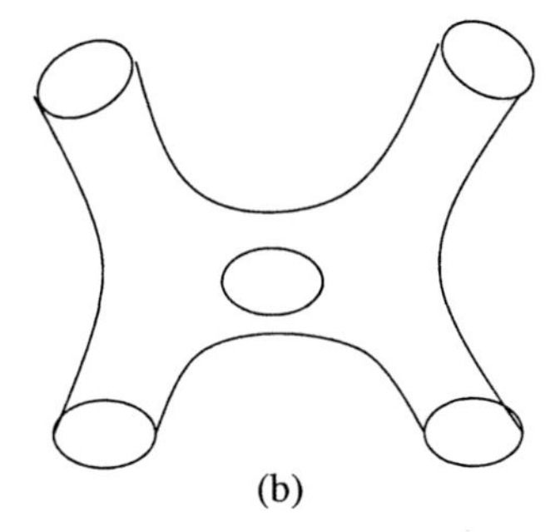

(b)

FIGURE 4. (a) A tree level scattering process in string theory. (b) A one loop level scattering process in string theory

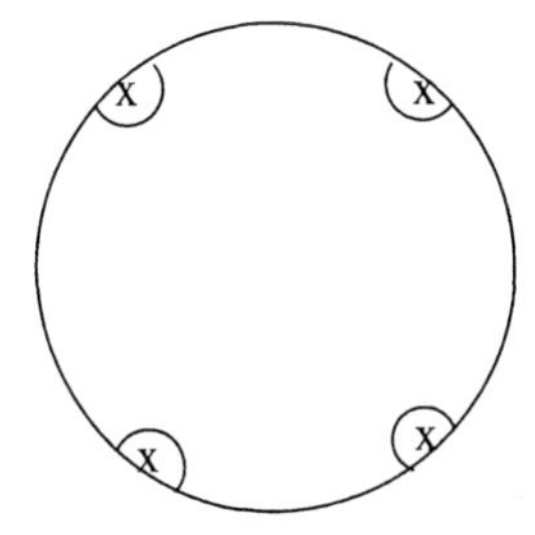

FIGURE 5. Conformal invariance makes it possible to compactify the world sheet, closing the holes corresponding to incoming and outgoing particles. The external string states are projected to points where local vertex operators must be inserted

once the relative normalization between these two diagrams is fixed, the relative normalization among all other diagrams is fixed. Thus besides the dimensionful parameter α', string theory has a single dimensionless coupling constant. We will see that both these parameters can be absorbed into definitions of various fields in the theory.

The more relevant quantity is the scattering amplitude where the external strings are in the eigenstates of the energy and momenta operators. There is an efficient way of doing this calculation: the so called vertex operators. Using the conformal symmetry of the theory it is possible to map the world sheets of figure (4) into compact surfaces where the external string states appear as vertex operators inserted at points (see figure 5). Every on shell physical state has associated a vertex operator with suitable properties. We have already discussed a heuristic argument to justify the ultraviolet finiteness of these diagrams. This property can be verified by explicit calculations. (See references [8] and [9] for details of these calculations).

Our main interest is in the scattering involving external massless states. The most convenient way to summarize the result of this computation is to specify the effective action. By definition the effective action is such that if we compute the tree level scattering amplitude using this action we should reproduce the S-matrix elements involving the massless states of string theory. In general such effective action will contain an infinite number of terms, but they can be organized according to the number of spacetime derivatives that appear. Terms with the lowest number of derivatives give the dominant contribution.

The low energy effective action of all five superstring theories have been found. The actions for the type IIA and type IIB strings correspond to those of IIA and IIB supergravity in ten dimensions respectively. The actions for the heterotic and type I theories correspond to $N = 1$ supergravity coupled to $N = 1$ super Yang Mills with gauge group $SO(32)$ for the type I and heterotic $SO(32)$ and with gauge group $E_8 \times E_8$ for the $E_8 \times E_8$ heterotic theory.

The emergence of the Einstein-Hilbert action for gravity in the effective action is the most striking result of string theory. This combined with the result of finiteness of the scattering amplitudes shows that string theory gives a finite theory of quantum gravity.

The effective actions for all five string theories are invariant under the simultaneous transformation of the dilaton and string coupling,

$$\Phi \to \Phi - 2C \quad g_s \to e^C g_s \tag{26}$$

therefore g_s can be absorbed into Φ and it has no physical significance. Similarly one can argue that even the string tension α' is physically meaningless. Indeed, since α' is the only dimensionful parameter in the theory, the effective action has an invariance under

$$\alpha' \to \lambda\alpha' \quad G_{\mu\nu} \to \lambda G_{\mu\nu} \tag{27}$$

and thus α' can be absorbed into the definition of $G_{\mu\nu}$.

C Compactification

All five superstring theories live in ten spacetime dimensions. Since our world is $(3 + 1)$ dimensional these are not realistic theories. String theories in less dimensions can be constructed using the idea of compactification, *i.e.* taking the $(9 + 1)$ background manifold as the product of $(3 + 1)$ spacetime times a 6 dimensional compact manifold with euclidean signature, *i.e.* $M^4 \times K_6$. The internal dimensions are highly curved and cannot be detected by the present day experiments, thus the world effectively appears to be four dimensional.

The internal manifold cannot be arbitrary, it has to satisfy the equations of motion of the effective field theory. One also considers only those manifolds that preserve part of the spacetime supersymmetry of the original ten dimensional theory, since this guarantees the vanishing of the cosmological constant and hence consistency of the corresponding string theory order by order in perturbation theory. There are many known examples of manifolds satisfying these restrictions, *e.g.* tori, K3, Calabi Yau manifolds, etc. There is an alternative procedure to determine the allowed background manifolds. One can modify the world sheet action (1) replacing $\eta_{\mu\nu}$ by $\eta_{ab} + g_{ij}$ $(a, b = 0, ..., 3; i, j = 4, ..., 9)$. This turns the free theory into a non linear σ-model. Requiring that this theory be conformally invariant leads to the same field equations for the background fields [10]. Therefore consistency of string theory restricts the kind of manifold on which the string can propagate. One can even abandon the geometrical picture and choose group manifolds for the internal space [11].

It is instructive to work out one simple example of compactification. Let us discuss the bosonic string theory in (1 + 24) dimensional Minkowski space times the circle S^1, *i.e.* $M^{25} \times S^1$. This is the simplest example of compactification and yet it illustrates many interesting features. In particular the compactification of the open string theory in this background led to the discovery of $D-branes$, extended objects which have played a crucial role in the recent developments of the theory.

Let us start with the closed string theory. Consider the contribution of the zero modes to the closed string fields (5)

$$X^\mu = X_L^\mu + X_R^\mu = x^\mu - i\sqrt{\frac{\alpha'}{2}}(\alpha_0^\mu + \tilde{\alpha}_0^\mu)\tau + \sqrt{\frac{\alpha'}{2}}(\alpha_0^\mu - \tilde{\alpha}_0^\mu)\sigma \tag{28}$$

For non compact dimensions, X^μ is periodic. Therefore the vanishing of δX^μ when $\sigma \to \sigma + 2\pi$ implies that $\alpha_0^\mu = \tilde{\alpha}_0^\mu$. For compact dimensions instead, for instance $X^{25} \to X^{25} + 2\pi\omega R$, R being the radius of compactification,

$$\alpha_0^{25} - \tilde{\alpha}_0^{25} = \omega R\sqrt{\frac{2}{\alpha'}} \tag{29}$$

Single valuedness of the wave function $e^{ik\cdot X}$ requires $\alpha_0^{25} + \tilde{\alpha}_0^{25} = \sqrt{2\alpha'}n/R$ and thus

$$\alpha_0^{25} = \sqrt{\frac{\alpha'}{2}}\left(\frac{n}{R} + \frac{\omega R}{\alpha'}\right) \quad ; \quad \tilde{\alpha}_0^{25} = \sqrt{\frac{\alpha'}{2}}\left(\frac{n}{R} - \frac{\omega R}{\alpha'}\right) \tag{30}$$

ω and n are integers denoted winding and Kaluza Klein modes respectively.

Now the mass spectrum in 25 dimensional spacetime gets modified to

$$M^2 = -k^\mu k_\mu = \frac{n^2}{R^2} + \frac{\omega^2 R^2}{\alpha'^2} + \frac{2}{\alpha'}(N-2) \tag{31}$$

where $N = \sum_n(\alpha_{-n}\cdot\alpha_n + \tilde{\alpha}_{-n}\cdot\tilde{\alpha}_n)$ is the oscillator number and now $\mu = 0, ..., 24$. It is interesting to note that the mass spectra of the theories at radius R and $\frac{\alpha'}{R}$ are identical after interchanging the winding and Kaluza Klein modes, $\omega \leftrightarrow n$, which amounts to $\alpha_0^{25} \to \alpha_0^{25}$, $\tilde{\alpha}_0^{25} \to -\tilde{\alpha}_0^{25}$.

Suppose the theory is formulated in terms of a new field $X'^{25} = X_L^{25} - X_R^{25}$. The energy momentum tensor and the correlation functions do not change since the signs enter in pairs. The only change is that the spectrum of zero modes in this new variable is that of the old theory at radius $R' = \alpha'/R$. The two theories are said to be T-dual and X'^{25} is the natural coordinate for the dual theory. It can be shown that T-duality is an exact symmetry of perturbative closed string theory.

Notice that as $R \to \infty$, the decompactification limit, the winding modes become infinitely massive while the Kaluza Klein modes go to a continuous spectrum. This behavior is, as expected, similar to having an additional noncompact dimension. Surprisingly, the same behavior is obtained when $R \to 0$, though now the infinitely massive states are the Kaluza Klein modes whereas the winding states approach a continuum. Thus as the radius goes to zero the spectrum again seems to approach that of a 26 dimensional theory. This equivalence between the $R \to 0$ limit and the $R \to \infty$ limit is completely different from the behaviour of point particle theories and is another indication that strings see geometry very differently at short distances. The space of inequivalent theories is the half line $R \leq \alpha'^{1/2}$ and thus there is no radius smaller than the self dual radius which is of the order of the string length. However when non perturbative effects are considered we will see that there appears structure at shorter distance.

The situation is different for the open string theory. Unlike closed strings, open strings cannot wind. Therefore when $R \to 0$ the Kaluza Klein modes become infinitely massive, but there is no continuum of states and the compact coordinate desappears, leaving a 25 dimensional theory. But the problem is that open strings couple to closed strings and the latter live in 26 dimensions when $R \to 0$. The apparent paradox is resolved in the T-dual formulation of the open string theory as follows.

Let us write the T-dual open string theory in terms of the new coordinate X'^{25}, namely

$$\begin{aligned} X'^{25}(\sigma, \tau) &= 2\alpha' k^{25}\sigma + oscillators \\ &= \frac{2\alpha' n}{R}\sigma + oscillators \end{aligned} \tag{32}$$

The oscillator terms vanish in the string endpoints $\sigma = 0$ and π, and noticing that there is no dependence of the zero modes on τ, we conclude that the endpoints do not move in the direction X'^{25}. This can also be seen from the boundary conditions

$$\begin{aligned} \partial_\sigma X^{25} &= \partial_z X^{25}\frac{\partial z}{\partial \sigma} + \partial_{\bar{z}} X^{25}\frac{\partial \bar{z}}{\partial \sigma} \\ &\sim \partial X'^{25}\frac{\partial z}{\partial \tau} + \partial_{\bar{z}} X'^{25}\frac{\partial \bar{z}}{\partial \tau} = \partial_\tau X'^{25} \end{aligned} \tag{33}$$

i.e. the Neumann boundary condition of the original coordinate X^{25} becomes Dirichlet boundary condition for the dual coordinate X'^{25}. Thus the endpoints in the dual theory are constrained to lie on a 25 dimensional hyperplane, they are fixed and the string can now wind.

Let us compute the compact coordinate of the endpoints

$$X'^{25}(\pi) - X'^{25}(0) = 2\pi\alpha' k^{25} = 2\pi\alpha'\frac{n}{R} = 2\pi n R' \tag{34}$$

The total change in X'^{25} between the two ends is an integral multiple of the periodicity of the dual dimension, so the endpoints lie on the same hyperplane in the periodic T-dual space as shown in figure (6).

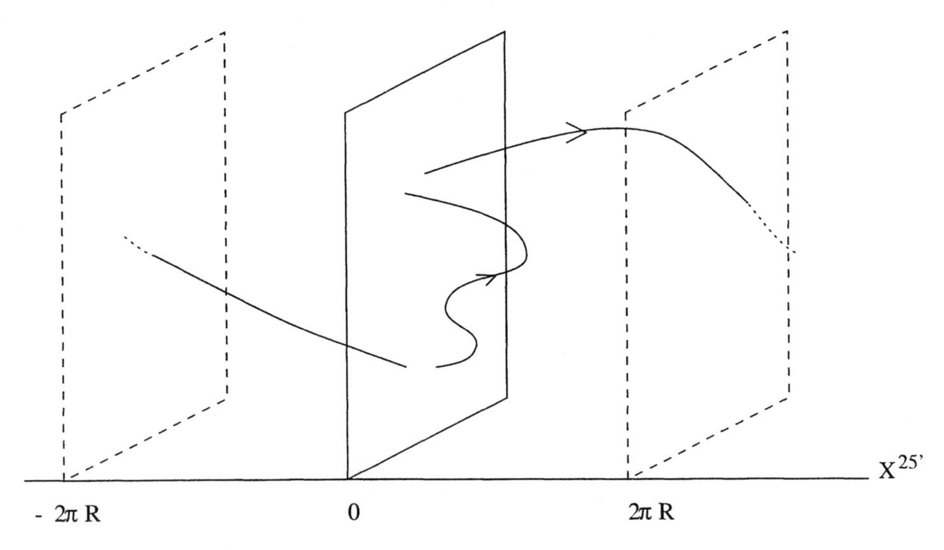

FIGURE 6. Schematic representation of the moduli space of a string

All this can be generalized to several periodic coordinates $X^m = \{X^{25}, X^{24}, ..., X^{p+1}\}$, by rewriting them in terms of the dual coordinates X'^m. In this case the endpoints are restricted to lie on $(p+1)$-dimensional hyperplanes. The Neumann boundary conditions become Dirichlet boundary conditions for the dual coordinates. The $(p+1)$-dimensional hypersurface is the world volume of an extended p dimensional object named a Dirichlet p-brane or $D-brane$.

Notice that although we have found the $D-branes$ through compactification these objects exist also in non compact spacetime: a Dp-brane is a $(p+1)$ dimensional hypersurface where the endpoints of the open string are fixed.

The $D-branes$ are dynamical objects and their properties, at weak coupling, can be determined with the same perturbative tools used elsewhere in string theory. It is interesting to compute the $D-brane$ tension, *i.e.* the mass per unit volume of a $D-brane$. A convenient way to do this is to consider the interaction between two parallel identical $D-branes$. In the bosonic string theory there is an attractive force between the branes due to the mutual exchange of gravitons and dilatons. The result of the calculation turns out to be proportional to the inverse string coupling, thus the $D-branes$ are very heavy at weak coupling.

Another important property of $D-branes$ is that they are smaller than strings. This follows from calculations of the relevant form factors.

The calculation of the force between two $D-branes$ is more interesting in the superstring theory where the attractive force due to the NS-NS states is cancelled by a repulsion due to the exchange of R-R states. In fact, the world volume of a $Dp-brane$ couples naturally to a $(p+1)$-form field and it can be shown that the $Dp-brane$ carries one unit of charge under the R-R $(p+1)$-form gauge field. Therefore in type IIA string theory there are branes with p even whereas in type IIB string they have p odd (recall the spectrum of R-R states in these theories). The net force between two parallel identical branes vanishes reflecting the supersymmetry of the spectrum and the fact that $D - branes$ are BPS states (see below). This property has important consequences for the non perturbative formulation of the theory which we discuss in the following Section. For a complete review of $D - branes$ see reference [12].

II DUALITY SYMMETRIES IN STRING THEORY

We have seen in the previous section that there are many consistent string theories. Not only there are five superstring theories in ten dimensions but one can also choose many possible background geometries as well as other background fields consistent with the string equations of motion. This large number of possibilities goes against the predictibility of the theory. In this Section we will see that the duality symmetries discovered in recent years give relations among different theories which reduce the degree of degeneracy. Moreover they suggest the existence of one theory in eleven dimensions, M theory, from which the five superstring theories appear to be just different limits. We have already seen an example of duality when we introduced the $D - branes$. For a complete review of this subject see reference [13].

Each of the possible string theories discussed above can be characterized by a set of parameters known as *moduli*. Examples of moduli are the string coupling constant, the shape and size of the compact manifold, the various other background fields, etc. There is a region of the moduli space of the theory where the string coupling is weak and perturbation theory is valid.

String duality provides an equivalence map between different regions in the moduli space of one theory or of two theories. In general a duality maps the weak coupling region of one theory to the strong coupling region of another theory and viceversa. Since we only have a perturbative formulation of string theory and the duality relations involve strongly coupled string theory, the duality equivalences remain at the level of conjectures and cannot be proved. One can only prove the consequences of the proposed dualities at weak coupling and in theories with enough supersymmetry where the non renormalization theorems allow non trivial tests of the duality conjectures.

A schematic representation of the moduli space of string theory is shown in figure (7.a). The black circle represents the weak coupling region and the rest corresponds to strong and intermediate coupling. Figure (7.b) represents a duality map between the strongly coupled region of theory A on a given background $\mathcal{M}$ and the weakly coupled region of another theory B on a background $\mathcal{M}'$ and viceversa.

Under duality typically perturbation expansions get mixed up. For example tree level results of one theory might correspond to perturbative and non perturbative contributions in the dual theory. Also under duality many of the elementary string states in one theory get mapped to solitons and their bound states in the dual

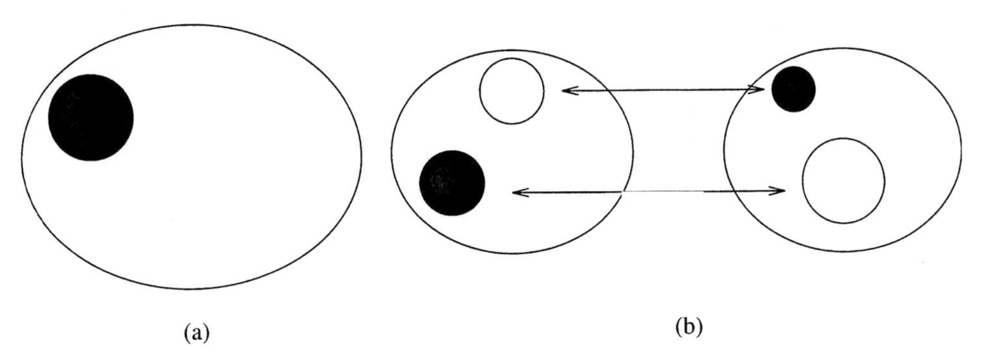

FIGURE 7. (a) Schematic representation of the moduli space of a string theory. The black circle denotes the weak coupling region. b) Duality map between the strongly coupled region of one theory and the weakly coupled region of another and viceversa

theory.

There are other special cases of duality. For example we have already discussed T-duality which maps the weak coupling region of one theory to the weak coupling region of another theory or of the same theory. Similar arguments as those that led us to conclude the self duality of the closed bosonic string theory can be followed to show that either of the two heterotic string theories compactified on a circle of radius R is dual to the same theory compactified on a circle of radius α'/R at the same value of the coupling constant. For the type II theories, it can be shown that type IIA string theory compactified on a circle of radius R is dual to type IIB string theory compactified on a circle of radius α'/R.

In the case of self duality, the duality transformations form a symmetry group that acts on the moduli space of the theory. For example type IIB superstring theory in ten dimensions is conjectured to have an $SL(2, Z)$ self-duality group.

In a generic situation duality can relate not just two theories but a whole chain of theories. For example type IIA superstring theory compactified on K3 is related to the heterotic string on T^4. On the other hand due to the equivalence of the $SO(32)$ heterotic string and type I superstring theories in ten dimensions, these theories are also equivalent when compactified on T^4.

From this discussion we see that the presence of duality in string theory has two important consequences. On one hand it reduces the degree of non uniqueness of the theory by relating various apparently unrelated (compactified) string theories. On the other hand it allows to study a strongly coupled string theory by mapping it to a weakly coupled dual theory.

In order to prove the duality conjectures we must be able to analyze the theories at strong coupling. But this seems impossible since we only know how to define the theory perturbatively at weak coupling. However fortunately string theory is supersymmetric. Supersymmetry gives rise to certain non renormalization theorems in string theory, due to which some of the weak coupling calculations can be trusted even at strong coupling. Thus we can focus our attention on such non renormalized quantities and ask if they are invariant under the proposed duality transformations. Testing the duality invariance of these quantities provides us with various tests of the duality conjectures.

The number of supersymmetries present in the theory determine the precise content of these non renormalization theorems. The maximum number of supersymmetry generators that can be present in a string theory is 32. This gives $N = 2$ supersymmetry in ten dimensions and $N = 8$ supersymmetry in four dimensions. Examples of such theories are type IIA or type IIB compactified on d dimensional tori T^d. There are also theories with 16 supersymmetry generators which give $N = 1$ supersymmetry in ten dimensions and $N = 4$ supersymmetry in four dimensions. Examples of these theories are type IIA or IIB compactified on $K3 \times T^d$, heterotic string compactified on T^d, etc. There is another class of theories with 8 supersymmetry generators, for example heterotic string on $K3 \times T^d$, type IIA or IIB on six dimensional Calabi Yau manifolds, etc.

For theories with 16 or more supersymmetry generators the non renormalization theorems are particularly powerful. In particular, the form of the low energy effective action involving the massless states of the theory is completely fixed by the spectrum and by the requirement of supersymmetry. Thus this effective action cannot get renormalized by string loop corrections. As a result, any valid symmetry of the theory must be a symmetry of this effective field theory. Moreover these theories contain a special class of states which are invariant under part of the supersymmetry transformations. They are known as BPS states, named after Bogomol'nyi, Prasad and Sommerfeld. The mass of a BPS state is completely determined in terms of its charge as a consequence of the supersymmetry algebra. Since this relation is derived purely from an analysis of the supersymmetry algebra, it is not modified by quantum corrections. Furthermore it can be argued that the degeneracy of BPS states of a given charge does not change as one moves in the moduli space even from the weak to the strong coupling region. Thus the spectrum of BPS states can be calculated from weak coupling analysis and the result can be continued to the strong coupling region. Since any valid symmetry of the theory must be a symmetry of the spectrum of BPS states, we can use this to design non trivial tests of duality.

Most of the duality conjectures in string theory were arrived at by analyzing the symmetries of the low energy effective action. However, once one arrives at a duality conjecture a much more precise test can be performed by analysing the spectrum of BPS states. Let us work out a concrete example: type I- $SO(32)$ heterotic duality.

We have already discussed the massless spectrum of the $SO(32)$ heterotic string. The low energy dynamics of these fields is governed by $N = 1$ supergravity coupled to $SO(32)$ super Yang Mills in ten dimensions. The action is

$$S^{(H)} = \frac{1}{(2\pi)^7(\alpha'_H)^4 g_H^2} \int d^{10}\, x\, \sqrt{-G^{(H)}}[R^{(H)} - \frac{1}{8}\partial_\mu \Phi^{(H)} \partial^\mu \Phi^{(H)}$$

$$- \frac{1}{4} e^{-\Phi^{(H)}/4} Tr(F^{(H)}_{\mu\nu} F^{(H)\mu\nu}) - \frac{1}{12} e^{-\Phi^{(H)}/2} H^{(H)}_{\mu\nu\rho} H^{(H)\mu\nu\rho}] \quad (35)$$

where $R^{(H)}$ is the Ricci scalar, $F^{(H)}_{\mu\nu}$ denotes the non-abelian gauge field strength

$$F^{(H)}_{\mu\nu} = \partial_\mu A^{(H)}_\nu - \partial_\nu A^{(H)}_\mu + \sqrt{\frac{2}{\alpha'_H}} [A^{(H)}_\mu, A^{(H)}_\nu], \quad (36)$$

Tr denotes the trace in the vector representation of $SO(32)$, and $H^{(H)}_{\mu\nu\rho}$ is the field strength associated with the antisymmetric tensor field $B^{(H)}_{\mu\nu}$:

$$H^{(H)}_{\mu\nu\rho} = \partial_\mu B^{(H)}_{\nu\rho} - \frac{1}{2} Tr(A^{(H)}_\mu F^{(H)}_{\nu\rho} - \frac{1}{3} \sqrt{\frac{2}{\alpha'_H}} A^{(H)}_\mu [A^{(H)}_\nu, A^{(H)}_\rho]) + \textit{cyclic permutations of } \mu, \nu, \rho \quad (37)$$

$2\pi\alpha'_H$ and g_H are the inverse string tension and the coupling constant of the heterotic string respectively. The rescalings that we discussed in the previous Section can be used to set α'_H and g_H equal to one.

Let us now turn to the type I string. The low energy dynamics is again described by $N = 1$ supergravity coupled to $SO(32)$ super Yang Mills. It is instructive to rewrite the effective action in terms of the type I variables. For suitable choice of the string tension and coupling constant, it is

$$S^{(I)} = \frac{1}{(2\pi)^7} \int d^{10}x \sqrt{-G^{(I)}} [R^{(H)} - \frac{1}{8} \partial_\mu \Phi^{(I)} \partial^\mu \Phi^{(I)} - \frac{1}{4} e^{-\Phi^{(I)}/4} Tr(F^{(I)}_{\mu\nu} F^{(I)\mu\nu}) - \frac{1}{12} e^{-\Phi^{(I)}/2} H^{(I)}_{\mu\nu\rho} H^{(I)\mu\nu\rho}] \quad (38)$$

The notation is as above. For both the type I and the $SO(32)$ heterotic string the low energy effective action is derived from the string tree level scattering amplitudes. However, to this order in derivatives, the form of the effective action is determined completely by the requirement of supersymmetry. Thus neither action can receive any quantum corrections.

It is straightforward to see that the actions $S^{(H)}$ and $S^{(I)}$ are identical under the identifications:

$$\begin{array}{lcl} \Phi^{(H)} = -\Phi^{(I)} & , & G^{(H)}_{\mu\nu} = G^{(I)}_{\mu\nu} \\ B^{(H)}_{\mu\nu} = B^{(I)}_{\mu\nu} & , & A^{a(H)}_\mu = A^{a(I)}_\mu \end{array} \quad (39)$$

Note the minus $(-)$ sign in the relation between $\Phi^{(H)}$ and $\Phi^{(I)}$. Recalling that $e^{\Phi/2}$ is the string coupling, we see that the strong coupling limit of one theory is related to the weak coupling region of the other theory and viceversa. This led to the hypothesis that the type I and the $SO(32)$ heterotic string theories in ten dimensions are equivalent. Stronger evidence for this hypothesis can be found by analysing the spectrum of supersymmetric states, but the equivalence of the two effective actions was the reason for proposing this duality in the first place.

Let us look at the BPS states. These states are invariant under part of the supersymmetry transformations and are characterized by two important properties: a) they belong to a supermultiplet which has typically less dimensions than a non-BPS state; b) the mass of a BPS state is completely determined by its charge as a consequence of the supersymmetry algebra.

Schematically, the supersymmetry algebra can be written as

$$\{Q, Q\} = H + I \tag{40}$$

where Q is the supersymmetry generator, H is the hamiltonian and I is the generator of an internal symmetry. We have ommited the indices that would distinguish the various supersymmetry and internal symmetry generators. Consider a one particle state $|\psi>$ which is invariant under part of the supersymmetry algebra, *i.e.* $Q|\psi>= 0$ for some Q's. This is a BPS state. Take the expectation value of the anticommutator

$$<\psi|\{Q, Q\}|\psi>=<\psi|H|\psi> + <\psi|I|\psi> \tag{41}$$

For those $Q's$ which annihilate $|\psi>$, the left hand side is zero. The two terms on the right are just the mass of the particle and its I-charge. Thus, the mass of any BPS state is determined entirely by its charge. This is a consequence of symmetry and does not depend on dynamics. In particular it remains true even after the coupling is large. In fact, the degeneracy of BPS states with a given set of charge quantum numbers is independent of the value of the moduli. This is the key property of PBS states that makes them so useful in testing duality. One can get different kinds of BPS states depending on the number of supersymmetry generators that leave the state invariant.

Given this result one can now adopt the following strategy to check the various duality conjectures using the spectrum of BPS states in the theory:

i. Identify BPS states in the spectrum of elementary string states. This spectrum can be trusted at all values of the coupling even though it is calculated at weak coupling.

ii. Make a conjectured duality transformation. This typically takes a BPS state in the spectrum of elementary string states to another BPS state, but with quantum numbers that are not present in the spectrum of elementary string states. Thus these states must arise as solitons or composite states.

iii. Try to explicitly verify the existence of these solitonic states with degeneracy as predicted by duality. This will provide a non trivial test of the corresponding duality conjecture.

$D-branes$ are BPS states, they are backgrounds preserving one half of the supersymmetries. Indeed, following the same procedure that we described in the bosonic case, one can compute the mass and the charge of the Dp-branes and they turn out to verify the relation $T_p = Q_p \propto g_s^{-1}$, *i.e.* the tension of a $D-brane$

equals its charge. These are the solitonic states that fill the duality multiplets. Different string theories contain different $D - branes$ and different bound states of $D - branes$. In particular, the duality between type I and $SO(32)$ heterotic string theories verifies the predictions for the spectrum of BPS states.

III M THEORY

There are two aspects of duality: elementary $\longleftrightarrow$ composite; and classical $\longleftrightarrow$ quantum. Since duality relates two apparently different theories, it gives a unified picture of all string theories. Moreover, duality can lead to discover new theories. For example, a conjectured theory living in eleven spacetime dimensions.

We have seen that the low energy effective action of Type IIA string theory is non-chiral $N = 2$ supergravity in ten dimensions. This theory can be obtained from dimensional reduction of $N = 1$ supergravity in eleven dimensions. Consider this latter theory. The bosonic fields are the metric $G^{(S)}_{MN}$ and an antisymmetric three form $C^{(S)}_{MNP}$ with $M, N = 0, ..., 10$. The dynamics is described by the action

$$S_{N=1\ SG} \sim \int d^{11}x[\sqrt{-G^{(S)}}(R^{(S)} - \frac{1}{48}F^{(S)2}) - \frac{1}{(12)^4}\epsilon^{\mu_0...\mu_{10}}C^{(S)}_{\mu_0\mu_1\mu_2}F^{(S)}_{\mu_3...\mu_6}F^{(S)}_{\mu_7...\mu_{10}}] \tag{42}$$

where $F^{(S)} \sim dC^{(S)}$ is the four form field strength associated with the rank three antisymmetric field $C^{(S)}$. The eleven dimensional Planck constant has been absorbed into the metric. Let us compactify this supergravity theory on a circle of radius $R \sim \sqrt{G^{(S)}_{10,10}}$. The dimensionally reduced theory agrees with the low energy effective theory of type IIA string theory, namely

$$\begin{aligned} S_{IIA} \propto \int d^{10}x\sqrt{-G}\ [\ & R - \frac{1}{8}\partial_\mu\Phi\partial^\mu\Phi - \frac{1}{12}e^{-\Phi/4}H_{\mu\nu\rho}H^{\mu\nu\rho} \\ & - \frac{1}{4}e^{3\Phi/4}F_{\mu\nu}F^{\mu\nu} - \frac{1}{48}e^{\Phi/4}F_{\mu\nu\rho\sigma}F^{\mu\nu\rho\sigma} \\ & - \frac{1}{(48)^2}(\sqrt{-G})^{-1}\epsilon^{\mu_0...\mu_9}B_{\mu_0\mu_1}F_{\mu_2...\mu_5}F_{\mu_6...\mu_9}] \end{aligned} \tag{43}$$

where

$$F_{\mu\nu\rho\sigma} = \partial_\mu C_{\nu\rho\sigma} + A_\mu H_{\nu\rho\sigma} + (-1)^P \cdot cyclic\ \ permutations, \tag{44}$$

under the identification

$$\begin{aligned} \sqrt{G^{(S)}_{10,10}} = e^{\Phi/3}, \qquad G^{(S)}_{\mu\nu} \sim e^{-\Phi/12}G_{\mu\nu}, \qquad G^{(S)}_{10\mu} \sim e^{\Phi/3}A_\mu, \\ C^{(S)}_{\mu\nu\rho} \sim C_{\mu\nu\rho}, \qquad C^{(S)}_{10\mu\nu} \sim B_{\mu\nu} \end{aligned} \tag{45}$$

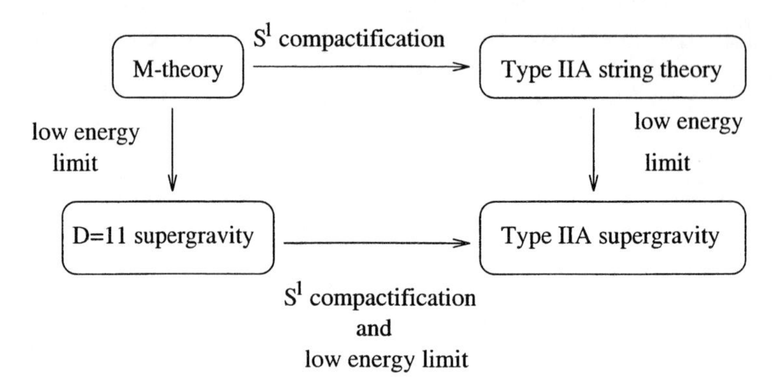

FIGURE 8. The relationship between M, type IIA and supergravity theories

where $\mu, \nu = 0, ..., 9$. Note that $\Phi \to \infty$ corresponds to $R \to \infty$. This is the strong coupling region of the type IIA string theory. This leads to the conjecture that type IIA string theory at strong coupling approaches an eleven dimensional theory whose low energy limit is eleven dimensional $N = 1$ supergravity. This theory has been called M-theory. The situation is illustrated in figure (8).

The evidence for the existence of an eleven dimensional theory is based on the low energy effective action. However there are also more precise tests involving the spectrum of BPS states. M-theory on S^1 has Kaluza Klein modes carrying momentum in the compact dimension. These are BPS states which can be shown to belong to the 256 dimensional representation of the supersymmetry algebra. The charge quantum number of these states is the momentum k/R along S^1. Thus for every integer k we should find such BPS states in type IIA string theory in ten dimensions. In M-theory these states carry k units of $G^{(S)}_{10\mu}$ charge. Since $G^{(S)}_{10\mu}$ gets mapped to A_μ under the M-theory-IIA duality, these states must carry k units of A_μ charge in type IIA theory. Recall that in Type IIA theory A_μ arises in the RR sector, therefore these states cannot come from elementary string states since elementary string excitations are neutral under RR sector gauge fields. However the D0-branes in this theory carry A_μ charge. In particular the state with $k = 1$ corresponds to a single D0-brane whose low energy dynamics is described by the dimensional reduction of $N = 1$ super Maxwell theory from $(9 + 1)$ to $(0 + 1)$ dimensions. This theory has sixteen fermion zero modes whose quantization leads to a 256 fold degenerate state. Thus there is effectively a multiplet with unit A_μ charge as predicted by the M-theory-IIA duality conjecture.

States with $k > 1$ must arise as bound states of k D0-branes. The dynamics of such states is given by the dimensional reduction of $N = 1$ supersymmetric $U(k)$ gauge theory from $(9 + 1)$ to $(0 + 1)$ dimensions. Finding these bound states is difficult, the main obstacle being that the energy of a charge k state is the same as that of k charge 1 states. For $k = 2$ such a bound state with the correct degeneracy has been found. The analysis for higher k still remains to be done.

Compactifying M-theory on different tori simplifies the analysis and the conjecture has been verified in several cases.

We have seen that M-theory can be defined as the strong coupling limit of type IIA string theory. However this is not enough to specify the theory. A non perturbative formulation of M-theory is still missing.

IV CONCLUSIONS

We have seen that duality conjectures relate two or more apparently different string theories. This leads to a unified picture which is schematically summarized in figure (9). According to this picture the apparently different string theories and their compactifications are just different limits of the same theory, which has been named U-theory. Starting with any of the five known superstring theories and going to strong coupling one can reach any of the others. M-theory is one corner of this unknown theory.

The string description is useful only near the corners of the figure where the coupling becomes weak. In the center of the parameter space and in the M-theory corner we do not know what degrees of freedom are supposed to appear. It has been suggested that instead of being the one-dimensional objects that we use in string theory they are matrices with the coordinates of the $D - branes$ as entries.

In fact, $D - branes$ play a fundamental role in the non-perturbative formulation of this unknown theory which we are seaking. An important aspect of these objects, which we have not discussed, is that they form black holes. The D-brane description of black holes has allowed the first microscopic counting of states giving rise to the Bekenstein-Hawking entropy, a longstanding problem in quantum gravity. This result led to a new duality, the Maldacena conjecture [14], which not only has important consequences for the black hole information paradox but also implies a novel relationship between some region of U-theory and ordinary quantum field theories. The Maldacena conjecture takes the form of an equivalence between a string theory or M-theory on a certain manifold and a quantum field theory. The precise form of this quantum field theory depends on the choice of background manifold as well as on which theory one is considering. Since quantum field theories have non-perturbative definition (even though explicit computations may be difficult) this conjecture has been used to give non-perturbative definitions of string theories on specific backgrounds. One can use the Maldacena conjecture to explore certain regions of U-theory and also, in the reverse direction, to study gauge theories using known results in string theory.

To summarize, we have seen that string theory gives a consistent quantum theory of gravity. Duality symmetries provide us with unification of many different string theories and of elementary and composite particles. Even though the picture is not complete, progress in string theory has improved our understanding of quantum field theory. Unfortunately, all this progress is theoretical and there are still no concrete new predictions of string theory at low energies.

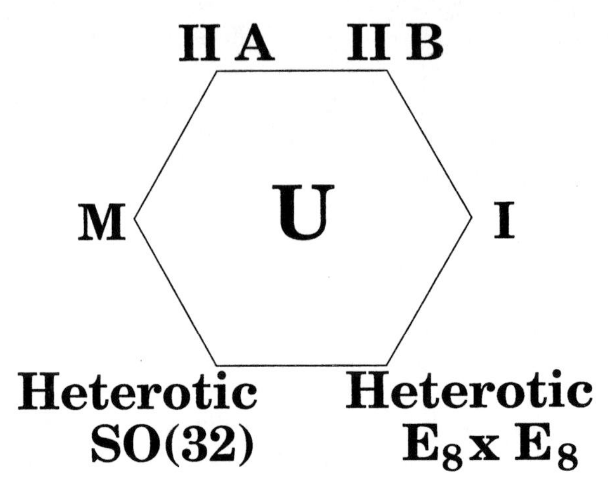

FIGURE 9. The relationship between M, type IIA and supergravity theories

REFERENCES

1. M. Green, J. Schwarz and E. Witten, *"Superstring Theory"*, Cambridge University Press, 1987
2. J. Polchinski, *"String Theory"*, Cambridge University Press, 1998
3. A. Neveu, J. Schwarz and C. Thorn, Phys. Lett. **B 35** (1971) 529
4. M. Green and J. Schwarz, Nucl. Phys. **B 181** (1981) 502
5. F. Gliozzi, J. Scherk and D. Olive, Nucl. Phys. **B 122** (1977) 253
6. J. Dai, R. Leigh and J. Polchinski, Mod. Phys. Lett. **A 4** (1989) 2073
7. D. Friedan, E. Martinec and S. Shenker, Nucl. Phys. **B 271** (1986) 93
8. E. D'Hocker and D. Phong, Rev. Mod. Phys. **60** (1988) 917
9. G. Aldazabal, M. Bonini, R. Iengo and C. Núñez, Nucl. Phys. **B 307** (1988) 291
10. C. Callan, D. Friedan, E. Martinec and M. Perry, Nucl. Phys. **B 262** (1985) 593
11. D. Gepner and E. Witten, Nucl. Phys. **B 278** (1986) 493
12. J. Polchinski, *Tasi lectures on D-branes*, hep-th-9611050
13. A. Sen, *An introduction to non-perturbative string theory*, hep-th-9802051
14. J. M. Maldacena, hep-th/9711200

Finite and Gauge-Yukawa Unified Theories: Theory and Predictions

T. Kobayashi[♣], J. Kubo[◇], M. Mondragón[♡], and G. Zoupanos[♠]

[♣] *Dept. of Physics, High Energy Physics Division, Univ. of Helsinki and Helsinki Institute of Physics, FIN-00014 Helsinki, Finland*
[◇] *Institute for Theoretical Physics, Kanazawa Univ., Kanazawa 920-1192, Japan*
[♡] *Instituto de Física, UNAM, Apdo. Postal 20-364, México 01000 D.F., México* [1]
[♠] *Nat. Technical University, Zografou, Greece, Athens* [2]

Abstract. All - loop Finite Unified Theories (FUTs) are very interesting $N = 1$ GUTs in which a complete reduction of couplings has been achieved. FUTs realise an old field theoretical dream and have remarkable predictive power. Reduction of dimensionless couplings in $N = 1$ GUTs is achieved by searching for renormalization group invariant (RGI) relations among them holding beyond the unification scale. Finiteness results from the fact that there exists RGI relations among dimensionless couplings that guarantee the vanishing of the β - functions in certain $N = 1$ supersymmetric GUTS even to all orders. Recent developments in the soft supersymmetry breaking (SSB) sector of $N = 1$ GUTs and FUTs lead to exact RGI relations also in this sector of the theories. Of particular interest is a RGI sum rule for the soft scalar masses holding to all orders. The characteristic features of SU(5) models that have been constructed based on the above tools are: a) the old agreement of the top quark prediction with the measured value remains unchanged, b) the lightest Higgs boson is predicted to be around 120 GeV, c) the s - spectrum starts above several hundreds of GeV.

I INTRODUCTION

In the recent years very interesting frameworks have been developed such as superstrings, non-commutative geometry and quantum groups aiming to describe in a unified manner all interactions including gravity. In fact even more recent progress has shown that these theoretical endeavours could be also understood in a unified way. It is very interesting that there exist frameworks to discuss quantum gravity, but we should not forget that the main result expected from a unified description of interactions by the particle physics community is to understand the

[1] Supported by the mexican projects CONACYT 3275-PE and PAPIIT-125298
[2] Presented by G. Zoupanos. Supported by the EU projects ERBFMRXCT960090, ERBIC17CT98309 and by the Greek project PENED95/1170;1981

CP490, *Particles and Fields: Eighth Mexican School,* edited by J. C. D'Olivo et al.

present day parameters of the SM in terms of a few fundamental ones or in other words to achieve *reduction of couplings* at a more fundamental level. Clearly all the above theoretical frameworks did not offer anything in the understanding of the free parameters of the SM, and in the best case they just manage to accommodate earlier ideas such as Grand Unified Theories (GUTs) and supersymmetry but without providing them with any predictive power.

In our recent studies [1]- [7], [14] we have developed a complementary strategy in searching for a more fundamental theory possibly at the Planck scale, whose basic ingredients are GUTs and supersymmetry, but its consequences certainly go beyond the known ones. Our method consists of hunting for renormalization group invariant (RGI) relations holding below the Planck scale, which in turn are preserved down to the GUT scale. This programme, called Gauge–Yukawa unification scheme, applied in the dimensionless couplings of supersymmetric GUTs, such as gauge and Yukawa couplings, had already noticable successes by predicting correctly, among others, the top quark mass in the finite and in the minimal N=1 supersymmetric SU(5) GUTs. An impressive aspect of the RGI relations is that one can guarantee their validity to all-orders in perturbation theory by studying the uniqueness of the resulting relations at one-loop, as was proven in the early days of the programme of *reduction of couplings* [15]. Even more remarkable is the fact that it is possible to find RGI relations among couplings that guarantee finiteness to all-orders in perturbation theory [16,18].

Although supersymmetry seems to be an essential feature for a successful realization of the above programme, its breaking has to be understood too, since it has the ambition to supply the SM with predictions for several of its free parameters. Indeed, the search for RGI relations has been extended to the soft supersymmetry breaking sector (SSB) of these theories [5,19], which involves parameters of dimension one and two. More recently a very interesting progress has been made [20]- [26] concerning the renormalization properties of the SSB parameters based conceptually and technically on the work of ref. [46]. In ref. [46] the powerful supergraph method [47] for studying supersymmetric theories has been applied to the softly broken ones by using the "spurion" external space-time independent superfields [48]. In the latter method a softly broken supersymmetric gauge theory is considered as a supersymmetric one in which the various parameters such as couplings and masses have been promoted to external superfields that acquire "vacuum expectation values". Based on this method the relations among the soft term renormalization and that of an unbroken supersymmetric theory have been derived. In particular the β-functions of the parameters of the softly broken theory are expressed in terms of partial differential operators involving the dimensionless parameters of the unbroken theory. The key point in the strategy of refs. [24]- [26] in solving the set of coupled differential equations so as to be able to express all parameters in a RGI way, was to transform the partial differential operators involved to total derivative operators. This is indeed possible to be done on the RGI surface which is defined by the solution of the reduction equations.

On the phenomenological side there exist some serious developments too. Previously an appealing "universal" set of soft scalar masses was asummed in the SSB sector of supersymmetric theories, given that apart from economy and simplicity (1) they are part of the constraints that preserve finiteness up to two-loops [12] [13], (2) they are RGI up to two-loops in more general supersymmetric gauge theories, subject to the condition known as $P = 1/3\ Q$ [19] and (3) they appear in the attractive dilaton dominated supersymmetry breaking superstring scenarios [51]. However, further studies have exhibited a number of problems all due to the restrictive nature of the "universality" assumption for the soft scalar masses. For instance (a) in finite unified theories the universality predicts that the lightest supersymmetric particle is a charged particle, namely the superpartner of the τ lepton $\tilde{\tau}$ (b) the MSSM with universal soft scalar masses is inconsistent with the attractive radiative electroweak symmetry breaking [54] and (c) which is the worst of all, the universal soft scalar masses lead to charge and/or colour breaking minima deeper than the standard vacuum [55]. Therefore, there have been attempts to relax this constraint without loosing its attractive features. First an interesting observation was made that in $N = 1$ Gauge–Yukawa unified theories there exists a RGI sum rule for the soft scalar masses at lower orders; at one-loop for the non-finite case [6] and at two-loops for the finite case [7]. The sum rule manages to overcome the above unpleasant phenomenological consequences. Moreover it was proven [26] that the sum rule for the soft scalar massses is RGI to all-orders for both the general as well as for the finite case. Finally the exact β-function for the soft scalar masses in the Novikov-Shifman-Vainstein-Zakharov (NSVZ) scheme [35] for the softly broken supersymmetric QCD has been obtained [26]. Armed with the above tools and results we are in a position to study the spectrum of the full finite and minimal supersymmetric SU(5) models in terms of few free parameters with emphasis on the predictions for the masses of the lightest Higgs and LSP and on the constraints imposed by having a large $\tan\beta$.

II THE REDUCTION OF COUPLINGS METHOD

Let us outline briefly the idea of reduction of couplings. Any RGI relation among couplings (which does not depend on the renormalization scale μ explicitly) can be expressed, in the implicit form $\Phi(g_1, \cdots, g_A) = \text{const.}$, which has to satisfy the partial differential equation (PDE)

$$\mu \frac{d\Phi}{d\mu} = \vec{\nabla} \cdot \vec{\beta} = \sum_{a=1}^{A} \beta_a \frac{\partial \Phi}{\partial g_a} = 0 \,, \tag{1}$$

where β_a is the β-function of g_a. This PDE is equivalent to a set of ordinary differential equations, the so-called reduction equations (REs) [15],

$$\beta_g \frac{dg_a}{dg} = \beta_a \,,\ a = 1, \cdots, A \,, \tag{2}$$

where g and β_g are the primary coupling and its β-function, and the counting on a does not include g. Since maximally $(A-1)$ independent RGI "constraints" in the A-dimensional space of couplings can be imposed by the Φ_a's, one could in principle express all the couplings in terms of a single coupling g. The strongest requirement is to demand power series solutions to the REs,

$$g_a = \sum_{n=0} \rho_a^{(n)} \, g^{2n+1} \;, \tag{3}$$

which formally preserve perturbative renormalizability. Remarkably, the uniqueness of such power series solutions can be decided already at the one-loop level [15]. To illustrate this, let us assume that the β-functions have the form

$$\begin{aligned} \beta_a &= \frac{1}{16\pi^2}\Big[\sum_{b,c,d\neq g} \beta_a^{(1)\,bcd} g_b g_c g_d + \sum_{b\neq g} \beta_a^{(1)\,b} g_b g^2\Big] + \cdots \;, \\ \beta_g &= \frac{1}{16\pi^2}\beta_g^{(1)} g^3 + \cdots \;, \end{aligned} \tag{4}$$

where $\cdots$ stands for higher order terms, and $\beta_a^{(1)\,bcd}$'s are symmetric in b, c, d. We then assume that the $\rho_a^{(n)}$'s with $n \leq r$ have been uniquely determined. To obtain $\rho_a^{(r+1)}$'s, we insert the power series (3) into the REs (2) and collect terms of $O(g^{2r+3})$ and find

$$\sum_{d\neq g} M(r)_a^d \, \rho_d^{(r+1)} = \text{lower order quantities} \;, \tag{5}$$

where the r.h.s. is known by assumption, and

$$M(r)_a^d = 3 \sum_{b,c\neq g} \beta_a^{(1)\,bcd} \, \rho_b^{(1)} \, \rho_c^{(1)} + \beta_a^{(1)\,d} - (2r+1)\, \beta_g^{(1)} \, \delta_a^d \;, \tag{6}$$

$$0 = \sum_{b,c,d\neq g} \beta_a^{(1)\,bcd} \, \rho_b^{(1)} \, \rho_c^{(1)} \, \rho_d^{(1)} + \sum_{d\neq g} \beta_a^{(1)\,d} \, \rho_d^{(1)} - \beta_g^{(1)} \, \rho_a^{(1)} \;. \tag{7}$$

Therefore, the $\rho_a^{(n)}$'s for all $n > 1$ for a given set of $\rho_a^{(1)}$'s can be uniquely determined if $\det M(n)_a^d \neq 0$ for all $n \geq 0$.

Our experience examining specific examples has taught us that the various couplings in supersymmetric theories have easily the same asymptotic behaviour. Therefore searching for a power series solution of the form (3) to the REs (2) is justified and moreover, one can rely that in applications keeping only the first terms a good approximation is obtained. This is not the case in non-supersymmetric theories.

An Instructive Example

Now that we have established the tools, let us illustrate an interesting example. Consider a gauge theory based on the gauge group $SU(N)$. The matter content is as follows: $\phi^i(\mathbf{N})$ and $\hat{\phi}_i(\overline{\mathbf{N}})$ are complex scalars, $\psi^i(\mathbf{N})$ and $\hat{\psi}_i(\overline{\mathbf{N}})$ are left-handed Weyl spinor, and $\lambda^a(a = 1, \ldots, N^2-1)$ is a right-handed Weyl spinor in the adjoint representation of $SU(N)$.

The Lagrangian is:

$$\mathcal{L} = -\frac{1}{4} F^a_{\mu\nu} F^{a\,\mu\nu} + i\sqrt{2}[\; g_Y \overline{\psi} \lambda^a T^a \phi - \hat{g}_Y \overline{\hat{\psi}} \lambda^a T^a \hat{\phi} + \text{h.c.}\;] - V(\phi, \overline{\phi}) \;, \tag{8}$$

$$V(\phi,\overline{\phi}) = \frac{1}{4}\lambda_1(\phi^i\phi_i^*)^2 + \frac{1}{4}\lambda_2(\hat{\phi}_i\hat{\phi}^{*\,i})^2 + \lambda_3(\phi^i\phi_i^*)(\hat{\phi}_j\hat{\phi}^{*\,j}) + \lambda_4(\phi^i\phi_j^*)(\hat{\phi}_i\hat{\phi}^{*\,j}) \;, \tag{9}$$

which is the most general renormalizable form of dimension four, consistent with the $SU(N) \times SU(N)$ global symmetry.

We find the following β - functions of the couplings of the theory

$$\begin{aligned}
\beta^{(1)}(g) &= (1 - 3N)g^3 \;, \\
\beta^{(1)}(g_Y) &= g_Y[\,-3(N + \frac{N^2-1}{2N})\,g^2 + (\frac{1}{2} + \frac{3}{2}\frac{N^2-1}{N})\,g_Y^2 + \frac{1}{2}\hat{g}_Y^2\;] \;, \\
\beta^{(1)}(\hat{g}_Y) &= \hat{g}_Y[\,-3(N + \frac{N^2-1}{2N})\,g^2 + (\frac{1}{2} + \frac{3}{2}\frac{N^2-1}{N})\,\hat{g}_Y^2 + \frac{1}{2}g_Y^2\;] \;, \\
\beta^{(1)}(\lambda_1) &= (4+N)\lambda_1^2 + 4\lambda_4^2 + 4N\lambda_3^2 + 8\lambda_3\lambda_4 + 3(N + 1 - \frac{4}{N} + \frac{2}{N^2})g^4 \\
&\quad -8(N - \frac{2}{N} + \frac{1}{N^2})g_Y^4 + \frac{N^2-1}{N}\lambda_1(4g_Y^2 - 6g^2) \;, && (10) \\
\beta^{(1)}(\lambda_2) &= (4+N)\lambda_2^2 + 4\lambda_4^2 + 4N\lambda_3^2 + 8\lambda_3\lambda_4 + 3(N + 1 - \frac{4}{N} + \frac{2}{N^2})g^4 \\
&\quad -8(N - \frac{2}{N} + \frac{1}{N^2})\hat{g}_Y^4 + \frac{N^2-1}{N}\lambda_2(4\hat{g}_Y^2 - 6g^2) \;, && (11) \\
\beta^{(1)}(\lambda_3) &= 4\lambda_3^2 + 2\lambda_4^2 + (1+N)\lambda_3(\lambda_1+\lambda_2) + \lambda_4(\lambda_1+\lambda_2) \\
&\quad +\frac{3}{2}(1 + \frac{2}{N^2})g^4 - 4(1 + \frac{1}{N^2})g_Y^2\hat{g}_Y^2 \\
&\quad +\frac{N^2-1}{N}\lambda_3(2g_Y^2 + 2\hat{g}_Y^2 - 6g^2) \;, && (12) \\
\beta^{(1)}(\lambda_4) &= 2N\lambda_4^2 + 8\lambda_3\lambda_4 + \lambda_4(\lambda_1+\lambda_2) + \frac{3}{2}(N - \frac{4}{N})g^4 \\
&\quad +8\frac{1}{N}g_Y^2\hat{g}_Y^2 + \frac{N^2-1}{N}\lambda_4(2g_Y^2 + 2\hat{g}_Y^2 - 6g^2) \;. && (13)
\end{aligned}$$

Searching for a power series solution (3) to the REs (2), we find in lowest order the following one:

$$g_Y = \hat{g}_Y = g \;,\; \lambda_1 = \lambda_2 = \frac{N-1}{N}g^2 \;,\; \lambda_3 = \frac{1}{2N}g^2 \;,\; \lambda_4 = -\frac{1}{2}g^2 \;, \tag{14}$$

which corresponds to an $N = 1$ supersymmetric gauge theory. Besides this interesting symmetric solution, the reduction equations have also non-symmetric solutions

[27]. It has been checked that in both kinds of solutions the coefficients in the power series can be uniquely determined according to the above described theorem [15]. However both kind of solutions have been examined for several but specific values of N.

III REDUCTION OF COUPLINGS AND FINITENESS IN $N = 1$ SUSY GAUGE THEORIES

Consider a chiral, anomaly free, $N = 1$ globally supersymmetric gauge theory based on a group G with gauge coupling constant g. The superpotential of the theory is given by

$$W = \frac{1}{2}\, m_{ij}\, \phi_i\, \phi_j + \frac{1}{6}\, C_{ijk}\, \phi_i\, \phi_j\, \phi_k \ , \tag{15}$$

where m_{ij} and C_{ijk} are gauge invariant tensors and the matter field ϕ_i transforms according to the irreducible representation R_i of the gauge group G. The renormalization constants associated with the superpotential (15), assuming that supersymmetry is preserved, are

$$\phi_i^0 = (Z_i^j)^{(1/2)}\, \phi_j \ , \tag{16}$$
$$m_{ij}^0 = Z_{ij}^{i'j'}\, m_{i'j'} \ , \tag{17}$$
$$C_{ijk}^0 = Z_{ijk}^{i'j'k'}\, C_{i'j'k'} \ . \tag{18}$$

The $N = 1$ non-renormalization theorem [10] ensures that there are no mass and cubic-interaction-term infinities and therefore

$$\begin{aligned} Z_{ijk}^{i'j'k'}\, Z_{i'}^{1/2\, i''}\, Z_{j'}^{1/2\, j''}\, Z_{k'}^{1/2\, k''} &= \delta_{(i}^{i''}\, \delta_{j}^{j''}\, \delta_{k)}^{k''} \ , \\ Z_{ij}^{i'j'}\, Z_{i'}^{1/2\, i''}\, Z_{j'}^{1/2\, j''} &= \delta_{(i}^{i''}\, \delta_{j)}^{j''} \ . \end{aligned} \tag{19}$$

As a result the only surviving possible infinities are the wave-function renormalization constants Z_i^j, i.e., one infinity for each field. The one-loop β-function of the gauge coupling g is given by [11]

$$\beta_g^{(1)} = \frac{dg}{dt} = \frac{g^3}{16\pi^2}\, [\sum_i l(R_i) - 3\, C_2(G)] \ , \tag{20}$$

where $l(R_i)$ is the Dynkin index of R_i and $C_2(G)$ is the quadratic Casimir of the adjoint representation of the gauge group G. The β-functions of C_{ijk}, by virtue of the non-renormalization theorem, are related to the anomalous dimension matrix γ_{ij} of the matter fields ϕ_i as:

$$\beta_{ijk} = \frac{dC_{ijk}}{dt} = C_{ijl}\, \gamma_k^l + C_{ikl}\, \gamma_j^l + C_{jkl}\, \gamma_i^l \ . \tag{21}$$

At one-loop level γ_{ij} is [11]

$$\gamma_{ij}^{(1)} = \frac{1}{32\pi^2} [\, C^{ikl}\, C_{jkl} - 2\, g^2\, C_2(R_i)\delta_{ij}\,], \tag{22}$$

where $C_2(R_i)$ is the quadratic Casimir of the representation R_i, and $C^{ijk} = C^*_{ijk}$. According to the discussion in Chapter 3, the non-renormalization theorem ensures there are no extra mass and cubic-interaction-term renormalizations, implying that the β-functions of C_{ijk} can be expressed as linear combinations of the anomalous dimensions γ_{ij} of ϕ^i. Therefore, all the one-loop β-functions of the theory vanish if $\beta_g^{(1)}$ and $\gamma_{ij}^{(1)}$, given in Eqs. (20) and (22) respectively, vanish, i.e.

$$\sum_i \ell(R_i) = 3C_2(G)\,, \tag{23}$$

$$C^{ikl}C_{jkl} = 2\delta^i_j g^2 C_2(R_i)\,, \tag{24}$$

A very interesting result is that the conditions (23,24) are necessary and sufficient for finiteness at the two-loop level [11].

In case supersymmetry is broken by soft terms, one-loop finiteness of the soft sector imposes further constraints on it [12]. In addition, the same set of conditions that are sufficient for one-loop finiteness of the soft breaking terms render the soft sector of they theory two-loop finite [13].

The one- and two-loop finiteness conditions (23,24) restrict considerably the possible choices of the irreps. R_i for a given group G as well as the Yukawa couplings in the superpotential (15). Note in particular that the finiteness conditions cannot be applied to the supersymmetric standard model (SSM), since the presence of a $U(1)$ gauge group is incompatible with the condition (23), due to $C_2[U(1)] = 0$. This naturally leads to the expectation that finiteness should be attained at the grand unified level only, the SSM being just the corresponding, low-energy, effective theory.

Another important consequence of one- and two-loop finiteness is that supersymmetry (most probably) can only be broken by soft breaking terms. Indeed, due to the unacceptability of gauge singlets, F-type spontaneous symmetry breaking [8] terms are incompatible with finiteness, as well as D-type [9] spontaneous breaking which requires the existence of a $U(1)$ gauge group.

A natural question to ask is what happens at higher loop orders. The answer is contained in a theorem [16] which states the necessary and sufficient conditions to achieve finiteness at all orders. Before we discuss the theorem let us make some introductory remarks. The finiteness conditions impose relations between gauge and Yukawa couplings. To require such relations which render the couplings mutually dependent at a given renormalization point is trivial. What is not trivial is to guarantee that relations leading to a reduction of the couplings hold at any renormalization point. As we have seen, the necessary, but also sufficient, condition for this to happen is to require that such relations are solutions to the REs (2) and hold at all orders. As we have seen, remarkably the existence of all-order solutions to (2) can be decided at the one-loop level.

A The all-loop finiteness theorem

Let us now turn to the all-order finiteness theorem [16], which states when a $N = 1$ supersymmetric gauge theory can become finite to all orders in the sense of vanishing β-functions, that is of physical scale invariance. It is based on (a) the structure of the supercurrent in $N = 1$ SYM [29–31], and on (b) the non-renormalization properties of $N = 1$ chiral anomalies [16,17]. Details on the proof can be found in refs. [16,28]. Here, following mostly ref. [28] we present a comprehensible sketch of the proof.

Consider a $N = 1$ supersymmetric gauge theory, with simple Lie group G. The content of this theory is given at the classical level by the matter supermultiplets S_i, which contain a scalar field ϕ_i and a Weyl spinor ψ_{ia}, and the gauge fields V_a, which contain a gauge vector field A_μ^a and a gaugino Weyl spinor λ_α^a.

Let us first recall certain facts about the theory:
(1) A massless $N = 1$ supersymmetric theory is invariant under a $U(1)$ chiral transformation R under which the various fields transform as follows

$$A'_\mu = A_\mu, \;\; \lambda'_\alpha = \exp(-i\theta)\lambda_\alpha \;\; \phi' = \exp(-i\frac{2}{3}\theta)\phi, \;\; \psi'_\alpha = \exp(-i\frac{1}{3}\theta)\psi_\alpha, \;\; \cdots \tag{25}$$

The corresponding axial Noether current $J_R^\mu(x)$ is

$$J_R^\mu(x) = \bar{\lambda}\gamma^\mu\gamma^5\lambda + \cdots \tag{26}$$

is conserved classically, while in the quantum case is violated by the axial anomaly

$$\partial_\mu J_R^\mu = r(\epsilon^{\mu\nu\sigma\rho}F_{\mu\nu}F_{\sigma\rho} + \cdots). \tag{27}$$

From its known topological origin in ordinary gauge theories [32], one would expect that the axial vector current J_R^μ to satisfy the Adler-Bardeen theorem [33] and receive corrections only at the one-loop level. Indeed it has been shown that the same non-renormalization theorem holds also in supersymmetric theories [17]. Therefore

$$r = \hbar\beta_g^{(1)}. \tag{28}$$

(2) The massless theory we consider is scale invariant at the classical level and, in general, there is a scale anomaly due to radiative corrections. The scale anomaly appears in the trace of the energy momentum tensor $T_{\mu\nu}$, which is traceless classically. It has the form

$$T_\mu^\mu \;=\; \beta_g F^{\mu\nu}F_{\mu\nu} + \cdots \tag{29}$$

(3) Massless, $N = 1$ supersymmetric gauge theories are classically invariant under the supersymmetric extension of the conformal group – the superconformal group. Examining the superconformal algebra, it can be seen that the subset of superconformal transformations consisting of translations, supersymmetry transformations, and axial R transformations is closed under supersymmetry, i.e. these transformations form a representation of supersymmetry. It follows that the conserved currents corresponding to these transformations make up a supermultiplet represented by an axial vector superfield called supercurrent [29] J,

$$J \equiv \{J'^{\mu}_{R},\ Q^{\mu}_{\alpha},\ T^{\mu}_{\nu}, ...\}, \tag{30}$$

where J'^{μ}_{R} is the current associated to R invariance, Q^{μ}_{α} is the one associated to supersymmetry invariance, and T^{μ}_{ν} the one associated to translational invariance (energy-momentum tensor).

The anomalies of the R current J'^{μ}_{R}, the trace anomalies of the supersymmetry current, and the energy-momentum tensor, form also a second supermultiplet, called the supertrace anomaly

$$\begin{aligned} S\ &= \{Re\ S,\ Im\ S,\ S_{\alpha}\} = \\ &\{T^{\mu}_{\mu},\ \partial_{\mu}J'^{\mu}_{R},\ \sigma^{\mu}_{\alpha\dot{\beta}}\bar{Q}^{\dot{\beta}}_{\mu}\ +\ \cdots\} \end{aligned} \tag{31}$$

where T^{μ}_{μ} in Eq.(29) and

$$\partial_{\mu}J'^{\mu}_{R} = \beta_{g}\epsilon^{\mu\nu\sigma\rho}F_{\mu\nu}F_{\sigma\rho} + \cdots \tag{32}$$

$$\sigma^{\mu}_{\alpha\dot{\beta}}\bar{Q}^{\dot{\beta}}_{\mu} = \beta_{g}\lambda^{\beta}\sigma^{\mu\nu}_{\alpha\beta}F_{\mu\nu} + \cdots \tag{33}$$

(4) It is very important to note that the Noether current defined in (26) is not the same as the current associated to R invariance that appears in the supermultiplet J in (30), but they coincide in the tree approximation. So starting from a unique classical Noether current $J^{\mu}_{R(class)}$, the Noether current J^{μ}_{R} is defined as the quantum extension of $J^{\mu}_{R(class)}$ which allows for the validity of the non-renormalization theorem. On the other hand J'^{μ}_{R}, is defined to belong to the supercurrent J, together with the energy-momentum tensor. The two requirements cannot be fulfilled by a single current operator at the same time.

Although the Noether current J^{μ}_{R} which obeys (27) and the current J'^{μ}_{R} belonging to the supercurrent multiplet J are not the same, there is a relation [16] between quantities associated with them

$$r = \beta_{g}(1 + x_{g}) + \beta_{ijk}x^{ijk} - \gamma_{A}r^{A} \tag{34}$$

where r was given in Eq. (28). The r^{A} are the non-renormalized coefficients of the anomalies of the Noether currents associated to the chiral invariances of the superpotential, and –like r– are strictly one-loop quantities. The γ_{A}'s are linear combinations of the anomalous dimensions of the matter fields, and x_{g}, and x^{ijk}

are radiative correction quantities. The structure of equality (34) is independent of the renormalization scheme.

One-loop finiteness, i.e. vanishing of the β-functions at one-loop, implies that the Yukawa couplings λ_{ijk} must be functions of the gauge coupling g. To find a similar condition to all orders it is necessary and sufficient for the Yukawa couplings to be a formal power series in g, which is solution of the REs (2).

We can now state the theorem for all-order vanishing β-functions.

Theorem: Consider an $N = 1$ supersymmetric Yang-Mills theory, with simple gauge group. If the following conditions are satisfied

1. There is no gauge anomaly.

2. The gauge β-function vanishes at one-loop

$$\beta_g^{(1)} = 0 = \sum_i l(R_i) - 3\, C_2(G). \tag{35}$$

3. There exist solutions of the form

$$\lambda_{ijk} = \rho_{ijk} g, \qquad \rho_{ijk} \in \mathbb{C} \tag{36}$$

to the conditions of vanishing one-loop matter fields anomalous dimensions

$$\gamma_j^{i\ (1)} = 0 = \frac{1}{32\pi^2} [\ C^{ikl}\, C_{jkl} - 2\ g^2\ C_2(R_i)\delta_{ij}]. \tag{37}$$

4. these solutions are isolated and non-degenerate when considered as solutions of vanishing one-loop Yukawa β-functions:

$$\beta_{ijk} = 0. \tag{38}$$

Then, each of the solutions (36) can be uniquely extended to a formal power series in g, and the associated super Yang-Mills models depend on the single coupling constant g with a β function which vanishes at all-orders.

It is important to note a few things: The requirement of isolated and non-degenerate solutions guarantees the existence of a formal power series solution to the reduction equations. The vanishing of the gauge β-function at one-loop, $\beta_g^{(1)}$, is equivalent to the vanishing of the R current anomaly (27). The vanishing of the anomalous dimensions at one-loop implies the vanishing of the Yukawa couplings β-functions at that order. It also implies the vanishing of the chiral anomaly coefficients r^A. This last property is a necessary condition for having β functions vanishing at all orders.

Proof: Insert β_{ijk} as given by the REs into the relationship (34) between the axial anomalies coefficients and the β-functions. Since these chiral anomalies vanish, we get for β_g an homogeneous equation of the form

$$0 = \beta_g(1 + O(\hbar)). \tag{39}$$

The solution of this equation in the sense of a formal power series in $\hbar$ is $\beta_g = 0$, order by order. Therefore, due to the REs (2), $\beta_{ijk} = 0$ too.

B Other criteria leading to all-loop finiteness

Another method of searching for all loop finite theories was suggested some time ago in refs. [18] and was revived recently in ref. [36] .It is based on the all orders relation among the gauge β-function β_g and the anomalous dimensions of the superfields γ which was first derived using instanton calculus [35] given by

$$\beta_g^{NSVZ} \propto [\sum_i \lambda(R_i) - 3C_2(G) - \sum_i \lambda(R_i)\gamma_i], \tag{40}$$

which is believed also to hold non-perturbatively [43]. In addition the second method is based on the general relation among the β-functions for the Yukawa couplings and anomalous dimensions of the superfields given in eq. (22) in accordance with the non-renormalization theorem [37].

It is worth noting that both β-functions are linear functionals of the anomalous dimensions of the matter superfields. These anomalous dimensions now are complicated functions of the couplings, however the relations among β's and γ's are very simple. The criterion for having a finite theory is that all β-functions vanish simultaneously. This puts n constraints on the n couplings g, C_{ijk}. In the case that these constraints are linearly independent, then one expects that their solutions are isolated points in the space of couplings. (Note that this case looks the same as the condition (4) of the theorem [16] mentioned above. The difference is that the theorem refers to one-loop β-functions.) However if only p constraints are linearly independent, then one is led to an $n-p$ dimensional manifold of fixed points. Thus, if some of the β-functions are linearly dependent,and there is at least one generic fixed point somewhere in the space of couplings, then the theory will have a manifold of fixed points. Of course it is always possible that the constraints have no solutions at all, as for instance when they put contradictory conditions on the anamalous dimensions.

In practice, when searching for all loop finite theories the second method gives a faster answer, since it does not require calculation of the anomalous dimensions even at one-loop. However in general one has to be careful. The first method is certainly better for real calculations in a given theory and gives unambiguous results, but requires more work. Before closing our reference to the second method it is worth mentioning the relatin among the β_g^{NSVZ} and the β_g and γ, when calculated in the

DRED scheme. The relationship between β_g^{NSVZ} and β_g^{DRED} has been explored [42], with the conclusion that there exists an analytic redefinition of g, $g \longrightarrow g'(g,C)$ which connects them, however the two schemes start to deviate at three loops.

C The supertrace anomaly in the presence of gravity

Having established how to construct an all-loop finite gauge theory, it is interesting to examine further the trace anomaly, T^μ_μ of the energy momentum tensor of such a theory in the presence of gravity. In that case one studies the contribution of graviton excitations to the trace anomaly. The latter is given by

$$T^\mu_\mu = (c/16\pi^2)W^2_{\mu\nu\rho\sigma} - (a/16)\pi^2 \tilde{R}_{\mu\nu\rho\sigma}\tilde{R}^{\mu\nu\rho\sigma}, \tag{41}$$

where $W_{\mu\nu\rho\sigma}$ is the Weyl tensor and $\tilde{R}^{\mu\nu\rho\sigma}$ is the dual of the curvature tensor, the second term being the Euler density. In a supersymmetric gauge theory with $N_V = dim\ G$ gauge multiplets and N_χ chiral multiplets the constants are [38]

$$c = 1/24(3N_V + N_\chi), \qquad a = 1/48(9N_V + N_\chi) \tag{42}$$

Similarly the supersymmetric partner of the trace anomaly, the divergence of the $J_R^{'\mu}$ is [61]

$$\delta_\mu(\sqrt{g}J_R^{'\mu}) = (c-a)/24\pi^2 R\tilde{R}, \tag{43}$$

where R and $\tilde{R}$ are the curvature tensor and its dual. In general a gravitational background breaks supersymmetry. In fact the only gravitational vacua compatible with supersymmetry are the flat Minkowski space-time M_{1+n} with the standard Poincaré supersymmetry and the anti-de Sitter one AdS_{2+n} with AdS supersymmetry [34].

Therefore knowing how to construct all-loop finite theories on M_4, we can examine all the potentially all-loop finite models and check to which extent the condition for cancellation of the graviton excitation contributions to the T^μ_μ and to the divergence of the $J_R^{'\mu}$, i.e.

$$c = a \tag{44}$$

can be satisfied. The latter is obeyed by models with $3N_V = N_\chi$.[3] Indeed we have checked all the $N=1$, chiral, all-loop finite models of ref. [40], which potentially can become all-loop finite and all the potentially interesting $N=2$, one- and therefore all-loop finite models based on $SU(5)$, $SO(10)$, E_6, E_7, E_8. We found that the

[3] In particular the isometries of the AdS_{2+n} space are described by the group $SO(2,n)$. The latter is actually the conformal group of the boundary Minkowski space M_{1+n}. According to recent ideas, the supergravity theory in the bulk of AdS is dual to a superconformal theory on the boundary [39].

only $N=1$, chiral model of ref. [40] which satisfies the condition (4) is an SO(10) gauge theory with chiral multiplets in the 2 **10**+1 **54**+1 **45**+**1** representations. We did not find any of the potentially interesting $N=2$ finite theories satisfying the condition (44), while on the contrary all $N=4$ gauge theories satisfy automatically the condition.

The general problem of studying superconformal anomalies of field theories on curved superspace has already been considered in ref. [52] for the massless Wess-Zumino model as well as for gauge theories [53]. Here we have limited ourselves in the case of considering finite gauge theories defined on M_4 including graviton excitations to the supertrace anomaly at one-loop. We found that apart from the phenomenologically uninteresting models, one $N=1$ based on $SO(10)$ and all $N=4$, the supertrace anomaly receives gravitational contributions. These contributions cannot be cancelled in non-finite theories.

IV REDUCTION OF DIMENSIONFUL PARAMETERS

The reduction of couplings was originally formulated for massless theories on the basis of the Callan-Symanzik equation [15]. The extension to theories with massive parameters is not straightforward if one wants to keep the generality and the rigor on the same level as for the massless case; one has to fulfill a set of requirements coming from the renormalization group equations, the Callan-Symanzik equations, etc. along with the normalization conditions imposed on irreducible Green's functions [49]. There has been progress in this direction [50]. Here, to simplify the situation, we would like to assume that a mass-independent renormalization scheme has been employed so that all the RG functions have only trivial dependencies of dimensional parameters.

To be general, we consider a renormalizable theory which contain a set of $(N+1)$ dimension-zero couplings, $\{\hat{g}_0,\hat{g}_1,\ldots,\hat{g}_N\}$, a set of L parameters with dimension one, $\{\hat{h}_1,\ldots,\hat{h}_L\}$, and a set of M parameters with dimension two, $\{\hat{m}_1^2,\ldots,\hat{m}_M^2\}$. The renormalized irreducible vertex function satisfies the RG equation

$$0=\mathcal{D}\Gamma[\,\mathbf{\Phi}'s;\hat{g}_0,\hat{g}_1,\ldots,\hat{g}_N;\hat{h}_1,\ldots,\hat{h}_L;\hat{m}_1^2,\ldots,\hat{m}_M^2;\mu\,]\,, \tag{45}$$
$$\mathcal{D}=\mu\frac{\partial}{\partial\mu}+\sum_{i=0}^{N}\beta_i\frac{\partial}{\partial\hat{g}_i}+\sum_{a=1}^{L}\gamma_a^h\frac{\partial}{\partial\hat{h}_a}+\sum_{\alpha=1}^{M}\gamma_\alpha^{m^2}\frac{\partial}{\partial\hat{m}_\alpha^2}+\sum_J\Phi_I\gamma^{\phi I}{}_J\frac{\delta}{\delta\Phi_J}\,.$$

Since we assume a mass-independent renormalization scheme, the γ's have the form

$$\gamma_a^h=\sum_{b=1}^{L}\gamma_a^{h,b}(g_0,\ldots,g_N)\hat{h}_b\,,$$
$$\gamma_\alpha^{m^2}=\sum_{\beta=1}^{M}\gamma_\alpha^{m^2,\beta}(g_0,\ldots,g_N)\hat{m}_\beta^2+\sum_{a,b=1}^{L}\gamma_\alpha^{m^2,ab}(g_0,\ldots,g_N)\hat{h}_a\hat{h}_b\,, \tag{46}$$

where $\gamma_a^{h,b}, \gamma_\alpha^{m^2,\beta}$ and $\gamma_a^{m^2,ab}$ are power series of the dimension-zero couplings g's in perturbation theory.

As in the massless case, we then look for conditions under which the reduction of parameters,

$$\hat{g}_i = \hat{g}_i(g) \ , \ (i = 1, \ldots, N) \ , \tag{47}$$

$$\hat{h}_a = \sum_{b=1}^{P} f_a^b(g) h_b \ , \ (a = P+1, \ldots, L) \ , \tag{48}$$

$$\hat{m}_\alpha^2 = \sum_{\beta=1}^{Q} e_\alpha^\beta(g) m_\beta^2 + \sum_{a,b=1}^{P} k_\alpha^{ab}(g) h_a h_b \ , \ (\alpha = Q+1, \ldots, M) \ , \tag{49}$$

is consistent with the RG equation (1), where we assume that $g \equiv g_0$, $h_a \equiv \hat{h}_a$ $(1 \le a \le P)$ and $m_\alpha^2 \equiv \hat{m}_\alpha^2$ $(1 \le \alpha \le Q)$ are independent parameters of the reduced theory. We find that the following set of equations has to be satisfied:

$$\beta_g \frac{\partial \hat{g}_i}{\partial g} = \beta_i \ , \ (i = 1, \ldots, N) \ , \tag{50}$$

$$\beta_g \frac{\partial \hat{h}_a}{\partial g} + \sum_{b=1}^{P} \gamma_b^h \frac{\partial \hat{h}_a}{\partial h_b} = \gamma_a^h \ , \ (a = P+1, \ldots, L) \ , \tag{51}$$

$$\beta_g \frac{\partial \hat{m}_\alpha^2}{\partial g} + \sum_{a=1}^{P} \gamma_a^h \frac{\partial \hat{m}_\alpha^2}{\partial h_a} + \sum_{\beta=1}^{Q} \gamma_\beta^{m^2} \frac{\partial \hat{m}_\alpha^2}{\partial m_\beta^2} = \gamma_\alpha^{m^2} \ , \ (\alpha = Q+1, \ldots, M) \ . \tag{52}$$

Using eq. (2) for γ's, one finds that eqs. (6)–(8) reduce to

$$\beta_g \frac{df_a^b}{dg} + \sum_{c=1}^{P} f_a^c [\gamma_c^{h,b} + \sum_{d=P+1}^{L} \gamma_c^{h,d} f_d^b] - \gamma_a^{h,b} - \sum_{d=P+1}^{L} \gamma_a^{h,d} f_d^b = 0 \ , \tag{53}$$

$$(a = P+1, \ldots, L; b = 1, \ldots, P) \ ,$$

$$\beta_g \frac{de_\alpha^\beta}{dg} + \sum_{\gamma=1}^{Q} e_\alpha^\gamma [\gamma_\gamma^{m^2,\beta} + \sum_{\delta=Q+1}^{M} \gamma_\gamma^{m^2,\delta} e_\delta^\beta] - \gamma_\alpha^{m^2,\beta} - \sum_{\delta=Q+1}^{M} \gamma_\alpha^{m^2,\delta} e_\delta^\beta = 0 \ , \tag{54}$$

$$(\alpha = Q+1, \ldots, M; \beta = 1, \ldots, Q) \ ,$$

$$\beta_g \frac{dk_\alpha^{ab}}{dg} + 2\sum_{c=1}^{P} (\gamma_c^{h,a} + \sum_{d=P+1}^{L} \gamma_c^{h,d} f_d^a) k_\alpha^{cb} + \sum_{\beta=1}^{Q} e_\alpha^\beta [\gamma_\beta^{m^2,ab} + \sum_{c,d=P+1}^{L} \gamma_\beta^{m^2,cd} f_c^a f_d^b$$

$$+2 \sum_{c=P+1}^{L} \gamma_\beta^{m^2,cb} f_c^a + \sum_{\delta=Q+1}^{M} \gamma_\beta^{m^2,\delta} k_\delta^{ab}] - [\gamma_\alpha^{m^2,ab} + \sum_{c,d=P+1}^{L} \gamma_\alpha^{m^2,cd} f_c^a f_d^b$$

$$+2 \sum_{c=P+1}^{L} \gamma_\alpha^{m^2,cb} f_c^a + \sum_{\delta=Q+1}^{M} \gamma_\alpha^{m^2,\delta} k_\delta^{ab}] = 0 \ , \tag{55}$$

$$(\alpha = Q+1, \ldots, M; a, b = 1, \ldots, P) \ .$$

If these equations are satisfied, the irreducible vertex function of the reduced theory

$$
\begin{aligned}
&\Gamma_R[\,\mathbf{\Phi}'s; g; h_1,\ldots,h_P; m_1^2,\ldots,\hat{m}_Q^2;\mu\,] \\
&\equiv \Gamma[\,\mathbf{\Phi}'s; g, \hat{g}_1(g),\ldots,\hat{g}_N(g); h_1,\ldots,h_P, \hat{h}_{P+1}(g,h),\ldots,\hat{h}_L(g,h); \\
&m_1^2,\ldots,\hat{m}_Q^2, \hat{m}_{Q+1}^2(g,h,m^2),\ldots,\hat{m}_M^2(g,h,m^2);\mu\,]
\end{aligned} \tag{56}
$$

has the same renormalization group flow as the original one.

The requirement for the reduced theory to be perturbative renormalizable means that the functions $\hat{g}_i$, f_a^b, e_α^β and k_α^{ab}, defined in eqs. (47-49), should have a power series expansion in the primary coupling g:

$$
\begin{aligned}
\hat{g}_i &= g\sum_{n=0}^{\infty}\rho_i^{(n)}g^n\;,\quad f_a^b = g\sum_{n=0}^{\infty}\eta_a^{b\;(n)}g^n\;, \\
e_\alpha^\beta &= \sum_{n=0}^{\infty}\xi_\alpha^{\beta\;(n)}g^n\;,\quad k_\alpha^{ab} = \sum_{n=0}^{\infty}\chi_\alpha^{ab\;(n)}g^n\;,
\end{aligned} \tag{57}
$$

To obtain the expansion coefficients, we insert the power series ansatz above into eqs. (50,53–55) and require that the equations are satisfied at each order in g. Note that the existence of a unique power series solution is a non-trivial matter: It depends on the theory as well as on the choice of the set of independent parameters.

Sum rule for Soft Breaking Terms and All-loop Results

The method of reducing the dimensionless couplings has been extended [5], as we have just seen above, to the soft supersymmetry breaking (SSB) dimensionful parameters of $N=1$ supersymmetric theories. In addition it was found [6] that RGI SSB scalar masses in Gauge-Yukawa unified models satisfy a universal sum rule. Here we will describe first how the use of the available two-loop RG functions and the requirement of finiteness of the SSB parameters up to this order leads to the soft scalar-mass sum rule [7].

Consider the superpotential given by (15) along with the Lagrangian for SSB terms

$$
-\mathcal{L}_{\rm SB} = \frac{1}{6}\,h^{ijk}\,\phi_i\phi_j\phi_k + \frac{1}{2}\,b^{ij}\,\phi_i\phi_j + \frac{1}{2}\,(m^2)^j_i\,\phi^{*\,i}\phi_j + \frac{1}{2}\,M\,\lambda\lambda + \text{H.c.}, \tag{58}
$$

where the ϕ_i are the scalar parts of the chiral superfields Φ_i , λ are the gauginos and M their unified mass. Since we would like to consider only finite theories here, we assume that the gauge group is a simple group and the one-loop β function of the gauge coupling g vanishes. We also assume that the reduction equations admit power series solutions of the form

$$
C^{ijk} = g\sum_{n=0}\rho^{ijk}_{(n)}g^{2n}\;, \tag{59}
$$

According to the finiteness theorem of ref. [16], the theory is then finite to all orders in perturbation theory, if, among others, the one-loop anomalous dimensions $\gamma_i^{j(1)}$ vanish. The one- and two-loop finiteness for h^{ijk} can be achieved by

$$h^{ijk} = -MC^{ijk} + \ldots = -M\rho^{ijk}_{(0)}\, g + O(g^5) \ . \tag{60}$$

Now, to obtain the two-loop sum rule for soft scalar masses, we assume that the lowest order coefficients $\rho^{ijk}_{(0)}$ and also $(m^2)^i_j$ satisfy the diagonality relations

$$\rho_{ipq(0)}\rho^{jpq}_{(0)} \propto \delta^j_i \text{ for all } p \text{ and } q \ \text{ and } \ (m^2)^i_j = m^2_j\delta^i_j \ , \tag{61}$$

respectively. Then we find the following soft scalar-mass sum rule

$$(\ m^2_i + m^2_j + m^2_k\ )/MM^\dagger = 1 + \frac{g^2}{16\pi^2}\,\Delta^{(1)} + O(g^4) \tag{62}$$

for i, j, k with $\rho^{ijk}_{(0)} \neq 0$, where $\Delta^{(1)}$ is the two-loop correction

$$\Delta^{(1)} = -2\sum_l [(m^2_l/MM^\dagger) - (1/3)]\ T(R_l), \tag{63}$$

which vanishes for the universal choice in accordance with the previous findings of ref. [13].

If we know higher-loop β-functions explicitly, we can follow the same procedure and find higher-loop RGI relations among SSB terms. However, the β-functions of the soft scalar masses are explicitly known only up to two loops. In order to obtain higher-loop results, we need something else instead of knowledge of explicit β-functions, e.g. some relations among β-functions.

The recent progress made using the spurion technique [47], [48] leads to the following all-loop relations among SSB β-functions, [20]- [25]

$$\beta_M = 2\mathcal{O}\left(\frac{\beta_g}{g}\right) \ , \tag{64}$$

$$\beta^{ijk}_h = \gamma^i{}_l h^{ljk} + \gamma^j{}_l h^{ilk} + \gamma^k{}_l h^{ijl}$$
$$-2\gamma^i_1{}_l C^{ljk} - 2\gamma^j_1{}_l C^{ilk} - 2\gamma^k_1{}_l C^{ijl} \ , \tag{65}$$

$$(\beta_{m^2})^i{}_j = \left[\Delta + X\frac{\partial}{\partial g}\right]\gamma^i{}_j \ , \tag{66}$$

$$\mathcal{O} = \left(Mg^2\frac{\partial}{\partial g^2} - h^{lmn}\frac{\partial}{\partial C^{lmn}}\right) \ , \tag{67}$$

$$\Delta = 2\mathcal{O}\mathcal{O}^* + 2|M|^2 g^2\frac{\partial}{\partial g^2} + \tilde{C}_{lmn}\frac{\partial}{\partial C_{lmn}} + \tilde{C}^{lmn}\frac{\partial}{\partial C^{lmn}} \ , \tag{68}$$

where $(\gamma_1)^i{}_j = \mathcal{O}\gamma^i{}_j$, $C_{lmn} = (C^{lmn})^*$, and

$$\tilde{C}^{ijk} = (m^2)^i{}_l C^{ljk} + (m^2)^j{}_l C^{ilk} + (m^2)^k{}_l C^{ijl} \ . \tag{69}$$

It was also found [25] that the relation

$$h^{ijk} = -M(C^{ijk})' \equiv -M\frac{dC^{ijk}(g)}{d\ln g} , \tag{70}$$

among couplings is all-loop RGI. Furthermore, using the all-loop gauge β-function of Novikov *et al.* [35] given by

$$\beta_g^{\rm NSVZ} = \frac{g^3}{16\pi^2}\left[\frac{\sum_l T(R_l)(1-\gamma_l/2) - 3C(G)}{1 - g^2C(G)/8\pi^2}\right] , \tag{71}$$

it was found the all-loop RGI sum rule [26],

$$m_i^2 + m_j^2 + m_k^2 = |M|^2\{ \frac{1}{1-g^2C(G)/(8\pi^2)}\frac{d\ln C^{ijk}}{d\ln g} + \frac{1}{2}\frac{d^2\ln C^{ijk}}{d(\ln g)^2} \}$$
$$+\sum_l \frac{m_l^2 T(R_l)}{C(G) - 8\pi^2/g^2}\frac{d\ln C^{ijk}}{d\ln g} . \tag{72}$$

In addition the exact-β function for m^2 in the NSVZ scheme has been obtained [26] for the first time and is given by

$$\beta_{m_i^2}^{\rm NSVZ} = \left[|M|^2\{ \frac{1}{1-g^2C(G)/(8\pi^2)}\frac{d}{d\ln g} + \frac{1}{2}\frac{d^2}{d(\ln g)^2} \}\right.$$
$$\left. +\sum_l \frac{m_l^2 T(R_l)}{C(G) - 8\pi^2/g^2}\frac{d}{d\ln g}\right] \gamma_i^{\rm NSVZ} . \tag{73}$$

Surprisingly enough, the all-loop result (72) coincides with the superstring result for the finite case in a certain class of orbifold models [7] if $d\ln C^{ijk}/d\ln g = 1$.

V $SU(5)$ FINITE AND GAUGE - YUKAWA UNIFIED MODELS

A Finite Unified Models

From the classification of theories with vanishing one-loop gauge β function [40], one can easily see that there exist only two candidate possibilities to construct $SU(5)$ GUTs with three generations. These possibilities require that the theory should contain as matter fields the chiral supermultiplets $\mathbf{5}$, $\overline{\mathbf{5}}$, $\mathbf{10}$, $\overline{\mathbf{5}}$, $\mathbf{24}$ with the multiplicities $(6,9,4,1,0)$ and $(4,7,3,0,1)$, respectively. Only the second one contains a **24**-plet which can be used to provide the spontaneous symmetry breaking (SB) of SU(5) down to $SU(3)\times SU(2)\times U(1)$. For the first model one has to incorporate another way, such as the Wilson flux breaking mechanism to achieve the desired SB of $SU(5)$ [1]. Therefore, for a self-consistent field theory discussion we would like to concentrate only on the second possibility.

It is clear, at least for the dimensionless couplings, that the matter content of a theory is only a necessary condition for all-order finiteness. Therefore, there exist, in principle, various finite models for a given matter content. However, during the early studies [41,44], the theorem [16] that guarantees all-order finiteness and requires the existence of power series solution to any finite order in perturbation theory was not known. The theorem introduces new constraints, in particular requires that the solution to the one-loop finiteness conditions should be non-degenerate and isolated. In most previous studies the freedom resulted as a consequence of the degeneracy in the one- and two-loop solutions was used to make specific ansätze that could lead to phenomenologically acceptable predictions. Note that the existence of such freedom is incompatible with the power series solutions [15,16].

Taking into account the new constraints an all-order finite $SU(5)$ model has been constructed [1], which among others successfully predicted the bottom and the top quark masses [1,4]. The latter is due to the Gauge-and-Yukawa-of-the-third-generation Unification [1]- [4] which has been achieved. In general the predictive power of a finite $SU(5)$ model depends on the structure of the superpotential and on the way the four pairs of Higgs quintets and anti-quintets mix to produce the two Higgs doublets of the minimal supersymmetric standard model (MSSM). Given that the finiteness conditions do not restrict the mass terms, there is a lot of freedom offered by this sector of the theory in mixing the four pairs of Higgs fields. As a result it was possible in the early studies (a) to provide the adequate doublet-triplet splitting in the pair of $\mathbf{5}$ and $\overline{\mathbf{5}}$ which couple to ordinary fermions so as to suppress the proton decay induced by the coloured triplets and (b) to introduce angles in the gauge-Yukawa relations suppressing in this way the strength of the Yukawa couplings. Concerning the requirement (b) one has to recall that at that time it was very unpleasant to have a top mass prediction at $O(150-200)$ GeV; the popular top quark mass was at $O(40)$ GeV. The above was most clearly stated in ref. [44] and has been revived [45] taking into account the recent data. However, it is clear that using the large freedom offered by the Higgs mass parameter space in requiring the condition (b) one strongly diminishes the beauty of a finite theory. Consequently, this freedom was abandoned in the recent studies of the all-loop finite SU(5) model [1] and only the condition (a) was kept as a necessary requirement.

A predictive Gauge-Yukawa unified SU(5) model which is finite to all orders, in addition to the requirements mentioned already, should also have the following properties.

1. One-loop anomalous dimensions are diagonal, i.e., $\gamma_i^{(1)\,j} \propto \delta_i^j$, according to the assumption (61).
2. Three fermion generations, $\overline{\mathbf{5}}_i$ $(i=1,2,3)$, obviously should not couple to $\mathbf{24}$. This can be achieved for instance by imposing $B-L$ conservation.
3. The two Higgs doublets of the MSSM should mostly be made out of a pair of Higgs quintet and anti-quintet, which couple to the third generation.

In the following we discuss two versions of the all-order finite model.

A: The model of ref. [1], whose matter content we just described above.
B: A slight variation of the model **A**, which can also be obtained from the class of the models suggested by Kazakov *et al.* [23] with a modification to suppress non-diagonal anomalous dimensions.

The quark mixing can be accommodated in these models, but for simplicity we neglect the intergenerational mixing and postpone the interesting problem of predicting the mixings to a future publication.

The superpotential which describe the two models takes the form [1,23]

$$W = \sum_{i=1}^{3} [\ \frac{1}{2} g_i^u\, \mathbf{10}_i \mathbf{10}_i H_i + g_i^d\, \mathbf{10}_i \overline{\mathbf{5}}_i\, \overline{H}_i\ ] + g_{23}^u\, \mathbf{10}_2 \mathbf{10}_3 H_4$$
$$+ g_{23}^d\, \mathbf{10}_2 \overline{\mathbf{5}}_3\, \overline{H}_4 + g_{32}^d\, \mathbf{10}_3 \overline{\mathbf{5}}_2\, \overline{H}_4 + \sum_{a=1}^{4} g_a^f\, H_a\, \mathbf{24}\, \overline{H}_a + \frac{g^\lambda}{3}\, (\mathbf{24})^3\ , \qquad (74)$$

where H_a and $\overline{H}_a$ $(a = 1, \dots, 4)$ stand for the Higgs quintets and anti-quintets. Given the superpotential W, we can compute now the γ functions of the model, from which we then compute the β functions. We find:

$$\gamma_{\mathbf{10}_1}^{(1)} = \frac{1}{16\pi^2} [-\frac{36}{5} g^2 + 3\,(g_1^u)^2 + 2\,(g_1^d)^2\,]\ ,$$
$$\gamma_{\mathbf{10}_2}^{(1)} = \frac{1}{16\pi^2} [-\frac{36}{5} g^2 + 3\,(g_2^u)^2 + 2\,(g_2^d)^2 + 3\,(g_{23}^u)^2 + 2\,(g_{23}^d)^2\,]\ ,$$
$$\gamma_{\mathbf{10}_3}^{(1)} = \frac{1}{16\pi^2} [-\frac{36}{5} g^2 + 3\,(g_3^u)^2 + 2\,(g_3^d)^2 + 3\,(g_{23}^u)^2 + 2\,(g_{32}^d)^2\,]\ ,$$
$$\gamma_{\overline{\mathbf{5}}_1}^{(1)} = \frac{1}{16\pi^2} [-\frac{24}{5} g^2 + 4\,(g_1^d)^2\,]\ ,$$
$$\gamma_{\overline{\mathbf{5}}_2}^{(1)} = \frac{1}{16\pi^2} [-\frac{24}{5} g^2 + 4\,(g_2^d)^2 + 4\,(g_{32}^d)^2\,]\ ,$$
$$\gamma_{\overline{\mathbf{5}}_3}^{(1)} = \frac{1}{16\pi^2} [-\frac{24}{5} g^2 + 4\,(g_3^d)^2 + 4\,(g_{23}^d)^2\,]\ ,$$
$$\gamma_{H_i}^{(1)} = \frac{1}{16\pi^2} [-\frac{24}{5} g^2 + 3\,(g_i^u)^2 + \frac{24}{5} (g_i^f)^2\,]\ \ i = 1,2,3\ , \qquad (75)$$
$$\gamma_{\overline{H}_i}^{(1)} = \frac{1}{16\pi^2} [-\frac{24}{5} g^2 + 4\,(g_i^d)^2 + \frac{24}{5} (g_i^f)^2\,]\ \ i = 1,2,3\ ,$$
$$\gamma_{H_4}^{(1)} = \frac{1}{16\pi^2} [-\frac{24}{5} g^2 + 6\,(g_{23}^u)^2 + \frac{24}{5} (g_4^f)^2\,]\ ,$$
$$\gamma_{\overline{H}_4}^{(1)} = \frac{1}{16\pi^2} [-\frac{24}{5} g^2 + 4\,(g_{23}^d)^2 + 4\,(g_{32}^d)^2 + \frac{24}{5} (g_4^f)^2\,]\ ,$$
$$\gamma_{\mathbf{24}}^{(1)} = \frac{1}{16\pi^2} [-10\, g^2 + \sum_{a=1}^{4} (g_a^f)^2 + \frac{21}{5}\, (g^\lambda)^2\,]\ .$$

The non-degenerate and isolated solutions to $\gamma_i^{(1)} = 0$ for the models $\{\mathbf{A}\ ,\ \mathbf{B}\}$ are:

$$
\begin{aligned}
&(g_1^u)^2 = \{\frac{8}{5} , \frac{8}{5})\}g^2 , \ (g_1^d)^2 = \{\frac{6}{5} , \frac{6}{5}\}g^2 , \ (g_2^u)^2 = (g_3^u)^2 = \{\frac{8}{5} , \frac{4}{5}\}g^2 , \\
&(g_2^d)^2 = (g_3^d)^2 = \{\frac{6}{5} , \frac{3}{5}\}g^2 , \ (g_{23}^u)^2 = \{0 , \frac{4}{5}\}g^2 , \ (g_{23}^d)^2 = (g_{32}^d)^2 = \{0 , \frac{3}{5}\}g^2 , \\
&(g^\lambda)^2 = \frac{15}{7}g^2 , \ (g_2^f)^2 = (g_3^f)^2 = \{0 , \frac{1}{2}\}g^2 , \ (g_1^f)^2 = 0 , \ (g_4^f)^2 = \{1 , 0 \}g^2 .
\end{aligned}
\tag{76}
$$

We have explicitly checked that these solutions (76) are also the solutions of the reduction equation (2) and that they can be uniquely extended to the corresponding power series solutions (3) [4]. Consequently, these models are finite to all orders.

After the reduction of couplings (76) the symmetry of W (74) is enhanced: For the model **A** one finds that the superpotential has the $Z_7 \times Z_3 \times Z_2$ discrete symmetry

$$
\begin{aligned}
&\overline{\mathbf{5}}_1 : (4,0,1) , \ \overline{\mathbf{5}}_2 : (1,0,1) , \ \overline{\mathbf{5}}_3 : (2,0,1) , \\
&\mathbf{10}_1 : (1,1,1) , \ \mathbf{10}_2 : (2,2,1) , \ \mathbf{10}_3 : (4,0,1) , \\
&H_1 : (5,1,0) , \ H_2 : (3,2,0) , \ H_3 : (6,0,0) , \\
&\overline{H}_1 : (-5,-1,0) , \ \overline{H}_2 : (-3,-2,0) , \ \overline{H}_3 : (-6,0,0) , \\
&H_4 : (0,0,0) , \ \overline{H}_4 : (0,0,0) , \ \mathbf{24} : (0,0,0) ,
\end{aligned}
\tag{77}
$$

while for the model **B** one finds $Z_4 \times Z_4 \times Z_4$ defined as

$$
\begin{aligned}
&\overline{\mathbf{5}}_1 : (1,0,0) , \ \overline{\mathbf{5}}_2 : (0,1,0) , \ \overline{\mathbf{5}}_3 : (0,0,1) , \\
&\mathbf{10}_1 : (1,0,0) , \ \mathbf{10}_2 : (0,1,0) , \ \mathbf{10}_3 : (0,0,1) , \\
&H_1 : (2,0,0) , \ H_2 : (0,2,0) , \ H_3 : (0,0,2) , \\
&\overline{H}_1 : (-2,0,0) , \ \overline{H}_2 : (0,-2,0) , \ \overline{H}_3 : (0,0,-2) , \\
&H_4 : (0,3,3) , \ \overline{H}_4 : (0,-3,-3) , \ \mathbf{24} : (0,0,0) ,
\end{aligned}
\tag{78}
$$

where the numbers in the parenthesis stand for the charges under the discrete symmetries.

The main difference of the models **A** and **B** is that three pairs of Higgs quintets and anti-quintets couple to the **24** for **B** so that it is not necessary [23] to mix them with H_4 and $\overline{H}_4$ in order to achieve the triplet-doublet splitting after SB of $SU(5)$. This enhances the predicitivity, because then the mixing of the three pairs of Higgsess are strongly constrained to fit the phenomenology of the first two generations [24].

B The Minimal Asymptotically Free SU(5) Model

The minimal N=1 supersymmetric SU(5) model [59] is particularly interesting, being the the simplest GUT supported by the LEP data [60]. Here we will consider

---

[4]) The coefficients in (76) are slightly different from those in models considered in refs. [23].

it as an attractive example of a partially reduced model. Its particle content is well defined and has the following transformation properties under SU(5): three $(\overline{\mathbf{5}}+\mathbf{10})$- supermultiplets which accommodate three fermion families, one $(\mathbf{5}+\overline{\mathbf{5}})$ to describe the two Higgs supermultiplets appropriate for electroweak symmetry breaking and a **24**-supermultiplet required to provide the spontaneous symmetry breaking of $SU(5)$ down to $SU(3)\times SU(2)\times U(1)$.

Since we are neglecting the dimensional parameters and the Yukawa couplings of the first two generations, the superpotential of the model is exactly given by

$$W=\frac{1}{2}\,g_t\mathbf{10}_3\,\mathbf{10}_3\,H+g_b\,\overline{\mathbf{5}}_3\,\mathbf{10}_3\,\overline{H}+g_\lambda\,(\mathbf{24})^3+g_f\,\overline{H}\,\mathbf{24}\,H\ , \tag{79}$$

where $H,\overline{H}$ are the $\mathbf{5},\overline{\mathbf{5}}$- Higgs supermultiplets and we have suppressed the $SU(5)$ indices. Let us now define

$$\tilde{\alpha}_i\equiv\frac{\alpha_i}{\alpha}\ ,\ i=1,\cdots,n\ , \tag{80}$$

According to this notation the reduction equations become

$$\begin{aligned}
\alpha\,\frac{d\tilde{\alpha}_t}{d\alpha}&=\frac{27}{5}\,\tilde{\alpha}_t-3\,\tilde{\alpha}_t^2-\frac{4}{3}\,\tilde{\alpha}_t\tilde{\alpha}_b-\frac{8}{5}\,\tilde{\alpha}_t\,\tilde{\alpha}_f\ ,\\
\alpha\,\frac{d\tilde{\alpha}_b}{d\alpha}&=\frac{23}{5}\,\tilde{\alpha}_b-\frac{10}{3}\,\tilde{\alpha}_b^2-\tilde{\alpha}_b\tilde{\alpha}_t-\frac{8}{5}\,\tilde{\alpha}_b\,\tilde{\alpha}_f\ ,\\
\alpha\,\frac{d\tilde{\alpha}_\lambda}{d\alpha}&=9\tilde{\alpha}_\lambda-\frac{21}{5}\,\tilde{\alpha}_\lambda^2-\tilde{\alpha}_\lambda\,\tilde{\alpha}_f\ ,\\
\alpha\,\frac{d\tilde{\alpha}_f}{d\alpha}&=\frac{83}{15}\,\tilde{\alpha}_f-\frac{53}{15}\,\tilde{\alpha}_f^2-\tilde{\alpha}_f\tilde{\alpha}_t-\frac{4}{3}\,\tilde{\alpha}_f\,\tilde{\alpha}_b-\frac{7}{5}\,\tilde{\alpha}_f\,\tilde{\alpha}_\lambda\ ,
\end{aligned} \tag{81}$$

in the one-loop approximation. Given the above equations describing the evolution of the four independent couplings $(\alpha_i\ ,\ i=t,b,\lambda,f)$, there exist $2^4=16$ non-degenerate solutions corresponding to vanishing ρ's as well as non-vanishing ones. The possibility to predict the top quark mass depends on a nontrivial interplay between the vacuum expectation value of the two $SU(2)$ Higgs doublets involved in the model and the known masses of the third generation $(m_b\ ,\ m_\tau)$. It is clear that only the solutions of the form

$$\rho_t\ ,\ \rho_b\ \neq\ 0 \tag{82}$$

can predict the top and bottom quark masses. Of these, there are only two solutions which satisfy the criteria of being asymptotically free.

In the one-loop approximation, the GYU solution yields

$$g_{t,b}^2\ =\ \sum_{m,n=1}^{\infty}\kappa_{t,b}^{(m,n)}\ h^m\ f^n\ g^2 \tag{83}$$

(h and f are related to the Higgs couplings), where h is allowed to vary from 0 to 15/7, while f may vary from 0 to a maximum which depends on h and vanishes at

$h = 15/7$. As a result, it was obtained [2]: $0.97\, g^2 \lesssim g_t^2 \lesssim 1.37\, g^2$, $0.57\, g^2 \lesssim g_b^2 = g_\tau^2 \lesssim 0.97\, g^2$. It was found [4,14] that consistency with proton decay requires g_t^2, g_b^2 to be very close to the left hand side values in the inequalities. Note furthermore that not only α_t but also α_b is predicted in this reduction solution.

VI PREDICTIONS OF LOW ENERGY PARAMETERS

In this section we will refine the predictions of the AFUT and FUT models for the third family quark masses, taking into account certain corrections and we will compare them with the experimental data.

Since the gauge symmetry is spontaneously broken below $M_{\rm GUT}$, the finiteness conditions do not restrict the renormalization property at low energies, and all it remains are boundary conditions on the gauge and Yukawa couplings (76) and the $h = -MY$ relation (60) and the soft scalar-mass sum rule (62) at $M_{\rm GUT}$. So we examine the evolution of these parameters according to their renormalization group equations at two-loop for dimensionless parameters and at one-loop for dimensional ones with these boundary conditions. Below $M_{\rm GUT}$ their evolution is assumed to be governed by the MSSM. We further assume a unique supersymmetry breaking scale M_s so that below M_s the SM is the correct effective theory.

A Top Mass Predictions

We recall that $\tan\beta$ is usually determined in the Higgs sector. However, it has turned out that in the case of GYU models it is convenient to define $\tan\beta$ by using the matching condition at M_s [56],

$$\begin{aligned} \alpha_t^{\rm SM} &= \alpha_t \sin^2\beta\ ,\ \alpha_b^{\rm SM} = \alpha_b \cos^2\beta\ ,\ \alpha_\tau^{\rm SM} = \alpha_\tau \cos^2\beta\ , \\ \alpha_\lambda &= \frac{1}{4}(\frac{3}{5}\alpha_1 + \alpha_2)\cos^2 2\beta\ , \end{aligned} \tag{84}$$

where $\alpha_i^{\rm SM}$ $(i = t, b, \tau)$ are the SM Yukawa couplings and α_λ is the Higgs coupling ($\alpha_I = g_I^2/4\pi^2$). With a given set of the input parameters [57],

$$M_\tau = 1.777\ {\rm GeV}\ ,\ M_Z = 91.188\ {\rm GeV}\ , \tag{85}$$

with [58]

$$\begin{aligned} \alpha_{\rm EM}^{-1}(M_Z) &= 127.9 + \frac{8}{9\pi}\log\frac{M_t}{M_Z}\ , \\ \sin^2\theta_{\rm W}(M_Z) &= 0.2319 - 3.03\times 10^{-5}T - 8.4\times 10^{-8}T^2\ , \\ T &= M_t/[{\rm GeV}] - 165\ , \end{aligned} \tag{86}$$

the matching condition (84) and the GYU boundary condition at $M_{\rm GUT}$ can be satisfied only for a specific value of $\tan\beta$. Here M_τ, M_t, M_Z are pole masses, and

TABLE 1. The predictions for different M_s for **A**

M_s [GeV]	$\alpha_{3(5f)}(M_Z)$	$\tan\beta$	M_{GUT} [GeV]	M_b [GeV]	M_t [GeV]
300	0.123	54.1	2.2×10^{16}	5.3	183
500	0.122	54.2	1.9×10^{16}	5.3	183
10^3	0.120	54.3	1.5×10^{16}	5.2	184

TABLE 2. The predictions for different M_s for **B**

M_s [GeV]	$\alpha_{3(5f)}(M_Z)$	$\tan\beta$	M_{GUT} [GeV]	M_b [GeV]	M_t [GeV]
800	0.120	48.2	1.5×10^{16}	5.4	174
10^3	0.119	48.2	1.4×10^{16}	5.4	174
1.2×10^3	0.118	48.2	1.3×10^{16}	5.4	174

the couplings above are defined in the $\overline{\text{MS}}$ scheme with six flavors. Under the assumptions specified above, it is possible without knowing the details of the scalar sector of the MSSM to predict various parameters such as the top quark mass [1]-[4]. We present them for the model **A** in table 1, for the model **B** in table 2, and for the minimal $SU(5)$ model in table 3.
Comparing, for instance, the M_t predictions above with the most recent experimental value

$$M_t = (173.8 \pm 5.2)\ \text{GeV}\ , \tag{87}$$

and recalling that the theoretical values for M_t given in the tables may suffer from a correction of less than $\sim 4\ \%$ [4], we see that they are consistent with the experimental data. For more details, see ref. [4], where various corrections on the predictions of GYU models such as the MSSM threshold corrections are estimated [5].

B Implications of the Sum Rule

Finite SU(5) Models: Here we will discuss the structure of the sum rule for the

[5] The GUT threshold corrections in the $SU(5)$ finite model are given in ref. [45].

TABLE 3. The predictions for different M_s for MIN $SU(5)$

M_s [GeV]	$\alpha_{3(5f)}(M_Z)$	$\tan\beta$	M_{GUT} [GeV]	M_b [GeV]	M_t [GeV]
300	0.123	47.9	2.2×10^{16}	5.5	178
500	0.122	47.8	1.8×10^{16}	5.4	178
1000	0.119	47.7	1.5×10^{16}	5.4	178

soft scalar masses for each of the finite models. According to (61), we recall that they are supposed to be diagonal. From the one-loop finiteness for the soft scalar masses, we obtain (there are $\{10\ ,\ 13\}$ equations for 15 unknown $\kappa^{(0)}$'s):

$$\kappa^{(0)}_{H_i} = 1 - 2\kappa^{(0)}_{\mathbf{10}_i}\ ,\ \kappa^{(0)}_{\overline{H}_i} = 1 - \kappa^{(0)}_{\mathbf{10}_i} - \kappa^{(0)}_{\overline{\mathbf{5}}_i}\ \ (i = 1,2,3)\ , \tag{88}$$
$$\kappa^{(0)}_{H_4} = \frac{2}{3} - \kappa^{(0)}_{\overline{H}_4}\ ,\ \kappa^{(0)}_{\mathbf{24}} = \frac{1}{3}\ \text{ for }\mathbf{A}\ ,$$

and

$$\kappa^{(0)}_{H_1} = 1 - 2\kappa^{(0)}_{\mathbf{10}_1}\ ,\ \kappa^{(0)}_{H_2} = \kappa^{(0)}_{H_3} = \kappa^{(0)}_{H_4} = 1 - 2\kappa^{(0)}_{\mathbf{10}_3}\ ,$$
$$\kappa^{(0)}_{\overline{H}_1} = 1 - \kappa^{(0)}_{\mathbf{10}_1} - \kappa^{(0)}_{\overline{\mathbf{5}}_1}\ ,\ \kappa^{(0)}_{\overline{H}_2} = \kappa^{(0)}_{\overline{H}_3} = \kappa^{(0)}_{\overline{H}_4} = -\frac{1}{3} + 2\kappa^{(0)}_{\mathbf{10}_3}\ , \tag{89}$$
$$\kappa^{(0)}_{\overline{\mathbf{5}}_2} = \kappa^{(0)}_{\overline{\mathbf{5}}_3} = \frac{4}{3} - 3\kappa^{(0)}_{\mathbf{10}_3}\ ,\ \kappa^{(0)}_{\mathbf{10}_2} = \kappa^{(0)}_{\mathbf{10}_3}\ ,\ \kappa^{(0)}_{\mathbf{24}} = \frac{1}{3}\ \text{ for }\mathbf{B}\ ,$$

where we have defined

$$\frac{m_i^2}{|M|^2} = \kappa_i^{(0)} + \frac{g^2}{16\pi^2}\kappa_i^{(1)} + \cdots\ ,\ i = \mathbf{10}_1, \mathbf{10}_2, \ldots, \mathbf{24}\ . \tag{90}$$

We then use the solution (76) to calculate the actual value for S', which express the two-loop correction to the sum rule. Surprisingly, it turns out for both models that

$$S' = 0\ . \tag{91}$$

That is, the one-loop sum rule in the present models is not corrected in two-loop order.

Next we would like to address the question of whether the sum rule (62) is the unique solution to the two-loop finiteness. The two-loop finiteness for the soft scalar masses follows if the following condition

$$\rho_{ipq(0)}\rho^{jpq}_{(0)}(\kappa_i^{(1)} + \kappa_p^{(1)} + \kappa_q^{(1)}) = -8C(i)\sum_l [\kappa_p^{(0)} - (1/3)]\ T(R_l) = -8C(i)S'\ , \tag{92}$$

is satisfied. There are 15 equations for 15 unknown $\kappa^{(1)}$'s. We find that the solution is not unique; it can be parameterized by $\{7,4\}$ parameters for a given S' which is zero for the present models. For instance,

$$\kappa^{(1)}_{H_i} = -2S' - 2\kappa^{(1)}_{\mathbf{10}_i}\ ,\ \kappa^{(1)}_{\overline{H}_i} = -2S' - \kappa^{(1)}_{\overline{\mathbf{5}}_i} - \kappa^{(1)}_{\mathbf{10}_i}\ (i = 1,2,3)\ , \tag{93}$$
$$\kappa^{(1)}_{H_4} = -\frac{4S'}{3} - \kappa^{(1)}_{\overline{H}_4}\ ,\ \kappa^{(1)}_{24} = -\frac{2S'}{3}\ \text{ for }\mathbf{A}\ ,$$

and

$$\kappa^{(1)}_{H_1} = -2S' - 2\kappa^{(1)}_{\mathbf{10}_1} \ , \ \kappa^{(1)}_{24} = -\frac{2S'}{3} \ , \ \kappa^{(1)}_{\mathbf{10}_2} = \kappa^{(1)}_{\mathbf{10}_3} \ ,$$

$$\kappa^{(1)}_{H_2} = \kappa^{(1)}_{H_3} = \kappa^{(1)}_{H_4} = -2S' - 2\kappa^{(1)}_{\mathbf{10}_3} \ , \ \kappa^{(1)}_{\overline{H}_2} = \kappa^{(1)}_{\overline{H}_3} = \kappa^{(1)}_{\overline{H}_4} = \frac{2S'}{3} + 2\kappa^{(1)}_{\mathbf{10}_3} \ , \tag{94}$$

$$\kappa^{(1)}_{\overline{H}_1} = -2S' - \kappa^{(1)}_{\overline{\mathbf{5}}_1} - \kappa^{(1)}_{\mathbf{10}_1} \ , \ \kappa^{(1)}_{\overline{\mathbf{5}}_2} = \kappa^{(1)}_{\overline{\mathbf{5}}_3} = -\frac{8S'}{3} - 3\kappa^{(1)}_{\mathbf{10}_3} \ \text{for } \mathbf{B} \ .$$

As one can easily see that $\kappa^{(1)}$'s satisfy

$$\kappa^{(1)}_i + \kappa^{(1)}_j + \kappa^{(1)}_k = -2S' = 0 \ , \tag{95}$$

which shows that the sum rule (62) in the present models is the unique solution to two-loop finiteness.

Thus, the sum rule gives us the following boundary conditions at the GUT scale [7]:

$$m^2_{H_u} + 2m^2_{\mathbf{10}} = m^2_{H_d} + m^2_{\overline{\mathbf{5}}} + m^2_{\mathbf{10}} = M^2 \ \text{ for } \mathbf{A} \ , \tag{96}$$

$$m^2_{H_u} + 2m^2_{\mathbf{10}} = M^2 \ , \ m^2_{H_d} - 2m^2_{\mathbf{10}} = -\frac{M^2}{3} \ ,$$

$$m^2_{\overline{\mathbf{5}}} + 3m^2_{\mathbf{10}} = \frac{4M^2}{3} \ \text{ for } \mathbf{B}, \tag{97}$$

where we use as free parameters $m_{\overline{\mathbf{5}}} \equiv m_{\overline{\mathbf{5}}_3}$ and $m_{\mathbf{10}} \equiv m_{\mathbf{10}_3}$ for the model **A**, and $m_{\mathbf{10}}$ for **B**, in addition to M.

The minimal supersymmetric $SU(5)$ model: In this model, the reduction of parameters implies that at the GUT scale the SSB terms are proportional to the gaugino mass, which thus characterizes the scale of supersymmetry breaking [5].

C Predictions for Higgs the S-spectrum

In the SSB sector, besides the constraints imposed by reduction of couplings and finiteness, we also look for solutions which are compatible with radiative electroweak symmetry breaking. As it has been mentioned, in the minimal $SU(5)$ model the SSB sector contains only one independent parameter, the gaugino mass M, which characterizes the scale of supersymmetry breaking. The lightest supersymmetric particle is found to be a neutralino of $\sim$ 220 GeV for $M(M_{GUT}) \sim 0.5$ TeV. In fig. 1 we present the dependence of the lightest Higgs mass m_h on the gaugino mass M.

Concerning the SSB sector of the finite theories **A** and **B**, besides the gaugino mass we have two and one more free parameters respectively, as previously mentioned. Thus, we look for the parameter space in which the lighter $\tilde{\tau}$ mass squared $m^2_{\tilde{\tau}}$ is larger than the lightest neutralino mass squared m^2_{χ} (which is the LSP). In the case where all the soft scalar masses are universal at the unfication scale, there is no region of $M_s = M$ below O(few) TeV in which $m^2_{\tilde{\tau}} > m^2_{\chi}$ is satisfied. But

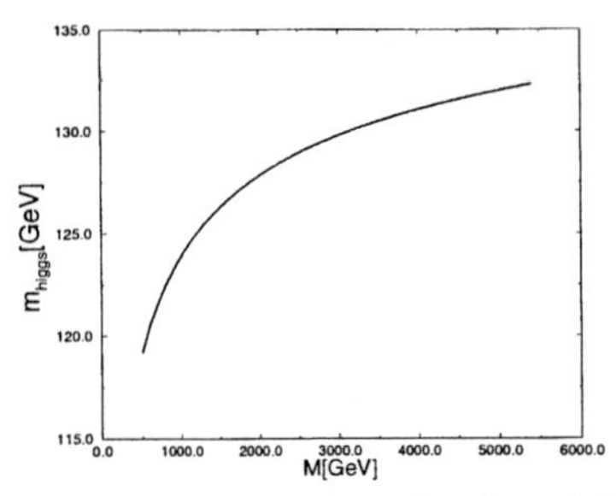

FIGURE 1. The M dependence of m_h for the minimal $SU(5)$ model.

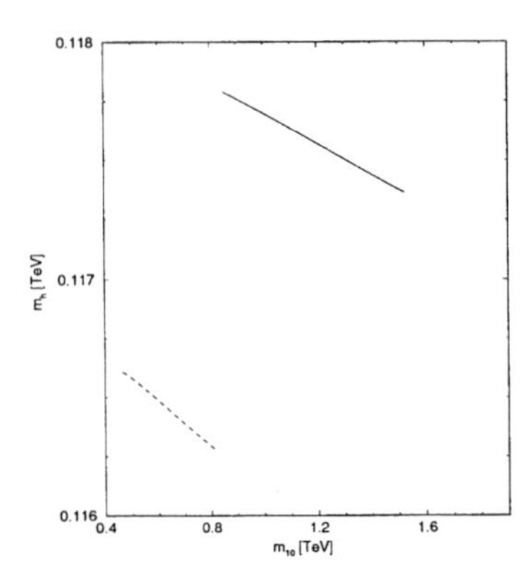

FIGURE 2. m_h as function of m_{10} for $M = 0.8$ (dashed) 1.0 (solid) TeV for the finite model **B**.

once the universality condition is relaxed this problem can be solved naturally (provided the sum rule). More specifically, using the sum rule (62) and imposing the conditions a) successful radiative electroweak symmetry breaking b) $m_{\tilde{\tau}^2} > 0$ and c) $m_{\tilde{\tau}^2} > m_{\chi^2}$, we find a comfortable parameter space for both models (although model **B** requires large $M \sim 1$ TeV).

In fig. 2 we present the m_{10} dependence of m_h for for $M = 0.8$ (dashed) 1.0 (solid) TeV for the finite Model **B**, which shows that the value of m_h is rather stable. Similar results hold also for Model **A**.

In Tables 4, 5, and 6 we present representative examples of the values obtained for the sparticle spectra in each of the models. The value of the lightest Higgs physical mass m_h has already the two-loop radiative corrections included [62,63].

The prediction of the Higgs mass for the three models is

$$m_h = 120 \pm 5 \pm 2 \;\; GeV \tag{98}$$

TABLE 4. A representative example of the predictions for the s-spectrum for the finite model **A** with $M = 1.0$ TeV, $m_{\bar{\mathbf{5}}} = 0.8$ TeV and $m_{\mathbf{10}} = 0.6$ TeV.

$m_\chi = m_{\chi_1}$ (TeV)	0.45	$m_{\tilde{b}_2}$ (TeV)	1.76
m_{χ_2} (TeV)	0.84	$m_{\tilde{\tau}} = m_{\tilde{\tau}_1}$ (TeV)	0.63
m_{χ_3} (TeV)	1.49	$m_{\tilde{\tau}_2}$ (TeV)	0.85
m_{χ_4} (TeV)	1.49	$m_{\tilde{\nu}_1}$ (TeV)	0.88
$m_{\chi_1^\pm}$ (TeV)	0.84	m_A (TeV)	0.64
$m_{\chi_2^\pm}$ (TeV)	1.49	$m_{H^\pm}$ (TeV)	0.65
$m_{\tilde{t}_1}$ (TeV)	1.57	m_H (TeV)	0.65
$m_{\tilde{t}_2}$ (TeV)	1.77	m_h (TeV)	0.122
$m_{\tilde{b}_1}$ (TeV)	1.54		

TABLE 5. A representative example of the predictions of the s-spectrum for the finite model **B** with $M = 1$ TeV and $m_{\mathbf{10}} = 0.65$ TeV.

$m_\chi = m_{\chi_1}$ (TeV)	0.45	$m_{\tilde{b}_2}$ (TeV)	1.70
m_{χ_2} (TeV)	0.84	$m_{\tilde{\tau}} = m_{\tilde{\tau}_1}$ (TeV)	0.47
m_{χ_3} (TeV)	1.30	$m_{\tilde{\tau}_2}$ (TeV)	0.67
m_{χ_4} (TeV)	1.31	$m_{\tilde{\nu}_1}$ (TeV)	0.88
$m_{\chi_1^\pm}$ (TeV)	0.84	m_A (TeV)	0.73
$m_{\chi_2^\pm}$ (TeV)	1.31	$m_{H^\pm}$ (TeV)	0.73
$m_{\tilde{t}_1}$ (TeV)	1.51	m_H (TeV)	0.73
$m_{\tilde{t}_2}$ (TeV)	1.73	m_h (TeV)	0.118
$m_{\tilde{b}_1}$ (TeV)	1.56		

where the first uncertainty comes from variations of the gaugino mass M and the soft scalar masses, and is smaller than above in the case of the FUTB and minimal model, due to the more restricted parameter space. The second uncertainty comes from the finite (i.e. not logarithmically divergent) corrections in going from the $\overline{\text{MS}}$ scheme to the pole scheme [64]. Our results are generally a few GeV smaller (≤ 3 GeV) than the ones obtained with the diagrammatic approach in the most refined approximation [65].

Finally, we calculate $BR(b \to s\gamma)$ [66], whose experimental value is $1 \times 10^{-4} < BR(b \to s\gamma) < 4 \times 10^{-4}$. The SM predicts $BR(b \to s\gamma) = 3.1 \times 10^{-4}$. This imposes a further restriction in our parameter space, namely $M \sim 1$ TeV if $\mu < 0$ for all three models. This restriction is less strong in the case that $\mu > 0$. For example, the minimal model with $M = 1$ TeV leads to $BR(b \to s\gamma) = 3.8 \times 10^{-4}$ for $\mu < 0$.

TABLE 6. A representative example of the predictions of the s-spectrum for the minimal $SU(5)$ model with $M = 1.0$ TeV.

$m_{\chi} = m_{\chi_1}$ (TeV)	0.45	$m_{\tilde{b}_2}$ (TeV)	1.88
m_{χ_2} (TeV)	0.84	$m_{\tilde{\tau}} = m_{\tilde{\tau}_1}$ (TeV)	0.92
m_{χ_3} (TeV)	1.73	$m_{\tilde{\tau}_2}$ (TeV)	1.10
m_{χ_4} (TeV)	1.73	$m_{\tilde{\nu}_1}$ (TeV)	1.43
$m_{\chi_1^{\pm}}$ (TeV)	0.84	m_A (TeV)	0.70
$m_{\chi_2^{\pm}}$ (TeV)	1.73	$m_{H^{\pm}}$ (TeV)	0.70
$m_{\tilde{t}_1}$ (TeV)	1.69	m_H (TeV)	0.70
$m_{\tilde{t}_2}$ (TeV)	1.89	m_h (TeV)	0.120
$m_{\tilde{b}_1}$ (TeV)	1.70		

VII CONCLUSIONS

The programme of searching for exact RGI relations among dimensionless couplings in supersymmetric GUTs, started few years ago, has now supplemented with the derivation of similar relations involving dimensionful parameters in the SSB sector of these theories. In the earlier attempts it was possible to derive RGI relations among gauge and Yukawa couplings of supersymmetric GUTs, which could lead even to all-loop finiteness under certain conditions. These theoretically attractive theories have been shown not only to be realistic but also to lead to a successful prediction of the top quark mass. The new theoretical developments include the existence of a RGI sum rule for the soft scalar masses in the SSB sector of $N = 1$ supersymmetric gauge theories exhibiting gauge-Yukawa unification. The all-loop sum rule substitutes now the universal soft scalar masses and overcomes its phenomenological problems. Of particular theoretical interest is the fact that the finite unified theories, which could be made all-loop finite in the supersymmetric sector can now be made *completely finite*. In addition it is interesting to note that the sum rule coincides with that of a certain class of string models in which the massive string modes are organized into $N = 4$ supermultiplets. Last but not least, in ref. [26] the exact β-function for the soft scalar masses in the NSVZ scheme was obtained for the first time. On the other hand the above theories have a remarkable predictive power leading to testable predictions of their spectrum in terms of very few parameters. In addition to the prediction of the top quark mass, which holds unchanged, the characteristic features that will judge the viability of these models in the future are: 1) the lightest Higgs mass is found to be around 120 GeV and 2) the s-spectrum starts beyond several hundreds of GeV. Therefore the next important test of Gauge-Yukawa and Finite Unified theories will be given with the measurement of the Higgs mass, for which these models show an appreciable stability, which is alarmingly close to the IR quasi fixed point prediction of the MSSM for large tan β [67].

ACKNOWLEDGEMENTS

It is a pleasure for one of us (G.Z.) to thank the Organizing Committee for the very warm hospitality offered. We acknowledge useful and interesting discussions with W. Hollik, G. Weiglein, M. Quirós, M. Carena and C. Wagner, and with A. Kehagias. We thank also W. Hollik and collaborators for use of the program FeynHiggs [68].

REFERENCES

1. D. Kapetanakis, M. Mondragón and G. Zoupanos, Zeit. f. Phys. **C60** (1993) 181; M. Mondragón and G. Zoupanos, Nucl. Phys. B (Proc. Suppl.) **37C** (1995) 98.
2. J. Kubo, M. Mondragón and G. Zoupanos, Nucl. Phys. **B424** (1994) 291.
3. J. Kubo, M. Mondragón, N.D. Tracas and G. Zoupanos, Phys. Lett. **B342** (1995) 155; J. Kubo, M. Mondragón, S. Shoda and G. Zoupanos, Nucl. Phys. **B469** (1996) 3.
4. J. Kubo, M. Mondragón, M. Olechowski and G. Zoupanos, Nucl. Phys. **B479** (1996) 25.
5. J. Kubo, M. Mondragón and G. Zoupanos, Phys. Lett. **B389** (1996) 523.
6. T. Kawamura, T. Kobayashi and J. Kubo, Phys. Lett. **B405** (1997) 64.
7. T. Kobayashi, J. Kubo, M. Mondragón and G. Zoupanos, Nucl. Phys. **B511** (1998) 45.
8. L. O'Raifeartaigh, *Nucl. Phys.* **B96** (1975) 331.
9. P. Fayet and J. Iliopoulos, *Phys. Lett.* **B51** (1974) 461.
10. J. Wess and B. Zumino, Phys. Phys. **B49** 52; J. Iliopoulos and B. Zumino, *Nucl. Phys.* **B76** (1974) 310; S. Ferrara, J. Iliopoulos and B. Zumino, *Nucl. Phys.* **B77** (1974) 413; K. Fujikawa and W. Lang, *Nucl. Phys.* **B88** (1975) 61.
11. A.J. Parkes and P.C. West, *Phys. Lett.* **B138** (1984) 99; *Nucl. Phys.* **B256** (1985) 340; P. West, *Phys. Lett.* **B137** (1984) 371; D.R.T. Jones and A.J. Parkes, *Phys. Lett.* **B160** (1985) 267; D.R.T. Jones and L. Mezincescu, *Phys. Lett.* **B136** (1984) 242; **B138** (1984) 293; A.J. Parkes, *Phys. Lett.* **B156** (1985) 73.
12. D.R.T. Jones, L. Mezincescu and Y.-P. Yao, Phys. Lett. **B148** (1984) 317.
13. I. Jack and D.R.T. Jones, Phys. Lett. **B333** (1994) 372.
14. For an extended discussion and a complete list of references see: J. Kubo, M. Mondragón and G. Zoupanos, Acta Phys. Polon. **B27** (1997) 3911.
15. W. Zimmermann, Com. Math. Phys. **97** (1985) 211; R. Oehme and W. Zimmermann, Com. Math. Phys. **97** (1985) 569.
16. C. Lucchesi, O. Piguet and K. Sibold, *Helv. Phys. Acta* **61** (1988) 321; *Phys. Lett.* **B201** (1988) 241; see also C. Lucchesi and G. Zoupanos, Fortsch. Phys. **45** (1997) 129.
17. O. Piguet and K. Sibold, *Int. Journ. Mod. Phys.* **A1** (1986) 913; *Phys. Lett.* **B177** (1986) 373; R. Ensign and K.T. Mahanthappa, *Phys. Rev.* **D36** (1987) 3148.
18. A.Z. Ermushev, D.I. Kazakov and O.V. Tarasov, Nucl. Phys. **281** (1987) 72; D.I. Kazakov, Mod. Phys. Lett. **A9** (1987) 663.

19. I. Jack and D.R.T. Jones, Phys. Lett. **B349** (1995) 294.
20. J. Hisano and M. Shifman, Phys. Rev. **D56** (1997) 5475.
21. I. Jack and D.R.T. Jones, Phys. Lett. **B415** (1997) 383.
22. L.V. Avdeev, D.I. Kazakov and I.N. Kondrashuk, Nucl. Phys. **B510** (1998) 289; D.I.Kazakov, hep-ph/9812513.
23. D.I. Kazakov, M.Yu. Kalmykov, I.N. Kondrashuk and A.V. Gladyshev, Nucl. Phys. **B471** (1996) 387.
24. D.I. Kazakov, Phys. Lett. **B412** (1998) 21.
25. I. Jack, D.R.T. Jones and A. Pickering, Phys. Lett. **B426** (1998) 73.
26. T. Kobayashi, J. Kubo and G. Zoupanos, Phys. Lett. **B427** (1998) 291.
27. N. Stamatopoulos, Master thesis.
28. O. Piguet, *Supersymmetry, Ultraviolet Finiteness and Grand Unification*, hep-th/9606045.
29. S. Ferrara and B. Zumino, *Nucl. Phys.* **B87** (1975) 207.
30. O. Piguet and K. Sibold, *Nucl. Phys.*. **B196** (1982) 428; **B196** (1982) 447.
31. O. Piguet and K. Sibold, *Renormalized Supersymmetry*, Birkhäuser Boston, 1986.
32. L. Alvarez-Gaumé and P. Ginsparg, *Nucl. Phys.* **B243** (1984) 449; W.A. Bardeen and B. Zumino, *Nucl. Phys.***B243** (1984) 421; B. Zumino, Y. Wu and A. Zee, *Nucl. Phys.* **B439** (1984) 477.
33. S.L. Adler and W.A. Bardeen, *Phys. Rev.* **182** (1969) 1517.
34. W. Nahm, Nucl. Phys. **B124** (1977) 121.
35. V. Novikov, M. Shifman, A. Vainstein and V. Zakharov, Nucl. Phys. **B229** (1983) 381; Phys. Lett. **B166** (1986) 329; M. Shifman, Int.J. Mod. Phys.**A11** (1996) 5761 and references therein.
36. R.G. Leigh and M.J. Strassler, *Nucl. Phys.* **B447** (1995) 95.
37. M. Grisaru, W. Siegel and M. Ročec, *Nucl. Phys.* **B159** (1979) 429.
38. S Christensen and M.J. Duff, Phys. Lett. **B76** (1978) 571.
39. J. Maldacena, Adv. Theor. Maths. Phys. **2** (1998) 231.
40. S. Hamidi, J. Patera and J.H. Schwarz, Phys. Lett. **B141** (1984) 349; X.D. Jiang and X.J. Zhou, Phys. Lett. **B197** (1987) 156; **B216** (1985) 160.
41. S. Hamidi and J.H. Schwarz, Phys. Lett. **B147** (1984) 301; D.R.T. Jones and S. Raby, Phys. Lett. **B143** (1984) 137; J.E. Bjorkman, D.R.T. Jones and S. Raby, Nucl. Phys. **B259** (1985) 503.
42. I. Jack, D. Jones and C. North, Nucl. Phys. **B473** (1996) 308; *Phys. Lett.* **B386** (1996) 138, Nucl.Phys. **B486** (1997) 479.
43. N. Seiberg, Phys. Rev. **D49** (1994) 6857.
44. J. León et al, Phys. Lett. **B156** (1985) 66.
45. K. Yoshioka, *Finite SUSY GUT Revisited*, hep-ph/9705449.
46. Y. Yamada, Phys.Rev. **D50** (1994) 3537.
47. R. Delbourgo, Nuovo Cim **25A** (1975) 646; A. Salam and J. Strathdee, Nucl. Phys. **B86** (1975) 142; K. Fujikawa and W. Lang, Nucl. Phys. **B88** (1975) 61; M.T. Grisaru, M. Rocek and W. Siegel, Nucl. Phys. **B159** (1979) 429.
48. L. Girardello and M.T. Grisaru, Nucl. Phys. **B194** (1982) 65; J.A. Helayel-Neto, Phys. Lett. **B135** (1984) 78; F. Feruglio, J.A. Helayel-Neto and F. Legovini, Nucl. Phys. **B249** (1985) 533; M. Scholl, Zeit. f. Phys. **C28** (1985)545.

49. O. Piguet and K. Sibold, Phys. Lett. **229B** (1989) 83.
50. D. Maison, unpublished; W. Zimmermann, *Reduction of Couplings in Massive Models of Quantum Field Theory*, talks given at 12th Max Born Symposium, Wroclow, Sept. 1998.
51. L.E. Ibáñez and D. Lüst, Nucl. Phys. **B382** (1992) 305; V.S. Kaplunovsky and J. Louis, Phys. Lett. **B306** (1993) 269; A. Brignole, L.E. Ibañez and C. Muñoz, Nucl. Phys. **B422** (1994) 125 [Erratum: **B436** (1995) 747].
52. J. Erdmenger, Ch. Rupp and K. Sibold, Nucl. Phys. **B 530** (1998) 501;
53. O.Piguet and S.Wolf, JHEP 04 (1998) 1. J. Erdmenger and Ch. Rupp, hep-th/9809090; H.Osborn, Annals Phys. 272 (1999) 24.
54. A. Brignole, L.E. Ibáñez and C. Muñoz, Phys. Lett. **B387** (1996) 305.
55. J.A. Casas, A. Lleyda and C. Muñoz, Phys. Lett. **B380** (1996) 59.
56. V. Barger, M.S. Berger and P. Ohmann, Phys. Rev. **D47** (1993) 1093.
57. Particle Data Group, L. Montanet *et al.*, Phys. Rev. **D50** (1994) 1173.
58. P.H. Chankowski, Z. Pluciennik and S. Pokorski, Nucl. Phys. **B439** (1995) 23.
59. S. Dimopoulos and H. Georgi, *Nucl. Phys.* **B193** (1981) 150; N. Sakai, *Zeit. f. Phys.* **C11** (1981) 153.
60. J. Ellis, S. Kelly and D.V. Nanopoulos, Phys. Lett. **B260** (1991) 131; U. Amaldi, W. de Boer and H. Fürstenau, Phys. Lett. **B260** (1991) 447
61. Anselmi et. al., Nucl. Phys. **B526** (1998) 543.
62. A.V. Gladyshev, D.I. Kazakov, W. de Boer, G. Burkart, R. Ehret, Nucl. Phys. **B498**(1997) 3.
63. M. Carena, J.R. Espinosa, M. Quirós, C.E.M. Wagner, Phys. Lett. **B355** (1995) 209.
64. W. Hollik, M.Quirós, M. Carena and C.E.M. Wagner, private communication.
65. By S. Heinemeyer, W. Hollik, G. Weiglein, Phys. Lett. **B440** (1998) 296; S. Heinemeyer, W. Hollik, G. Weiglein, Phys. Rev. **D58** (1998) 091701.
66. S. Bertolini, F. Borzumati, A. Masiero and G. Ridolfi, Nucl. Phys. **B353** (1991) 591.
67. M. Jurčišin and D.I. Kazakov, hep-ph/9902290.
68. S. Heinemeyer, W. Hollik, G. Weiglein, hep-ph/9812320.

SEMINARS

Recent results on the operation of a Cherenkov detector prototype for the Pierre Auger Observatory

M. Alarcón, M. Medina, L. Villaseñor

Instituto de Física y Matemáticas, Universidad Michoacana de San Nicolás de Hidalgo, Apdo. Postal 2-82, Morelia, Mich., 58040, México

A. Fernández, H. Salazar

Facultad de Ciencias, BUAP, 72 000 Puebla, Pue., México

J.F. Valdés-Galicia

Instituto de Geofísica, UNAM, 04510 México, D.F., México

J.C. D'Olivo, L. Nellen

Instituto de Ciencias Nucleares, UNAM, Apdo. Postal 70-543, 04510 México, D.F., México

and A. Zepeda

Departamento de Física, Cinvestav-IPN, 07000 México, D.F., México

Abstract. A full-sized water Cherenkov detector (WCD) prototype (cylinder 3.57 m diameter filled with purified water up to a height of 1.2 m) was used to obtain experimental results that validate the concept of remote calibration and monitoring of WCDs based on the use of the natural flux of cosmic ray muons. Two types of events can be used to monitor and to calibrate each of the WCDs: through-going muons (i.e., isolated muons) and decay electrons from muons stopped inside each detector. The different triggers that will be used to obtain these events and the on-line calibration and monitoring histograms are discussed along with the way these data can be used to diagnose component failures of any of the surface stations of the Auger Observatory.

CP490, *Particles and Fields: Eighth Mexican School,* edited by J. C. D'Olivo et al.

INTRODUCTION

The main purpose of the Pierre Auger Observatory (PAO) is to study the origin and nature of the cosmic rays reaching the earth with energies above 10^{19} eV and to measure their energy spectra, their arrival direction and their composition [1]. These cosmic rays carry macroscopic energies on microscopic particles and their acceleration mechanism remains as one of the biggest and oldest mysteries in astrophysics. At these extreme energies, the cosmic rays are constrained to travel distances shorter than about 100 Mpc by the GZK effect [2,3]; as a consequence, their trajectories are deflected only a few degrees in the typical intergalactic magnetic fields of a few nanogauss; therefore, their arrival direction provides important information about their source location in the sky. The Auger Observatory surface detector is described elsewhere [1]. For a surface array as big as this one it is indispensable that the continuous calibration and monitoring of each WCD be done remotely. In the present paper we describe one way in which these tasks can be performed on the basis of the flux of secondary cosmic ray muons reaching the surface detectors. This flux is about $250 muons/m^2 s$ at sea level and even higher at the Auger sites [4]. Muons with energies greater than about 300 MeV cross the detector completely and in the process leave a Cherenkov signal which is, on the average, constant in time; lower-energy muons can stop and decay inside the water volume of the detectors, in the latter case, the decay electron will give rise to a Cherenkov signal which is also constant in time. These signals can be used on-line to provide an absolute calibration of the energy scale of the surface stations and to obtain a reliable way to monitor and diagnose the performance of each of the 1600 stations in a remote way.

DISCUSSION

The WCD prototype we used consisted of a full-sized prototype made of a polyethylene cylinder with a cross section of 10 m^2 filled with purified water up to a height of 1.2 m; the tank had three 8″ PMTs looking downwards at the tank volume from the water surface. The inner surface of the tank was covered with a highly reflective tyvek sheet (reflectivity of about 90% in the ultraviolet region of the EM spectrum) cut to the cylindrical shape and kept in place by circular PVC hoses tightly stretched against the inner walls of the tank. It is useful to take the Cherenkov signal from muons that cross the detector vertically as a reference point; for this purpose we used the experimental setup shown in Figure 1. A 1" slab of steel was placed between the bottom of the tank and the lower hodoscope in order to harden the energy spectrum of the triggering muons. For the purpose of acquiring events in which the muons either cross the detector or stop and decay inside it we used the experimental setup shown in Figure 2.

The trigger for the first case, see Figure 1, is simply given by the time coincidence of the PMT pulses from two scintillation hodoscopes placed vertically, one above and the other below the tank. The trigger for the second case, see Figure 2, is given by the presence of an isolated signal (i.e., no further PMT activity in a time window of 20 μs) in the sum of the three PMTs above 30 mV or by the occurrence

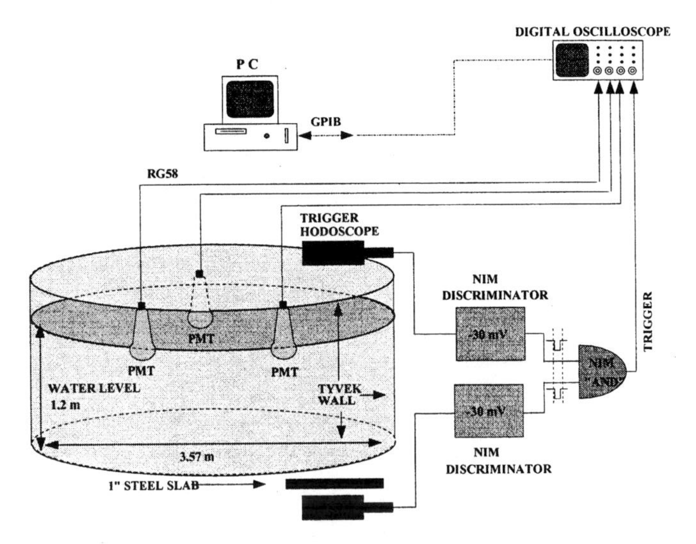

FIGURE 1. Schematic diagram of the experimental array used to collect events with muons that cross the detector vertically.

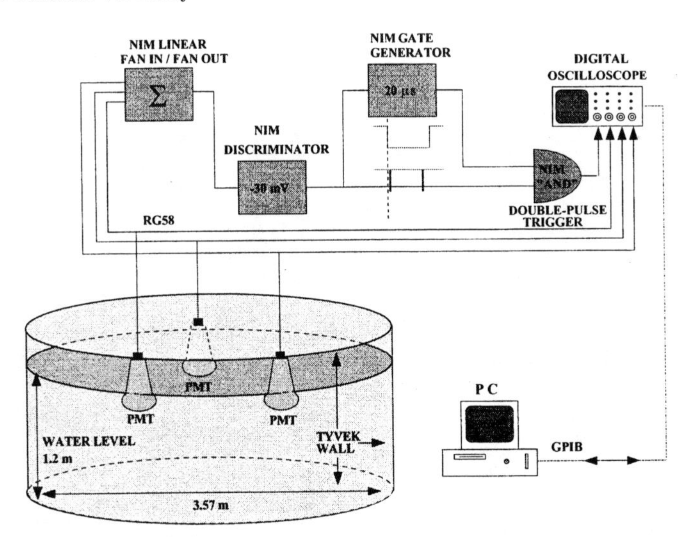

FIGURE 2. Schematic diagram of the experimental array used to collect events with muons that either stop and decay inside the detector, or cross it at any angle.

of two consecutive signals above 30 mV in the sum of the pulses from the three PMTs within a time window of 20 μs, respectively. The CAMAC controller used was the LeCroy 8901; it was connected to a National Instruments GPIB port on a pentium PC running at 133 MHz. The DAQ program was written in LabView, which is a graphic programming package from National Instruments. A typical muon-decay event is shown in Figure 3 which displays the signal from each of the three PMTs. The calibration procedure using crossing muons on one hand and decaying muons on the other is described in detail elsewhere [5,6] for a tank of reduced dimensions with a single PMT; in the case of full-sized WCDs with three PMTs the procedure is similar with the additional advantage that the trigger can be defined to require double or triple coincidences on the PMTs (see Figure 3) to enhance the signal to noise ratio. Results on the stability of the water purity have also been reported elsewhere [7]. The simulation of vertical, crossing and decaying muons for a full-sized WCD is described in [8].

Triggers

The following level-2 triggers will be used to perform the continuous calibration and monitoring of the Auger WCDs in a remote way:

1) Isolated Muon. Defined as Sum(Signals from PMTs) > 15% VEM and no further PMT activity in 20 microsecond, where 1 VEM is the signal corresponding to one vertical muon.

2) Double-Coincidence Isolated Muon. Defined as 1) and two PMTs in coincidence.

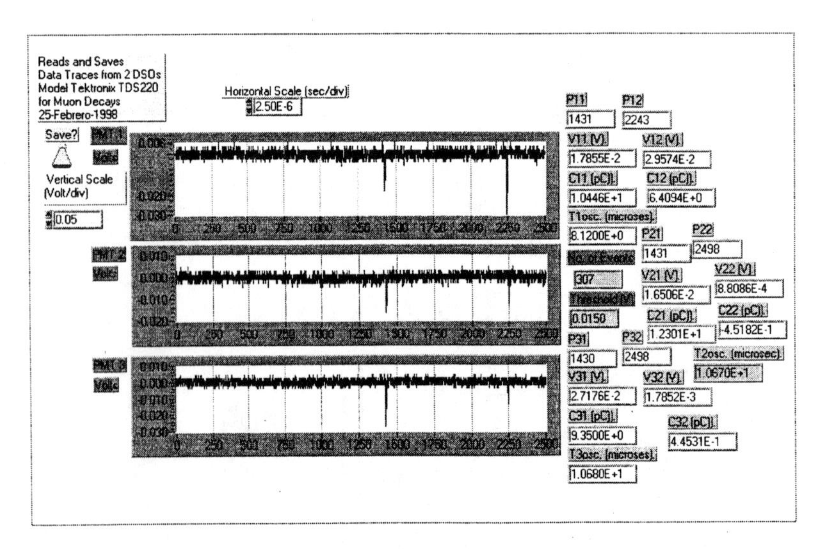

FIGURE 3. Front view of the LabView program used to acquire muon-decay events. Presumably the first pulse corresponds to a stopping muon and the second to a decay electron.

3) Triple-Coincidence Isolated Muon. Defined as 1) and three PMTs in coincidence.

4) Double-Pulse. Defined as two pulses in a time window of 20 microsecond with variable threshold with Sum(Signals from PMTs) for first pulse $> x_1$VEM and Sum(Signals from PMTs) for second pulse$> x_2$VEM, where x_1 and x_2 will be of the order of 15% and will be optimized for muon-decay discrimination.

5) LED. Obtained by flashing an LED inside the detectors.

Calibration histograms

A set of histograms will be used to obtain an absolute calibration of each station on the basis of the average energy deposition of through-going muons associated with Isolated-Muon events, and decay electrons associated with Double-Pulse events. The LED events will be used to calibrate the gains of the ADCs used for the the most significant bits of the dynamic range by comparing with overlapping bits of lower significance. The following histograms will be used for each case: Isolated Muons events (IM) Q, V. Q/V and event rate; LED events (LED) Q, V, and Q/V; Double-Pulse events (D-P) Q1,V1,Q1/V1, rate, Q2, V2, Q2/V2, and t12.

Monitoring histograms

In addition of the above, the following histograms will be used specifically to monitor the performance of each station on-line and in a remote way.

At least the following 9 temperatures: Water, air inside tank, air outside tank, three PMTs, front-end electronics, batteries, tank top,

Voltage and current for each PMT,

Input current, output current and voltage for solar panel,

Cloud detectors,

Pressure monitor at tank bottom to calculate water level,

Atmospheric pressure,

Others up to 32 channels.

Diagnosis

The following diagnoses will be done on-line on the basis of the monitoring data.

Failure: Dirt water. Manifestation: Q, V, and Q/V decrease for IM and LED events correlated on all PMTs.

Failure: PMT HV. Manifestation: Q and V change, Q/V constant on affected PMT correlated with variations on HVs for IM and LED events.

Failure: PMT window. Manifestation: Same as above but uncorrelated with HVs for IM and LED events.

Failure: Tank liner. Manifestation: Similar to Dirt water, LED events show less variation depending on LED location.

Failure: Afterpulsing. Manifestation: D-P rate increases, correlated with peaks on t12.

Failure: Ice. Manifestation: Water temperature.

Failure: Water leaks. Manifestation: Q, V decrease on all PMTs correlated with water height.

Failure: PMT signal reflections, overshooting, rings, etc. Manifestation: Abnormal PMT traces, rates increase for D-P events. Transfer of the full PMT traces (500 ns for 4 ADCs at 10 bits/ADC and 25 ns sampling period gives 1 Kbit which can be transferred in about 5 s at 200 bit/s) can help elucidate the exact nature of the failure in these cases.

In addition, test leads will exist at each station to perform a more thorough diagnosis of its performance in situm.

CONCLUSIONS

We have discussed the way in which the natural flux of background muons can be used to calibrate and monitor each of the 1600 WCDs of the Pierre Auger Observatory in a remote way. Two types of events can be used: crossing and decaying muons. The triggers to be used to collect the events from which the calibration and monitoring histograms can be obtained on-line have been listed. The calibration and monitoring procedures are similar to the ones described elsewhere [5,6] for a tank of reduced dimensions with a single PMT; in the case of full-sized WCDs with three PMTs the procedures have the additional advantage that the trigger can be defined to require double or triple coincidences of the PMT signals. A detailed list of the different diagnoses that can be performed on-line on the basis of the monitoring histograms was also discussed.

REFERENCES

1. Auger collaboration, "The Pierre Auger Project – Design Report", http://www-td-auger.fnal.gov:82, March 1997.
2. Greisen, K., *Phys. Rev Letters* **16**, 748 (1965)
3. Zatsepin, G.T. and Kuz'min, V.A., *JETP Letters* **4**, 78 (1966)
4. Aguilar-Benitez, M. et al, *Review of Particle Properties, Phys. Rev.* **D50** (1994).
5. Villaseñor, L. et al., *Proc. of the 25th. Intl. Cosmic Ray Conference (Durban)*, Vol. 7, 197 (1997).
6. Alarcón, M. et al., *Nucl. Instr. and Meths. in Phys. Res. A* **420**, 39 (1999).
7. Alarcón, M. et al., *Rev. Mex. Fis.* **44-5**, 479 (1998).
8. Medina, M., L. Nellen and L. Villaseñor, *these proceedings.*

Neutrino magnetic moment and supernovae

Alejandro Ayala [a], Juan Carlos D'Olivo [a], and Manuel Torres [b]

[a] *Instituto de Ciencias Nucleares, Universidad Nacional Autónoma de México Apdo. Postal 70-543. México D.F. 04510, México.*
[b] *Instituto de Física, Universidad Nacional Autónoma de México, Apdo. Postal 20-364, 01000 México, D.F., México.*

Abstract. For neutrinos with a magnetic moment, we calculate the production rate of right handed neutrinos in a hot and dense plasma via the quirality flip ($\nu_L \to \gamma^* \nu_R$) and the plasmon decay ($\gamma^* \to \nu_L \nu_R$) processes. The rate for for these processes is computed in terms of a resummed photon propagator which consistently incorporates the background effects. Applying the results to the case of supernova collapse, our results can be used to place an upper limit on the neutrino magnetic moment $\mu_\nu < (0.1 - 0.4) \times 10^{-11} \mu_B$

The existence of a neutrino magnetic moment would have interesting astrophysical and cosmological implications. In turn the analysis of astrophysical and cosmological phenomena can be used to infer bounds on the possible values of the neutrino magnetic moment [1]. For example, is has been known for long time that a value $\mu_\nu \approx 1 \times 10^{-10} \mu_B$ (μ_B is the Bohr magneton) may provide a solution to the solar neutrino puzzle. The previous value should be compared with the one obtained if the standard model is extended to include Dirac neutrino masses: $\mu_\nu \approx 3 \times 10^{-19} \mu_B$. In recent work we studied the production of right handed neutrinos in the core of a supernova, deriving the following limit [2,5]

$$\mu_\nu < (0.1 - 0.4) \times 10^{-11} \mu_B \,, \tag{1}$$

in this paper we summarized the main results.

Consider a QED plasma in thermal equilibrium at a temperature T and with an electron chemical potential $\tilde{\mu}_e$ in the limit $T\,, \tilde{\mu}_e \gg m_e$. The production rate Γ of a right-handed neutrino with total energy E and momentum $\vec{p}$ can be conveniently expressed in terms of the imaginary part of the ν_R self-energy as

$$\Gamma\,(E) = \frac{1}{2E} n_F \ \mathrm{Tr}\,[\not{P} R \,\mathrm{Im}\,\Sigma] \,, \tag{2}$$

where, $P^\mu = (E, \vec{p})$; $L, R = \frac{1}{2}\,(1 \pm \gamma_5)$ and n_F is the Fermi distribution for the ν_R's.

CP490, *Particles and Fields: Eighth Mexican School,* edited by J. C. D'Olivo et al.

The use of the thermal cutting rules allow us to relate $\Gamma(E)$ to the amplitudes of various processes. The most important are: the spin flip ($\nu_L \rightarrow \gamma^* \nu_R$) and the plasmon decay ($\gamma^* \rightarrow \nu_L \nu_R$). A consistent formalism for the computation of the ν_R production rate requires the use of an effective photon propagator that resums the leading temperature corrections according to the Braaten -Pisarski method [3]. For the intermediate neutrino propagator and for the neutrino-photon vertex ($\mu_\nu \sigma_{\alpha\beta} K^\beta$) their bare expressions can be taken. The properties of the effective photon propagator can written in terms of the spectral decomposition for the longitudinal and transverse modes

$$\rho_{L,T}(k_0\,, k) = Z_{L,T}\,(k)\,[\delta\,(k_0 - \omega_{L,T}(k))] + \beta_{L,T}(k_0, k)\theta\left(k_0^2 - k^2\right)\,. \tag{3}$$

The spectral function includes the contribution of the transverse and longitudinal time-like collective excitations with a mass given by the photon thermal mass

$$m_\gamma^2 = \frac{e^2}{2\pi^2}\left(\tilde{\mu}_e^2 + \frac{\pi^2 T^2}{3}\right), \tag{4}$$

and $Z_{L,T}\,(k)$ are the residues of the longitudinal and transverse propagator at the position of the pole. As a consequence of Landau damping there is also an "effective photon" that propagates below the light cone with an amplitude proportional to β, see Ref. [4] for the explicit expressions.

ν_R production through plasmon decay

The contribution to the plasmon decay process $\gamma^* \rightarrow \bar{\nu}_L \nu_R$ process arises from the decay of the collective plasma excitations. The production of ν_R through plasmon decay has been obtained as [5]

$$\begin{aligned} \Gamma(E) = \frac{\mu_\nu^2}{16\pi E^2} \sum_{i=L,T} \int_0^\infty & k dk\, \theta(\omega^+ - E)\theta(E - \omega^-) f(\omega_i(k)) \\ & [1 - \tilde{n}_F(\omega_i - E])]\, C_i(E, -\omega_i) Z_i(k)\,, \end{aligned} \tag{5}$$

where f and n_F represent the boson and fermion distribution respectively, θ is the step function and C_i are kinematical factors obtained carrying out the Dirac traces. The energy carried by the produced right-handed neutrino is computed in terms of the ν_R emissivity (luminosity per unit volume)

$$\mathcal{E}_{\nu_R} = \int \frac{d^3p}{(2\pi)^3}\; E\;\Gamma(E)\,. \tag{6}$$

When applied to a supernova this contribution is two orders of magnitude smaller than the corresponding quirality flip contribution, so we can safely ignore it.

$\nu_L \to \nu_R$ quirality flip

The quirality flip arises from the absorption (emission) of an effective space-like photon ($\gamma^* \nu_L \to \nu_R$) originated in the Landau damping mechanism. The rate of production of right-handed neutrinos from the $\nu_L \to \nu_R$ flip can be computed utilizing the effective photon propagator as [2]

$$\Gamma(E) = \frac{\mu_\nu^2}{16\pi E^2} \int_0^\infty kdk \int_{-k}^{k} d\omega\, \theta(2E + \omega - k) \\ (1 + f(\omega))\, n_F(E + \omega) \sum_{i=L,T} C_i(E, K)\beta_i(\omega, k)\,. \tag{7}$$

The computation of $\Gamma(E)$ requires in general numerical integration. However, following the method introduced by Braaten and Yuan [6] we can explicitly extract the leading logarithmic terms to order e, leading to a very good analytical approximation. The method is valid in the limit $T, \tilde{\mu}_\nu \lesssim E$ and introduces an intermediate cutoff q^* that separates: (i) the hard momentum region contributions $k > q^*$, where the tree-level approximation is justified for the virtual photon and (ii) the soft momentum region $k < q^*$, where the full resumed propagators must be used, but approximations such as $1 + f(\omega) \approx T/\omega$ are justified. Adding the hard and soft contributions, the dependence on the arbitrary scale q^* cancels. After a lengthy calculation, the final result at leading order in (T/E) has been worked out as [5]

$$\Gamma(E) = \frac{\mu_\nu^2 m_\gamma^2 T}{2\pi} n_F(E) \Bigg[1.88 + \ln\left(\frac{T}{\sqrt{2}m_\gamma}\right) + \frac{\tilde{\mu}_\nu - E}{4T} \\ + \frac{1}{2}\ln\left(\frac{\cosh^2\left(\frac{E-\tilde{\mu}_\nu}{2T}\right)\sinh\left(\frac{E}{2T}\right)}{\cosh\left(\frac{\tilde{\mu}_\nu}{2T}\right)}\right)\Bigg], \tag{8}$$

where $\tilde{\mu}_\nu$ is the neutrino chemical potential. The emissivity for ν_R emission from the quirality flip process is obtained inserting $\Gamma(E)$ from the exact Eq. (7) or the approximated Eq. (8) expressions in Eq. (6) obtaining very similar results.

The ν_R emission in a supernova

We consider the emission of right-handed neutrinos immediately after a supernova core collapse. The large mean free path of the right handed neutrinos compared to the core radius implies that the ν_R's would freely fly away from the supernova. Therefore, the core luminosity for ν_R emission can be simply computed as $Q_{\nu_R} = V\,\mathcal{E}$, where V is the volume of the supernova core and $\mathcal{E}$ is the ν_R emissivity.

To make a numerical estimate, we shall adopt a simplified picture of the inner core, corresponding to the average parameters of SN1987A [7]. Consequently, we take a constant density $\rho \approx 8 \times 10^{14}\,\mathrm{g/cm^3}$, a volume $V \approx 8 \times 10^{18}\,\mathrm{cm^3}$, an electron to baryon ratio $Y_e \simeq Y_p \simeq 0.3$, and temperatures in the range $T = 30 \sim 60\,\mathrm{MeV}$. This corresponds to a degenerate electron gas with a chemical potential $\tilde{\mu}_e$ ranging from 307 to 280 MeV. For the left-handed neutrino we take $\tilde{\mu}_\nu \approx 160\,\mathrm{MeV}$. Using this values, we obtain by numerical integration

$$Q_{\nu_R} = \left(\frac{\mu_\nu}{\mu_B}\right)^2 (0.7 - 4.3) \times 10^{76}\,\mathrm{ergs/s}\,, \tag{9}$$

for T ranging from 30 to 60 MeV. The main contribution to this result arises from the $\nu_L \to \nu_R$ flip process, whereas the contribution from the $\gamma \to \bar{\nu}_L \nu_R$ decay is smaller by two orders of magnitude. Moreover, an estimate of the ν_R luminosity derived from the approximated solution obtained in Eq. (8) is surprisingly accurate, it differs from the result in Eq. (9) by less than 2%. Assuming that the emission of ν_R's lasts approximately 1 s the luminosity bound is $Q_{\nu_R} \leq 10^{53}$ ergs/s. Thus, comparing with the result in Eq. (9), we obtain the upper limit on the neutrino magnetic moment quoted in Eq. (1).

Barbieri and Mohapatra [8] derived the limit $\mu_\nu < (0.2 - 0.8) \times 10^{-11}\mu_B$ considering the helicity flip scattering $\nu_L e \to \nu_R e$ to order e^4 introducing the Debye mass in the photon propagator as an infrared regulator. Our result for μ_ν in Eq. (1) slightly improves their result; moreover puts the analysis of possible bounds derived from the chirality flip process in supernovae on a firm basis. We notice that our result in Eq. (1) is very similar to the one that has been derived by Raffelt [9] from the analysis of plasmon decay in globular-custer stars which however includes the Majorana case. In conclusion, we have derived the result in Eq. (1) that represents one of the most stringent constraints for the neutrino magnetic moment.

REFERENCES

1. For a review see: G. Raffelt, "Particle Physics from Stars", in Annual Review of Nuclear and Particle Science (1999); hep-ph/9903472.
2. A. Ayala, J.C. D'Olivo and M. Torres, hep-ph/9804230; Phys. Rev. D (rapid communication in press).
3. E. Braaten and R. D. Pisarski, Nuc. Phys. **B337**, 569
4. M. Le Bellac, *Thermal Field Theory.* Cambridge University Press (1996).
5. A. Ayala, J.C. D'Olivo and M. Torres, " Right handed neutrino production in a dense and hot plasma"; to be published.
6. E. Braaten and T. C. Yuan, Phys. Rev. **D66**, 2183 (1991).
7. A. Burrows, Annu. Rev, Nucl. Part. Sci. **40**, 181 (1990); S. E. Woosley and T. A. Weaver, Annu. Rev, Astron. Astrophys. **24**, 205 (1986).
8. R. Barbieri and R. N. Mohapatra, Phy. Rev. Lett. **61**, 27 (1988).
9. G. Raffelt, Phys. Rev. Lett. **64**, 2856 (1990); Ap.J. **365**, 559 (1990).

Dynamical chiral symmetry breaking in Yukawa and Wess-Zumino models

A. Bashir[1] and J. L. Diaz Cruz [2]

1) Instituto de Fisica y Matematicas, Universidad Michoacana de San Nicolas de Hidalgo Ap. Postal 2-82, Morelia, Mich., Mexico
2) Instituto de Fisica "Luis Rivera Terrazas", Benemérita Universidad Autónoma de Puebla, Apartado Postal J-48, 72570 Puebla, Pue., México.

Abstract. We study dynamical mass generation for fermions in the Yukawa and the supersymmetric Wess-Zumino (WZ) models. It is found that above a critical coupling fermion mass can be generated dinamicaly in the Yukawa model, whereas in the WZ model the fermion does not acquire mass. We also show that the supersymmetry (SUSY) preserving solution is permitted and hence the scalars may not acquire mass either.

INTRODUCTION

Despite the success of quantum field theory in the description of elementary particles in the perturbative regime, it still remains a challenge to understand the non-perturbative domain satisfactorily. One of the methods which has gained attention in this regard in recent years is the study of Schwinger-Dyson equations (SDE) [1]. This approach has been very successful in addressing issues like dynamical mass generation for fundamental fermions in strong interactions [2]. Moreover, recent attempts, e.g., [3–5] to improve the reliability of the approximations used have increased the credibility of the results obtained through such studies.

Application of Schwinger-Dyson formalism to supersymmetric (SUSY) models has been less unambiguous. For example, in supersymmetric QED (SQED), it was first thought impossible to obtain dynamical mass generation for fermions [6]. However, some more recent studies [7] have shown that it is probably possible to break chiral symmetry dynamically in SQED. In this paper, we take the SUSY Wess-Zumino model, and attempt to solve the corresponding SDE for the fermion and scalar propagators. We believe this exercise will provide us with a deeper insight into what role supersymmetry plays in the context of dynamical mass generation. Such a study should provide us with a better starting point for other SUSY theories such as SQED and SQCD.

Since our discussion of the WZ model shall rely on previous work for the Yukawa model [8], it is convenient to summarize the main results. The Yukawa model

CP490, *Particles and Fields: Eighth Mexican School,* edited by J. C. D'Olivo et al.

with one real scalar and one Majorana fermion, can be considered as a truncated Wess-Zumino model. We used the quenched approximation. Keeping in mind the perturbative expansion of the 3-point vertex beyond the lowest order and its transformation under charge conjugation symmetry, we proposed an *ansatz* for the full effective vertex. One of the advantages of using that vertex is that the equations for the mass function $\mathcal{M}(p^2)$ and the wavefunction renormalization $F(p^2)$ decouple completely in the neighbourhood of the critical coupling, α_c above which mass is generated for the fermions. We solved both the equations to find analytical expressions for $F(p^2)$ and the anomalous mass dimensions in the neighbourhood of α_c. The results show that non-perturbative interaction of fermions with fundamental scalars can give masses to fermions in a dynamical way provided the interaction is strong enough. We used numerical calculation to draw Euclidean mass of the fermions as a function of the coupling, and confirm that it obeys Miransky scaling. We also evaluated $F(p^2)$ numerically.

THE WESS-ZUMINO MODEL

We now discuss the massless Wess-Zumino model, characterized by the following Lagrangian:

$$\mathcal{L} = \frac{1}{2}(\partial_\mu A)^2 + \frac{1}{2}(\partial_\mu B)^2 + \frac{1}{2}(i\bar{\psi}\gamma^\mu\partial_\mu\psi) - \frac{1}{2}g^2(A^2+B^2)^2 - g\bar{\psi}(B - iA\gamma_5)\psi\,. \quad (1)$$

The SDE for the fermion propagator in this model can be derived using standard methods. Before we embark on solving the resulting DS equation, we must propose an ansatz for the 3-point scalar-fermion-fermion vertex: $\Gamma_A(k,p)$. Any *ansatz* for the 3-point vertex must satisfy: $\Gamma_A(k,p) = 1 + \mathcal{O}(g^2)$, because of the perturbative limit, and it must be symmetric in k and p. However, it is easy to see that in the presence of both the scalars A and B, none of the 3-point vertices gets modified at $\mathcal{O}(\alpha)$ in perturbation theory. Therefore, it is reasonable to use the bare vertex here. Though one would expect to solve coupled integral equations for $F(p^2)$ and $\mathcal{M}(p^2)$, a miraculous cancellation of terms takes place as evident from the following SDE for the fermion propagator:

$$\frac{\not{p} - \mathcal{M}(p^2)}{iF(p^2)} = \frac{\not{p}}{i} - \frac{\alpha}{\pi^3}\int d^4k \left[\frac{F(k^2)}{\not{k} - \mathcal{M}(p^2)}\frac{1}{q^2}\right] + \frac{\alpha}{\pi^3}\int d^4k \left[\gamma_5 \frac{F(k^2)}{\not{k} - \mathcal{M}(p^2)}\gamma_5 \frac{1}{q^2}\right]. \quad (2)$$

Taking the trace of this equation, we get: $\mathcal{M}(p^2) = 0$. As the cancellation of terms takes place at the very beginning, it is easy to see that it is not only a feature of the crude approximations made, but goes beyond it. Dynamical mass generation will remain an impossibility for the full vertex and the full scalar propagator as long as they are identical for both the scalars. The vertex corrections for A and B have been proven to be equal up to $\mathcal{O}(\alpha^2)$ [9]. We shall shortly see that the same is true for the full scalar propagator at least upto $\mathcal{O}(\alpha)$.

As far as wavefunction renormalization $F(p^2)$ is concerned, its leading log behaviour modifies slightly, by the inclusion of the other scalar, to:

$$F(p^2) = 1 + \frac{\alpha}{\pi} \ln \frac{p^2}{\Lambda^2} \quad . \tag{3}$$

Although it is an interesting conclusion in its own right that supersymmetry prevents dynamical mass generation for fermions in the Wess-Zumino model, another important issue to probe will be whether supersymmetry itself remains intact, i.e., whether the scalars can also be kept massless. This can be discussed with the SDE for the scalar (for example A). A scalar propagator, unlike a fermion, needs only one unknown function to describe it. But we shall prefer to split it into two parts and write the full scalar propagator as follows:

$$S_A(p) = \frac{F_A(p^2)}{p^2 - \mathcal{M}_A^2(p^2)} \quad . \tag{4}$$

The non-zero value of the mass function $\mathcal{M}_A(p^2)$ will be responsible for shifting the pole from $p^2 = 0$ to some finite value, generating the mass for the scalar dynamically. $F_A(p^2)$ on the other hand is the scalar wavefunction renormalization. The SDE for the scalar propagator in Euclidean space can now be written as:

$$\begin{aligned} \frac{p^2 + \mathcal{M}_A^2(p^2)}{F_A(p^2)} &= p^2 + \frac{3\alpha}{2\pi^3} \int d^4k \, \frac{F_A(k^2)}{k^2 + \mathcal{M}_A^2(k^2)} + \frac{\alpha}{2\pi^3} \int d^4k \, \frac{F_B(k^2)}{k^2 + \mathcal{M}_B^2(k^2)} \\ &\quad - \frac{2\alpha}{\pi^3} \int d^4k \, \frac{F(k^2)F(q^2)}{k^2 q^2} \, \Gamma_A(k,p) \; k \cdot q \end{aligned} \tag{5}$$

where we have used the fact that the fermions do not acquire mass. If we want to preserve supersymmetry and do not want the scalars to acquire mass, we must have:

$$\mathcal{M}_A(p^2) = \mathcal{M}_B(p^2) = 0 \quad . \tag{6}$$

We are then left with:

$$\frac{1}{F_A(p^2)} = 1 + \frac{\alpha}{2\pi^3 p^2} \int \frac{d^4k}{k^2} \left[3F_A(k^2) + F_B(k^2) - 4\frac{k \cdot q}{q^2} \, F(k^2)F(q^2)\Gamma_A(k,p) \right] \tag{7}$$

and there is a similar equation for the scalar B. These equations should yield a solution for $F_A(p^2)$ and $F_B(p^2)$ such that it does not change the position of the pole for the scalar propagator and that the quadratic divergences cancel. It is well-known that it does happen in perturbation theory to $\mathcal{O}(\alpha)$. In fact, one can evaluate $F_A(p^2)$ and $F_B(p^2)$. The leading log expression for these functions to $\mathcal{O}(\alpha)$ is

$$F_A(p^2) = F_B(p^2) = 1 + \frac{\alpha}{\pi} \ln \frac{p^2}{\Lambda^2} \tag{8}$$

which is exactly the same expression as that for $F(p^2)$ for the fermion propagator. This result indicates that supersymmetry needs not be broken.

CONCLUSIONS

We have studied the SDE for the Yukawa and the Wess-Zumino models. In the simple Yukawa model, we propose a vertex *ansatz* which we argue should perform better than the bare vertex. In the quenched approximation, we find dynamical mass generation for fermions above a critical value of the coupling $\alpha_c = \pi/4$. The generated Euclidean mass obeys Miransky scaling. When we extend this Yukawa model to equate the scalar and fermionic degrees of freedom (Wess-Zumino model), we find that a neat cancellation of terms occurs and there is no mass generation for the fermions. This fact remains true beyond the rainbow approximation and is supported by perturbative calculations available for the 3-point vertex and of the scalar propagator. If supersymmetry has to be preserved, the scalars should also acquire no mass dynamically. Studying the SDE for the scalars, we observe that such a solution is allowed and in fact leads to the wavefunction renormalization function for the scalars which is exactly the same as that for the fermion.

It is interesting to see the role of supersymmetry in more complicated theories such as SQED, SQCD or the resulting DSE for the gravitino that arises in Supergravity, as well as the need to explore further connections between the Holomorphic approach and that of the SDE [10], which we plan to present in a future publication.

ACKNOWLEDGEMENTS

This work was partly supported by a TWAS-AIC award and CONACYT-SNI (México).

REFERENCES

1. For a review see: C.D. Roberts and A.G. Williams, Prog. Part. Nucl. Phys. **33** 477 (1994).
2. V.A. Miransky, Nuovo Cim. **90A** 149 (1985) ; Sov. Phys. JETP **61** 905 (1985) ; P.I. Fomin, V.P. Gusynin, V.A. Miransky and Yu.A. Sitenko, La rivista del Nuovo Cim. **6**, numero5, 1 (1983).
3. D.C. Curtis and M.R. Pennington, Phys. Rev. **D42** 4165 (1990).
4. Z. Dong, H.J. Munczek and C.D. Roberts, Phys. Lett. **B333** 536 (1994).
5. A. Bashir and M.R. Pennington, Phys. Rev. D50 7679 (1994).
6. T.E. Clark and S.T. Love, Nucl. Phys. **B310** 371 (1988)
7. A. Kaiser and S.B. Selipsky, hep-th/9708087, Yale preprint YCTP-P14-97.
8. A. Bashir and J.L. Diaz-Cruz, submited to J. Phys.
9. I. Jack, D.R.T. Jones and P. West, Phys. Lett. **B258** 382 (1991).
10. T. Appelquist, A. Nyffeler and S.B. Belipsky, hep-th/9709177, Yale preprint YCTP-P12-97.

Bose-Einstein correlations and the Dalitz plot of hadronic meson decays[1]

E. Cuautle* and G. Herrera*†

*Centro de Investigación y de Estudios Avanzados
Apdo. Postal 14 740, México 07000, DF, México.

†Institut für Physik, Universität Dortmund
Otto-Hahn-Str.4
44221 Dortmund, Germany

Abstract. We show that the presence of residual Bose-Einstein correlations may affect the non-resonant contribution of hadronic decays where two identical pions appear in the final state. The distortion of the phase-space of the reaction would be visible in the Dalitz plot. We discuss the decay $K^+ \to \pi^-\pi^+\pi^+$ which contain two identical pions in the final state.

INTRODUCTION

Hadronic decays of D and K mesons are an important source of information. The lifetime difference between the D^+ and the D^0 mesons as well as the role of final state interactions may be understood by carefully studying their hadronic decays. The Bose-Einstein correlation among identical mesons has been used to probe the space-time structure of the intermediate state right before hadrons appear [1,2] in high energy and nuclear collisions.
The Dalitz plot analysis is a powerful technique widely used in the study of resonances substructures on meson decays. The plot represents the phase space of the decay and it is weighted by the squared amplitude of the reaction. Therefore, it contains information on both kinematics and dynamics. A three body decay $D \to A + B + C$ involving spin 0 particles has two degrees of freedom. This mean that two real variables are necessary to describe any such decay uniquely. One possible choice of variables is the so-called Dalitz variables $m^2_{AB} = (P_A + P_B)^2$ and $m^2_{AC} = (P_A + P_C)^2$, where P_i is the four momentum of particle i. The m^2_{AB} and m^2_{AC} are natural variables in terms of which two body intermediate resonances are described. The kinematical allowed region of the Dalitz plot depends upon the parent mass.

1) Supported by the Alexander von Humboldt Foundation and by CONACyT

CP490, *Particles and Fields: Eighth Mexican School,* edited by J. C. D'Olivo et al.

The Dalitz plot has been used to get information about meson decays e.g. the spin of D^+.
The estimation of the size of CP-violation effects in $K \to 3\pi$ and the strong interaction effects in the same decay [3] are studied through Dalitz plot too.
The three-body amplitude is normally assumed to be constant over the Dalitz plot but, decays like $D^+ \to K^-\pi^+\pi^+$ and $D^+ \to \pi^-\pi^+\pi^+$ contain identical charged pions in the final state and their interference produce a non-constant behaviour of the phase-space [4–6].
In this work we present a study of the $K^+ \to 3\pi$ decay from the perspective of Bose-Einstein interference among the identical bosons in the final state and the residual correlation produced. This may help to understand the Dalitz plot of this decay and the effect in any decay with several bosons in general.

BOSE-EINSTEIN CORRELATION

The study of Bose-Einstein correlations (BEC) in high energy physics reactions became important when it was realized that such correlations may affect measurements of the standard model parameters [7,8].
When the particles in a decay are produced via fragmentation the process may not be coherent. The incoherence present in the decay could give rise to the Bose Einstein interference.
One parametrizes the effect assuming a set of point-like sources emitting bosons. These point like sources are distributed according to a density $\rho(r)$. The correlation function can then be written as,

$$R_{BE}(\vec{p_1}, \vec{p_2}) = \int \rho(\vec{r_1})\rho(\vec{r_2}) \mid \psi_{BE}(\vec{p_1}, \vec{p_2}) \mid^2 d^3r_1 d^3r_2, \qquad (1)$$

where $\vec{p_1}, \vec{p_2}$ are the momenta of the two bosons, ψ_{BE} represents the Bose-Einstein symmetrized wave function of the bosons system and $\int_V \rho(\vec{r})d^3r = 1$. Taking plane waves to describe the bosons one obtains:

$$R_{BE}(\vec{p_1}, \vec{p_2}) = 1 + \mid \mathcal{F}(\rho(\vec{r})) \mid^2, \qquad (2)$$

where $\mathcal{F}(\rho)$ represents the Fourier transform of the density function $\rho(\vec{r})$.
The Golhaber's parametrization [1] is the most commonly used and is given by

$$R_{BE}(\vec{p_1}, \vec{p_2}) = 1 + \lambda e^{-Q^2\beta}, \qquad (3)$$

where $Q^2 = -(p_1 - p_2)$ is the Lorentz invariant, which can be written also as, $Q^2 = m^2 - 4m_\pi^2$, where m is the invariant mass of the two pions and $m_\pi = 0.139$ GeV the mass of the pion. The parameter λ lies between 0 and 1 and reflects the degree of coherence in the pion production. The radius of the pions source will be given by $r = \hbar c\sqrt{\beta}[fm]$.

The presence of Bose-Einstein correlations will modify not only the invariant mass spectrum of like charged but also that of unlike charged pions. This reflection of BEC is known as residual correlation.

SIMULATION OF BOSE-EINSTEIN CORRELATIONS

In this work we have analysed the $K^+ \to \pi^-\pi_1^+\pi_2^+$ decay. We will take the approach used in [9] where the BEC are simply simulated by weighting each event, according to:

$$W = \prod_{i,j}(1 + \lambda e^{-\beta Q_{ij}^2}), \tag{4}$$

where the product was taken over all pairs (i, j) of like charged pions. In our case the product is reduced to one pair of like charged pions

$$W = 1 + \lambda e^{-\beta Q^2}. \tag{5}$$

We did simulations with different values for λ and β. When we plot R_{BE} as a function of Q^2 according to Eq. 3, we can see that one obtains what one puts in the simulation.

EFFECTS ON THE DALITZ PLOT

Figs.1a,b show the invariant mass espectra of $m^2(\pi^-\pi_1^+)$ and $m^2(\pi_1^+\pi_2^+)$ before and after the BEC has been incorporated. Fig.1b shows the BEC directly since

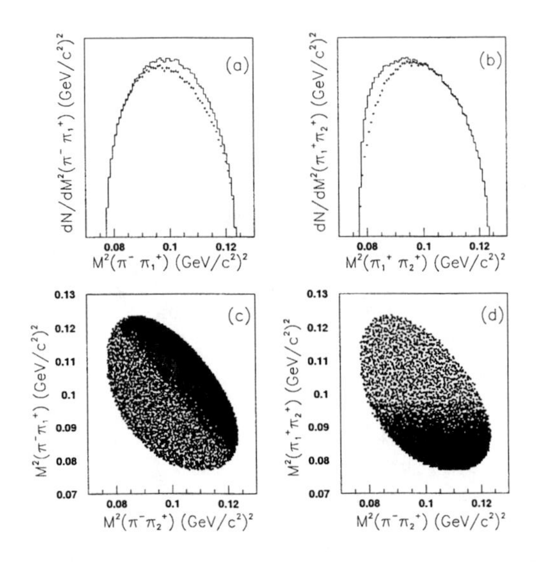

FIGURE 1. Invariant mass spectrum for the decay $K^+ \to \pi^-\pi_1^+\pi_2^+$ showing the residual (a) and direct (b) with (histogram) and without BEC. The Dalitz plot (c) show the residual effect of BEC and (d) show the direct effect. The parameters used were $\lambda = 0.7$ and $\beta = 200$ GeV^{-2}.

it shows the invariant mass of like charged pion combination. The distribution in Fig.1a exhibits residual effects only. Figs.1c,d show the Dalitz plots for $K^+ \rightarrow \pi^- \pi_1^+ \pi_2^+$. In this case Fig.1d shows the BEC directly in one of the axis while only residual effects are present in Fig.1c where the two axis are combination of unlike pions.
The phase space available to pions in the decay $D \rightarrow 3\pi$ [5] is much larger than in the decay $K^+ \rightarrow 3\pi$. The Q values $Q_{K,D} = M_{K,D} - 3m_\pi$ are $Q_K = 76$ MeV, $Q_D = 1450$ MeV and for a similar B meson decay $Q_B = 4863$ MeV. As a result of a small Q_K value the pion are strongly restricted in phase space and the interference is less visible.
We had used $\beta = 200$ MeV^{-2} ($r \approx 2.79$ fm) and $\lambda = 0.7$ values, just to make the effect more visible. Experimental distributions with this shape are not observed and this indicates a much smaller source size (β) and probably a larger coherence (λ). The value $\lambda = 1$ indicate a totally chaotic production in the decay and $\lambda = 0$ mean a completely coherent process in which Bose-Einstein interference would not be present. The fragmentation may introduce some degree of incoherence given to λ a value between 0 and 1. The exact value, however, would be obtained fitting the correlation function as in the case of the source radius to measured distributions.

ACKNOWLEDGEMENTS

We would like to thank the organizers for inviting us to present this work. GH wants to thank to Prof. D. Wegener (U. of Dortmund) for his kind hospitality during GH sabbatical stay at the Physics Institute.

REFERENCES

1. G. Goldhaber *et al.*, *Phys. Rev.* **120** (1960)300
2. R. Hernández and G. Herrera, *Phys. Lett.* **B332** (1994)448
   A. Gago and G. Herrera, *Mod. Phys. Lett.* **A10**(1995)1435
3. A.A. Bel'kov, G. Bohm, D. Ebert, A.V. Layov and A. Schaale, *Phys. Lett.* **196B** (1987)107; A.A. Bel'kov, G. Bohm, A.V. Layov and A. Schaale, hep-ph/9311295
4. E. Cuautle and G. Herrrera, *Phys. Lett.* **B434** (1998)153-157
5. J. Adler *et al.*, *Phys. Lett.* **196B** (1987)107
6. P. L. Frabetti *et al.*, *Phys. Lett.* **B407** (1997)79
7. A. Bialas and A. Krzywicki, *Phys. Lett.* **B 354** (1995)134
8. L. Lonnblad and T. Sjostrand, *Phys. Lett.* **B 351** (1995)293
9. G.D. Lafferty, *Z. Phys.* **C60** (1993)659

Jacobi Elliptic Solutions of $\lambda\phi^4$ Model in a Finite Domain [1]

J. A. Espichán Carrillo [2], A. Maia Jr. [3] and V. M. Mostepanenko [4]

[2] *Instituto de Física "Gleb Wathagin", University of Campinas (UNICAMP) - 13.081-970 - Campinas (SP), Brazil.*
[3] *Instituto de Matemática, University of Campinas (UNICAMP) - 13.081-970 - Campinas (SP), Brazil.*
[4] *Friedmann Laboratory for Theoretical Physics (Russia) and Department of Physics, Federal University of Paraíba (UFPB) - C.P. 5008; 58059-970 - João Pessoa (PB), Brazil.*

Abstract. We calculate the general static solutions of the scalar field equations for the potential $V(\phi) = -\frac{1}{2}M^2\phi^2 + \frac{\lambda}{4}\phi^4$ for a finite domain in $(1+1)$ dimensions. Solutions with vacuum-vacuum boundary conditions at $x = \pm\infty$ were firstly obtained by Dashen et all (DHN) [1]. We define expressions for the "topological charge", "total energy" and "energy density" for general solutions in a finite domain. Also we calculate the quantum fluctuations and radiative corrections for general solutions.

I GENERAL SOLUTIONS

We consider the Lagrangian density $\mathcal{L}$ of a scalar field ϕ in $(1+1)$ dimensions,

$$\mathcal{L}(\phi, \partial_\mu\phi) = \frac{1}{2}\dot{\phi}^2 - \frac{1}{2}(\partial_x\phi)^2 - V(\phi),$$

where $V(\phi) = -\frac{1}{2}M^2\phi^2 + \frac{\lambda}{4}\phi^4$.

The classical equation of motion (static case) is

$$\partial_{xx}\phi - \frac{\partial V}{\partial \phi} = 0. \tag{1}$$

A first integral of eq.(1), after a change of variable, is given by

$$x - x_0 = \pm\frac{1}{M}\int dz\, f(z), \tag{2}$$

---

1) Presented by J. A. Espichán Carrillo
2) e-mail: espichan@ifi.unicamp.br
3) e-mail: maia@ime.unicamp.br
4) e-mail: mostep@fisica.ufpb.br

CP490, *Particles and Fields: Eighth Mexican School,* edited by J. C. D'Olivo et al.

where $f(z) = \sqrt{z^4 - z^2 + c/2}$, $z = \frac{\sqrt{\lambda}}{\sqrt{2}M}\phi$, and x_0, c are constants of integration.

We are looking for real solutions of Eq. (2). So we must have $(z^4 - z^2 + c/2) > 0$.

The stationary points of the function $f(z)$, are given by $z = 0$ and $z = \pm 1/2$. For $c = 0$ and $c = 1/2$, the integral (2) is not defined at these points.

Below we present the main results.

CASE $0 < c < \frac{1}{2}$

$(z^4 - z^2 + c/2) > 0$ is satisfied for $|z| \leq \sqrt{\frac{1-\sqrt{1-2c}}{2}} \quad \Rightarrow \quad |\phi_c(x)| < \frac{M}{\sqrt{\lambda}}$.

The solution for ϕ, will be given by

$$\phi_c(x) = \pm \frac{M}{\sqrt{\lambda}} \frac{\sqrt{2c}}{\sqrt{1+\sqrt{1-2c}}} \operatorname{sn}\left(\frac{M\sqrt{1+\sqrt{1-2c}}}{\sqrt{2}}(x - x_0), m\right), \tag{3}$$

where $m = \frac{1}{-1+\frac{1+\sqrt{1-2c}}{c}}$ and sn is the Jacobi Elliptic Function [3]. Also, observe that for $c \to \frac{1}{2}$, the Kink solution [1] is re-obtained.

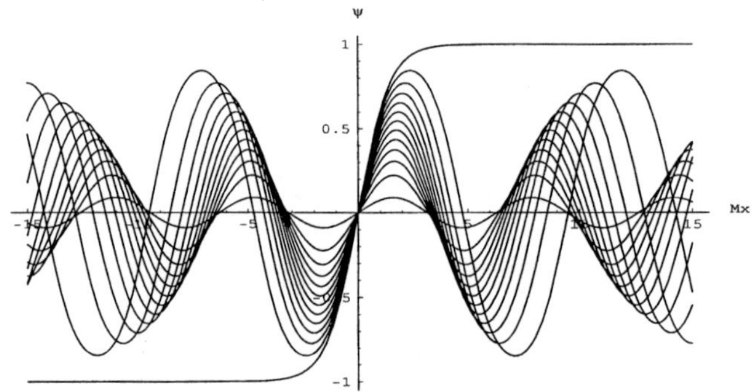

FIGURE 1. Family of sn-type solutions (JE-sn) given by eq.(3), for c = 0.008, 0.049,...,0.459, 0.5. The solution for c = 0.5 is the DHN's Kink. Here and below we have defined $\psi = \frac{M}{\sqrt{\lambda}}\phi$.

CASE $c \geq \frac{1}{2}$

In this case, the solution for ϕ, will be given by

$$|\phi_c(x)| = \frac{M\sqrt[4]{2c}}{\sqrt{\lambda}} \sqrt{\frac{1 - \operatorname{cn}\left(\sqrt[4]{8c}\,M(x - x_0), \frac{1}{2}(1 + \frac{1}{\sqrt{2c}})\right)}{1 + \operatorname{cn}\left(\sqrt[4]{8c}\,M(x - x_0), \frac{1}{2}(1 + \frac{1}{\sqrt{2c}})\right)}}. \tag{4}$$

II JE-SN IN AN 0NE-DIMENSIONAL BOX

In this section we study the solutions (3) constrained to a box $[-\frac{L}{2}, \frac{L}{2}]$. Dirichlet boundary conditions at $x = \pm\frac{L}{2}$, in the JE-sn (3), gives

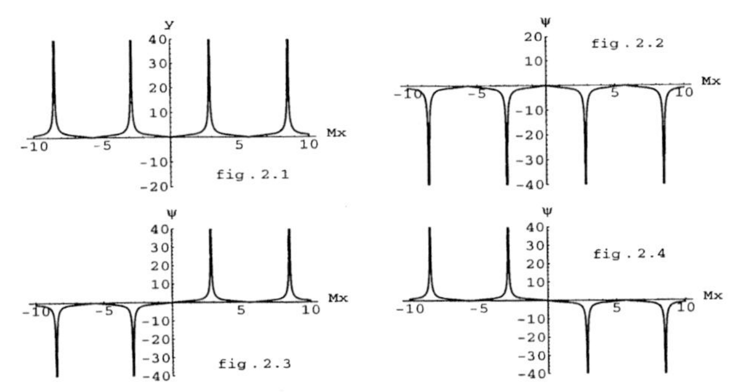

FIGURE 2. Solutions given by eq. (4) for $c = 1$. Here we took $x_0 = 0$. For a given length L, there exists just one solution satisfying vacuum-vacuum condition at $\pm\frac{L}{2}$.

$$ML = \frac{4\sqrt{2}}{\sqrt{1+\sqrt{1-2c}}}K(m),$$

where $4K(m)$ is a period of the Jacobi Elliptic Functions [3].

On the other hand, it is easy to show that $\frac{d(ML)}{dc} > 0$. So, "ML" increasing monotonically with the parameter c. Besides, in the limit $c \to 0$ we get $ML = 2\pi$.

This relation shows that the product "ML" has a minimal value 2π. It is easy to see that if we fix M as the mass of our field, it can not exist in a cavity with length smaller than $\frac{2\pi}{M}$.

Now we have calculated the following quantities in a finite domain

a) "Topological charge"

$$Q(L) = \frac{1}{2}\int_{-\frac{L}{2}}^{\frac{L}{2}} \frac{\sqrt{\lambda}}{M}\, d\phi = \pm\frac{\sqrt{2c}}{\sqrt{1+\sqrt{1-2c}}}\,\mathrm{sn}\left(\frac{ML}{2\sqrt{2}}\sqrt{1+\sqrt{1-2c}}, m\right),$$

where for $L = \infty$ or $c = \frac{1}{2}$, we have that $Q = \pm 1$.

b) "Total energy" (*classical mass*) is given by

$$\mathcal{M}(L) = \int_{-L/2}^{L/2} dx\, \epsilon_c(x) \tag{5}$$

where $\epsilon_c(x)$ is the **"energy density"**

$$\epsilon_c(x) = \frac{M^4}{\lambda}\left(\frac{c}{2}\mathrm{cn}^2\!\left(\frac{Mx}{\sqrt{2}}\sqrt{1+\sqrt{1-2c}}, m\right)\times \mathrm{dn}^2\!\left(\frac{Mx}{\sqrt{2}}\sqrt{1+\sqrt{1-2c}}, m\right) - \frac{c\,\mathrm{sn}^2\!\left(\frac{Mx}{\sqrt{2}}\sqrt{1+\sqrt{1-2c}}, m\right)}{(1+\sqrt{1-2c})} + \frac{c^2\,\mathrm{sn}^4\!\left(\frac{Mx}{\sqrt{2}}\sqrt{1+\sqrt{1-2c}}, m\right)}{(1+\sqrt{1-2c})^2} + \frac{1}{4}\right). \tag{6}$$

It is not difficult to obtain an explicit expression for the "total energy" (5), using the previous relation (6). For $c = \frac{1}{2}$, we obtain DHN's, result that is, $\epsilon(x) = \frac{M^4}{2\lambda}\, sech^4(\frac{Mx}{\sqrt{2}})$ and $\mathcal{M} = \frac{2\sqrt{2}M^3}{3\lambda}$.

III JE-SN FLUCTUATIONS (WORK IN PROGRESS)

To determine the quantum corrections we must solve the following eigenvalue equation [1]:

$$\left(-\frac{d^2}{dx^2} - M^2 + 3\lambda\phi_c^2\right)\psi_n(x) = \omega_n^2\psi_n(x), \tag{7}$$

where $\psi_n(x)$ is an eigenfunction of the differential operator and ω_n is its eigenvalue.

Substitutitng (3) in (7), as well the changes variables, $\alpha = \frac{Mx}{\sqrt{2}}\sqrt{1+\sqrt{1-2c}}$ and $\omega_n^2 = \frac{(E-2)M^2}{2}$, we get the Lamé equation [2], i.e.,

$$\frac{d^2}{d\alpha^2}\psi(\alpha) = \left(6\,m\,\mathrm{sn}^2(\alpha,m) - \frac{E(1+m)}{2}\right)\psi(\alpha).$$

Now, imposing Dirichlet boundary conditions, on Lamé eigenfunctions we can get the eigenvalues ω_n. So, finally, we obtain the non-renormalized mass correction, that is,

$$\Delta\mathcal{M}(L) = \sum\!\!\!\!\!\!\int \omega_n = \frac{\hbar}{2}\sum_{n=1}\sqrt{\frac{3}{m_n(L)+1}}M + \frac{\hbar}{2}\sum_{n=1}\sqrt{\frac{3\,m_n(L)}{m_n(L)+1}}M + 0$$

$$+\frac{\hbar}{2}\sqrt{\frac{1+m(L)+2\sqrt{m^2(L)-m(L)+1}}{1+m(L)}}M + CP,$$

where CP denotes the contribution of the continuum part of the spectrum. This is yet under investigation.

A test of consistence of our calculations is that the DHN's mass correction [1] (renormalized) should be obtained in the limit $L \to \infty$, that is

$$\lim_{L\to\infty}(\Delta\mathcal{M})_R(L) = (\Delta\mathcal{M}_R).$$

ACKNOWLEDGMENTS

This work was supported in part by FAPESP (Fundação de Amparo à Pesquisa do Estado de São Paulo), Brazil.

REFERENCES

1. R. Dashen, B. Hasslacher and A. Neveu. *Phys. Rev.* **D10**, *4131 (1974)*
2. Z. X. Wang and D. R. Guo. *Special Functions*, World Scientific (1989)
3. M. Abramowitz and I. A. Stegun. *Handbook of Mathematical Functions* (Dover Publications, INC., New York, 1972)

Stretched Horizon for Non–Supersymmetric Black Holes

C. Espinoza and M. Ruiz–Altaba

Departamento de Física Teórica
Instituto de Física
Universidad Nacional Autónoma de México
Apartado Postal 20-364
01000 México, D.F.

Abstract. We review the idea of stretched horizon for extremal black holes in supersymmetric string theories, and we compute it for non-supersymmetric black holes in four dimensions. Only when the angluar momentum is large is the stretched horizon bigger than the event horizon.

I INTRODUCTION

It was long ago suggested that black holes should be treated as elementary particles [1], because both are parametrized only by their mass, angular momentum (or spin) and gauge charges. String theory has made a significant contribution towards putting this assertion on a firm basis. A common feature of black holes and elementary string states is that the degeneracy of states with a given mass increases with it. Unfortunately, for elementary string states the logarithm of the degeneracy of states increases linearly with mass, whereas the Bekenstein-Hawking entropy of the black hole increases as the square of the mass. There are some cases in which the discrepancy between the two entropies can be removed appealing to the large renormalization of the mass of a black hole [2]. There are, however, some particular states in string theory, called BPS states, which do not receive any mass renormalization [3]. Whereas the logarithm of the degeneracy of BPS states grows linearly with the mass, the area of the event horizon of a BPS or extremal black hole actually vanishes. This motivates the argument [4] that the entropy of an extremal black hole is not exactly equal to the area of the event horizon, but to the area of a surface close to the event horizon called "stretched horizon". By carefully defining the location of the stretched horizon in a consistent way, the Bekenstein-Hawking formula for the black hole entropy can receive corrections in such a way that it correctly reproduces the logarithm of the density of elementary string states.

CP490, *Particles and Fields: Eighth Mexican School,* edited by J. C. D'Olivo et al.

The stretched horizon of black hole is defined as the surface where the space-time curvature (in the string metric) becomes large. It is also the surface where the local Unruh temperature for a stationary observer (constant r) is the Hagedorn temperature of the string theory. (The local Unruh temperature becomes infinite at the event horizon). Specifically, we define the stretched horizon as the surface where the scalar curvature $C = (\text{Riemann})^2$ is equal to one in Planck's units.

II STRETCHED HORIZON FOR THE CLASSICAL SOLUTIONS

Throughout, we shall be thinking of black holes as the classical description of a quantum object. The nature of this object is well approximated by (classical) general relativity far away from the horizon. At any rate, we should not expect the metric which solves Einstein's equations to make any sense at distances to the origin smaller than the de Broglie wavelength $\lambda_B = M^{-1}$ for a black hole of mass M. Since we are stringy, it is perhaps more appropriate to use Veneziano's generalized uncertainty principle [5] $\Delta x \geq \frac{\ell_s^2}{2\hbar}\Delta p + \frac{\hbar}{\Delta p}$, where ℓ_s is the string scale, g is the string coupling constant and $\ell_p = g\ell_s$ is Planck's length. So, in Planck units, the metric certainly is not expected to make any sense at radii smaller than

$$\lambda_V = \frac{1}{2}M + \frac{1}{M} \tag{1}$$

Where is the stretched horizon? If the place where the curvature becomes big (unity in Planck units) is inside the event horizon, then clearly there is no need to stretch it at all. Similarly, if it falls at a radius smaller than λ_V, there is no point in talking about it. We are interested in finding under what circumstances the stretched horizon is a meaningful and useful concept for non-supersymmetric four-dimensional black holes. In other words, we must find when the stretched horizon is bigger than the event horizon and also bigger than λ_V (or λ_B).

Consider the general solution to the Einstein-Maxwell equations

$$\begin{aligned} ds^2 = &-\left(\frac{\Delta - a^2\sin^2\theta}{\Sigma}\right)dt^2 - \frac{2a\sin^2\theta\,(r^2+a^2-\Delta)}{\Sigma}dtd\phi \\ &+\left[\frac{(r^2+a^2)^2 - \Delta a^2\sin^2\theta}{\Sigma}\right]\sin^2\theta d\phi^2 + \frac{\Sigma}{\Delta}dr^2 + \Sigma d\theta^2, \end{aligned} \tag{2}$$

$$A_\mu = -\frac{er}{\Sigma}\left[(dt)_\mu - a\sin^2\theta(d\phi_\mu)\right], \tag{3}$$

where

$$\Sigma = r^2 + a^2 cos^2\theta, \quad \Delta = r^2 + a^2 + e^2 - 2Mr, \tag{4}$$

Using Mathematica, we have computed the square of the Riemann tensor for this solution, C, and evaluated where it becomes one.

For the Schwarszchild black hole ($e = a = 0$) the scalar curvature is simply

$$C = \frac{48M^2}{r^6} \tag{5}$$

whereby the stretched horizon radius is

$$r_s = 48^{\frac{1}{6}} M^{\frac{1}{3}} \tag{6}$$

Thus, the stretched horizon is always inside the event horizon, except for such ridiculously small masses that the de Broglie wavelength is actually bigger than both.

This continues to be the case for any charged static black hole: in Fig. 1 we plot the critical mass (in Planck units) at which the stretched horizon crosses the event horizon in terms of the charge e. Only for masses below the line is the stretched horizon bigger than the event horizon. The curve is essentially flat just below $M \sim 0.95$ when it begins to grow. When e reaches its extremal value $e = 1$, M_c reaches 1.088.

So both for Schwarzschild and for Reissner-Nordström black holes, the stretched horizon is useless.

When $a \neq 0$, i.e. for rotating black holes, the situation changes drastically. The stretched horizon is again bigger than the event horizon only for small masses, but the critical mass below which the stretched horizon is relevant (bigger than the event horizon) is now moderately big, and grows with a. Fig. 2 shows this critical mass when $e = 0$ (the plot was computed in the axial direction $\theta = 0$). In the extremal limit $a \to 1$, the critical mass approaches the value 2.68×10^6. The charge, in the general Kerr-Newman case, remains rather irrelevant.

III CONCLUSIONS

In this work we have shown two things. First, that outside the event horizon of a static black hole, the curvature is always small. Secondly, that when a four-dimensional black hole rotates, the event horizon is effectively hidden from an outside observer by a *stretched horizon*, where classical physics breaks down. We have found numerically the critical mass above which the event horizon is bigger than the stretched horizon. In the extremal limit, it ranges from 1.088 when $a = 0$, to 2.68×10^6 when $e = 0$.

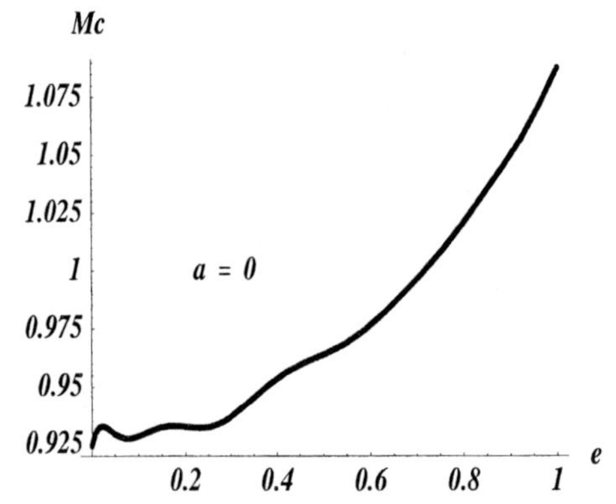

FIGURE 1. Critical mass Mc for a nonrotating charged black hole.

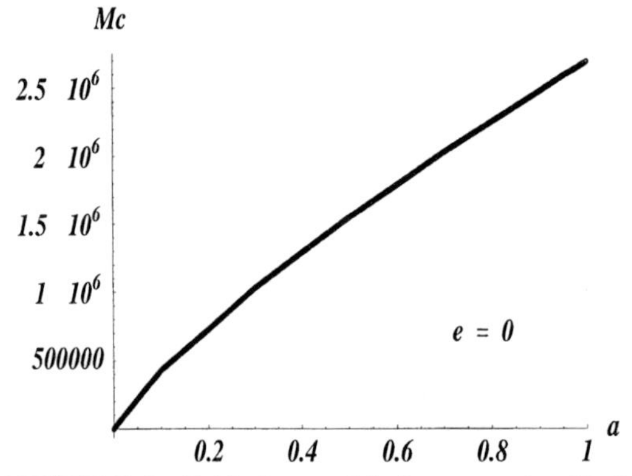

FIGURE 2. Critical mass Mc for a rotating black hole.

Acknowledgements. This work is supported in part by CONACYT 25504E, DGAPA-UNAM IN103997. C.E. enjoys a scholarship from DGEP-UNAM.

REFERENCES

1. G.'tHooft, Nucl. Phys. **B335** (1990) 138.
2. L. Susskind, hep-th/9309145.
3. E. Witten, D. Olive, Phys. Lett. **B78** (1978) 97.
4. A. Sen, Mod. Phys. Lett. **A10** (1995) 2081-2084; hep-th/9504147.
5. G. Veneziano, CERN-TH.5366/89

Description of Λ^0 polarization

Julián Félix

Instituto de Física, Universidad de Guanajuato, León, GTO 37000, México

Abstract. Reanalyzing two inclusive Λ^0 polarization data samples, created from *proton-nucleus* collisions, we found that Λ^0 polarization is well described, in the average, by $\mathcal{P}(x_F, P_T) = (-0.450 \pm 0.018)x_F P_T + (0.060 \pm 0.007)$, with $-1 < x_F < P_T$ GeV/c. This function fits also $\bar{\Lambda}^0$ polarization created in $p\bar{p} \to \bar{\Lambda}^0 X$ reactions, and Ξ^- polarization.

INTRODUCTION

Since Λ^0 polarization was discovered in unpolarized inclusive pp collisions[1], it is customary to characterize Λ^0 polarization as function of both x_F and P_T separately, leaving unresolved the problem of the simultaneous dependence of Λ^0 polarization on x_F and P_T. We can extract, from experimental evidences, some hints about this functionality. Hitherto our experimental knowledge on Λ^0 polarization produced in unpolarized pp or *p-nucleus* reactions, is as follows[1-7]: 1. Λ^0 polarization is negative with respect to the creation plane. 2. At fixed x_F, Λ^0 polarization decreases linearly with P_T for $0 < P_T < 1.2$ GeV/c; the slope of the straight line fit decreases (increases in absolute value) as x_F increases. 3. At fixed P_T, Λ^0 polarization decreases linearly with x_F for $-1 < x_F < +1$; the slope of the straight line fit decreases (increases in absolute value) as P_T increases.

From items 2 and 3, it is evident that Λ^0 polarization must depend on both x_F and P_T simultaneously, and that both variables must play identical role.
To establish the dependence of Λ^0 polarization on x_F and P_T we analyzed two different experimental data previously published by other authors[3,7]. Those data collections were created using a beam of 400 GeV protons hitting on beryllium targets, and employing the same experimental apparatus to detect them; some differences between data sets were the production angles, and therefore the deduced different x_F and P_T values. From this analysis, we deduced an analytic expression that fits both the x_F and the P_T Λ^0 polarization dependence, as we show in the next section.

CP490, *Particles and Fields: Eighth Mexican School,* edited by J. C. D'Olivo et al.

Λ^0 POLARIZATION

Fig. 1(a) and 1(b) show Λ^0 polarization distribution, from Ref. 3, and as function of both P_T and x_F respectively. Each distribution were fitted to a straight line; the results are indicated in each figure, also x_F and P_T average, and χ^2/dof. Dividing the slope of the Fig. 1(a) equation by the average x_F and the slope of the Fig. 1(b) equation by the average P_T, we obtained two equations that the average is:

$$\mathcal{P}(x_F, P_T) = (-0.403 \pm 0.015)x_F P_T + (0.050 \pm 0.004), \qquad (1)$$

showing that Λ^0 polarization depends on both the x_F and the P_T linearly. This equation represents the most simple bi-linear form extracted from Ref. 3 data. Refitting the above distributions, using this equation, we obtain 0.82 and 1.07 for the χ^2/dof respectively, indicating that this equation fits well those distributions.

Fig. 1(c) and 1(d) show Λ^0 polarization distribution as function of both P_T and x_F respectively, from Ref. 7, and in similar x_F and P_T regions that those treated before. Following the above procedure, we determined that Λ^0 polarization is fitted by the equations indicated in the Fig. 1(c) and 1(d); also the χ^2/dof. Dividing Fig. 1(c) equation slope by the average x_F and Fig. 1(d) equation slope by average P_T, we obtained two equations that in the average give:

$$\mathcal{P}(x_F, P_T) = (-0.517 \pm 0.024)x_F P_T + (0.074 \pm 0.009). \qquad (2)$$

This equation represents the most simple bi-linear form extracted from Ref. 7 data. Refitting the above two distributions, using this equation, we obtain 3.44 and 2.71 for the χ^2/dof respectively; hence Λ^0 polarization is a bi-linear function of both x_F and P_T.

If we average both Eq. 1 and Eq. 2 we obtain Eq. 3:

$$\mathcal{P}(x_F, P_T) = (-0.450 \pm 0.018)x_F P_T + (0.060 \pm 0.007). \qquad (3)$$

Using this equation to fit the previous four Λ^0 polarization distributions (from Ref. 3 and Ref. 7, and as function of P_T and x_F), we obtain 1.89, 1.13, 4.95, and 2.65 for the χ^2/dof respectively. Therefore Λ^0 polarization is function of both x_F ($-1 < x_F < +1$) and P_T ($0 < P_T < 1.2$ GeV/c).

We must note that the presented Λ^0 polarization distributions lack data around zero in both x_F and P_T, see Fig. 1 for the sample was created inclusively with limited x_F and P_T, resulting in an interception different from zero, see Eqs. 1, 2, and 3; it is known, from arguments of symmetry that the interception must be zero, however the parameterization presented (Eq. 3) is the best one obtained from the data. In Ref. 7 it is suggested the expression $\mathcal{P}(x_F, P_T) = -(C_1 x_F + C_2 x_F^3)(1 - e^{C_3 P_T^2})$ to fit Λ^0 polarization; but, this expression fits badly the Λ^0 polarization distributions at low x_F-P_T values -the χ^2/dof's are very big-. In the other hand -using exclusive reactions, where it is possible to get Λ^0 polarization values in the regions of x_F and P_T close to zero, it is shown that in the region of $-1 < x_F < +1$ and $P_T \leq 1.5$ GeV/c Λ^0 polarization follows accurately the linear dependence on

both x_F and P_T[8]; the slope of the Eq. 3 and the slope of the one presented in Ref. 8 agree statistically.
We tested Eq. 3 using data from Ref. 5. The experiment from where this last data come from is entirely dissimilar from the previous one, in energy mainly. The Λ^0 polarization distribution, as function of x_F, is shown in Fig. 1(e); the average P_T is 1.14 GeV/c. Using this value in Eq. 3, when we fitted the Ref. 5 Λ^0 polarization distribution as function of x_F, we got 1.87 for the χ^2/dof. This value proves that the Eq. 3, as function of both x_F and P_T, fits well Λ^0 polarization distribution.

We can draw two more not so strong conclusions: a) Λ^0 polarization is independent of the beam energy, and b) Λ^0 polarization is independent of the target nature -implying that Λ^0 polarization is not a nuclear effect-.

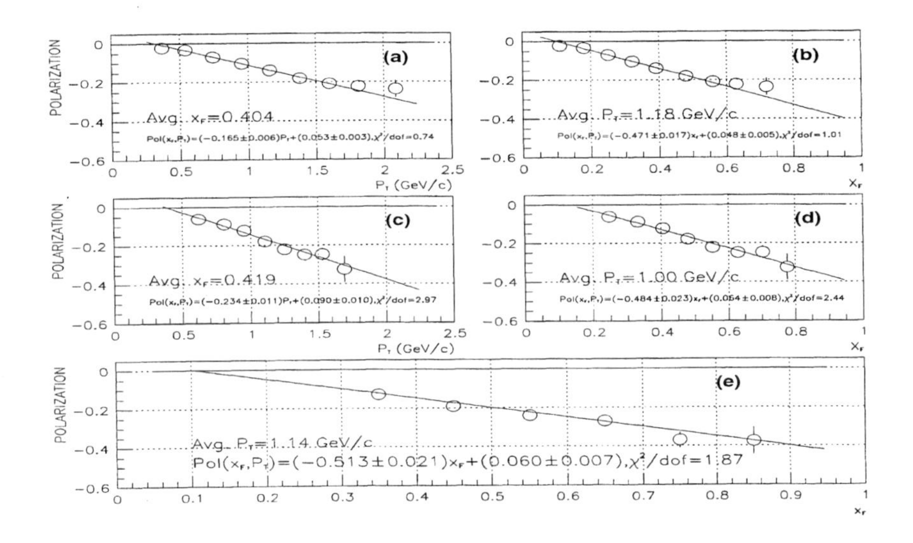

FIGURE 1. Λ^0 polarization as function of x_F. We explain the meaning inside the text.

HYPERON POLARIZATION

Probably the hyperon polarization is universal; no only its production in *hadron-hadron* collisions but in its distribution as function of both x_F and P_T. It is known that the polarization of Λ^0, Ξ^-, and $\bar{\Lambda}^0$ is function of both x_F and P_T; however, the functional dependence has not been stated definitely. Also it is known that Σ^+, $\bar{\Sigma}^-$, and $\bar{\Xi}^+$ are produced polarized [9], depending on P_T.

The $\bar{\Lambda}^0$ polarization created in $p\bar{p} \rightarrow \bar{\Lambda}^0 X$ is also function of both x_F and P_T[10], and this polarization is fitted well by Eq. 3. The obtained values indicate that $\bar{\Lambda}^0$ polarization depends on both x_F and P_T in the same way that Λ^0 polarization depends on those variables.

The polarization of Ξ^-, created from a beam of Σ^- at 330 GeV/c [11], also follows the x_F and P_T dependence of Λ^0 polarization, as function of both P_T and x_F.

CONCLUSIONS

In résumé, we have analyzed two different samples of Λ^0 polarization as function of both x_F and P_T, to find that, in the average, Λ^0 polarization follows Eq. 3. This equation represents the most simple bi-linear form as function of both x_F and P_T. We state that this form is universally valid to describe Λ^0 polarization distribution, and also to describe $\bar{\Lambda}^0$ polarization and Ξ^- polarization. Also that Λ^0 polarization must be independent of the beam energy and of the target nature.

ACKNOWLEDGMENTS

This work was supported in part by CoNaCyT of México under Grant 458100-5-4009PE.

REFERENCES

1. G. Bunce *et al.*, *Phys. Rev. Lett.* **36**, 1113 (1976).
2. F. Lomanno *et al.*, *Phys. Rev. Lett.* **43**, 1905 (1979); S. Erhan *et al.*, *Phys. Lett.* **82B**, 301 (1979); F. Abe *et al.*, *Phys. Rev. Lett.* **50**, 1102 (1983); K. Raychaudhuri *et al.*, *Phys. Lett.* **90B**, 319 (1980).
3. K. Heller *et al.*, *Phys. Rev. Lett.* **41**, 607 (1978).
4. F. Abe *et al.*, *J. of the Phys. S. of Japan.* Vol. **52**, 12 (1983) 4107-4117; P. Aahlin *et al.*, *Lettere al Nuovo Cimento* Vol. **21**, No. 7, (1978).
5. A.M. Smith *et al.*, *Phys. Lett.* **185B**, 209 (1987).
6. V. Blobel *et al.*, *Nuclear Physics B* **122**, 429 (1977); T. Henkes *et al.*, *Phys. Lett.* **283B**, 155 (1992); K. Heller *et al.*, *Phys. Lett.* **68B**, 480 (1977); J. Félix, *Ph.D. thesis*, **U. de Guanajuato, México** 1994; J. Félix *et al.*, *Phys. Rev. Lett.* **76**, 22 (1996).
7. B. Lundberg *et al.*, *Phys. Rev. D* **40**, 3557 (1989).
8. J. Félix *et al.*, *resummitted to PRL, Jan. 1999.*
9. A. Morelos *et al.*, *Phys. Rev. Lett.* **71**, 2175 (1993); P.M. Ho *et al.*, *Phys. Rev. Lett.* **65**, 1713 (1990).
10. S.A. Gourlay *et al.*, *Phys. Rev. Lett.* **56**, 2244 (1986).
11. M.I. Adamovich *et al.*, *Z. Phys. A* **350**, 379-386 (1995).

Couplings between generalized gauge fields

J. Antonio Garcia [a] and Bernard Knaepen [b]

[a] *Instituto de Ciencias Nucleares Universidad Nacional Autónoma de México Apartado Postal 70-543, 04510 México, D.F.*
[b] *Physique Théorique et Mathématique, Université Libre de Bruxelles, Campus Plaine C.P. 231, B-1050 Bruxelles, Belgium.*

Abstract. We analyze the BRST field-antifield construction for generalized gauge fields consisting of massless mixed representations of the Lorentz Group and we calculate all the strictly gauge invariant interactions between them. All these interactions are higher derivative terms constructed out from the derivatives of the curl of the field strength.

Generalized gauge fields consisting of free massless integer spin mixed representations of the Lorentz group (mixed tensors) can be used as models for higher spin covariant bosonic theories and to study the long standing problem of interactions for massless higher spin particles.

Free consistent massless theories whose field content is this class of mixed tensors have attracted some attention in the past. In particular, possible lagrangians, compatible with gauge invariance, have been proposed in [1–3] while the complete ghost spectrum and the BRS operator are described in [1,2,4].

In this article we calculate the cohomology of the "longitudinal exterior derivative" in order to obtain all the consistent, strictly gauge invariant interactions that can be added to the free theory.

For simplicity, we will concentrate on tensors with three indices which satisfy the following identities:

$$T_{[ab]c} = -T_{[ba]c}, \quad T_{[ab]c} + T_{[ca]b} + T_{[bc]a} = 0; \tag{1}$$

we comment at the end of the paper on how to generalize our results to other mixed tensors. In terms of Young diagrams the fields (1) are represented by $\begin{array}{|c|c|}\hline a & c \\ \hline b \\ \cline{1-1}\end{array}$.

Using the 'Hook' formula [5,6], one easily calculates that the tensors (1) have $\frac{1}{3}D(D-1)(D+1)$ components in D dimensions.

The lagrangian of the theory is,

CP490, *Particles and Fields: Eighth Mexican School,* edited by J. C. D'Olivo et al.

$$\mathcal{L} = -\frac{1}{12}\left(F_{[abc]d}F^{[abc]d} - 3F_{[abx]}{}^{x}F^{[aby]}{}_{y}\right), \tag{2}$$

where $F_{[abc]d} = \partial_a T_{[bc]d} + \partial_b T_{[ca]d} + \partial_c T_{[ab]d}$. The corresponding action is, up to a total derivative, invariant under the gauge transformations,

$$\delta_{\epsilon,\eta} T_{[ab]c} = \partial_a \epsilon_{bc} - \partial_b \epsilon_{ac} + \partial_a \eta_{bc} - \partial_b \eta_{ac} - 2\partial_c \eta_{ab}, \tag{3}$$

where ϵ_{ab} are symmetric and η_{ab} are antisymmetric gauge parameters.

By performing the Dirac constraint analisys we can show that the Lagrangian theory is consistent and that the number of degrres of freeodom is $\frac{1}{3}D(D-2)(D-4)$.

The theory is reducibile and the reducible identities among the gauge transformations are responsible for the presence of ghosts of ghosts in the BRS ghost spectrum. Indeed, the gauge variations vanish for the choices, $\epsilon_{ab} = 3(\partial_a C_b + \partial_b C_a)$ and $\eta_{ab} = \partial_a C_b - \partial_b C_a$, where C_a are D arbitrary functions.

I BRST FORMALISM AND GUAGE INVARIANT FUNCTIONS

According to the general rules of the BRST field-antifield formalism [7], the BRST differential s is constructed as follows.

First, one defines a differential δ called the Koszul-Tate differential whose role is to implement the equations of motion in cohomology. We are interested in study the strictly gauge invariant interactions. In the BRST construction is the longitudinal exterior derivative γ which takes into account the gauge invariance of the model. In our case, we first need to introduce the ghosts S_{ab} and A_{ab} in place of each gauge parameter according to the definition,

$$\gamma T_{[ab]c} = \partial_a S_{bc} - \partial_b S_{ac} + \partial_a A_{bc} - \partial_b A_{ac} - 2\partial_c A_{ab}, \tag{4}$$

where S_{ab} and A_{ab} are respectively symmetric and antisymmetric in ab.

Because the gauge transformations are reducible we also need the ghosts of ghosts C_a which satisfy, $\gamma S_{ab} = 3(\partial_a C_b + \partial_b C_a)$ and $\gamma A_{ab} = \partial_a C_b - \partial_b C_a$. With these definitions, we have $\gamma^2 = 0$. A grading called the 'pureghost' number is associated to the ghost fields and we have: $pureghost(S_{ab}) = pureghost(A_{ab}) = 1$, $pureghost(C_a) = 2$ and also $pureghost(\gamma) = 1$. Note that the fields and their derivatives are of antighost and pureghost number 0 and that $\gamma(antifields) = \delta(ghosts) = 0$.

For the model we consider, the full BRST differential is simply given by the sum of the Koszul-Tate differential and the longitudinal exterior derivative: $s = \delta + \gamma$. The grading of s is called the 'ghost' number and is given by $ghost = pureghost - antighost$.

The procedure we use is based on the following result [7,9]:

Lemma 1 *If $\mathcal{A}$ is a free graded-commutative differential algebra with differential D, then it is possible to find generators of $\mathcal{A}$ which satisfy the following relations:*

$$Dx^i = f(x^i), \quad Dy^\alpha = z^\alpha. \tag{5}$$

Furthermore, if $f \equiv 0$, then the cohomology of D in $\mathcal{A}$, $H(D) \equiv \frac{Ker\ D}{Im\ D}$, is given by the polynomials in the x^i.

In our case the algebra $\mathcal{A}$ is the algebra generated by the fields, the antifields, the ghosts, the ghosts of ghosts and all their derivatives. Our task is thus to redefine all our generators in such a way that they obey (6).

First of all, let us note that the antifields and their derivatives are all γ-closed and do not appear in the γ variations of the other fields. This implies that they are automatically part of the x^i variables.

For the other variables, we denote by V^k the vector space spanned by $\partial_{s_1 \ldots s_k} C_a$, $\partial_{s_1 \ldots s_{k-1}} A_{ab}$, $\partial_{s_1 \ldots s_{k-1}} S_{ab}$, $\partial_{s_1 \ldots s_{k-2}} T_{[ab]c}$. Our whole algebra is $\mathcal{A} = \oplus_k V^k$ and γ has a well defined action in each V^k. We will therefore look for new coordinates in each V^k.

According to the general theory of the representations of the symmetric group, we asociate to the algebraic expresion $\partial_{d_1 \ldots d_k} T_{[ab]c}$ the Young diagram $\begin{ytableau} d_1 & \cdot & \cdot & \cdot & \cdot & d_k \end{ytableau} \otimes \begin{ytableau} a & c \\ b \end{ytableau}$ .. This tensor product decomposes into the following irreducible components under a general invertible transformation,

$$\begin{ytableau} d_1 & \cdot & \cdot & \cdot & \cdot & d_k \end{ytableau} \otimes \begin{ytableau} a & c \\ b \end{ytableau} \simeq \begin{ytableau} a & c & d_1 & \cdot & \cdot & \cdot & \cdot & d_k \\ b \end{ytableau} \oplus \begin{ytableau} a & c & d_1 & \cdot & \cdot & \cdot & \cdot \\ b & d_k \end{ytableau}$$
$$\oplus \begin{ytableau} a & c & d_1 & \cdot & \cdot & \cdot & \cdot \\ b \\ d_k \end{ytableau} \oplus \begin{ytableau} a & c & d_1 & \cdot & \cdot & \cdot \\ b & d_l \\ d_k \end{ytableau},$$

where $l = k - 1$.

The combinations of $\partial_{d_1 \ldots d_k} T_{[ab]c}$ represented by the above diagrams are respectively denoted by $R^T_{abcd_1 \ldots d_k}$, $H^T_{abcd_1 \ldots d_k}$, $F^T_{abcd_1 \ldots d_k}$, and $E^T_{abcd_1 \ldots d_k}$. They are obtained by first symmetrizing $\partial_{d_1 \ldots d_k} T_{[ab]c}$ according to every line and then antisymmetrizing the result according to every column. By convention, for every symmetrization or antisymmetrization, we divide the corresponding sum of terms by a factorial term. The variables R, H, F and E form a new basis for the space spanned by $\partial_{d_1 \ldots d_k} T_{[ab]c}$

In exactly the same way, the $(k+1)$-th order derivatives of the ghosts A_{ab} are decomposed according to,

$$\begin{ytableau} d_1 & \cdot & \cdot & \cdot & \cdot & d_k & c \end{ytableau} \otimes \begin{ytableau} a \\ b \end{ytableau} \simeq \begin{ytableau} a & c & d_1 & \cdot & \cdot & \cdot & \cdot & d_k \\ b \end{ytableau} \oplus \begin{ytableau} a & c & d_1 & \cdot & \cdot & \cdot & \cdot \\ b \\ d_k \end{ytableau}.$$

The diagrams of the rhs of the above decomposition are denoted respectively by $R^A_{abcd_1 \ldots d_k}$ and $F^A_{abcd_1 \ldots d_k}$. Note that in the case $k = 0$ the above notation is not well adapted because the diagram $\begin{array}{|c|}\hline a\\\hline b\\\hline c\\\hline\end{array}$ is missing. In that case, we denote the corresponding combination of $\partial_c A_{ab}$ by F^A_{abc}.

For the $(k+1)$-th order derivatives of the S_{ac} we have,

$$\begin{array}{|c|c|c|c|c|c|c|}\hline d_1&\cdot&\cdot&\cdot&\cdot&d_k&b\\\hline\end{array} \otimes \begin{array}{|c|c|}\hline a&c\\\hline\end{array} \simeq \begin{array}{|c|c|c|c|c|c|c|c|c|}\hline a&c&b&d_1&\cdot&\cdot&\cdot&\cdot&d_k\\\hline\end{array} \oplus \begin{array}{|c|c|c|c|c|c|c|c|}\hline a&c&d_1&\cdot&\cdot&\cdot&\cdot&d_k\\\hline b\\\cline{1-1}\end{array}$$

$$\oplus \begin{array}{|c|c|c|c|c|c|c|}\hline a&c&d_1&\cdot&\cdot&\cdot&\cdot\\\hline b&d_k\\\cline{1-2}\end{array} \; .$$

The components of the decomposition are denoted respectively, $L^S_{abcd_1\ldots d_k}$, $R^S_{abcd_1\ldots d_k}$, $H^S_{abcd_1\ldots d_k}$.

Finally, the $(k+2)$-th derivatives of the C_a decompose according to,

$$\begin{array}{|c|c|c|c|c|c|c|c|}\hline d_1&\cdot&\cdot&\cdot&\cdot&d_k&b&c\\\hline\end{array} \otimes \begin{array}{|c|}\hline a\\\hline\end{array} \simeq \begin{array}{|c|c|c|c|c|c|c|c|c|}\hline a&c&b&d_1&\cdot&\cdot&\cdot&\cdot&d_k\\\hline\end{array} \oplus \begin{array}{|c|c|c|c|c|c|c|c|}\hline a&c&d_1&\cdot&\cdot&\cdot&\cdot&d_k\\\hline b\\\cline{1-1}\end{array} \; .$$

The two different components are denoted $L^C_{abcd_1\ldots d_k}$ and $R^C_{abcd_1\ldots d_k}$.

With the above definitions, an explicit calculation shows that we have the following relations among the variables:

$$\gamma R^T_{abcd_1\ldots d_k} = 3R^A_{abcd_1\ldots d_k} + \frac{k+3}{2} R^S_{abcd_1\ldots d_k}, \quad \gamma H^T_{abcd_1\ldots d_k} = \frac{k+2}{2} H^S_{abcd_1\ldots d_k}, \tag{6}$$

$$\gamma F^T_{abcd_1\ldots d_k} = 3F^A_{abcd_1\ldots d_k}, \quad \gamma L^S_{abcd_1\ldots d_k} = 6L^C_{abcd_1\ldots d_k}, \tag{7}$$

$$\gamma R^S_{abcd_1\ldots d_k} = 3R^C_{abcd_1\ldots d_k}, \quad \gamma E^T_{abcd_1\ldots d_k} = 0, \quad \gamma F^A_{abc} = 0. \tag{8}$$

From here we conclude that the cohomology of γ is generated by the variables C_a, F^A_{abc}, $E^T_{abcd_1\ldots d_k}$, the antifields and their derivatives.

REFERENCES

1. C.S. Aulakh, I.G. Koh and S. Ouvry, *Phys. Lett.* **B173** (1986) 284.
2. K.S. Chung, C.W. Han, J.K. Kim and I.G. Koh, *Phys. Rev.* **D37** (1988) 1079.
3. T. Curtright, *Phys. Lett.* **165B** (1985) 304.
4. J.M.F. Labastida and T.R. Morris, *Phys. Lett.* **B180** (1986) 101.
5. N. V. Dragon, *Tensor Algebra and Young Tableaux*, HD-THEP-81-16.
6. M. Hamermesh, *Group Theory*, Addison Wesley, (1962).
7. M. Henneaux and C. Teitelboim, *Quantization of Gauge Systems*, Princeton University Press, (1992).
8. M. Henneaux, *Consistent Interactions Between Gauge Fields: The Cohomological Approach*, hep-th/9712226, International Conference on *Secondary calculus and cohomological Physics*, Moscow, August 1997.
9. D. Sullivan, *Publ. IHES* **47** (1997) 269.

Non-divergent Formula for the Beta Energy Spectrum in the Four Body Decay of Hyperons

F. Guzmán A. and S.R. Juárez W.

Escuela Superior de Física y Matemáticas, Instituto Politécnico Nacional
Edif.9, Unidad Prof. Adolfo López Mateos
Col. Lindavista, C.P. 07738 México D.F., MEXICO.

Abstract. The bremsstrahlung in the region of the Dalitz plot that covers exclusively the four-body events of unpolarized semileptonic decays of charged and neutral hyperons is described by very accurate analytical expressions. These expressions contain a logarithmic singularity at the upper edge of the plot. We perform the analytical integration and obtain divergentless formulas for the energy spectrum of the produced charged lepton. With these formulas for the four body region and the analogous for the three body region, the whole spectrum of physically possible events are covered.

The precise description of the semileptonic weak decays is relevant in several areas of particle physics. The precise formulas that contain terms of the order α multiplied by the momentum transfer, for the energy spectrum of a decay product (charged lepton) in the hyperon semileptonic decay, require the knowledge of the radiative corrections in the four body region of the Dalitz plot. The subject of this paper is to show how the logarithmic divergences that are contained in previous results, which are obtained in an analytical way (for the radiative corrections to baryon beta decays in the four body region of the Dalitz plot) are cancelled after performing the integration over the energy of the final hyperon that emerges in the process. These new analytical results are valid for photon bremsstrahlung calculations in any charged or neutral hyperon decay. The total decay rate can now be computed directly through these analytical formulas for the radiative corrections, because these formulas do not contain any divergences. Such formulas are suitable for a direct evaluation of the radiative corrections for any event in the allowed physical region.

We consider the four body decay with the emission of a real photon

$$A(p_1) \rightarrow B(p_2) + e(\ell) + \bar{\nu}_e(p_{\bar{\nu}}) + \gamma(k). \tag{1}$$

The four-momenta and masses of the particles involved in the hyperon semileptonic decay are denoted by, $p_1 = (E_1, \vec{p_1})$, $p_2 = (E_2, \vec{p_2})$, $\ell = (E, \vec{\ell})$, $p_{\bar{\nu}} = (E_{\bar{\nu}}, \vec{p_{\bar{\nu}}})$, and

CP490, *Particles and Fields: Eighth Mexican School,* edited by J. C. D'Olivo et al.

$k=(k_0,\vec{k})$, and by M_1, M_2, m, $m_{\bar{\nu}}$ and m_k, respectively. The four body region for the process in Eq. (1), in the rest frame of A, is defined by

$$M_2 \le E_2 \le E_2^{\min} = \frac{M_1 - E - E\beta}{2} + \frac{M_2^2}{2\left(M_1 - E - E\beta\right)}, \quad \left|\vec{\ell}\right| = E\beta, \tag{2}$$

$$m \le E \le E_B = \frac{\left(M_1 - M_2\right)^2 + m^2}{2\left(M_1 - M_2\right)}. \tag{3}$$

The result for the bidimensional distribution of energies in the charged process (CHSD) is compactly given by [1],

$$d\Gamma_{BC}(E,E_2) = \frac{\alpha}{\pi} d\Omega \sum_{i=0}^{16} H_i' \theta_i^T, \quad d\Omega = \frac{G_v^2}{2} \frac{2M_1 d\Omega_\ell d\phi_2}{\left(2\pi\right)^5} dE_2 dE. \tag{4}$$

Similarly for the neutral process decay (NHSD), [2],

$$d\Gamma_{BN}(E,E_2) = \frac{\alpha}{\pi} d\Omega \left[\left(H_0' + N_0'\right)\theta_0^T + H_1'\theta_1^{nT} + \sum_{i=2}^{16}\left(H_i' + N_i'\right)\theta_i^T + N_{17}'\theta_{17}^{nT}\right], \tag{5}$$

the H_i''s, N_i''s and θ_i^T's are shown in the preceding references [1,2].
To obtain the electron energy spectrum, the following integration is required

$$d\Gamma_B\left(E\right) = \int_{M_2}^{E_2^{\min}} \frac{d\Gamma_B\left(E,E_2\right)}{dE}.$$

For both processes, the CHSD and NHSD, the $d\Gamma\left(E,E_2\right)$ depends on a parameter y_0 through the θ_1^T, and the θ_1^{nT}, in the following way

$$\theta_1^{T,nT} \propto \ln\left|\frac{y_0+1}{y_0-1}\right|, \qquad y_0 = \frac{\left(E_\nu^0\right)^2 - E^2\beta^2 - \left|\vec{p}_2\right|^2}{2\left|\vec{p}_2\right| E\beta}, \quad E_\nu^0 = M_1 - E_2 - E, \tag{6}$$

where y_0 is the cosine of the angle between the electron and the residual baryon when $\left|\vec{k}\right| = 0$. One finds that $y_0 \to 1$, for collinear events, case in which $E_2 = E_2^{\min}$. In this special situation one has to deal with a logarithmical divergence. This divergence is an obstacle to perform a direct numerical integration. This difficulty disappears when an analytical integration is performed.
The integral to be solved for both processes, according to Eqs. (4) and (5), is

$$d\Gamma_B^D\left(E\right) = \frac{\alpha}{\pi}\frac{G_v^2}{2}\frac{M_1}{2\pi^3} \sum_{k=0}^{2} \int_{M_2}^{E_2^{\min}} \varepsilon_k E_2^k \ln\left|\frac{y_0+1}{y_0-1}\right| dE_2. \tag{7}$$

To simplify, we consider the replacements:

$$a_0 = \frac{M_1^2 + M_2^2 + m^2 - 2M_1 E}{2M_2 \left|\vec{\ell}\right|} \quad \text{and } b_0 = \frac{(E - M_1)}{\beta E},$$

then one has to evaluate integrals of the form

$$R_k = \int_{z_b}^{z_t} z^k \ln \left| \frac{a_0 + b_0 z + \sqrt{z^2 - 1}}{a_0 + b_0 z - \sqrt{z^2 - 1}} \right| dz \quad \text{with} \quad z = \frac{E_2}{M_2}, \; z_t = \frac{E_2^{\min}}{M_2}, \; z_b = 1. \tag{8}$$

After performing a very subtle analysis [3], we obtain the following non-divergent analytical result

$$\begin{aligned} R_k = & -\frac{1}{k+1} \sum_{r=0}^{k} \left(z_s^r \sqrt{z_s^2 - 1} - z_t^r \sqrt{z_t^2 - 1} \right) \int_{z_b}^{z_t} \frac{z^{k-r}}{\sqrt{z^2 - 1}} dz \\ & - \frac{z_t^{k+1}}{k+1} \ln \left| z_t - b_0 \sqrt{z_t^2 - 1} \right| \mp \frac{z_s^{k+1}}{k+1} \ln \left| \frac{z_s z_t - 1 \mp \sqrt{z_s^2 - 1} \sqrt{z_t^2 - 1}}{z_s - z_t} \right| . \end{aligned} \tag{9}$$

Gathering and refining previous results, we obtain the whole spectrum of events in the four body region [4].

For the charged process $\Sigma^- \rightarrow n e \bar{\nu} \gamma$,

$$d\Gamma_{BC}(E) = d\Gamma_{BC}^{D}(E) + d\Gamma_{BC}^{ND}(E), \qquad \text{where} \tag{10}$$

$$d\Gamma_{BC}^{D}(E) = \frac{\alpha}{\pi} \frac{G_v^2}{2} \frac{M_1}{2\pi^3} \sum_{r=0}^{2} \varepsilon_r^C M_2^{r+1} R_r, \; , \qquad \text{and} \tag{11}$$

$$d\Gamma_{BC}^{ND}(E) = \frac{\alpha}{\pi} \frac{G_v^2}{2} \frac{M_1}{2\pi^3} \int_{M_2}^{E_2^{\min}} \left(H_0' \theta_0^T + \sum_{i=2}^{16} H_i' \theta_i^T \right) dE_2. \tag{12}$$

For the neutral process $\Lambda^0 \rightarrow p e \bar{\nu} \gamma$,

$$d\Gamma_{BN}(E) = d\Gamma_{BN}^{D}(E) + d\Gamma_{BN}^{ND}(E), \qquad \text{where} \tag{13}$$

$$d\Gamma_{BN}^{D}(E) = \frac{\alpha}{\pi} \frac{G_v^2}{2} \frac{M_1}{2\pi^3} \sum_{r=0}^{2} \varepsilon_r^N M_2^{r+1} R_r, \qquad \text{and} \tag{14}$$

$$\begin{aligned} d\Gamma_{BN}^{ND}(E) \; = & \frac{\alpha}{\pi} \frac{G_v^2}{2} \frac{M_1}{2\pi^3} \int_{M_2}^{E_2^{\min}} \Big[(H_0' + N_0') \, \theta_0^T + H_1' \theta_1^{nTND} \\ & + \sum_{i=2}^{16} (H_i' + N_i') \, \theta_i^T + N_{17}' \theta_{17}^{nT} \Big] \, dE_2. \end{aligned} \tag{15}$$

TABLE 1. $\Sigma^-(p_1) \rightarrow n(p_2) + e^-(\ell) + \bar{\nu}_e(p_\nu) + \gamma(k)$

$x = E/E_m$	0.1	0.2	0.3	0.4	0.5
% in Ref. [5]	7.8	1.5	0.5	0.1	0.02
% from Eq. (10)	7.776	1.533	0.458	0.135	0.019

TABLE 2. $\Lambda(p_1) \rightarrow p^+(p_2) + e^-(\ell) + \bar{\nu}_e(p_\nu) + \gamma(k)$

$x = E/E_m$	0.1	0.2	0.3	0.4	0.5
% in Ref. [5]	9.5	2.3	0.8	0.25	0.02
% from Eq. (13)	9.277	2.200	0.765	0.242	0.024

In Tables 1 and 2 we compare the data given in [5] for the radiative corrections in percents, caused by bremsstrahlung events in the four body region of the Dalitz plot, with the numerical values obtained from our analytical results.

The precise formulas are suitable to be evaluated numerically without any ambiguity, at any energy in which the charged lepton is emitted. They contain the bremsstrahlung in the four body decay region where the three body decay does not take place and they include events in which the electron is collinear to the produced hadron (at the edge of Dalitz plot). These results for unpolarized decays are of high precision, are model independent and are useful for processes where the momentum transfer is not small and therefore cannot be neglected. The analytical results are useful to obtain precise information, about the underlaying interactions in the decay processes and the internal structure of hadrons through the determination of the values of the form factors involved in the effective interaction. Let us mention that the radiative corrections were also computed by other authors using a Monte Carlo method [6].

We acknowledge the financial support from CONACYT (México). S.R.J.W. also thanks to "Comisión de Operación y Fomento de Actividades Académicas" (COFAA) from Instituto Politécnico Nacional.

REFERENCES

1. S.R. Juárez W., *Phys. Rev.* **D53**, 3746 (1996).
2. S.R. Juárez W., *Phys. Rev.* **D55**, 2889 (1997).
3. F. Guzmán A., M.S. Thesis, Escuela Superior de Física y Matemáticas, I.P.N. (unpublished), (1998).
4. S.R. Juárez W. and F. Guzmán A., in preparation (1999).
5. K. Tóth, K Szego and A. Margaritis, *Phys. Rev.* **D33**, 3306 (1986).
6. F. Glück and I. Joo, *Comp. Phys. Com.* **107**, 92 (1997).

New Properties of the Renormalization Group Equations of the Yukawa Couplings and the CKM Matrix[1]

S.R. Juárez[†2], P. Kielanowski* and G. Mora*

*Centro de Investigación y de Estudios Avanzados, Apdo. Postal 14 740, México 07000, DF, México.
[†]Escuela Superior de Física y Matemáticas, Instituto Politécnico Nacional, Edif. 9, Unidad Prof. Adolfo López Mateos, Col. Lindavista, C. P. 07738 México D. F., MEXICO.

Abstract. We solve the one-loop Renormalization Group Equations (RGE) and show that each of the Quark Yukawa Couplings (QYC) matrices for the up and down quarks depends on two functions of energy. The energy dependence is such that the diagonalizing matrices of the up QYC and the ratio m_c/m_u are constant while for the down QYC the diagonalizing matrices are energy dependent. From this it follows that the evolution of the CKM matrix depends on energy through one known function.

I YUKAWA COUPLINGS AND HIERARCHY

One of the most important problems of the Standard Model (SM) of the electroweak and strong interactions is the mass generation of the particles and the flavor mixing which have their origin in the Yukawa couplings

$$\sum_{i,j=1}^{3}\left(f_{ij}^{(e)}\bar{e}_L^i\phi e_R^j + y_{ij}^{(u)}\bar{u}_L^i\tilde{\phi}u_R^j + y_{ij}^{(d)}\bar{d}_L^i\phi d_R^j + h.c.\right). \tag{1}$$

Here, $f^{(e)}$, y_u and y_d are the Yukawa coupling matrices of leptons, up and down quarks to the scalar Higgs field ϕ, respectively. In this paper we consider only the quark sector.

The quark masses are the eigenvalues of the Yukawa couplings obtained after its diagonalization by the biunitary transformations $(U_{u,d})_{L,R}$

$$\mathrm{Diag}(m_u, m_c, m_t) = (U_u)_L y^{(u)} (U_u)_R^\dagger, \qquad \mathrm{Diag}(m_d, m_s, m_b) = (U_d)_L y^{(d)} (U_d)_R^\dagger. \tag{2}$$

---

[1] Supported by the Conacyt, México.
[2] Partially supported by the Comisión de Operación y Fomento de Actividades Académicas (COFAA) del Instituto Politécnico Nacional.

CP490, *Particles and Fields: Eighth Mexican School,* edited by J. C. D'Olivo et al.

Furthermore, from the diagonalizing matrices we obtain the flavor mixing in the charged current described by the familiar Cabibbo-Kobayashi-Maskawa (CKM) matrix

$$V_{ckm} = (U_u)_L (U_d)^{\dagger}_L. \tag{3}$$

The quark masses and the CKM matrix keep a close relation, and both show a hierarchical structure that can be expressed in terms of the λ parameter, $\lambda \equiv \sin\theta_c \sim 0.22$, by the following relations [1]:

$$\frac{m_u}{m_c} \sim \lambda^4, \quad \frac{m_c}{m_t} \sim \lambda^4, \quad \frac{m_d}{m_s} \sim \lambda^2, \quad \frac{m_s}{m_b} \sim \lambda^2, \quad \frac{m_b}{m_t} \sim \lambda^2, \tag{4}$$

and by the very useful Wolfenstein's parameterization [2] of the CKM matrix:

$$V_{ckm} = \begin{bmatrix} 1-\lambda^2/2 & \lambda & A\lambda^3(\rho - i\eta) \\ -\lambda & 1-\lambda^2/2 & A\lambda^2 \\ A\lambda^3(1-\rho-i\eta) & -A\lambda^2 & 1 \end{bmatrix}. \tag{5}$$

II RENORMALIZATION GROUP EQUATIONS

The RGE are an important tool for the search of the properties of the quark masses and the CKM matrix at different energy scales. The RGE have been worked out in earlier papers [3]. We consider the equations obtained by Grzadkowski et al., which after neglecting terms of order λ^4 and higher become:

$$\frac{d}{dt}y_u(t) = \left\{\alpha_1(t) + \alpha_2 Tr\left(y_u(t)y_u^{\dagger}(t)\right) + \alpha_4 y_u(t)y_u^{\dagger}(t)\right\} y_u(t) \tag{6}$$

$$\frac{d}{dt}y_d(t) = \left\{\beta_1(t) + \beta_2 Tr\left(y_u(t)y_u^{\dagger}(t)\right) + \beta_4 y_u(t)y_u^{\dagger}(t)\right\} y_d(t), \tag{7}$$

where $t \equiv \ln|\frac{\mu}{m_t}|$ is the energy scale parameter, the α_i and β_i are different parameters for the Standard Model (SM), the Minimal Standard Supersymmetric Model (MSSM) and the two Doublets Higgs Model (DHM) (see the Appendix).
Eqs. (6) and (7) form a system of coupled non-linear differential equations. Eq. (6) depends only on y_u and can be solved independently of Eq. (7). Once y_u is known from Eq. (6) it is plugged in Eq. (7) which becomes a linear equation for y_d. We solve Eq. (6) for the up sector in two steps. First we show that the diagonalizing biunitary matrices of the Yukawa couplings do not depend on the energy. We then use this fact to diagonalize and decouple the equations. The diagonalizing biunitary matrices of the up QYC, $(U_u)_{L,R}$ are unitary matrices that diagonalize the hermitian matrices $H_u^1 = y_u y_u^{\dagger}$ and $H_u^2 = y_u^{\dagger} y_u$. The differential equations for $H_u^{1,2}$ are:

$$\frac{d}{dt}H_u^i(t) = 2\left\{\alpha_1(t) + \alpha_2 Tr\left(H_u^i(t)\right) + \alpha_4 H_u^i\right\} H_u^i(t) \qquad i = 1,2. \tag{8}$$

$H_u^i(t)$ which is the solution of Eq. (8) can be written as a power series of $(t - t_0)$. Each coefficient of this series is a function of $H_u^i(t_0)$, so it can be diagonalized by the matrix that diagonalizes $H_u^i(t_0)$ since it is hermitian. It thus follows that $H_u^i(t)$ and $H_u^i(t_0)$ can be diagonalized by the same matrix, so the diagonalizing matrices are energy independent.

If $(U_u)_L$ and $(U_u)_R$ are the diagonalizing matrices of $H_u^1(t)$ and $H_u^2(t)$, respectively, then

$$y_u(t) = (U_u)_L^\dagger \Delta_u (U_u)_R, \tag{9}$$

with $\Delta_u = \mathrm{Diag}(m_u, m_c, m_t)$. Substituting this relation in Eq. (8) we get

$$\frac{d}{dt}\Delta_u = \left\{\alpha_1(t) + \alpha_2 Tr\left[\Delta_u^2\right] + \alpha_4 \Delta_u^2\right\}\Delta_u. \tag{10}$$

For the $u-$ and $c-$quarks we obtain

$$m_{u,c} = m_{u,c}(t_0) r_g^{1/2}(t) \exp\left(\alpha_2 \int_{t_0}^t m_t^2(\tau) d\tau\right), \tag{11}$$

with

$$r_g(t) = \exp\left(2\int_{t_0}^t \alpha_1(\tau) d\tau\right), \tag{12}$$

and for $m_t(t)$,

$$m_t(t) = \frac{m_t(t_0) r_g^{1/2}(t)}{\sqrt{1 - 2(\alpha_2 + \alpha_4) m_t^2(t_0) \int_{t_0}^t r_g(\tau) d\tau}}. \tag{13}$$

Thus, we have the complete solution for $y_u(t)$. Also we notice that from Eq. (11) it follows that m_u/m_c is energy independent.

Eq. (7) for the Yukawa couplings of the down sector can be transformed using the fact that the diagonalizing matrices of y_u do not depend on the energy. With the substitution,

$$y_d(t) = (U_u)_L^\dagger W_1, \tag{14}$$

we find that W_1 fulfills the following equation

$$\frac{dW_1}{dt} = \left\{\beta_1(t) + \beta_2 Tr\left[\Delta_u^2\right] + \beta_4 \Delta_u^2\right\} W_1. \tag{15}$$

The solution of this equation is:

$$W_1(t) = \left(r'_g(t)\right)^{1/2} \exp\left(\beta_2 \int_{t_0}^{t} m_t^2(\tau)d\tau\right) Z(t) \cdot W_1(0), \tag{16}$$

where

$$r'_g(t) = \exp\left(2 \int_{t_0}^{t} \beta_1(\tau)d\tau\right), \tag{17}$$

and

$$Z(t) = \begin{bmatrix} 1 & 0 & 0 \\ 0 & 1 & 0 \\ 0 & 0 & h(t) \end{bmatrix}, \qquad h(t) = \exp\left(\int \beta_4 m_t^2 dt\right). \tag{18}$$

Finally, the solution is:

$$y_d(t) = \left(r'_g(t)\right)^{1/2} \exp\left(\beta_2 \int_{t_0}^{t} m_t^2(\tau)d\tau\right) (U_u)_L^\dagger Z(t)(U_u)_L y_d(0). \tag{19}$$

In summary, we have shown the following facts for the RG evolution of QYC. Each of the $y_u(t)$ and $y_d(t)$ depends on two functions of the energy, as can be seen from Eqs. (11), (13) and (18), (20). The diagonalizing matrices of the up-QYC and the ratio m_u/m_c remain constant. Given the form [Eq. (20)] of the solution for $y_d(t)$ we conclude that the diagonalizing matrix $(U_d)_L$ depends only on one function of energy $h(t)$. From Eq. (3) it then follows that also the CKM matrix is a function of $h(t)$ only.

III APPENDIX

$$\alpha_1(t) = -\frac{1}{(4\pi)^2} G_u(t), \quad \alpha_2 = \frac{3}{(4\pi)^2}, \quad \alpha_4 = \frac{1}{(4\pi)^2}\frac{3b}{2}, \tag{20}$$

$$\beta_1(t) = -\frac{1}{(4\pi)^2} G_d(t), \quad \beta_2 = \frac{3}{(4\pi)^2} a, \quad \beta_4 = \frac{1}{(4\pi)^2}\frac{3c}{2}, \tag{21}$$

where $G_u(t)$ and $G_d(t)$ contain the coupling constants (see Ref. [3]) and are model dependent. (a, b, c) are equal to $(1, 1, -1)$, $(0, 2, 2/3)$ and $(0, 1, 1/3)$ in the SM, MSSM and DHM, respectively.

REFERENCES

1. H. González et al., Phys. Lett. **B 440** 94 (1998); H. González et al., *A New symmetry of Quark Yukawa Couplings*, page 755, International Europhysics Conference on High Energy Physics, Jerusalem 1997, Eds. Daniel Lellouch, Giora Mikenberg, Eliezer Rabinovici, Springer-Verlag 1999.
2. L. Wolfenstein,, Phys. Rev. Lett. **51** 1945 (1983).
3. B. Grzadkowski, M. Lindner and S. Theisen, Phys. Lett., **B 198** 64 (1987); M. Olechowski and S. Pokorski, Phys. Lett. **B 257** 388 (1991).

Making glue in high energy nuclear collisions

Alex Krasnitz* and Raju Venugopalan†

* *UCEH, Campus de Gambelas, Universidade do Algarve, Faro, P-8000, Portugal*[1]
† *Physics Department, BNL, Upton, NY 11973*[2]

Abstract.
We discuss a real time, non–perturbative computation of the transverse dynamics of gluon fields at central rapidities in very high energy nuclear collisions.

INTRODUCTION

This year (1999) the Relativistic Heavy Ion Collider (RHIC) at Brookhaven will begin colliding beams of gold ions at center of mass energies of 200 GeV/nucleon. In slightly over 5 years from now, the LHC collider at CERN will do the same at center of mass energies of about 5.5 TeV/nucleon. At these energies, the appropriate basis to describe the colliding nuclei is that of partons, the quarks and gluons that constitute a nucleon, rather than a hadronic basis. An objective of the above mentioned experiments is to determine whether the partons, confined in nucleons prior to the collision, are liberated after the collision to form fleetingly, in the relatively large nuclear volume, an equilibrated state of matter popularly known as the quark gluon plasma.

Clearly, the space–time evolution, and possible equilibration, of matter produced in a nuclear collision must depend on the initial conditions. These are given by the parton distributions inside each of the nuclei. In perturbative QCD, the "factorized" expression for the multiplicity or energy distribution of a high p_t jet, with $p_t \approx \sqrt{s}$, is obtained by convolving the parton distributions in each of the nuclei, at the hard scale of interest, with the elementary parton–parton cross section. If $x \equiv p_t/\sqrt{s}$ is not too small, the factorized expression is reliable. With the measured nuclear structure functions, one can then compute the multiplicity and energy distributions of the jets produced [1].

---

1) AK's work supported by Portuguese Fundação para Ciênca e a Technologia, grants CERN/S/FAE/1111/96 and CERN/P/FAE/1177/97

2) Invited talk by RV at the VIII Mexican School of Particles and Fields. RV's work supported by DOE Nuclear Theory at BNL.

CP490, *Particles and Fields: Eighth Mexican School,* edited by J. C. D'Olivo et al.

However, for $x << 1$ (corresponding to the transverse momentum range $\Lambda_{QCD} << p_t << \sqrt{s}$) the factorization formula for energy and multiplicity distributions breaks down. Simply put, this is because partons in one nucleus can resolve more than one parton in the other [2]. The parton densities in the nuclei become very large and may even saturate at sufficiently small x. The regime of high parton densities is a novel regime in QCD where, although the coupling constant may be small, the fields strengths are large enough for the physics to be non–perturbative [3].

The precise x value at which the above mentioned leading twist factorization breaks down is not clear. There are hints from from HERA that parton saturation is already seen in the data for $x \approx 10^{-4}$ and $Q^2 \approx 4$ GeV^2 [4]. If this result is robust, similar effects may be seen in nuclei at larger values of x, even $x \sim 10^{-2}$. Their relevance for RHIC (and especially LHC) cannot then be ignored.

The effects of high parton densities in the central rapidity region of nuclear collisions can be studied in a model which is based on an effective field theory (EFT) approach to QCD at small x [5]. The model describes the time evolution of gauge fields in a nuclear collision. It takes into account, self-consistently, interference effects (which are also responsible for shadowing in deeply inelastic scattering) that become important at small x. Another nice feature is that it provides a space–time picture of the nuclear collision. This feature would be extremely useful if the gauge fields at late times were to provide the initial conditions for a parton cascade [6] or for hydrodynamic evolution if it can be determined that the matter produced has equilibrated [7].

The model is formulated in the infinite momentum frame $P^+ \to \infty$ and light cone gauge $A^+ = 0$. It contains a dimensionful parameter μ^2, defined to be the color charge squared per unit area,

$$\mu^2 = \frac{A^{1/3}}{\pi r_0^2} \int_{x_0}^{1} dx \left(\frac{1}{2N_c} q(x, Q^2) + \frac{N_c}{N_c^2 - 1} g(x, Q^2) \right) . \tag{1}$$

Here q, g stand for the *nucleon* quark and gluon structure functions at the resolution scale Q of the physical process of interest. Also, above $x_0 = Q/\sqrt{s}$. Using the HERA structure function data, Gyulassy and McLerran [8] estimated that $\mu \leq 1$ GeV for LHC energies and $\mu \leq 0.5$ GeV at RHIC. Thus a window of applicability for weak coupling techniques does exist, and higher order calculations will tell us if it is smaller or larger than the simple classical estimate.

An interesting property of the light cone gauge is that final state interactions are absent! Kovchegov and Mueller [9] showed that the effects of final state interactions, as seen in a covariant gauge computation, are already contained in the nuclear wavefunction in light cone gauge. This non–trivial observation is at the heart of the approach described in this talk. Finally, we should alert the reader to alternative approaches to the one described here pursued in Refs. [10]– [11].

CLASSICAL MODEL OF GLUON PRODUCTION

At very high energies, the hard valence quark (and gluon) modes are highly Lorentz contracted, static sources of color charge for the wee parton, Weizsäcker–Williams, modes in the nuclei. The sources are described by the current

$$J^{\nu,a}(r_t) = \delta^{\nu+}\rho_1^a(r_t)\delta(x^-) + \delta^{\nu-}\rho_2^a(r_t)\delta(x^+)\,, \tag{2}$$

where $\rho_{1(2)}$ correspond to the color charge densities of the hard modes in nucleus 1 (nucleus 2) respectively. The classical field describing the small x modes in the EFT is obtained by solving the Yang–Mills equations in the presence of the two sources. We have then

$$D_\mu F^{\mu\nu} = J^\nu\,. \tag{3}$$

The small x glue distribution is simply related to the Fourier transform $A_i^a(k_t)$ of the solution to the above equation by $< A_i^a(k_t)A_i^a(k_t) >_\rho$.

The above averaging over the classical charge distributions is defined by

$$\langle O\rangle_\rho = \int d\rho_1 d\rho_2\, O(\rho_1,\rho_2)\exp\left(-\int d^2r_t \frac{\mathrm{Tr}\left[\rho_1^2(r_t)+\rho_2^2(r_t)\right]}{2g^4\mu^2}\right)\,. \tag{4}$$

We have assumed identical nuclei with equal Gaussian weights $g^4\mu^2$.

Before the nuclei collide ($t<0$), a solution of the equations of motion is

$$A^\pm = 0\,;\; A^i = \theta(x^-)\theta(-x^+)\alpha_1^i(r_t) + \theta(x^+)\theta(-x^-)\alpha_2(r_t)\,, \tag{5}$$

where $\alpha_q^i(r_t)$ ($q=1,2$ denote the labels of the nuclei and $i=1,2$ are the two transverse Lorentz indices) are *pure gauge* fields defined through the gauge transformation parameters $\Lambda_q(\eta,r_t)$ [8]

$$\alpha_q^i(r_t) = \frac{1}{i}\left(Pe^{-i\int_{\pm\eta_{\rm proj}}^0 d\eta'\Lambda_q(\eta',r_t)}\right)\nabla^i\left(Pe^{i\int_{\pm\eta_{\rm proj}}^0 d\eta'\Lambda_q(\eta',r_t)}\right)\,. \tag{6}$$

Here $\eta = \pm\eta_{\rm proj} \mp \log(x^\mp/x^\mp_{\rm proj})$ is the rapidity of the nucleus moving along the positive (negative) light cone with the gluon field $\alpha^i_{1(2)}$. The $\Lambda_q(\eta,r_t)$ are determined by the color charge distributions $\Delta_\perp\Lambda_q = \rho_q$ (q=1,2) with $\Delta_\perp$ being the Laplacian in the perpendicular plane.

For $t>0$ the solution is no longer pure gauge. Working in the Schwinger gauge $A^\tau \equiv x^+A^- + x^-A^+ = 0$, Kovner, McLerran and Weigert [12] found that with the ansatz

$$A^\pm = \pm x^\pm\alpha(\tau,r_t)\;;\; A^i = \alpha_\perp^i(\tau,r_t)\,, \tag{7}$$

where $\tau=\sqrt{2x^+x^-}$, Eq. 3 could be written more simply in terms of α and $\alpha_\perp$. Note that these fields are independent of η-the solutions are explicitly boost invariant in the forward light cone!

The initial conditions for the fields $\alpha(\tau, r_t)$ and $\alpha_\perp^i$ at $\tau = 0$ are obtained by matching the equations of motion (Eq. 3) at the point $x^\pm = 0$ and along the boundaries $x^+ = 0, x^- > 0$ and $x^- = 0, x^+ > 0$. Remarkably, for such singular sources, there exist a set of non–singular initial conditions for the smooth evolution of the classical fields in the forward light cone. One obtains

$$\alpha_\perp^i|_{\tau=0} = \alpha_1^i + \alpha_2^i \; ; \; \alpha|_{\tau=0} = \frac{i}{2}[\alpha_1^i, \alpha_2^i] \,. \tag{8}$$

Gyulassy and McLerran have shown [8] that even when the fields $\alpha_{1,2}^i$ before the collision are smeared out in rapidity, to properly account for singular contact terms in the equations of motion, the above boundary conditions remain unchanged. Further, since as mentioned the equations are very singular at $\tau = 0$, the only condition on the derivatives of the fields that would lead to regular solutions are $\partial_\tau \alpha|_{\tau=0}, \partial_\tau \alpha_\perp^i|_{\tau=0} = 0$.

Perturbative solutions of the Yang–Mills equations to order ρ^2 in the color charge density (or equivalently to second order in $\alpha_S \mu / k_t$) were found, and at late times, after averaging over the Gaussian sources, the number distribution of classical gluons was found to be [12,8,13,14]

$$\frac{dN}{dy d^2 k_t} = \pi R^2 \frac{2 g^6 \mu^4}{(2\pi)^4} \frac{N_c(N_c^2 - 1)}{k_t^4} L(k_t, \lambda) \,, \tag{9}$$

where $L(k_t, \lambda)$ is an infrared divergent function at the scale λ. This result agrees with the quantum bremsstrahlung formula of Gunion and Bertsch [15]. Also, Guo has shown that the above result is equivalent to the perturbative QCD factorized result for the process $qq \to qqg$ [16].

From the above expression, it is clear that distributions are very sensitive to $L(k_t, \lambda)$. Usually, as in Gunion and Bertsch, this divergence is absorbed in a non–perturbative form factor. What is novel about the classical approach is that, at sufficiently high energies, the non–linearities in the Yang–Mills fields self–consistently regulate this infrared divergence. To confirm this claim, one needs to solve the Yang–Mills equations to all orders in $\alpha_S \mu / k_t$. A non–perturbative solution of the Yang–Mills equations on a two dimensional transverse lattice was performed by us [17] and is described below.

REAL TIME SIMULATIONS OF YANG–MILLS I: LATTICE FORMULATION

We have seen above that the Yang–Mills equations are boost invariant. This is a consequence of the sources being δ–functions on the light cone– the nuclei are assumed to move with the speed of light! Since this is not the case, boost invariance is only approximate. It should, however, be a good assumption at the energies of interest–especially at central rapidities.

The boost invariance assumption simplifies our numerical work considerably. We now have a 2+1–dimensional theory and all the dynamics is restricted to the transverse plane (we assume $\eta = 0$). The Yang–Mills equations are most conveniently solved by fixing $A^\tau = 0$ gauge and solving Hamilton's equations. Gauge invariance is ensured by defining the theory, in the usual way, on a 2-dimensional transverse lattice. The lattice Hamiltonian is the Kogut-Susskind Hamiltonian for gauge fields coupled to an adjoint scalar

$$\begin{aligned} H_L &= \frac{1}{2\tau} \sum_{l\equiv(j,\hat{n})} E_l^a E_l^a + \tau \sum_{\Box} \left(1 - \frac{1}{2}\mathrm{Tr} U_\Box\right) , \\ &+ \frac{1}{4\tau} \sum_{j,\hat{n}} \mathrm{Tr} \left(\Phi_j - U_{j,\hat{n}} \Phi_{j+\hat{n}} U_{j,\hat{n}}^\dagger\right)^2 + \frac{\tau}{4} \sum_j \mathrm{Tr}\, p_j^2, \end{aligned} \tag{10}$$

where E_l are generators of right covariant derivatives on the group and $U_{j,\hat{n}}$ is a component of the usual SU(2) matrices corresponding to a link from the site j in the direction $\hat{n}$. The first two terms correspond to the contributions to the Hamiltonian from the chromoelectric and chromomagnetic field strengths respectively. Also, above $\Phi \equiv \Phi^a \sigma^a$ is the adjoint scalar field with its conjugate momentum $p \equiv p^a \sigma^a$.

Lattice equations of motion follow directly from H_L of Eq. 10. For any dynamical variable v, with no explicit time dependence, $\dot{v} = \{H_L, v\}$, where $\dot{v}$ is the derivative with respect to τ, and $\{\}$ denote Poisson brackets. We take E_l, U_l, p_j, and Φ_j as independent dynamical variables, whose only nonvanishing Poisson brackets are $\{p_i^a, \Phi_j^b\} = \delta_{ij}\delta_{ab}$; $\{E_l^a, U_m\} = -i\delta_{lm} U_l \sigma^a$; $\{E_l^a, E_m^b\} = 2\delta_{lm}\epsilon_{abc} E_l^c$ (no summing of repeated indices). The equations of motion are consistent with a set of local constraints (Gauss' laws). Their evolution in τ after the nuclear collision is determined by Hamilton's equations and their values at the initial time $\tau = 0$.

The initial conditions on the lattice are the constraints on the longitudinal gauge potential $A^\pm$ and the transverse link matrices $U_\perp$ at $\tau = 0$. The longitudinal gauge potentials are zero outside the light cone and satisfy the Schwinger gauge condition $A^\tau = 0$ inside the light cone $x_\pm > 0$. They can be written, as in the continuum case (see Eq. 7), as

$$A^\pm = \pm x^\pm \theta(x^+)\theta(x^-)\alpha(\tau, x_t) . \tag{11}$$

The transverse link matrices are, for each nucleus, pure gauges before the collision. This fact is reflected by writing

$$U_\perp = \theta(-x^+)\theta(-x^-)I + \theta(x^+)\theta(x^-)U(\tau) + \theta(-x^+)\theta(x^-)U_1 + \theta(x^+)\theta(-x^-)U_2 , \tag{12}$$

where $U_{1,2}$ are pure gauge.

The pure gauges are defined on the lattice as follows. To each lattice site j we assign two SU(N_c) matrices $V_{1,j}$ and $V_{2,j}$. Each of these two defines a pure gauge lattice gauge configuration with the link variables $U_{j,\hat{n}}^q = V_{q,j} V_{q,j+n}^\dagger$ where

$q = 1, 2$ labels the two nuclei. Also, as in the continuum, the gauge transformation matrices $V_{q,j}$ are determined by the color charge distribution $\rho_{q,j}$ of the nuclei, normally distributed with the standard deviation $g^4\mu_L^2$:

$$P[\rho_q] \propto \exp\left(-\frac{1}{2g^4\mu_L^2}\sum_j \rho_{q,j}^2\right). \tag{13}$$

Parametrizing $V_{q,j}$ as $\exp(i\Lambda_j^q)$ with Hermitean traceless Λ_j^q, we then obtain Λ_j^q by solving the lattice Poisson equation

$$\Delta_L \Lambda_j^q \equiv \sum_n \left(\Lambda_{j+n}^q + \Lambda_{j-n}^q - 2\Lambda_j^q\right) = \rho_{q,j}. \tag{14}$$

The correct continuum solution (Eqs. 5 and 7) for the transverse fields $A_\perp$ is recovered by taking the formal continuum limit of Eq. 12.

Using the general representation of the gauge fields in Eqs. 11 and 12, we shall now state the initial conditions for them at $\tau = 0$.

$$\begin{aligned} U_\perp|_{\tau=0} &= (U_1 + U_2)(U_1^\dagger + U_2^\dagger)^{-1} \; ; \; E_l|_{\tau=0} = 0 \, . \\ p_j|_{\tau=0} &= 2\alpha \; ; \; \Phi_j = 0 \, , \end{aligned} \tag{15}$$

where $U_\perp$ is defined as $\exp(ia_\perp\alpha_\perp)$. The above initial conditions are obtained by matching the lattice equations of motion in the four light cone regions at $\tau = 0$. For details we refer the reader to Ref. [17].

REAL TIME SIMULATIONS OF YANG–MILLS II: DISCUSSION OF RESULTS

In Ref. [17], we reported results of simulations of the time evolution of classical fields in a 2+1–dimensional SU(2) gauge theory described by the Hamiltonian in Eq. 10. The simulations were carried out on transverse lattices ranging from 20×20 sites to 160×160 sites. The lattice results depend on one dimensionless parameter, $g^2\mu L$ [18]. The parameter L corresponds to the size of the nucleus; μ defined in Eq. (1) is determined by the size of the nucleus, the energy of the nucleus, and the hard scale Q of interest; the strong coupling g runs as a function of either Q or μ depending on which is greater.

Thus, once given the energy and size of the incoming nuclei, and the hard scale of interest, we can determine the evolution of gauge fields in the central rapidity region. Consider, for instance, colliding two gold nuclei (A$\sim$ 200) at RHIC and LHC energies. Approximating $L^2 = \pi R^2$, we obtain $L = 11.6$ fm. For the hard scales of interest, $Q \approx 1$–2 GeV, $\mu \sim 0.5$ GeV at RHIC ($\mu \sim 1$ GeV at LHC). One then obtains

$$\begin{aligned} g^2\mu L &\approx 120 \;\; \text{(RHIC)} \\ &\approx 240 \;\; \text{(LHC)} \, . \end{aligned} \tag{16}$$

Above we have chosen $g = 2$ (or equivalently, $\alpha_S \approx 0.3$).

On the lattice, the lattice coupling is $g^2(\mu a)(L/a) \equiv g^2\mu_L N$. The continuum limit is obtained by keeping $g^2\mu_L N$ fixed (to the physical value of interest-as in Eq. 16) and taking μ_L to zero. It appears from our simulations that we are in the weak coupling regime for $\mu_L = 0.017, 0.035$ in lattice units. For the physical values of $g^2\mu L$ above, these would correspond to lattices an order of magnitude larger than those considered so far. Detailed simulations on the above physical scenario will be reported at a later date.

We now turn to an issue of some concern; whether quantities of interest have a continuum limit (in the above defined sense) in the classical theory. For instance, in thermal field theories, it is not clear that dynamical quantities such as auto–correlation functions have a well defined limit as the lattice spacing $a \to 0$. In the EFT described here, there is reason to be more optimistic.

Consider the following gauge invariant quantity; the energy density $p^a p^a = E_k/N^2$ of the scalar field on the lattice (in units of μ_L^4). It is plotted as a function of the lattice size N (in units of the lattice spacing) in Fig. 1 for $\mu_L = 0.0177, 0.035$.

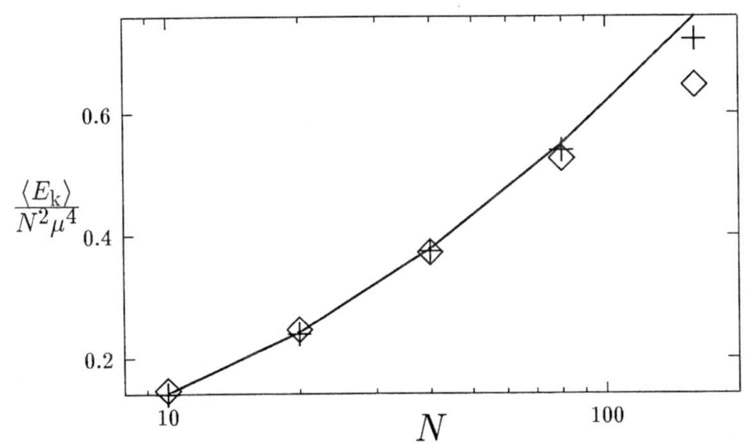

FIGURE 1. The lattice size dependence of the scalar kinetic energy density, expressed in units of μ^4 for $\mu = 0.0177$ (pluses) and $\mu = 0.035$ (diamonds). The solid line is the LPTh prediction. The error bars are smaller than the plotting symbols.

The solid line is the prediction from lattice perturbation theory (LPTh.). It is given by

$$p^a p^a = 6\left(\frac{\mu}{N}\right)^4 \sum_{n,n'}\left[\left(\sum_{\vec{k}} \frac{\sin(l_n)\sin(l_{n'})}{\Delta^2(l)}\right)^2 + 16\left(\sum_{\vec{k}} \frac{\sin^2(\frac{l_n}{2})\sin^2(\frac{l_{n'}}{2})}{\Delta^2(l)}\right)^2\right], \quad (17)$$

where $l_n = 2\pi n/N$ and $\Delta(l) = 2\sum_{n=1,2}(1-\cos(l_n))$ is the usual lattice Laplacian. The continuum limit of the above equation has the form $p^a p^a \longrightarrow A + B\log^2(L/a)$, where A and B are constants that can be determined from the above equation.

From Fig. 1, it can be seen that the kinetic energy density diverges as $\log^2(L/a)$– i.e., it does not have a well–defined continuum limit. This divergence is softer than in a thermal theory, where the energy density diverges as an inverse power of the lattice spacing. If we subtract the perturbative result in Eq. 17 at some scale $\Lambda_{nonpert}$, the resulting expression will have a continuum limit but will depend on $\Lambda_{nonpert}$. The trick is to find a $\Lambda_{nonpert}$ for which the contribution to the "non–perturbative" scalar kinetic energy density converges as $\mu_L \to 0$. In Table. 1, we show the results, at $\tau = 0$, of simulations where $g^2\mu L = 33.6$ is kept fixed (note: here we fix $g = 1$). The lattice cut–offs (recall that $k_t = 2\pi n/L$) of $n_{cut} = 10, 15, 20$ correspond to $\Lambda_{nonpert} = 1.08, 1.62, 2.16$ GeV respectively. For $n_{cut} =$

TABLE 1. Scalar k.e. density $\frac{E_k}{N^2\mu_L}$ vs μ_L for fixed $g^2\mu L = 33.6$. Columns 3–5 are the scalar.ke.d after subtracting LPTh. contribution for $n > n_{cut}$ in Eq. 17.

μ_L	$\frac{E_k}{N^2\mu^4}$	n_{cut}=10	n_{cut}=15	n_{cut}=20
.84	07.24e-02	.0341	.0562	.0702
.56	10.20e-02	.0407	.0636	.0791
.42	12.38e-02	.0426	.0694	.0852
.21	17.60e-02	.0268	.0702	.0945

15, the non–perturbative contribution to the scalar.ke.density appears to converge to a constant value as μ_L is decreased, keeping $g^2\mu L$ fixed. This result must of course be confirmed for larger lattices and other values of $g^2\mu L$. Caveat aside, our results appear to suggest that for a particular $\Lambda_{nonpert}$ there is a non–perturbative contribution to the energy density that survives in the continuum limit.

Presumably, the subtraction described above may also be performed for static quantities in a thermal field theory. It is unlikely that this procedure there is reliable for dynamic quantities. In our case, the scalar kinetic energy is a dynamic quantity evolving in time. If the above mentioned procedure is to be use, it must also be valid at later times. We are optimistic that this may be the case because our simulations suggest that the time evolution of hard modes decouples from that of the soft modes. This is illustrated very clearly by Fig. 2 of the most recent of our papers cited in Ref. [17].

The parameter $\Lambda_{nonpert}$ may be given a physical interpretation. At sufficiently high energies, one may be able to relate it to the physical scale at which one begins to see deviations from perturbative predictions (note: at very high energies one expects $\Lambda_{nonpert} >> \Lambda_{QCD}$) due to high parton density effects.

We are currently studying whether the non–perturbative contribution to the scalar energy density also has a robust continuum limit at late times [19].

REFERENCES

1. K. Kajantie, P. V. Landshoff, and J. Lindfors, *Phys. Rev. Lett.* **59** (1987) 2527; K. J. Eskola, K. Kajantie, and J. Lindfors, *Nucl. Phys.* **323** (1989) 37; J.-P. Blaizot and A. H. Mueller, *Nucl. Phys.* **B289** (1987)847; K. J. Eskola, *Comments in Nucl. and Part. Phys.* **22** (1998) 185.
2. R. Venugopalan, *Comments in Nucl. and Part. Physics*, **22** (1998) 113.
3. L. V. Gribov, E. M. Levin, and M. G. Ryskin, *Phys. Repts.* 1 (1983); A. H. Mueller and J. Qiu, *Nucl. Phys.* **B268** 427 (1986).
4. A. H. Mueller, hep-ph/9904404, hep-ph/9902302.
5. L. McLerran and R. Venugopalan, *Phys. Rev.* **D49** 2233 (1994); **D49** 3352 (1994); **50** 2225 (1994);J. Jalilian–Marian, A. Kovner, L. McLerran and H. Weigert, *Phys. Rev.* **D55** 5414 (1997); J. Jalilian–Marian, A. Kovner, A. Leonidov and H. Weigert, *Nucl. Phys.* **B504** (1997) 415; *Phys. Rev.* **D59** (1999) 034007; J. Jalilian–Marian, A. Kovner and H. Weigert, *Phys. Rev.* **D59** (1999) 014015.
6. K.. Geiger, *Phys.Rep.* **258** 237 (1995); X.-N. Wang, *Phys. Rep.* **280** 287 (1997).
7. J. D. Bjorken, *Phys.Rev.* **D27** 140 (1983); J. Sollfrank, P. Houvinen, M. Kataja, P. V. Ruuskanen, M. Prakash and R. Venugopalan, *Phys. Rev.* **C55** 392 (1997).
8. M. Gyulassy and L. McLerran, *Phys. Rev.* **C56** (1997) 2219.
9. Y. Kovchegov and A. H. Mueller *Nucl. Phys.* **B529** 451 (1998).
10. S. A. Bass, B. Müller, and W. Poschl, nucl-th/9808011; B. Müller and W. Poschl, nucl-th/9808031, nucl-th/9812066.
11. A. Makhlin and E. Surdutovich, *Phys. Rev.* **C58** 389 (1998).
12. A. Kovner, L. McLerran and H. Weigert, *Phys. Rev* **D52** 3809 (1995); **D52** 6231 (1995).
13. Y. V. Kovchegov and D. H. Rischke, *Phys. Rev.* **C56** (1997) 1084.
14. S. G. Matinyan, B. Müller and D. H. Rischke, *Phys. Rev.* **C56** (1997) 2191; *Phys. Rev.* **C57** (1998) 1927.
15. J. F. Gunion and G. Bertsch, *Phys. Rev.* **D25** 746 (1982).
16. X. Guo, *Phys. Rev.* **D59** 094017 (1999).
17. A. Krasnitz and R. Venugopalan, hep-ph/9706329, hep-ph/9808332, hep-ph/9809433.
18. R. V. Gavai and R. Venugopalan, *Phys. Rev.* **D54** 5795 (1996).
19. A. Krasnitz and R. Venugopalan, in progress.

The doubly compactified Schwinger model

R. Linares, L.F. Urrutia and J.D. Vergara

Departamento de Física de Altas Energías
Instituto de Ciencias Nucleares
Universidad Nacional Autónoma de México
Circuito Exterior, C.U., 04510 México, D.F.

Abstract. In this note we summarize the exact solution of the doubly compactified Schwinger model (CSM), defined by the condition that the domain of the electromagnetic degree of freedom $c = \frac{1}{L}\int_0^L dx A_1$ is $-\bar{c} < c < +\bar{c}$. The results are contrasted with the standard situation, where $-\infty < c < +\infty$, which we call the non-compact case (NCSM). Both theories are also compactified in a circle of length L for the space variable x.

INTRODUCTION

The motivation for having a compact domain in the electromagnetic variable c arises from the loop-space formulation of gauge theories [1], applied to this simple case [2]. In fact, the natural variable in this formulation is the holonomy $T = \exp ie \oint A_1(z)dz = \exp iecL$, which exhibits the property that the choice of c in the range $-\bar{c} < c < \bar{c}$, with $\bar{c} = \frac{\pi}{eL}$, is sufficient to describe the physics of the problem. As we will s how in the following, the properties of the compactified model are different from those arising in the non-compact case. Nevertheless, the loop space program is also able to encompass the latter results by *enlarging* the corresponding loop Hilbert space [3]. Among the numerous papers on the NCSM [4], we heavily relay on the work of Iso and Murayama [5]. A detailed discussion of the CSM can be found in Ref. [6].

QUANTIZATION AND SOLUTION

The Schwinger model is QED in $1+1$ dimensions. In a standard notation the Lagrangian is

$$\mathcal{L} = -\frac{1}{4}F_{\mu\nu}F^{\mu\nu} + \bar{\psi}\gamma^{\mu}\left(i\partial_{\mu} - eA_{\mu}\right)\psi, \tag{1}$$

CP490, *Particles and Fields: Eighth Mexican School,* edited by J. C. D'Olivo et al.

where we adopt periodic (antiperiodic) boundary conditions for the fields: A_μ (ψ). The above Lagrangian is invariant under the gauge transformations $\psi \rightarrow e^{ie\alpha(x,t)}\psi$, $A_\mu \rightarrow A_\mu - \partial_\mu \alpha(x,t)$. There are two families of gauge transformations: (i) those continuously connected to the identity, called small gauge transformations and characterized by the function $\alpha = b(t)e^{i2\pi nx/L}$. (ii) the second family corresponds to the so called large gauge transformations, which are generated by the non-periodic functions $\alpha = \frac{2\pi n}{eL}x$, $n = \pm1, \pm2, \ldots,$. The compactification condition upon c means that two values of c differing by $\frac{2\pi}{eL}$ must be identified. In this way c is invariant under both types of gauge transformations, as opposed to the non-compact case where $c \rightarrow c + \frac{2\pi}{eL}$ under large gauge transformations.

Some questions naturally arise now: (i) is the compactified model exactly solvable?, (ii) does it make any further difference whether or not we compactify the range of the remaining modes of $A(x)$? and (iii) in which way, if any, the imposed compactification propagates to the remaining degrees of freedom of the problem ?.

The quantization proceeds in a way completely similar to the standard case. We work in the Coulomb gauge $\frac{\partial A_1}{\partial x} = 0$, which means that $c = A_1(t)$. On the other hand, A_0 is given by the charge density via the instantaneous Coulomb interaction. The implementation of the zero-mode of the Gauss law, as a constraint determining the physical sates, implies that these must be states of zero electric charge.

The first step in the quantization is the construction of the fermionic Fock space in an external electromagnetic field. Here a main difference arises, because in the CSM it is possible to introduce the conserved and *gauge invariant* axial charge $\bar{Q}_5 = \sum_{n=-\infty}^{+\infty}\left(a_n^\dagger a_n - b_n^\dagger b_n\right) + \frac{ecL}{\pi}$, with eigenvalues $2N$, $N = 0, \pm1, \ldots,$. The one-particle energy eigenvalues of h_F,where $\mathcal{H}_F = \psi^\dagger h_F \psi$ is the fermionic Hamiltonian density, are $\epsilon_n = \frac{2\pi}{L}\left(n + \frac{1}{2} - \frac{eL}{2\pi}c\right) \equiv \frac{2\pi n}{L} + \mathcal{C}$, which are fully gauge invariant in virtue of the compactification process. Here, a_n and b_n denote generically the fermionic annihilation operators. Under both type of gauge transformations, they change as $a_n \rightarrow e^{ie\alpha(0)}a_n, b_n \rightarrow e^{ie\alpha(0)}b_n$. This establishes another difference with respect to the NCSM. In the latter situation, large gauge transformations lead to $a_n \rightarrow a_{n+1}, b_n \rightarrow b_{n+1}$.

States of minimum energy are Dirac-type vacuums with all negative energy levels filled, given by

$$|\mathcal{E}_N, 2N\rangle = \prod_{n=-\infty}^{N-1} a_n^\dagger|0\rangle \otimes \prod_{m=N}^{\infty} b_m^\dagger|0\rangle, \tag{2}$$

each having energy $\mathcal{E}_N(c) = \frac{2\pi}{L}\left\{\left(N - \frac{ecL}{2\pi}\right)^2 - \frac{1}{12}\right\}$ and axial charge $2N$. Our compactification condition implies that the state with $N = 0$ is the true vacuum at this stage. The excited states are constructed by using the following current operators

$$\psi_1{}^\dagger(x)\psi_1(x) = \frac{1}{L}\sum_{n=-\infty}^{+\infty} e^{-\frac{2\pi in}{L}x}\, j_+{}^n, \quad \psi_2{}^\dagger(x)\psi_2(x) = \frac{1}{L}\sum_{n=-\infty}^{+\infty} e^{+\frac{2\pi in}{L}x}\, j_-{}^n. \tag{3}$$

In terms of the corresponding zero modes, we have $Q = j_+{}^0 + j_-{}^0$, $\bar{Q}_5 = j_+{}^0 - j_-{}^0 + \frac{ecL}{\pi}$. The operators $j_\pm{}^n, n \geq 1$ annihilate the states (2). Finally, the fermionic Fock space in the background electromagnetic field will consist of all the local vacuums (2), together with all possible states constructed from them by the application of an arbitrary number of the current operators $(j_\pm{}^n)^\dagger, n = 1, 2, \ldots$, defined above.

The regularized current algebra on this fermionic Fock space is given by [5]

$$[j_+{}^n, (j_+{}^m)^\dagger] = n\delta_{m,n}, \qquad [j_-{}^n, (j_-{}^m)^\dagger] = n\delta_{m,n}, \qquad [j_+{}^n, j_-{}^m] = 0, \tag{4}$$

and corresponds to the *anomalous* Schwinger commutators [7]. In this Fock space, the fermionic Hamiltonian in the external field can be written as

$$H_F = \mathcal{E}_N(c) + \frac{2\pi}{L} \sum_{n>0} ((j_+^n)^\dagger j_+^n + (j_-^n)^\dagger j_-^n). \tag{5}$$

Next we include the quantization of the electromagnetic Fourier modes A_m, E_m, $m = 0, \pm 1, \ldots,$. The commutator $[E(x), \psi_\alpha(y)] = 0$ leads to

$$[E_m, a_n] = \frac{ie}{2\pi m} (a_n - a_{n+m}), \quad m \neq 0, \qquad [E_0, a_n] = 0, \tag{6}$$

together with an analogous relation for the b's. The above commutators arise because the expansion of the fermionic fields in Fourier modes makes use of the wave functions in the electromagnetic field, instead of pure plane waves. The remaining commutators are the expected ones. The required commutation relations are satisfied by choosing

$$E_m = \frac{1}{iL}\frac{\partial}{\partial A_m} - \frac{e}{2\pi i m}\left(j_+{}^m + (j_-{}^m)^\dagger\right), m \neq 0, \quad E_0 = \frac{1}{iL}\frac{\partial}{\partial c}. \tag{7}$$

The solution of Eq.(6) leads to the following dependence of the fermionic operators upon the non-zero electromagnetic modes

$$a_m = \exp\left(-\frac{eL}{2\pi}\sum_{k\neq 0}\frac{1}{k}A_k\right)\bar{a}_m, \tag{8}$$

where $\bar{a}_m$ are new fermionic operators which are independent of the gauge field A_k. With the realization (7), the non-zero modes of the Gauss law reduce to $\mathcal{G}_m = \frac{2\pi m}{L}\frac{\partial}{\partial A_m} \approx 0, m \neq 0$. Then we conclude that the wave functions of the system must be of zero electric charge and also independent of the modes A_m, $m \neq 0$. Our expression for the Gauss law is somewhat different from the one obtained in [5], though the final results are equivalent.

In this way we conclude that any further compactification in the electromagnetic modes A_m is irrelevant and also that the compactification of c does not have any further effect in the remaining fermionic degrees of freedom of the model.

The next step is to write the full Hamiltonian making use of the Gauss law. This allows to rewrite the operators $E_m, m \neq 0$, in terms of the currents introduced in (3). The resulting Hamiltonian is subsequently diagonalized by a Bogoliubo v transformation [5], leading to

$$H = \frac{\pi}{2L}\left(\bar{Q}_5 - \frac{ecL}{\pi}\right)^2 - \frac{1}{2L}\left(\frac{\partial}{\partial c}\right)^2 + \sum_{n>0} \frac{E_n}{n}\left((\tilde{j}_+^n)^\dagger \tilde{j}_+^n + (\tilde{j}_-^n)^\dagger \tilde{j}_-^n\right), \qquad (9)$$

up to an infinite constant. Here $E_n = \sqrt{\left(\frac{2\pi n}{L}\right)^2 + M^2}$, where $M = \frac{e}{\sqrt{\pi}}$. The Bogoliubov rotation is a unitary transformation U_B, which commutes with the electric and axial charge operators. We will denote by an overall tilde the rotated operators and states. To construct the Hilbert space of the full theory we will start from the states $|\widetilde{N}\rangle = U_B\,|\mathcal{E}_N, 2N\rangle$, arising from Eq. (2).

As in the non-compact case, each mode decouples in such a way that the Schroedinger equation $H\Delta = E\Delta$ is solved by $\Delta = \Pi_n \Delta_n$, where $H_n\ \Delta_n = E_n\ \Delta_n$ and $E = \sum E_n$. The general strategy to construct the Hilbert space will be to start from the zero modes $F_N(c) \times |\widetilde{N}\rangle$ and to subsequently apply all possible combinations of the raising operators $(\tilde{j}_\pm{}^m)^\dagger$.

The zero modes correspond to the choice $|\text{ground}; N\rangle = F_N(c) \times |\widetilde{N}\rangle$ and the wave functions $F_N(c)$ satisfy the following Schroedinger equation

$$\left(-\frac{1}{2L}\left(\frac{\partial}{\partial c}\right)^2 + \frac{e^2L}{2\pi}\left(\frac{2\pi N}{eL} - c\right)^2\right) F_N(c) = E_{N,0} F_N(c), \qquad (10)$$

subjected to the boundary conditions

$$F_N|_{c=-\bar{c}} = F_N|_{c=+\bar{c}}, \qquad \frac{\partial F_N}{\partial c}|_{c=-\bar{c}} = \frac{\partial F_N}{\partial c}|_{c=+\bar{c}}, \qquad (11)$$

arising from the compactification procedure. These should be contrasted with those of Manton, written in Eqs.(3.15) of Ref. [8]. The above Schroedinger equation (10) together with the boundary conditions (11) need to be solved numerically and lead to energies which are not any more given by the characteristic equally spaced harmonic oscillator spectrum, as it is the case in the NCSM. In general, the above energies and wave functions will have an additional label α to enumerate them. ¿From the numerical calculation we find that the absolute minimum value of the zero-modes energy correspond to the ground state of the $N = 0$ case. Thus, in the CSM the physical, non-degenerated, vacuum of the theory is $|\text{ground}; 0, 0\rangle = F_{0,0}(c) \times |\tilde{0}\rangle$.

The excited states are obtained by applying the creation operators $(\tilde{j}_\pm{}^m)^\dagger$ to each zero mode $|\text{ground}; N, \alpha\rangle$. Each individual action raises the energy by E_m. The excited states will be labeled by $|N,\ \alpha,\ N_1,\ \ldots, N_k, \ldots\rangle$, where N_k is the total number of times that the operators $(\tilde{j}_\pm{}^k)^\dagger$ have been applied to the corresponding zero mode. This is the occupation number of the k-level.

The fact that $\tilde{Q}_5 = \bar{Q}_5$ is conserved in the full Hilbert space of the model is a direct consequence of the way in which the Hilbert space has been constructed and can be proved accordingly.

FINAL COMMENTS

By providing the explicit construction of the Hilbert space of the model, we have shown the the CSM is exactly solvable, in analogy with the standard case. The CSM still possesses a chiral anomaly since the current $\bar{J}_5^\mu = J_5^\mu + \frac{e}{\pi}\epsilon^{\mu\nu}A_\nu, \epsilon^{01} = 1$, is conserved but not gauge invariant. Nevertheless, due to the compactification condition, the associated charge $\bar{Q}_5$ is invariant under small and large gauge transformations, thus providing an additional quantum number to label the Hamiltonian eigenstates. The model exhibits a non-degenerated vacuum. No theta-vacuums need to be introduced since gauge invariance is automatically preserved. As in the standard case, the Hamilto nian decouples in non-interacting modes. The zero modes are characterized by fermionic excitations just filling the negative energy states, together with electromagnetic states corresponding to a harmonic oscillator potential with periodic boundary conditions. In this way, the resulting spectra is not any more equally spaced and cannot be described as a superposit ion of quanta with a given mass. Our boundary conditions (11) are not continuously connected to those of Manton [8], which implies that the CSM does not have to reproduce the NCSM in the case $eL \to 0$.

Acknowledgments: Partial support from CONACyT grant 3141-PE to LFU and JDV is acknowledged. R.L., LFU and JDV are also supported by the grant DGAPA-UNAM-IN100397.

REFERENCES

1. R. Gambini and J. Pullin, *Loops, knots, gauge theories and quantum gravity* (Cambridge: Gambridge Univ. Press 1996).
2. R. Gambini, H. Morales, L. F. Urrutia and J.D. Vergara, Phys. Rev. (1998).
3. H. Fort and R. Gambini, *The U(1) and strong CP problems from the loop formulation perspective*, hep-th/9711174, nov. 1997.
4. J. Schwinger, Phys. Rev. **125**, 397 (1962); *ibid* **128**,2425 (1962). For a recent review see for example C. Adam, *Anomaly and Topological aspects of two-dimensional quantum electrodynamics*, Dissertation, Universitat Wien, october 1993. C. Adam, R. A. Bertlmann and P. Hofer, Riv. Nuovo Cim.**16**(1993)No 8.
5. S. Iso and H. Murayama, Prog. Theo. Phys. **84**,142 (1990).
6. R. Linares, L.F. Urrutia and J.D. Vergara, *The Hamiltonian solution of the doubly compactified Schwinger model*, in preparation.
7. J. Schwinger, Phys. Rev. Letts. **3**, 296 (1959).
8. N.S. Manton, Ann. Phys. **159**, 220 (1985).

Magnetic dipole moment of vector mesons

G. López Castro and G. Toledo Sánchez

Departamento de Física, Cinvestav del IPN,
Apartado Postal 14-740, México 07000 D.F., México

Abstract. We analyze the sensitivity to the vector-meson magnetic dipole moment of radiative processes involving the production and decay of vector mesons. These studies assume that vector mesons are stable particles. We then discuss how to incorporate the finite-width effects in the calculations without spoiling the electromagnetic gauge invariance of the scattering amplitudes.

The electromagnetic vertex of charged vector mesons (VM) have not been measured yet. In this paper we explore the possibility that the magnetic dipole moment (MDM) of light VM can be extracted from the effects produced in radiative processes that involve the production and decay of these particles. We shall first consider the approximation where vector mesons are stable particles. Then, we will discuss the whole radiative process involving the production *and* decay of the unstable vector meson and will propose a mechanism to obtain a gauge-invariant amplitude when the finite width is included.

Since vector mesons are quark-antiquark bound states, one expects that a measurement of their electromagnetic structure can provide useful information about the underlying QCD dynamics. As is well known, the comparison of theory and experiment for the magnetic dipole moment of the electron provides the golden precision test of the standard model [1]. In a similar way, forthcoming more accurate measurements of the magnetic dipole moment and electric quadrupole moment (EQM) of the $W^{\pm}$ bosons at LEP II and Tevatron colliders are expected to provide a test of the gauge sector of the electroweak theory. Finally, it is also well known that measurements of the MDM and EQM of the deuteron, confirms the picture that this particle can be viewed as a tightly bound state of a neutron and a proton. Following a similar reasoning, the measurement of the MDM or EQM of vector mesons can serve as useful tests for phenomenological quark models [2].

Vector mesons, as massive gauge bosons, are highly unstable particles (lifetimes of $O(10^{-24})$ sec). Therefore, neither electron–vector-meson elastic scattering nor spin precession techniques [3] are useful to measure their MDM's or EQM's. The only possibility to extract their properties is the study of radiative processes involving

CP490, *Particles and Fields: Eighth Mexican School,* edited by J. C. D'Olivo et al.

the production and/or decay of vector mesons [4]. The basic idea is that the photon emitted off the charged vector meson carries information on the MDM and the EQM.

The first assumption we make when considering the production *or* decay of vector mesons is that they are stable particles. Then we have to isolate the contribution of the MDM from other potential contributions as model-dependent effects, bremsstrahlung or the EQM. These additional contributions are small in the decay rate of $V^+ \to P^+P^0\gamma$ decays [5] (V (P) a vector (pseudoscalar meson)). Similarly, they can be rendered negligible in the energy and angular distribution of photons by looking at small angle emission of photons with respect to the charged particle in the final state [6,7].

In units of $e/2m_V$, the MDM of the vector meson can be written as $\mu = 1 + \kappa$, where $\kappa = 1$ can be considered the *canonical* value [8] and m_V is the mass of the VM. The magnitude of the ρ (K^*) MDM has been estimated using phenomenological quark models, with the result $1 \leq \mu \leq 3.7$ ($1.0 \leq \mu \leq 3.7$) [2].

Let us first consider the effects of the MDM in the *decays* of light vector mesons using as an example the decay $\rho^\pm \to \pi^\pm\pi^0\gamma$ (the procedure for the decays $K^{*\pm} \to K^\pm\pi^0\gamma, K^0\pi^\pm\gamma$ and $D^{*\pm} \to D^\pm\pi^0\gamma, D^0\pi^\pm\gamma$ is completely analogous [6]). We choose these decay modes of VM's because the corresponding non-radiative processes completely dominate the decay rates. The energy and angular distribution of photons can be written as [9] $P(\omega, \theta) = A\sin^2\theta/\omega^2 + B(\kappa, \theta)\omega^0 + C$, where ω denotes the photon energy in the rest frame of the VM and θ is the photon angle with respect to the P^+ meson. The function $B(\kappa, \omega)$ also contains model-dependent and EQM contributions, while C denotes the non-leading (model-dependent) terms in ω [6,7]. Observe that if we choose small values of θ, the contribution of order ω^0 can become the dominant term. It also happens that the model-dependent and EQM contributions in $B(\kappa, \theta)$ are also suppressed for this special kinematical configuration [6]. In Figure 1 we plot the $P(\omega, \theta = 10^0)$ distribution as a function of the dimensionless variable x ($x \equiv 2\omega/m_V$), for the $\rho^+ \to \pi^+\pi^0\gamma$ decay. The meaning of the different curves is explained in the Figure caption. Notice that in the vertical axis $P(\omega, \theta)$ has been represented by the double differential decay rate normalized to the non-radiative rate.

As we observe, the small angle distribution of photons is sensitive to the different values of the ρ^+ MDM κ in the intermediate photon energy region ($0.45 \leq x \leq 0.65$). As discussed in Ref. [6], a measurement of the angular and energy distribution of photons in this region with an accuracy of 25 %, would furnish a determination of the MDM within $\Delta\kappa \approx \pm 0.5$.

Now we focus on the decay $\tau^- \to \rho^-\nu_\tau\gamma$, which is an interesting example of a *production* process [7] (a similar analysis can be done for the production of a K^{*-} vector meson). In Figure 2 we plot the energy an angular distribution of photons for this decay. Now, we define the dimensionless variable $x = 2\omega/m_\tau$ where ω is the photon energy in the τ rest frame and θ is the angle of th e photon measured with respect to the direction of the ρ meson three-momentum. In these plots, the small angle is fixed to $\theta = 10^0$, which is a configuration that suppresses photon

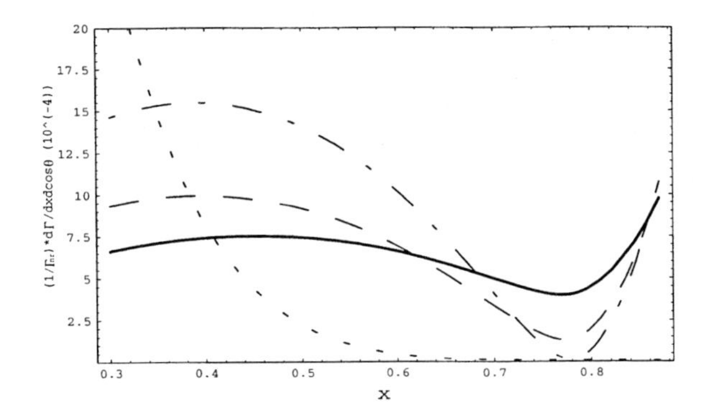

FIGURE 1. Photon energy distribution for $\theta = 10^0$ in $\rho^+ \to \pi^+\pi^0\gamma$. The short-dashed plot corresponds to the term of order ω^{-2}. The solid, long-dashed and long-short-dashed plots are the terms of order ω^0 when $\kappa = 0, 1$ and 2, respectively.

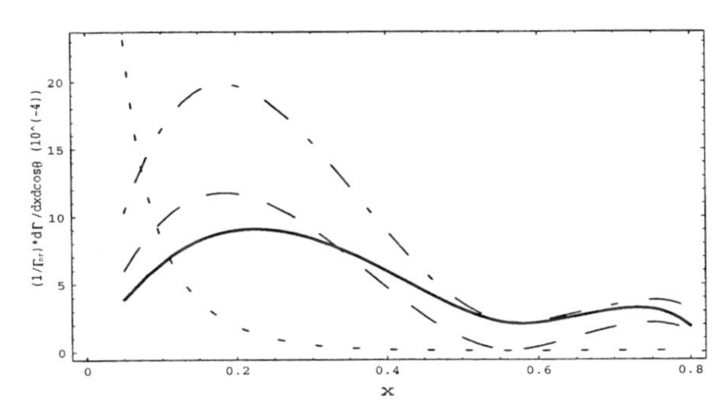

FIGURE 2. Photon energy distribution for $\theta = 10^0$ in $\tau^- \to \rho^- \nu_\tau \gamma$. The description of different plots is identical to Figure 1.

emission from the electric charges of τ and ρ and off the ρ EQM. In this case, the τ is assumed to have a *normal* MDM ($\kappa_\tau = 1$). As in the previous case, the intermediate photon energy region $0.15 \leq x \leq 0.5$ is more sensitive to the effects of the anomalous $\rho^\pm$ MDM and could provided a measurement of this property with good accuracy [7]. Notice also that these plots exhibit a dip at the end of the spectrum. This corresponds to the kinematical configuration where the vector meson is produced at rest and it is defined by the relation $x_{\rm dip} = 1 - \sqrt{m_V/m_\tau}$.

In the two examples considered above, the vector meson is assumed to be stable. However, unstable particles can not be described by asymptotic states in S-matrix amplitudes. Actually, unstable particles can appear only as intermediate states, as for example the vector meson V in the decay chain $\tau^- \to V^-\nu\gamma \to P^+P'^0\nu\gamma$. In general, unstable particles can not be detached from its production and decay mechanisms and be described by truncated amplitudes as in the previous examples. Therefore, a resonant propagator must be used to describe the finite propagation of the particle from its production to its decay vertex locations. However, this procedure must be done in a consistent way in order to preserve the electromagnetic gauge invariance of the full S-matrix amplitude [10].

As it has been stressed in Refs. [10], the electromagnetic gauge-invariance of the whole radiative amplitude can be preserved if the propagators and electromagnetic vertices of unstable particles are chosen so that the corresponding Ward identities remain valid order by order in perturbation theory. According to Refs. [10], the violation of the electromagnetic gauge invariance induced by the finite width effects can be cured by resumming the absorptive pieces of one loop corrections to the propagator and electromagnetic vertex of the unstable particles. In the case of the $W^\pm$ gauge bosons, it can be easily checked that only the fermion loop corrections generate an absorptive part for low virtualities ($q^2 \leq m_W^2$) of the charged gauge bosons. In a similar way, it can be proved that, for the relevant virtualities (off-shellness) of the vector mesons in the $\tau^- \to P^-P'^0\nu_\tau\gamma$ decays, only the loop corrections involving two-pseudoscalar mesons contribute to the absorptive corrections. If we consider massless pseudoscalar mesons (Goldstone bosons) in loop corrections, the propagator and electromagnetic vertex of the charged unstable vector meson can be written as:

$$D^{\mu\nu}(q) = \frac{-i}{q^2 - m_V^2 + iq^2\frac{\Gamma_V}{m_V}} \left[g^{\mu\nu} - \frac{q^\mu q^\nu}{m_V^2}\left(1 + \frac{i\Gamma_V}{m_V}\right)\right] \tag{1}$$

and

$$-ie\Gamma^{\alpha\beta\mu} = ie\left(1 + \frac{i\Gamma_V}{m_V}\right)\Gamma_0^{\alpha\beta\mu} \tag{2}$$

where Γ_V denotes the decay width of the vector meson and $\Gamma_0^{\alpha\beta\mu}$ corresponds to the tree-level electromagnetic vertex of the vector meson.

As it can be easily verified, Eqs. (1) and (2) satisfy the electromagnetic Ward identity. Therefore, using these Feynman rules in the evaluation of the different contributions to the whole decay process $\tau^- \to V^-\nu_\tau\gamma \to P^-P'^0\nu_\tau\gamma$ automatically furnish a gauge-invariant amplitude. At present, we are following this scheme in order to obtain information about the MDM of vector mesons from the whole production *and* decay process.

Finally let us comment that although the theoretical calculations presented in this paper indicate that the double differential distribution of photons is sensitive to the effects of the MDM, it remains the difficulty associated with the experimental feasibility to reconstruct small angle configuration of photons.

REFERENCES

1. Particle Data Group, Barnett R. M. et. al., *Phys. Rev. D* **54**,1 (1996).
2. Hecht M. B., and Mackellar B.H.J., *Phys. Rev. C.* **57**, 2638 (1998); Hawes, F. T. and Pichowsky, M. A. , e-print nucl-th/9806025.
3. Bargmann V., Michel L. and Teledgi V. L., *Phys. Rev. Lett.* **2**, 433 (1959).
4. Zakharov V. I., Kondratyuk L. A. and Ponomarev L. A. *Sov. J. Nucl. Phys.* **8**, 456(1969).
5. Bramon A., Díaz Cruz J. L. and López Castro G. *Phys. Rev. D* **47**, 5181 (1993).
6. López Castro G. and Toledo Sánchez G., *Phys. Rev. D* **56**, 4408 (1997).
7. López Castro G. and Toledo Sánchez G., *Vector-meson magnetic dipole moment effects in radiative τ decays.* To appear in *Phys. Rev. D.*
8. Ferrara S., Porrati M. and Telegdi V. L., *Phys. Rev. D* **46**, 3529 (1992).
9. Burnett T. H. and Kroll N. M. *Phys. Rev. Lett.* **20** 86(1968).
10. López Castro, G., Lucio M., J. L. and Pestieau, J., *Mod. Phys. Lett. A* **6**, 3679 (1991); Baur U. and Zeppenfeld D. *Phys. Rev. Lett.* **75**, 1002 (1995); Beuthe M., Gonzalez Felipe R., López Castro G. and Pestieau J. *Nucl. Phys. B* **498**, 55 (1997); Beenakker, W. et al., *Nucl. Phys. B* **500**, 255 (1997).

The decay $t \to bWZ$ in models with extended Higgs sector

D. A. López Falcón[1] and J. L. Díaz Cruz

Instituto de Física "Luis Rivera Terrazas", Benemérita Universidad Autónoma de Puebla, Apartado Postal J-48, 72570 Puebla, Pue., México.

Abstract. We study the contribution of charged Higgs boson to the rare decay of the quark top $t \to bWZ$ in models with extended Higgs sector that include doblets and triplets. Higgs doublets are needed to couple charged Higgs with quarks, whereas the Higgs triplet component is used to obtaing a vertex H^+W^-Z. It is found that this decay mode has a large branching ratio that may allow to test the custodial $SU(2)_c$ symmetry.

From a pure phenomenological point of view, the decay $t \to bWZ$ is interesting because it can provide a test for the possible existence of physiscs beyond the standard model (SM), which so far seems to agree quite well with experiment.On theoretical ground, this decay is interesting because it could indicate that a Higgs boson lies in a higher representation, beyond the doublet Higgs of the SM [1], which may or not satisfy a custodial symmetry, therefore, it is also a useful test for this symmetry.

The value of the BR($t \to bWZ$) predicted in SM is [2] 5.4 x 10^{-7} and then this decay is well beyond the sensitivity of Tevatron Run II or even LHC. Indeed, the rare decays of top quark are usually very suppressed [3]; thus its observation would imply new physics. In this paper, we present the results for BR($t \to bWZ$) obtained using first a general amplitude to describe the contribution of an intermediate charged Higgs to the decay, allowing some constants to describe the precise value of couplings, then, we improve this results, considering a model with two doublets and one triplet, which fix the values of the constants in the amplitude.

The amplitude in general is

$$\mathcal{M} = [\overline{u}(p_b)(a + b\gamma_5)u(p_t)]\,(\frac{-i}{p_H^2 - \hat{m}_H^2})\,[A g_{\mu\nu}\epsilon_W^{*\mu}\epsilon_Z^{*\nu}] \tag{1}$$

where a, b and A are the constants previously mentioned and $\hat{m}_H \equiv m_H + \frac{i}{2}\Gamma_H$. The masses of top quark, the bottom quark, the charged Higgs, the W boson and the Z boson are denoted by m_t, m_b, m_H, m_W and m_Z, respectively.

---

[1] Supported by a fellow of the Consejo Nacional de Ciencia y Tecnología (México)

CP490, *Particles and Fields: Eighth Mexican School,* edited by J. C. D'Olivo et al.

In order to calculate the partial width of this decay, we shall perform a numerical integration for the expression of the squared amplitude, over the three-body phase space, namely:

$$\Gamma(t \to bWZ) = \frac{1}{(2\pi)^3} \frac{1}{32m_t^3} \int | \overline{\mathcal{M}} |^2 \, ds \, dt \tag{2}$$

where $| \overline{\mathcal{M}} |^2$ denotes the squared amplitude, averaged over spins and summed over polarizations, that has the general form:

$$| \overline{\mathcal{M}} |^2 = \frac{| A |^2 \, (a^2 + b^2)(m_t^2 + m_b^2 - s)}{(s - m_H^2)^2 + m_H^2 \Gamma_H^2} \left[2 + \left(\frac{s - m_W^2 - m_Z^2}{2m_W m_Z} \right)^2 \right] \tag{3}$$

The integration limits are written as

$$(m_W + m_Z)^2 \leq s \leq (m_t - m_b)^2 \tag{4}$$

and

$$t^- \leq t \leq t^+ \tag{5}$$

where

$$t^{\pm} = m_t^2 + m_Z^2 - \frac{1}{2s}[(s + m_t^2 - m_b^2)(s + m_Z^2 - m_W^2)$$
$$\mp \lambda^{1/2}(s, m_t^2, m_b^2) \lambda^{1/2}(s, m_Z^2, m_W^2)] \tag{6}$$

and $\lambda(x, y, z) = (x + y - z)^2 - 4xy$.

We have considered that the branching ratio of this decay is given by the ratio of the Eq. (2) to the total width, which includes the SM dominant decay, whose expression is given by [4]

$$\Gamma(t \to bW^+) = \frac{G_F m_t^3}{8\pi\sqrt{2}} \left(1 - \frac{m_W^2}{m_t^2} \right)^2 \left(1 + 2\frac{m_W^2}{m_t^2} \right) \left[1 - \frac{2\alpha_s}{3\pi} \left(\frac{2\pi^2}{3} - \frac{5}{2} \right) \right] \tag{7}$$

with $\alpha_s = 0.118$, and the decay mode $t \to bH$ (whenever is kinematically allowed), whose expression is given by [5]

$$\Gamma(t \to bH^+) = \frac{g^2}{128\pi} \frac{V_{tb}}{m_W^2 m_t} [\Sigma_{tb} + \Delta_{tb}] \, \lambda^{1/2} \left(1, \frac{m_b^2}{m_t^2}, \frac{m_H^2}{m_t^2} \right) \tag{8}$$

where

$$\Sigma_{tb} = (m_b \tan\beta + m_t \cot\beta)^2 \left[(m_t^2 + m_b^2) - m_W^2 \right]$$

and

$$\Delta_{tb} = (m_b \tan\beta - m_t \cot\beta)^2 \left[(m_t^2 - m_b^2) - m_W^2 \right]$$

. Moreover, we assume that the Higgs width is dominated by the modes $H \to c\bar{s}$ and $H \to \tau\nu_\tau$, therefore [5]

$$\Gamma_H = \frac{g^2 m_H}{32\pi m_W^2} \left[3(m_c^2 \cot^2\beta + m_s^2 \tan^2\beta) + m_\tau^2 \tan^2\beta \right]. \tag{9}$$

In the so-called first approximation, the constants in the amplitude are taken as:

$$(a^2 + b^2) = 1 \qquad \text{and} \qquad | A |^2 = g^4 m_Z^2$$

for this case we get the branching ratio shown in Figure 1.

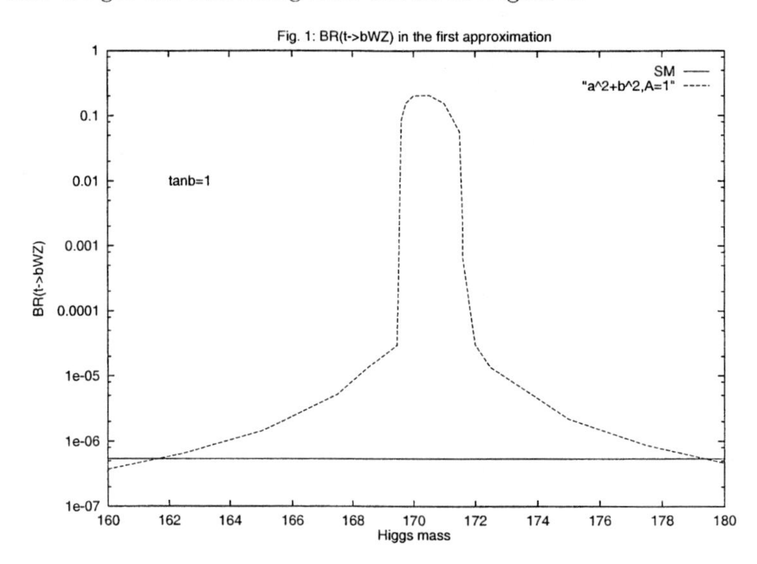

Now, we improve this results, by considering a particular model with two doublets and one triplet. To get a sizeable coupling tbH, we can use the two-doublet model, which gives for the vertex tbH, the expression [6]

$$\frac{ig}{2\sqrt{2}m_W}\left[(m_t \cot\beta + m_b \tan\beta) + (m_t \cot\beta - m_b \tan\beta)\gamma_5\right] \tag{10}$$

where $\tan\beta$ is the ratio of the vacuum expectation values of the two scalar doublets. Moreover, we require a representation higher than the doublet, to get a sizeable coupling HWZ, for simplicity let us consider a triplet, for which the coupling will be [1]

$$-\frac{igm_W}{\cos\theta_w} g_{\mu\nu}. \tag{11}$$

Finally, we parametrize the effect of mixing between the doublets and triplet representation with an angle defined by

$$\begin{pmatrix} D^\pm \\ T^\pm \end{pmatrix} = \begin{pmatrix} \cos\alpha & \sin\alpha \\ -\sin\alpha & \cos\alpha \end{pmatrix} \begin{pmatrix} G^\pm \\ H^\pm \end{pmatrix}. \tag{12}$$

Therefore, the previous couplings are modified by a factor $\sin\alpha$ and $\cos\alpha$ respectively. Therefore, the constants in the amplitude take the values

$$a = m_t \cot\beta + m_b \tan\beta, \qquad b = m_t \cot\beta - m_b \tan\beta \qquad \text{and} \qquad A = \frac{g^2 \sin\alpha \cos\alpha}{2\sqrt{2}\cos\theta_w}$$

We find that the Yukawa coupling Eq. (10) is the same for both values $\tan\beta = 1$ and $\tan\beta = m_t/m_b$, which are the limits implied by GUT [7], so we only use the first of this values here, on the other hand, the limits on the factor $\sin\alpha\cos\alpha$, which is part of the constant A, range from 0 to $\frac{1}{2}$.

With all this considerations, we shown in the Figure 2, the branching ratio with two values for $\sin\alpha\cos\alpha$, the first one (equal to $\frac{1}{2}$) is the minimum suppresion or the bigger $SU(2)_c$ breaking, and the last one (equal to 0.1) can be considered a sizeable suppresion or smaller $SU(2)_c$ breaking.

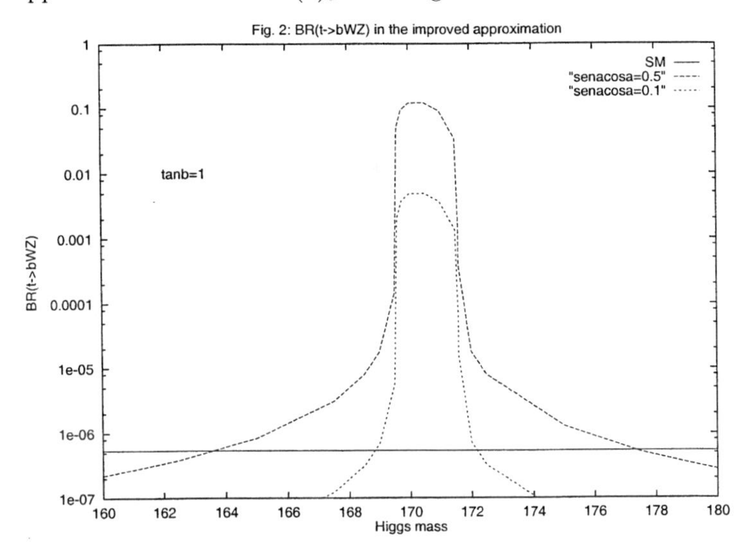

We conclude from Figures 1 and 2 that there exist a relatively small region in which it is factible to observe this decay. From $m_H = 166$ to $m_H = 176$ in the best case ($\sin\alpha\cos\alpha = 1/2$), and from $m_H = 169$ to $m_H = 172$ in the worst ($\sin\alpha\cos\alpha = 0.1$). On the other hand, because of the generality in the calculation of $BR(t \to bWZ)$, this decay mode of the quark top is a useful test of custodial $SU(2)_c$ symmetry.

REFERENCES

1. Cheung, K., Phillips, R. J. N. and Pilaftsis, A., *Phys. Rev. D* **51**, 4731 (1995).
2. Mahlon, G., "Theoretical Expectations in Radiative Top Decays", presented at the Thinkshop on Top-Quark Physics for Run II, held at the FermiLab, October 16-18, 1998.
3. Atwood, D. and Sher, M., *Phys. Lett. B* **411**, 306 (1997).
4. Caso, C. *et al., Eur. Phys. J. C* **3**, 1 (1998).
5. Barger, V. and Phillips, R. J. N., *Phys. Rev. D* **41**, 884 (1990).
6. Gunion, J.F., Haber, H. E., Kane, G., and Dawson, S., *"The Higgs Hunter's Guide"*, Addison-Wesley, 1990, ch. 4, pp. 203.
7. Bagger, V., Dimopoulos, S. and Masso, E., *Phys. Rev. Lett.* **55**, 920 (1985).

Constraint on the magnetic moment of the top quark

R. Martínez, J.-Alexis Rodríguez and M. Vargas

Depto. de Física, Universidad Nacional, Bogotá, Colombia

Abstract. We derive a bound on the magnetic dipole moment of the top quark in the context of the effective Lagrangian approach by using the ratios $R_b = \Gamma_b/\Gamma_h$, $R_l = \Gamma_h/\Gamma_l$ and the Z width. We take into account the vertex and oblique corrections. We obtain the allowed region for the magnetic dipole moment of top quark as $-2.94 \leq \delta\kappa \leq 1.9$.

The most recent analyses of precision measurements at the Large Electron Positron (LEP) collider lead to the conclusion that the predictions of the Standard Model (SM) of electroweak interactions, based on the gauge group $SU(2)_L \otimes U(1)_Y$ are in excellent agreement with the experimental results. Recently the discovery of the top quark has been announced by the Collider Detector at Fermilab (CDF) and D0 collaborations [1]. The direct measurement of the top quark mass m_t is in agreement with the indirect estimates derived by confronting the SM m_t dependent higher order corrections with the LEP and other experimental results.

The aim of the present work is to extract indirect information on the magnetic dipole moment of the top quark from LEP data, specifically we use the ratios R_b and R_l defined by

$$R_b = \frac{\Gamma(Z \to b\bar{b})}{\Gamma(Z \to hadron)}, \qquad R_l = \frac{\Gamma(Z \to hadron)}{\Gamma(Z \to l\bar{l})} \tag{1}$$

and the Z width, in the context of an effective Lagrangian approach. The oblique and QCD corrections to the b quark and hadronic Z decay widths cancel off in the ratio R_b. This property makes R_b very sensitive to direct corrections to the $Zb\bar{b}$ vertex, specially those involving the heavy top quark [2], while Γ_Z and R_l are more sensitive to oblique corrections.

In order to define an effective Lagrangian it is necessary to specify the symmetry and the particle content of the low-energy theory. In our case, we require the effective Lagrangian to be CP-conserving, invariant under SM symmetry $SU(2)_L \otimes U(1)_Y$, and to have as fundamental fields the same ones appearing in the SM spectrum. In the present work, we consider the following dimension six and CP-conserving operators,

CP490, *Particles and Fields: Eighth Mexican School,* edited by J. C. D'Olivo et al.

$$O_{uW}^{ab} = \bar{Q}_L^a \sigma^{\mu\nu} W_{\mu\nu}^i \tau^i \tilde{\phi} U_R^b \, , \qquad O_{uB}^{ab} = \bar{Q}_L^a \sigma^{\mu\nu} Y B_{\mu\nu} \tilde{\phi} U_R^b \, , \tag{2}$$

where Q_L^a is the quark isodoublet, U_R^b is the up quark isosinglet, a , b are the family indices, $B_{\mu\nu}$ and $W_{\mu\nu}$ are the $U(1)_Y$ and $SU(2)_L$ field strengths, respectively, and $\tilde{\phi} = i\tau_2\phi^*$. We use the notation introduced by Buchmüller and Wyler [3]. In the case of the operators O_{uB}^{ab} and O_{uW}^{ab}, some degrees of family mixing is made explicit (corresponding to $a \neq b$) without breaking SM gauge invariance. After spontaneous symmetry breaking, these fermionic operators generate also effective vertices proportional to the anomalous magnetic moments of quarks. The above operators for the third family give rise to the anomalous $t\bar{t}\gamma$ vertex and the unknown coefficients ϵ_{uB}^{ab} and ϵ_{uW}^{ab} are related respectively with the anomalous magnetic moment of the top quark through

$$\delta\kappa_t = -\sqrt{2}\frac{m_t}{m_W}\frac{g}{eQ_t}(s_W \epsilon_{uW}^{33} + c_W \epsilon_{uB}^{33}) \, , \tag{3}$$

where s_W denotes the sine of the weak mixing angle.

The expression for R_b is given by

$$R_b = R_b^{SM}(1 + (1 - R_b^{SM})\delta_b) \, , \tag{4}$$

where R_b^{SM} is the value predicted by the SM and δ_b is the factor which contains the new physics contribution, and it is defined as follows

$$\delta_b = \frac{2\left(g_V^{SM} g_V^{NP} + g_A^{SM} g_A^{NP}\right) + \left(g_V^{NP}\right)^2 + \left(g_A^{NP}\right)^2}{(g_V^{SM})^2 + (g_A^{SM})^2} \tag{5}$$

and g_V^{SM} and g_A^{SM} are the vector and axial vector couplings of the $Zb\bar{b}$ vertex. The contributions from new physics, eq. (2), to R_l and Γ_Z are of two classes. One from vertex correction to $Zb\bar{b}$ in the Γ_{hadr} and the other from the oblique correction through $\Delta\kappa$ in the $\sin^2\theta_W$. These can be written as

$$\begin{aligned} R_l &= R_l^{SM}(1 - 0.1851\ \Delta\kappa + 0.2157\ \delta_b) \, , \\ R_b &= R_b^{SM}(1 - 0.03\ \Delta\kappa + 0.7843\ \delta_b) \, , \\ \Gamma_Z &= \Gamma_Z^{SM}(1 - 0.2351\ \Delta\kappa + 0.1506\ \delta_b) \end{aligned} \tag{6}$$

where $\Delta\rho$ is equal to zero for the operators that we are considering.

Now we have various posibilities to explore the space of the parameters ε_{uW}^{33}, ε_{uB}^{33} and $\delta\kappa_t$, which are related by eq.(3). Further we have that the parameters ε_{uW}^{33}, ε_{uB}^{33} are involved with the measured quantities $Q_i = (\Gamma_Z, R_b, R_l)$ trought the eq.(6) and they form a surface in the respective space $(\varepsilon_{uW}^{33}, \varepsilon_{uB}^{33}, Q_i)$. Thus the experimental planes for the quantities Q_i cut the surface defined in that space and it is the allowed region in the plane $\varepsilon_{uW}^{33} - \varepsilon_{uB}^{33}$. If we do not neglect the term of the order $\left(g^{NP}\right)^2$ in eq.(7) we get the following expressions

$$-0.057\varepsilon_B + 0.058\varepsilon_B^2 + 0.053\varepsilon_W + 0.0015\varepsilon_W^2 + 0.01\varepsilon_B\varepsilon_W = 1 - (\Gamma_Z^{\exp}/\Gamma_Z^{SM}),$$
$$-0.023\varepsilon_B + 0.026\varepsilon_B^2 + 0.016\varepsilon_W + 0.0007\varepsilon_W^2 + 0.0048\varepsilon_B\varepsilon_W = 1 - (R_b^{\exp}/R_b^{SM}),$$
$$-0.656\varepsilon_B + 0.686\varepsilon_B^2 + 0.536\varepsilon_W + 0.018\varepsilon_W^2 + 0.1264\varepsilon_B\varepsilon_W = 1 - (R_l^{\exp}/R_l^{SM}),$$

where we have omitted the superindex 33, subindex u and each expression define two planes, with the upper and lower experimental limit respectively. The SM values for the parameters that we have used are $\Gamma_Z = 2.4972$ GeV, $R_l = 20.747$, $R_b = 0.2157$, $\Gamma_{hadr} = 1743.4$ MeV and $\Gamma_l = 84.03$ MeV; with the input parameters: $m_t = 175$ GeV, $\alpha_s(m_Z) = 0.118$, $m_Z = 91.1861$ GeV, $m_H = 100$ GeV and $\Lambda = 1$ TeV. And the experimental values are $\Gamma_Z = 2.4946 \pm 0.0027$ GeV, $R_l = 20.778 \pm 0.029$, $R_b = 0.2178 \pm 0.0011$.

In figure 1 we plot these curves in the plane ε_{uW}^{33}– ε_{uB}^{33} corresponding to the cut with the experimental values of Γ_Z. We are plotting in figures 2 and 3 the same as figure 1 for R_b and R_l respectively. In this kind of scenarios new physics is explored assuming that its effects are smaller than the SM effects, consequently one expect that $\left|g_{V,A}^{NP}/g_{V,A}^{SM}\right| \ll 1$ and, then we get from $g_{V(A)}^{SM}$ and the expressions (10), the inequalities $|\varepsilon_{uW}^{33}| \leq 0.11$, $|\varepsilon_{uB}^{33}| \leq 0.48$, which are plotted in figures 1, 2 and 3 as vertical and horizontal lines, respectively. The allowed region in the plane ε_{uW}^{33}– ε_{uB}^{33} in the figures 1, 2 and 3 is figure out as the intersected area between the curves and the limits of the above equations.

We try to get other constraints to the parameter ε_{uW}^{33} using other measured processes. For instance, we calculate the contribution to the top quark decay $t \to bW$ but with the available measurements by CDF and D0 of $B(t \to bW)$, even is not possible to get a better constraint for ε_{uW}^{33}, obtaining the value $|\varepsilon_{uW}^{33}| \leq 0.7$. We also inspect the system $B^0 - \overline{B}^0$ where we get a contribution identical to zero for the operators under consideration.

On the other hand, the allowed regions given in figure 2, 3 and 4 give maximal and minimal bounds for the the parameters ε_{uW}^{33} and ε_{uB}^{33}. By using the eq. (4) and the bounds showed in the figures, we obtain for $\delta\kappa_t$ the following values $-2.94 \leq \delta\kappa_t \leq 1.3$, $-0.76 \leq \delta\kappa_t \leq 1.9$ and $-1.3 \leq \delta\kappa_t \leq 1.7$, which correspond to Γ_Z, R_b and R_l, respectively. Therefore for these observables Γ_Z, R_b and R_l, we get the allowed region -2.94$\leq \delta\kappa \leq$ 1.9.

We thank COLCIENCIAS for financial support.

REFERENCES

1. CDF Collaboration, F. Abe et. al., Phys. Rev. Lett. **74**, 2626 (1995); D0 Collaboration, S. Abachi et. al., Phys. Rev. Lett. **74**, 2632 (1995).
2. S. Mrenna and C.- P. Yuan, Phys. Lett. **B 367**, 188 (1996); E. Ma, Phys. Rev. **D 53**, 2276 (1996); P. Bammert, et. al., Phys. Rev. **D 54**, 4275 (1996).
3. W. Buchmüller and D. Wyler, Nucl. Phys. **B 268**, 621 (1986); Phys. Lett. **B 197**, 379 (1987).
4. G. Altarelli, hep-ph/9611239.

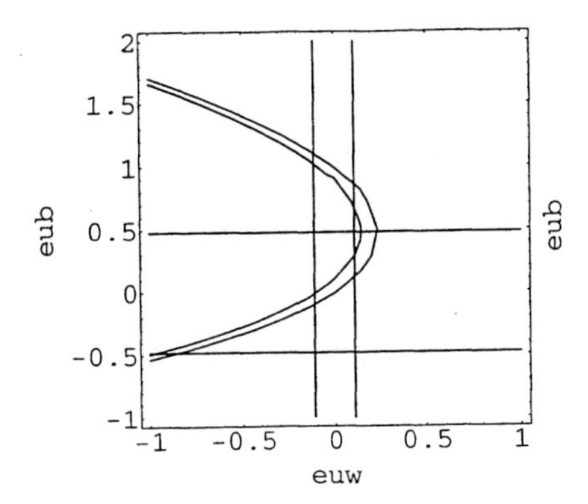

Figure 1. The projection on the plane $\varepsilon^{33}_{uW} - \varepsilon^{33}_{UB}$ of the first expression of eq.(10), *i.e.* the cut with the experimental values of Γ_Z . The horizontal and vertical lines are the constraints from eq.(11).

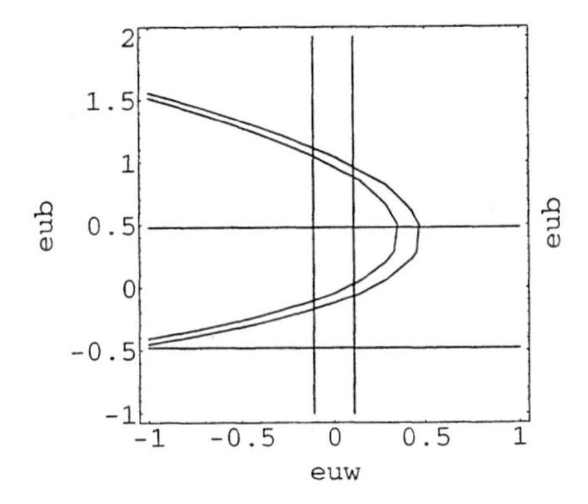

Figure 2. The same as figure 2 for the ratio R_b.

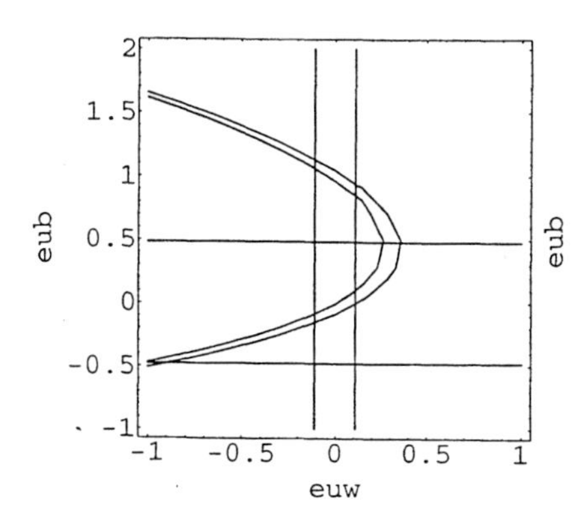

Figure 3.Same as fig. 2 for R_l.

From Superstrings Theory to the Dark Matter in Galaxies

Tonatiuh Matos[1]

Departamento de Física,
Centro de Investigación y de Estudios Avanzados del IPN,
PO. Box 14-740, 07000 México D. F., MEXICO

Abstract. Starting from the effective action of the low energy limit of superstrings theory, I find an exact solution of the field equations which geodesics behavie exactly as the trajectories of stars arround of a spiral galaxy. Here dark matter is of dilatonic origin. It is remarkable that the energy density of this space-time is the same as the used by astronomers to model galaxy stability. Some remarks about a universe dominated by dilatons are pointed out.

Till some years ago it was believed that in our era, matter is made of leptons, quarks and gauge bosons. Only theories beyond the Standard Model predict other exotic particles which can exist at very higher energies, maybe near of the origin of the universe. But the discovery of the existence of a great amount of dark matter in galaxies and galaxy clusters could change our building of how is matter constructed, furthermore it could be possible that we do not know of what 90% of the matter of the universe is made. Let me explain in some lines this affirmation. Since the discovery of Zwiky and Smith of the necessity of a great amount of wanting matter in the Coma and Virgo clusters in 1933 [1] in order that these clusters remain stable, the astronomers have discovered that a great amount of luminous matter is absent in the galaxies in order to understand their stability and age (for a better explanation of the dark matter problem in galaxies, see the G. Raffelt contribution in this volume). Astronomers have discovered even a greater amount of wanting luminous matter in most of the galaxy clusters, since these clusters have also shown to be very stable. In terms of the reason $\Omega_x = \rho_x/\rho_{crit}$ between a x matter contribution and the critical density ρ_{crit} which closes the universe, we can express the contribution of the luminous matter in the universe by $\Omega_{obs} = (0.003 \pm 0.002)/h$ (see for example [2]), which depends on the value of the Hubble constant h in units of 100 km/sec/ Mpc. If we consider the matter needed in the halos of the galaxies in order to conserve their stability, the mass density of the universe is $\Omega_{halos} \sim 0.05$, and considering the matter needed in order to have stability of the galaxies clusters,

1) e-mail:tmatos@fis.cinvestav.mx

CP490, *Particles and Fields: Eighth Mexican School,* edited by J. C. D'Olivo et al.

the density of the universe grows to $\Omega_{clos} = 0.25 \pm 0.10$. These two last densities do not depend on the value of the Hubble constant. Neutrinos can contribute to the total density of the universe, nevertheless due to their recently discovered low mass, their contribution cannot be much grater than the luminous one (see the contribution of R. Peccei in this volume).

Our actual understanding of the universe is sustained by the Standard Model of cosmology, namely the Freedman-Robertson-Weaker (FRW) cosmological model. The predictions of the FRW model is supported for important observations; the universe expansion, the microwave background and the observations in the early elements composition in the era of nucleosynthesis. All these three predictions are supported for a extraordinary coincidences with observations, therefore it could be very difficult to construct another cosmological model with so nice features. Remarkable is the fact that the theoretical predictions of nucleosynthesis do not permit a great amount of baryonic matter. If the value of the Hubble constant $h = 1$, the permitted values for the baryonic density implies $0.06 < \Omega_{baryon} < 0.02$. If the Hubble constant $h = 0.4$ these values could increase to $0.05 < \Omega_{baryon} < 0.12$ (see for example [2]). In any case this limits do not permit sufficient baryons for explaining the needed matter in clusters. **This fact implies that there must exist exotic matter in the universe**, *i.e.*, there is a great amount of non-baryonic matter in the cosmos and we do not know its nature. There are many hypothesis about the nature of this exotic matter. In this lines I want to explain one of this hypothesis, namely; the scalar field as dark matter in galaxies and in the universe [3] [4].

The FRW model contains some problems related with the origin of the universe, when the universe was quantum mechanical. Some of the most important problems of the FRW model are; the horizon, the flatness problem, galaxy formation, etc. Some of these problems can be resolved using an inflationary model of the universe, *i.e.*, introducing a scalar field by hand into the Einstein field equations. This procedure is preferred by theoretical physicist because it is elegant and simple. In general the inflationary model implies that $\Omega = 1$, where most of the matter is due to the scalar field. It is quit remarkable that all the actual most important unification theories, like the Standard Model of particles, the Kaluza-Klein and the Superstrings theories predict the existence of scalar fields. Scalar fields are needed in order to maintain consistence in the respective theory. Therefore the question arrays; is it possible that the wanting matter could have a scalar nature? In this lines I will show that this seems to be the case, this fact puts the **scalar fields as a good candidate to be the dark matter in the universe**. I will start supposing that scalar fields are the dark matter in spiral galaxies. Observational data show that the galaxies are composed by almost 90% of dark matter. Nevertheless the halo contains a larger amount of dark matter, because otherwise the observed dynamics of particles in the halo is not consistent with the predictions of Newtonian theory, which explains well the dynamics of the luminous sector of the galaxy. So we can suppose that luminous matter does not contribute in a very important way to the total energy density of the halo of the galaxy at least in the mentioned region, instead the scalar matter will be the main contributor to it. Luminous matter in galaxies posses a Newtonian behavior, we expect that only gravitational interactions are important in them. So, we can perfectly neglect all the other interactions, I will suppose that only gravitation and scalar interactions are present. So, the model I am dealing with will be given by the gravitational interaction modified by a scalar field and a scalar potential. Then, I start with the effective low energy action of superstrings theories with cosmological constant Λ in the Einstein frame

$$S = \int d^4x\sqrt{-g}[-\frac{R}{\kappa_0} + 2(\nabla\Phi)^2 + e^{-2\alpha\phi}\Lambda], \tag{1}$$

where R is the scalar curvature, Φ is the scalar field, $\kappa_0 = \frac{16\pi G}{c^3}$ and $\sqrt{-g}$ is the determinant of the metric. I have carried out a conformal transformation in order to have a more simple form of the field equations. Action (1) actually states that an exponential potential appears in a natural way in this theory.

On the other hand, the exact symmetry of the halo is stills unknown, but it is reasonable to suppose that the halo is symmetric with respect to the rotation axis of the galaxy. Here I let the symmetry of the halo as general as I can, so I choose it to be axial symmetric. Furthermore, the rotation of the galaxy do not affect the motion of test particles around the galaxy, dragging effects in the halo of the galaxy should be too small to affect the tests particles (stars) traveling around the galaxy. Hence, in the region of interest we can suppose the space-time to be static, given that the circular velocity of stars (like the sun) of about 230 Km/s seems not to be affected by the rotation of the galaxy and we can consider a time reversal symmetry of the space-time. The most general static and axial symmetric metric compatible with this action, written in the Papapetrou form is

$$ds^2 = \frac{1}{f}[e^{2k}(dzd\bar{z}) + W^2d\phi^2] - f\ c^2dt^2, \tag{2}$$

where $z := \rho + i\zeta$ and $\bar{z} := \rho - i\zeta$ and the functions f, W and k depend only on ρ and ζ. This metric represents the symmetries posted above.

An exact solution of the field equations derived from the action (1) in Boyer-Lindquist coordinates $\rho = \sqrt{r^2 + b^2}\sin\theta$, $\zeta = r\cos\theta$ reads [8] [9]

$$\begin{aligned} ds^2 = &\frac{1 + \frac{b^2\cos^2\theta}{r^2}}{f_0 r_0}(\frac{dr^2}{1 + \frac{b^2}{r^2}} + r^2\ d\theta^2) + \\ &\frac{r^2 + b^2\sin^2\theta}{f_0 r_0}d\phi^2 - f_0 c^2\frac{r^2 + b^2\sin^2\theta}{r_0}dt^2 \end{aligned} \tag{3}$$

The effective energy density μ_{DM} of (3) is given by the expression

$$\mu_{DM} = \frac{1}{2}V(\Phi) = \frac{4f_0 r_0}{\kappa_0(r^2 + b^2\sin^2\theta)} \tag{4}$$

The energy density (4) coincides with that required for a galaxy to explain the rotation curves of test particles in its halo, but in our model, this energy density is product of the scalar field and the scalar field potential, that is, this dark matter is produced by a Φ particle.

In what follows I study the circular trajectories of a test particle on the equatorial plane taking the space-time (2) as the background. The motion equation of a test particle in the space-time (2) can be derived from the Lagrangian

$$\mathcal{L} = \frac{1}{f}[e^{2k}(\left(\frac{d\rho}{d\tau}\right)^2 + \left(\frac{d\varsigma}{d\tau}\right)^2) + W^2\left(\frac{d\phi}{d\tau}\right)^2] - f\, c^2\left(\frac{d\,t}{d\tau}\right)^2. \tag{5}$$

This Lagrangian contains two constants of motion, the angular momentum per unit of mass B and the total energy of the test particle A. In terms of the metric components and the test particle velocity $v = (\dot{\rho}, \dot{\varsigma}, \dot{\phi})$ I obtain $A^2 = c^4 f^2/(f - \frac{v^2}{c^2})$. For a circular trajectory at the equatorial plane $\dot{\varsigma} = \dot{\rho} = 0$ the equation of motion is $B^2\, f/W^2 - A^2/c^2 f = -c^2$. This last equation determines the circular trajectories of the stars of the galaxy. Using these equations I obtain an expression for B in terms of v^2, $B^2 = v^2/(f - \frac{v^2}{c^2})W^2/f \sim v^2 W^2/f^2$, since $v^2 << c^2$. From this equation one concludes that for our solution (3) $v^2 = f_0^2 B^2$, *i.e.*

$$v_{DM} = f_0 B, \tag{6}$$

where I call $v \longrightarrow v_{DM}$ the circular velocity due to the dark matter.

Let us model the circular velocity profile of a spiral galaxy by the function

$$v_L^2 = v^2(R_{opt})\beta\frac{1.97\, x^{1.22}}{(x^2 + 0.78^2)^{1.43}} \tag{7}$$

which is the approximate model for the Universal Rotation Curves proposed by Persic *et.al.* [5] where $\beta = v_L(R_{opt})/v(R_{opt.})$. One obtains a typical profile of the circular velocity due to the luminous matter of a spiral galaxy [6]. With this velocity it is now easy to calculate the angular momentum (per unity of mass) of the test particle $B = v_L D$, where D is the distance between the center of the galaxy and the test particle. For our metric, $D = \int ds = \sqrt{(r^2 + b^2)/f_0 r_0}$. Finally, using equation (6) I find the profile of the dark matter velocity. The results are shown in fig. 1 for some galaxies. We see that the correspondence with typical circular velocities profiles given in the literature [6] [7] is excellent.

The crucial point for having the circular velocity $v_{DM} = f_0 B$ is that $f \sim W$ in the solution (3). But this fact remains unaltered after conformal transformations in the metric $\hat{ds}^2 = \Omega(\Phi)ds^2$, so that the circular velocity v_{DM} remains the same for all theories and frames related with metric (2) by conformal transformations. This point is very important. In order to derive action (1) from the effective action of the low energy limit of superstrings theory, I have carried out a conformal transformation, so this result is valid also for this last action. But then the result is valid for any theory conformaly equivalent to action (1).

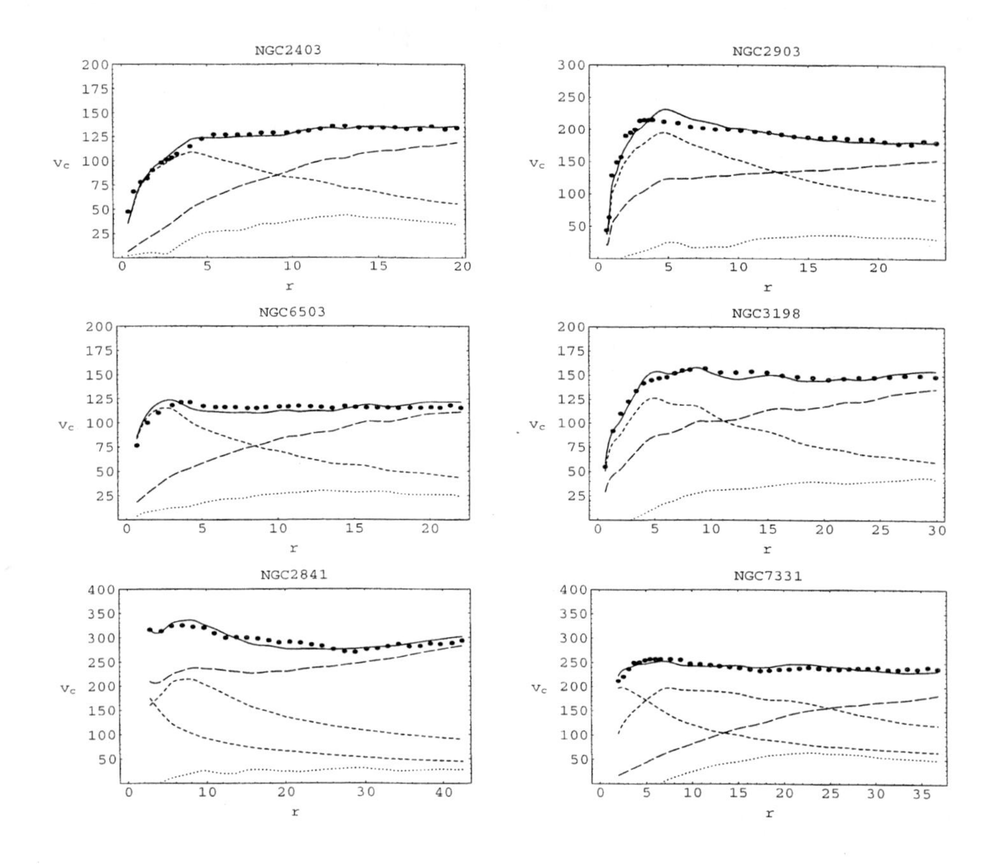

FIGURE 1. Curves of v_c in km/s vs. the distance r (kpc) for the observational data (dots) and the fits. Contributions from luminous matter v_L (short dashed), the dilaton v_{DM} (long dashed) and gas (dotted) are also shown.

This result has some very interesting consequences [10]. Scalar fields not only exist, but they represent 90% of the matter in the universe. This result and the inflationary models tell us that scalar fields are the most important part of matter in nature, they determine the structure of the universe. After the big bang, they inflated the universe; soon after they gave mass to the particles; later they concentrate maybe because of scalar field condensation, provoking that baryonic matter density fluctuations and forming stars, galaxies and galaxy clusters. Scalar fields can clarify why galaxies formed so soon after the recombination era, they condensate during the radiation era forming the arena which formed the galaxies. The question why nature use only the spin 1 and spin 2 fundamental interactions over the simplest spin 0 interactions becomes clear here. This result tell us that in fact nature have preferred the spin 0 interaction over the other two ones because scalar field interactions determine the cosmos structure. This result give also a limit for the validity of the Einstein's equations, they are valid at local level; planets, stars, star-systems, but they are not more valid at galactic or cosmological level where a scalar field interaction must be added to the original equations. This result could be the first contact of higher dimensional theories like the Kaluza-Klein or the Superstrings one with reality, furthermore, it could be the first trace to demonstrate the existence of extra dimensions in nature.

This work was partially supported by CONACyT México, grant 3697-E.

REFERENCES

1. Zwicky, F. Helv. Phys. Acta, **6,** (1933), 110
2. Schramm D. N. In "Nuclear and Particle Astrophysics", ed. J. G. Hirsch and D. Page, Cambridge Contemporary Astrophysics, (1998).
3. Matos, T. and Guzman, S. F. gr-qc/9810028
4. Guzman, S. F., Matos, T. and Villegas, B. H. astro-ph/9811143
5. Persic, M. Salucci, P and Stel, S. *MNRAS*, **281**, (1996), 27.
6. K.G. Begeman, A.H. Broeils and R.H. Sanders, *MNRAS,* **249,** (1991), 523-537.
7. Vera C. Rubin, W. Kent Ford,Jr., and Norbert Thonnard. *Astrophys. Journal*, **238**, (1980), 471-487.
8. T. Matos. *Ann. Phys. (Leipzig)* **46**, (1989), 462-472. T. Matos. *J. Math. Phys.*, **35**, (1994), 1302-1321. T. Matos. Exact Solutions of G-invariant Chiral Equations. *Math. Notes,* **58**, (1995), 1178-1182.
9. T. Matos and F. S. Guzmán. Modeling Dilaton Stars with Arbitrary Electromagnetic Field. In *Proceedings of the VIII Marcel Grossman Meeting, Jerusalem, Israel,* Ed. by D. Ruffini et. al., Word Scientific Singapore, (1998), in press.
10. T. Matos, F.S. Guzmán and H. Villegas B. *in preparation* 1999.

Simulations of the surface detector of the Pierre Auger Observatory — Calibration and Monitoring —

Martin Medina*, Lukas Nellen†, Luis Villaseñor*

*Instituto de Física y Matematicas, Universidad Michoacana, Apdo. Postal 2-82, Morelia Mich., 58040 México
†Instituto de Ciencias Nucleares, UNAM, Circuito Exterior C.U, A.Postal 70-543, 04510 México, D.F.

Abstract. Once completed, the *Pierre Auger Observatory* will be the cosmic ray detector with the largest aperture ($2 \times 7000\,\mathrm{km}^2\mathrm{sr}$). The design of this observatory and, in the future, the data analysis depend, at least partially, on simulations of the response of the detector to an extended air shower. In the following, we present briefly the *Pierre Auger Observatory* and discuss simulations of the surface detector. As a concrete example we discuss the simulation of calibration and monitoring methods.

INTRODUCTION

Soon after the detection of the cosmic microwave background (CMB) it was pointed out independently by Greisen [1] and by Zatsepin and Kuz'min [2] that interactions of ultra high energetic cosmic rays (UHECR) with the CMB form an important part of the energy loss processes of UHECR above a certain energy (GZK cutoff). For protons, this energy is $E_{\mathrm{GZK}} \approx 5 \times 10^{19}\,\mathrm{eV}$. This energy is the threshold energy for the reaction

$$p + \gamma_{\mathrm{CMB}} \rightarrow \Delta^+ \rightarrow \begin{cases} p + \pi^0 \\ n + \pi^+ \end{cases} .$$

Above this threshold, the mean free path for protons in the CMB is $\approx 10\,\mathrm{Mpc}$. This, together with the high inelasticity of this process, implies that the sources of cosmic rays above the GZK cutoff are not further than about 100 Mpc from us (see figure 1 (b)), a small distance on cosmological scales. Nevertheless, about 20 events with energies beyond the GZK cutoff have been reported in the literature, the first by Linsley in 1963 [3]. After careful examination of the data it is clear that those events are real and not a result of energy miss-assignment. Neither the nature nor the origin of the UHECR is known, though.

CP490, *Particles and Fields: Eighth Mexican School,* edited by J. C. D'Olivo et al.

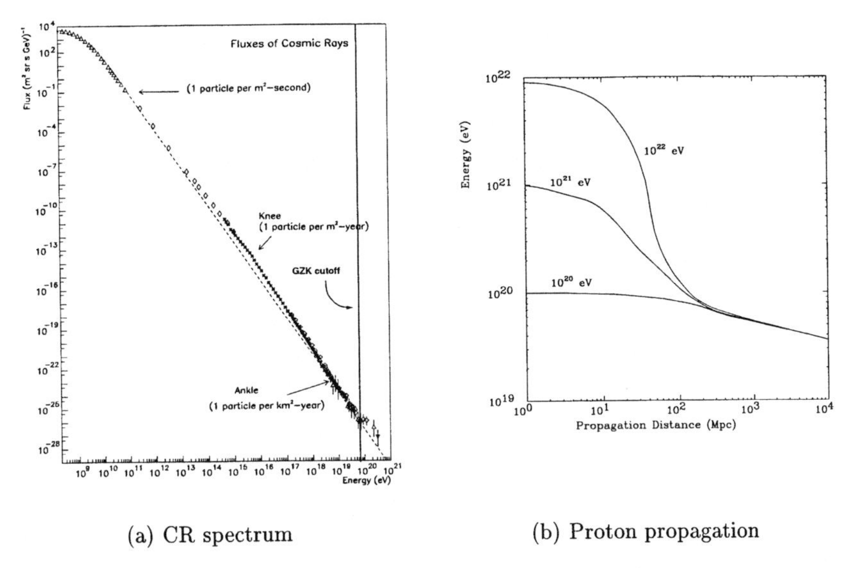

(a) CR spectrum

(b) Proton propagation

FIGURE 1. The observed spectrum of cosmic rays extends beyond the GZK cutoff without any sign of a significant change in the power law (a). Protons with an initial energy beyond the GZK cutoff reach the GZK threshold after $\approx 100\,\mathrm{Mpc}$, with little dependence on the initial energy (b).

THE *PIERRE AUGER OBSERVATORY*

Even though the number of particles detected so far beyond the GZK cutoff is sufficient to establish the existence of trans-GZK cosmic rays, we do not have enough data to identify the type of particle or the source. The problem here is the low flux of about one particle per year per hundred km^2. To be able to gather a statistically significant amount of events, an international collaboration decided to construct the *Pierre Auger Observatory*, consisting of two sites of $3000\,\mathrm{km}^2$ each, one in the southern hemisphere, in the province of Mendoza in Argentina, and one in the northern hemisphere, in Utah, USA. With this split of observing sites it is possible to obtain an almost even exposure of the whole sky.

Each site has two subsystems, a ground array and an air fluorescence detector, covering the same area. The ground array will consist of roughly 1600 water Čerenkov detectors layed out on a triangular grid with a spacing of 1.5 km. It will sample the front of an extensive air shower (EAS) as it crosses the ground level. The fluorescence detector will consist of about 36 telescopes with a field of view of $30° \times 30°$ and an angular resolution of $1.5°$, overlooking the whole area of the ground array. This subsystem will detect the nitrogen fluorescence produced by an EAS as it develops in the atmosphere, adding longitudinal information to the lateral information obtained by the ground array [4].

SIMULATIONS OF THE SURFACE DETECTOR

As part of the design process, a full set of simulation programs for the surface array, including all relevant physical processes in the WCD and its electronics, has been developed [5,6] and used to evaluate the performance of the the surface detector [7]. For this purpose, libraries of simulated air showers [8] using MOCCA [9] and CORSIKA [10] have been used. It is worth noting that CORSIKA and some of the software used for the simulation of the WCD uses standard HEP packages like GEANT, EGS4 and GEISHA. This, and the successful application to full detector simulations, demonstrates that the simulation of the surface detector is under control and reliable. It turned out to be easy to adapt the existing software to the simulation of calibration and monitoring methods.

CALIBRATION AND MONITORING

The sheer size of the surface array as well the desertic locations chosen for the installation of the *Pierre Auger Observatory* do not allow for frequent access to individual detector stations. Therefore, adequate remote calibration and monitoring capabilities are of great importance.

Calibration and monitoring of the detectors of the surface array use both sensors, like voltages, currents and temperatures, read out by the station controller and the analysis of signals produced by reference sources. Given the goals of low price and low maintenance, the sources of choice are signals produced by relativistic muons crossing the WCD and by the electrons produced in the decay of a stopping muon [11,12]. The background of secondary cosmic ray particles provides a constant flux of muons, eliminating the need for a separate calibration source. Changes in the distribution of charge collected, peak voltage and Q/V are indicative of various possible problems like contamination of the water or the PMT windows.

The method of using crossing muons as a reference signal has been tested experimentally and by simulations and is considered the baseline method for the *Pierre Auger Observatory* [11]. The idea of using electrons from muon decay was developed using a reduced prototype WCD and compared successfully against simulations [12]. In order to guide the experimental work testing the decay electron method on a full size tank, the simulations were extended to this configuration.

Figure 2 shows that both the crossing muon charge distribution and the decay electron charge distribution show a clear peak. The crossing muon peak is slightly above the vertical muon peak; the electron peak is at about 20% of a vertical muon. The signal in the full size tank about 75% of that in the reduced prototype due to the reduction of the ratio of wall to photo cathode area. This reduction does not affect the viability of the decay electron method. A potential problem which remains to be analysed is the increase of random muon coincidences in the bigger tank, increasing the background for the muon followed by decay electron double pulse trigger.

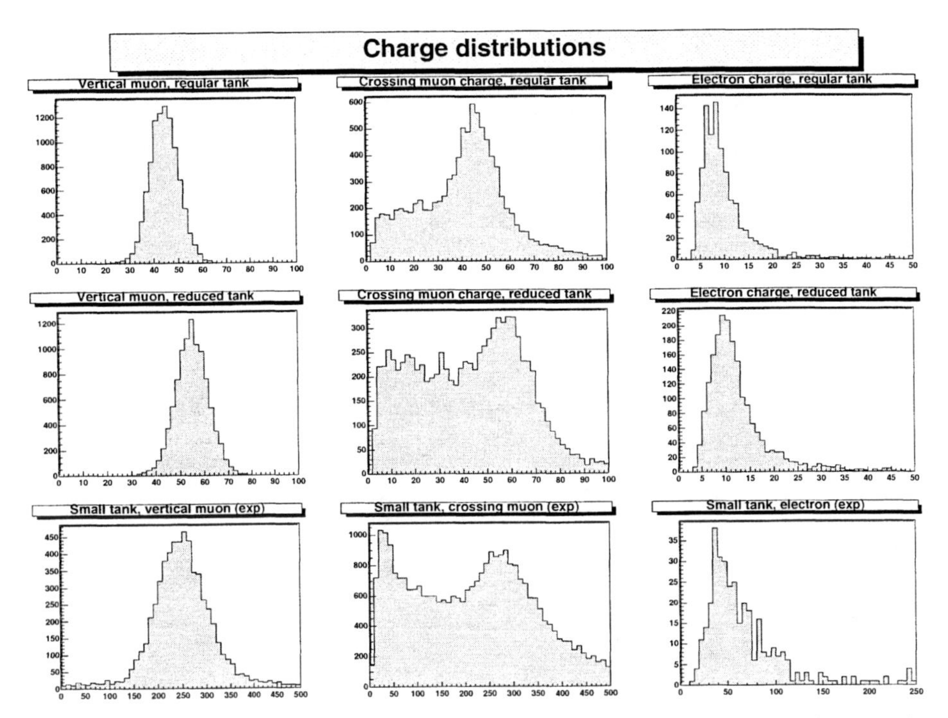

FIGURE 2. Charge distributions for vertical muons (column 1), crossing muons (column 2) and decay electrons (column 3). Experimental distributions (row 3) and simulation (row 2) for the reduced prototype show good agreement. Simulations for a full size tank (row 1) show somewhat smaller charges collected due to the relatively reduced photo cathode area.

CONCLUSIONS

The techniques for remote calibration and monitoring of the surface array of the *Pierre Auger Observatory* are developed and tested, both in experiments and in simulations. The future work is on refining the methods outlined above with an emphasis on understanding how different failure modes show up in the monitoring data. Simulations play an essential part in that work in order to identify and prepare the relevant experiments.

We would like to thank our Mexican colleagues in the Auger project for their collaboration.

REFERENCES

1. Greisen K., Phys. Rev. Letts. **16** (1966) 748.
2. Zatsepin G. Z., and Kuz'min V. A., JETP Letters **4** (1966) 78.

3. Linsley J., Phys. Rev. Letts. **10** (1963) 146.
4. The Auger Collaboration, *The Pierre Auger Observatory: Design Report*, 2nd ed., 1997. (http://www.auger.org/admin/DesignReport/). Updated information provided in the technical notes availble at http://www.auger.org/admin/GAP_Notes/.
5. Pryke C.L.S., "*Instrumentation development and experimental design for a next generation detector of the highest energy cosmic rays*", PhD Thesis, The University of Leeds, England 1996.
6. de Mello Neto J.R.T, GAP-1998-020. Bauleo P. and J.C. Rodriguez Martinez, GAP-1998-050.
7. Dawson B.R. and C.L. Pryke, GAP-1997-044. Pryke, C.L., GAP-1997-037; GAP-1997-005; GAP-1997-004.
8. http://aupc1.uchicago.edu/~pryke/auger/showsim/
9. A. M. Hillas, "*The MOCCA program: MOnte-Carlo CAscades*". In proceedings of the 24th International Cosmic Ray Conference, Rome, Italy, volume 1, 1995.
10. Heck D., J. Knapp, J.N. Capdevielle, G. Schatz, T. Thouw, GAP-98-013.
11. Kutter T. and C.L. Pryke, GAP-1997-025. Pryke C.L., GAP-1997-026.
12. Alcaráz F., J. Barrera, E. Cantoral, A. Fernández, M. Medina, L. Nellen, C. Pacheco, S. Roman, H. Salazar, J. Valdés-Galicia, M. Vargas, L. Villaseñor and A. Zepeda, Nucl. Instr. and Meth. **A420** (1999) 39–47. Medina M., L. Villaseñor *et al.*, *Recent results on the operation of a Cherenkov detector prototype for the Pierre Auger Observatory*, these proceedings.

Breaking of flavor permutational symmetry and the CKM matrix

A. Mondragón and E. Rodríguez-Jáuregui

Instituto de Física, Universidad Nacional Autónoma de México
Apdo. Postal 20-364, 01000 México, D. F., México. [1]

Abstract. Different ansätze for the breaking of the flavor permutational symmetry according to $S_L(3) \otimes S_R(3) \supset S_L(2) \otimes S_R(2)$ give different Hermitian mass matrices which differ in the symmetry breaking pattern. In this work we obtain a clear and precise indication on the preferred symmetry breaking pattern. The preferred pattern allows us to compute the CKM mixing matrix, the Jarlskog invariant J, and the three inner angles of the unitarity triangle in terms of four quark mass ratios and the CP violating phase Φ. Excellent agreement with the experimentally determined absolute values of the entries in the CKM matrix is obtained for $\Phi = 90°$. The corresponding computed values of the Jarlskog invariant and the inner angles are $J = 3.00 \times 10^{-5}$, $\alpha = 84°$, $\beta = 24°$ and $\gamma = 72°$ in very good agreement with current data on CP violation in the neutral kaon-antikaon system and oscillations in the B_s°-$\bar{B}_s^\circ$ system.

MASS MATRICES FROM THE BREAKING OF $S_L(3) \otimes S_R(3)$

Under exact $S_L(3) \otimes S_R(3)$ symmetry, the mass spectrum for either up or down quark sectors consists of one massive particle in a singlet irreducible representation and a pair of massless particles in a doublet irreducible representation, the corresponding quark mass matrix is $\mathbf{M}_{3q}$. In order to generate masses for the first and second families, we add the terms $\mathbf{M}_{2q}$ and $\mathbf{M}_{1q}$ to $\mathbf{M}_{3q}$. The term $\mathbf{M}_{2q}$ breaks the permutational symmetry $S_L(3) \otimes S_R(3)$ down to $S_L(2) \otimes S_R(2)$ and mixes the singlet and doublet representation of $S(3)$. $\mathbf{M}_{1q}$ transforms as the mixed symmetry term in the doublet complex tensorial representation of $S_{diag}(2) \subset S_L(2) \otimes S_R(2)$. Putting the first family in a complex representation allows us to have a CP violating phase. Then, in a symmetry adapted basis , $\mathbf{M}_q$ takes the form

[1] Presented by E. Rodríguez-Jáuregui

CP490, *Particles and Fields: Eighth Mexican School,* edited by J. C. D'Olivo et al.

$$M_q = m_{3q}\left[\begin{pmatrix} 0 & A_q e^{-i\phi_q} & 0 \\ A_q e^{i\phi_q} & 0 & 0 \\ 0 & 0 & 0 \end{pmatrix} + \begin{pmatrix} 0 & 0 & 0 \\ 0 & -\Delta_q+\delta_q & B_q \\ 0 & B_q & \Delta_q-\delta_q \end{pmatrix}\right] + m_{3q}\begin{pmatrix} 0 & 0 & 0 \\ 0 & 0 & 0 \\ 0 & 0 & 1-\Delta_q \end{pmatrix} \quad (1)$$

The entries in the mass matrix may be readily expressed in terms of the mass ratios $\tilde{m}_{1q} = m_{1q}/m_{3q}$ and $\tilde{m}_{2q} = m_{2q}/m_{3q}$: $A_q^2 = \tilde{m}_{1q}\tilde{m}_{2q}(1-\delta_q)^{-1}$, $\Delta_q = \tilde{m}_{2q} - \tilde{m}_{1q}$, $B_q = \delta_q((1-\tilde{m}_{1q}+\tilde{m}_{2q}-\delta_q) - \tilde{m}_{1q}\tilde{m}_{2q}(1-\delta_q)^{-1})$. If each possible symmetry breaking pattern (SBP)is now characterized by the ratio ${Z_q}^{1/2} = B_q/(-\Delta_q+\delta_q)$, the small parameter δ_q is obtained as the solution of the cubic equation

$$\delta_q\left[(1+\tilde{m}_{2q}-\tilde{m}_{1q}-\delta_q)(1-\delta_q) - \tilde{m}_{1q}\tilde{m}_{2q}\right] - Z_q(-\tilde{m}_{2q}+\tilde{m}_{1q}+\delta_q)^2 = 0 \quad (2)$$

which vanishes when Z_q vanishes. In the symmetry adapted basis, the term $\mathbf{M_{2q}}$ is decomposed as a linear combination of two linearly independent numerical matrices, $\mathbf{M_{2A}}$ and $\mathbf{M_{2S}}$,

$$\mathbf{M}_{2q} = m_{3q}(\delta_q - \Delta_q)\left\{N_{Aq}\begin{pmatrix} 0 & 0 & 0 \\ 0 & 1 & -\sqrt{8} \\ 0 & -\sqrt{8} & -1 \end{pmatrix} + N_{Sq}\begin{pmatrix} 0 & 0 & 0 \\ 0 & 1 & \frac{1}{\sqrt{8}} \\ 0 & \frac{1}{\sqrt{8}} & -1 \end{pmatrix}\right\} \quad (3)$$

the matrices, $\mathbf{M_{2A}}$ and $\mathbf{M_{2S}}$, are of the same form as $\mathbf{M_{2q}}$ with mixing parameters $Z_A = -\sqrt{8}$ and $Z_S = 1/\sqrt{8}$ respectively. From Eq.(3), it is evident that there is a corresponding decomposition of the mixing parameter $Z_q^{1/2}$,

$${Z_q}^{1/2} = N_{Aq}Z_A^{1/2} + N_{Sq}Z_S^{1/2} \quad with \quad 1 = N_{Aq} + N_{Sq}, \quad (4)$$

in this way a unique linear combination of ${Z_A}^{1/2}$ and ${Z_S}^{1/2}$ is associated to the SBP. The pair of numbers (N_A, N_S) are a convenient mathematical label of the SBP. The parameter ${Z_q}^{1/2} = M_{2q23}/M_{2q22}$ is a measure of the amount of mixing of singlet and doublet irreducible representations of $S_L(3) \otimes S_R(3)$. It will be assumed that, the up and down mass matrices are generated following the same SBP: $Z_u^{1/2} = Z_d^{1/2} = Z^{*1/2}$. Elsewhere [1], we found that the $S_L(3) \otimes S_R(3)$ flavour symmetry is broken down to $S_L(2) \otimes S_R(2)$ according to a mixed SBP characterized by $Z^{*1/2} = 1/2\left(Z_S^{1/2} - Z_A^{1/2}\right)$. Then, the mass matrix with the modified Fritzsch texture takes the form

$$\mathbf{M}^*_{q,H} = m_{3q}\begin{pmatrix} 0 & \sqrt{\frac{\tilde{m}_{1q}\tilde{m}_{2q}}{1-\delta_q^*}}e^{-i\phi_q} & 0 \\ \sqrt{\frac{\tilde{m}_{1q}\tilde{m}_{2q}}{1-\delta_q^*}}e^{i\phi_q} & -\tilde{m}_{2q}+\tilde{m}_{1q}+\delta_q^* & \frac{9\sqrt{2}}{8}(-\tilde{m}_{2q}+\tilde{m}_{1q}+\delta_q^*) \\ 0 & \frac{9\sqrt{2}}{8}(-\tilde{m}_{2q}+\tilde{m}_{1q}+\delta_q^*) & 1-\delta_q^* \end{pmatrix}_H, \quad (5)$$

where δ_q^* is the solution of the cubic equation obtained from Eq. (2) when $\sqrt{81/32}$ is subtituted for $Z^{*1/2}$.

THE MIXING MATRIX

The Hermitian mass matrix $\mathbf{M}_q$ may be written in terms of a real matrix $\bar{\mathbf{M}}_q$and a diagonal matrix of phases $\mathbf{P}_q$ as $\mathbf{P}_q \bar{\mathbf{M}}_q {\mathbf{P}_q}^{\dagger}$. Then, the mixing matrix $\mathbf{V}$ is given by

$$\mathbf{V} = {\mathbf{O}_u}^T \mathbf{P}^{u-d} \mathbf{O}_d \tag{6}$$

where $\mathbf{P}^{u-d} = diag[1, e^{i(\phi_u - \phi_d)}, e^{i(\phi_u - \phi_d)}]$ is the diagonal matrix of the relative phases and $\mathbf{O}_q$ is the orthogonal matrix that diagonalizes $\bar{\mathbf{M}}_q$

$$\mathbf{O}_q = \begin{pmatrix} (\tilde{m}_{2q} f_1/D_1)^{1/2} & -(\tilde{m}_{1q} f_2/D_2)^{1/2} & (\tilde{m}_{1q}\tilde{m}_{2q} f_3/D_3)^{1/2} \\ ((1-\delta_q)\tilde{m}_{1q} f_1/D_1)^{1/2} & ((1-\delta_q)\tilde{m}_{2q} f_2/D_2)^{1/2} & ((1-\delta_q) f_3/D_3)^{1/2} \\ -(\tilde{m}_{1q} f_2 f_3/D_1)^{1/2} & -(\tilde{m}_{2q} f_1 f_3/D_2)^{1/2} & (f_1 f_2/D_3)^{1/2} \end{pmatrix} \tag{7}$$

where $f_1 = 1 - \tilde{m}_1 - \delta_q^*$, $f_2 = 1 + \tilde{m}_2 - \delta_q^*$, $f_3 = \delta_q^*$, $D_1 = (1-\delta_q^*)(1-\tilde{m}_1)(\tilde{m}_{2q} + \tilde{m}_{1q})$, $D_2 = (1-\delta_q^*)(1+\tilde{m}_{2q})(\tilde{m}_{2q} + \tilde{m}_{1q})$ and $D_3 = (1-\delta_q^*)(1+\tilde{m}_{2q})(1-\tilde{m}_{1q})$.
From Eqs. (6) and (7), we derived closed, explicit expressions for all entries in the matrix $\mathbf{V}$ written in terms of four mass ratios $(\tilde{m}_u, \tilde{m}_c, \tilde{m}_d, \tilde{m}_s)$ and one free real parameter $\Phi = \phi_u - \phi_d$ [1]. The CP violating phase Φ measures the mismatch in the $S_L(2) \otimes S_R(2)$ symmetry breaking in the u- and d-sectors. We made a χ^2 fit of the exact expressions for the absolute values of the the entries in the mixing matrix $|V^{th}|$ and the Jarlskog invariant J to the experimentally determined values of $|V^{exp}|$ and J^{exp}. We took the values of the running quark masses evaluated at the scale of m_t from H. Fritzsch [2], [3], we left the mass ratios $\tilde{m}_c$ and $\tilde{m}_s$ fixed at their central values $\tilde{m}_c = 0.0048$ and $\tilde{m}_s = 0.03437$ but we took the values of $\tilde{m}_u = 0.000042$ and $\tilde{m}_d = 0.00148$ close to the upper and lower bounds of $\tilde{m}_u$ and $\tilde{m}_d$ respectively. A detailed account of the computation may be found in Mondragón and Rodríguez-Jáuregui [1]. We found the following best value for Φ,

$$\Phi^* = 90^\circ. \tag{8}$$

Therefore, the theoretical expressions for the entries in the mixing matrix $\mathbf{V}$ are functions of the four mass ratios $(\tilde{m}_u, \tilde{m}_c, \tilde{m}_d, \tilde{m}_s)$ with the CP violating phase $\Phi = 90^\circ$. The quark mixing matrix V^{th} computed from the theoretical expresions is

$$V^{th} = \begin{pmatrix} 0.9754 e^{i1^\circ} & 0.2208 e^{i156^\circ} & 0.0037 e^{i85^\circ} \\ 0.2207 e^{i114^\circ} & 0.9745 e^{i89^\circ} & 0.0396 e^{i90^\circ} \\ 0.0084 e^{i270^\circ} & 0.0388 e^{i270^\circ} & 0.9992 e^{i90^\circ} \end{pmatrix} \tag{9}$$

The Jarlskog invariant , J, may be computed directly from the commutator of the mass matrices [4]

$$J = -\frac{det\{-i[\mathbf{M}_{u,H}, \mathbf{M}_{d,H}]\}}{F} \tag{10}$$

where $F=(1+\tilde{m}_c)(1-\tilde{m}_u)(\tilde{m}_c+\tilde{m}_u)(1+\tilde{m}_s)(1-\tilde{m}_d)(\tilde{m}_s+\tilde{m}_d)$.

Substitution of the expression (5) for $\mathbf{M}_u$ and $\mathbf{M}_d$, in Eq. (10), gives

$$J=\frac{Z\sqrt{(\tilde{m}_u/\tilde{m}_c)/(1-\delta_u^*)}\sqrt{(\tilde{m}_d/\tilde{m}_s)/(1-\delta_d^*)}}{(1+\tilde{m}_c)(1-\tilde{m}_u)(1+\tilde{m}_u/\tilde{m}_c)(1+\tilde{m}_s)(1-\tilde{m}_d)(1+\tilde{m}_d/\tilde{m}_s)}$$
$$\times\Big\{-(\tilde{m}_u\tilde{m}_c/(1-\delta_u^*))\,(\delta_d^*-(\tilde{m}_s-\tilde{m}_d))^2-(\tilde{m}_d\tilde{m}_s/(1-\delta_d^*))(\delta_u^*-(\tilde{m}_c-\tilde{m}_u))^2$$
$$+\left[(\delta_u^*-(\tilde{m}_c-\tilde{m}_u))(1-\delta_d^*)-(\delta_d^*-(\tilde{m}_s-\tilde{m}_d))(1-\delta_u^*)\right]^2\Big\} \qquad (11)$$

The non-vanishing of J is a necessary and sufficient condition for the violation of CP [4]. From Eq. (11), it is apparent that J vanishes when Z, $sin\Phi$ and $\tilde{m}_u$ or $\tilde{m}_d$ vanish. Therefore, the violation of CP and the consequent non-vanishing of Z necessarily implies a mixing of singlet and doublet representations of $S_L(3)\otimes S_R(3)$. We also computed the three inner angles α, β and γ of the unitarity triangle,

$$\alpha\approx\Phi,\quad \beta\approx arctan\sqrt{(\tilde{m}_u\tilde{m}_s)/(\tilde{m}_c\tilde{m}_d)},\quad \gamma\approx arctan\sqrt{(\tilde{m}_d\tilde{m}_c)/(\tilde{m}_s\tilde{m}_u)}. \qquad (12)$$

The values obtained are $J=3x10^{-5},\alpha=84°,\beta=24°,\gamma=72°$ in good agreement with current data on CP violation in the $K^o-\bar{K}^o$ mixing system [5], [6] and oscillations in the B_s°-$\bar{B}_s^\circ$ system [5], [7].

ACKNOWLEDGMENTS

This work was partially supported by DGAPA-UNAM under contract No. PAPIIT-IN125298 and by CONACYT (México) under contract 3909P-E9607.

REFERENCES

1. A. Mondragón, E. Rodríguez-Jáuregui *The breaking of the flavour permutational symmetry: Mass textures and the CKM matrix. hep-ph/9807214, to be published in Phys. Rev. D*; see also A. Mondragón and E. Rodríguez-Jáuregui, *Rev. Mex. Fis.* **44(S1)**, 33 (1998), hep-ph/9804267
2. H. Fritzsch, *Mass hierarchies, Hidden Symmetry and Maximal CP-violation", hep-ph/9807551* See also H. Fritzsch, *The symmetry and the Problem of Mass Generation. Proceedings of the XXI International Colloquium on Group Theoretical Methods in Physics (Group 21), Goslar, Germany,* (1996), edited by H.-D. Doebner, W. Scherer, and C. Schutte (World Scientific, Singapore, 1997), Vol. II, p. 543.
3. H. Fusaoka and Y. Koide *Phys. Rev. D* **57**, 3986 (1998).
4. C. Jarlskog, *Phys. Rev. Lett.* **55**, 1039 (1985).
5. Particle Data Group, C. Caso *et al.*, *Eur. Phys. J. C***3**, 1 (1998).
6. Salvatore Mele *"Indirect Measurements of the Vertex Angles of the Unitarity Triangle", CERN-EP/98-133, hep-ph/9810333* (1998).
7. Ahmed Ali and Boris Kayser, *"Quark Mixing and CP Violation", hep-ph/9806230* (1998).

L-R asymmetries and signals for new bosons [1]

J. C. Montero, V. Pleitez and M. C. Rodriguez [2]

Instituto de Física Teórica
Universidade Estadual Paulista
Rua Pamplona, 145
01405-900- São Paulo, SP
Brazil

Abstract. Several left-right parity violating asymmetries in lepton-lepton scattering in fixed target and collider experiments are considered as signals for doubly charged vector bosons (bileptons).

The left-right asymmetry when only one of the lepton is polarized is defined as follows [1]

$$A_{RL}(ll \to ll) = \frac{d\sigma_R - d\sigma_L}{d\sigma_R + d\sigma_L}, \tag{1}$$

where $d\sigma_{R(L)}$ is the differential cross section for one right (left)-handed lepton l scattering on an unpolarized lepton l. That is

$$A_{RL}(ll \to ll) = \frac{(d\sigma_{RR} + d\sigma_{RL}) - (d\sigma_{LL} + d\sigma_{LR})}{(d\sigma_{RR} + d\sigma_{RL}) + (d\sigma_{LL} + d\sigma_{LR})}, \tag{2}$$

where $d\sigma_{ij}$ denotes the cross section for incoming leptons with helicity i and j, respectively, and they are given by

$$d\sigma_{ij} \propto \sum_{kl} |M_{ij;kl}|^2, \quad i,j;k,l = L,R. \tag{3}$$

Another interesting possibility is the case when both leptons are polarized. We can define an asymmetry $A_{R;RL}$ in which one beam is always in the same polarization state, say right-handed, and the other is either right- or left-handed polarized (similarly we can define $A_{L;LR}$):

---

1) Presented by M. C. Rodriguez.
2) e-mails: montero@axp.ift.unesp.br; vicente@axp.ift.unesp.br; mcr@axp.ift.unesp.br

CP490, *Particles and Fields: Eighth Mexican School,* edited by J. C. D'Olivo et al.

$$A_{R;RL} = \frac{d\sigma_{RR} - d\sigma_{RL}}{d\sigma_{RR} + d\sigma_{RL}}, \qquad A_{L;RL} = \frac{d\sigma_{LR} - d\sigma_{LL}}{d\sigma_{LL} + d\sigma_{LR}}. \tag{4}$$

We can define also an asymmetry when one incident particle is right- handed and the other is left-handed and the final states are right- and left or left- and right-handed:

$$A_{RL;RL,LR} = \frac{d\sigma_{RL;RL} - d\sigma_{RL;LR}}{d\sigma_{RL;RL} + d\sigma_{RL;LR}} \tag{5}$$

or similarly, $A_{LR;RL,LR}$. These asymmetries, in Eqs.(4) and (5), are dominated by QED contributions. However, this will not be the case if a bilepton resonance does exist at typical energies of the NLC. To show this fact is the goal of this paper. These asymmetries can be calculated for both fixed target (FT) and collider (CO) experiments.

We can integrate in the scattering angle and define the asymmetry $\overline{A}_{RL}$ as

$$\overline{A}_{RL}(ll \to ll) = \frac{(\int d\sigma_{RR} + \int d\sigma_{RL}) - (\int d\sigma_{LL} + \int d\sigma_{LR})}{(\int d\sigma_{RR} + \int d\sigma_{RL}) + (\int d\sigma_{LL} + \int d\sigma_{LR})}, \tag{6}$$

where $\int d\sigma_{ij} \equiv \int_{5^o}^{175^o} d\sigma_{ij}$. All these asymmetries can be studied in future accelerators [2,3].

The importance of these sort of parity breaking asymmetries in fixed target experiments in lepton-lepton scattering was first pointed out in Ref. [1]. For the case of electron-electron scattering the mass of the electrons can be neglected. For an energy of $E = 50$ GeV and for a scattering angle (in the center of mass frame) of $\theta = 90^o$ the left-right asymmetry, defined in Eq. (1) has a value $\approx -3 \times 10^{-7}$ in the standard model. Radiative corrections reduce this value about $40 \pm 3\,\%$ [4]. It is expected that fixed target experiments like those at SLAC [5] can measure this asymmetry [4]. For the muon-muon elastic scattering this asymmetry is $\approx 5.4 \times 10^{-5}$ [6]. We have studied also the non-diagonal elastic scattering $e\mu \to e\mu$. In the last case we obtain a value of -5.9×10^{-8} for a muon energy of 50 GeV and -2.9×10^{-7} for muon energy of 190 GeV. At these energies the muon mass cannot be neglected [7]. This type of asymmetry can be measured using the high-energy muon beam M2 of the CERN SPS as in the NA47 experiment [8].

The relevance of these asymmetries in collider experiments was first pointed out in Refs. [6,9]. In fixed target experiments the cross sections are large ($\sim$ mb) and the asymmetries small ($\sim 10^{-7}$). On the other hand, in collider experiments the cross sections are small ($\sim 10^{-3}$nb) but the asymmetries large (~ 0.1 for the muon-muon case). Explicitly we have that at energies $\sqrt{s} = 300$ GeV and $\theta \approx 90^o$ the asymmetry is

$$A_{RL}^{\text{CO,ESM}}(ee \to ee) \approx -0.05, \tag{7}$$

for the electron-electron case and

$$A_{LR}^{\mathrm{CO,ESM}}(\mu\mu \to \mu\mu) \approx -0.1436, \tag{8}$$

for the muon-muon case. Future colliders with polarized lepton-lepton scattering can have the appropriate luminosity to measure these parameters.

If a muon-electron collider is constructed in the future, it would be possible to measure the $A_{RL}^{CO;ESM}(\mu e) = -0.024$ for $E_\mu = 190$ GeV ($\sqrt{s} \sim 380$ GeV) and $\theta = 90^o$. At high energies the mass effects are not important.

So far all the results were obtained in the standard model. In certain kind of models there are doubly charged scalars (H^{--}) or/and vector (U^{--}) bileptons [10]. As expect the asymmetry is larger in the U-pole. For instance,

$$A_{RL}^{\mathrm{CO,ESM+U}}(ee \to ee) = -0.099, \tag{9}$$

and

$$A_{RL}^{\mathrm{CO,ESM+U}}(\mu\mu \to \mu\mu) = -0.1801, \tag{10}$$

when we add to the standard model asymmetry the contributions due to the the bilepton U with $M_U = 300$ GeV and $\Gamma_U = 36$ MeV. In Fig. 1 we show the behaviour of the asymmetry $A_{RL}^{\mathrm{CO,ESM+U}}$ as a function of the mass of the boson U^{--} with $\theta = \pi/2$, $\sqrt{s} = 300$ GeV and the same Γ_U.

For the electron-electron case we can define the quantity

$$\delta\,\overline{A}_{RL}(ee \to ee) \equiv (\overline{A}_{RL}^{\mathrm{CO,ESM+U}} - \overline{A}_{RL}^{\mathrm{CO,ESM}})/\overline{A}_{RL}^{\mathrm{CO,ESM}}, \tag{11}$$

where $\bar{A}$'s are define in Eq. (6). Although $\delta\,\overline{A}_{RL}$ is large (near 50 for $\sqrt{s} = 300$) at the U-resonance we would like to stress that it remains appreciably large even far from the U-peak. That particular behavior suggests that this quantity could be the one to be considered in the search for new physics, like the bilepton U^{--}, in future colliders. On the other hand, the asymmetry is insensitive to the contributions of the doubly charged scalars.

We have used also the asymmetries defined in Eq. (4). In this case it is interesting to note that

$$A_{R;RL}^{\mathrm{CO,ESM+U}}(ee \to ee) \approx -A_{R;RL}^{\mathrm{CO,ESM}}(ee \to ee), \tag{12}$$

and we see that such a difference on sign is a good signature for the discovery of the vector bilepton.

The contributions of an extra neutral vector boson Z' has also been considered for the case $A_{RL}(\mu e)$. In this case the asymmetry is considerably enhanced and it will be appropriate in searching for extra neutral vector bosons with mass up to 1 TeV. Since in the 331 model the Z' couplings with the leptons are flavor conserving we do not have additional suppression factors coming from mixing [10]. Hence, the μe elastic scattering can be very helpful, even with the present experimental capabilities, for looking for non-standard physics effects.

This work was supported by Fundação de Amparo à Pesquisa do Estado de São Paulo (FAPESP), Conselho Nacional de Ciência e Tecnologia (CNPq) and by Programa de Apoio a Núcleos de Excelência (PRONEX).

REFERENCES

1. Derman, E., Marciano, W., Ann. Phys. (N.Y.) **121**, 147 (1979
2. Kuhman, K., *et al.*, (The NLC Accelerator Design Group and The NLC Physics Working Group), *Physics and Technology of the Next Linear Collider*, FERMILAB-PUB96/112. See also `http://nlc.physics.upenn.edu/nlc/nlc.html` and `http://pss058.psi.ch/cuypers/e-e-.htm`.
3. Gunion, J. F., *Muon Colliders: The Machine and the Physics*, hep-ph/9707379. See also `http//www.cap.bnl.gov/mumu` and `http://www.fnal.gov/projects/muon_collider`.
4. Czarnecki, A., Marciano, W., *Phys. Rev. D* **53**, 1066 (1996).
5. See `htpp://www.salc.stanford.edu/FIND/explist.html`.
6. Montero, J. C., Pleitez, V., Rodriguez, M. C., *Phys. Rev. D*, 094026 (1998); hep-ph/9802313.
7. Montero, J. C., Pleitez, V., Rodriguez, M. C., *Phys. Rev. D*, 097505 (1998); hep-ph/9803450.
8. Adams, D., *et al.* (SMC Collaboration), *Phys. Rev. D* **56**, 5330 (1997).
9. Czarnecki, A., Marciano W., *Int. J. Mod. Phys.* A13, 2235 (1998); hep-ph/9801394.
10. F. Pisano and V. Pleitez, *Phys. Rev. D* **46**, 410 (1992).

Real sector of self-dual Gravity with only first class constraints

M. Montesinos*, H. A. Morales†, L. F. Urrutia ‡, and J. D. Vergara‡

**Departamento de Física, CINVESTAV, Av. I.P.N. No. 2508, 07000 México, D.F.*
†*Departamento de Física, UAM-I, A. Postal 55-534, 09340 México, D.F.*
‡*Instituto de Ciencias Nucleares, UNAM, A. Postal 70-543, 04510 México D.F.*

Abstract. General Relativity in terms of Ashtekar variables describe complex gravity. To identify the real sector of the theory, reality conditions are implemented as second class constraints, leading to two real configurational degrees of freedom per space point. Nevertheless, this realization makes non-polynomial some of the constraints. For the sake of preserving the simplicity of the constraints, an alternative method preventing the use of Dirac brackets, is discussed. It consists of converting all second class constraints into first class by adding extra variables.

INTRODUCTION

In spite of its simplicity, the constraints of GR in terms of Ashtekar variables describe complex gravity [1]. Ashtekar himself considered reducing to the real sector, through the introduction of an inner product designed to make hermitian the physical operators. However, this strategy has not worked up to now, except for some particular cases. Other alternatives have been presented to avoid the use of reality conditions, at the price of a more cumbersome form of the constraints [2].

There still remains the possibility of keeping the self-dual canonical formalism and trying to envisage how to select the real sector of the theory. Indeed this is possible, as it has been shown for pure gravity in [3], at the classical level. Reality conditions are implemented as second class constraints. The present work is devoted to show that in the case of pure gravity it is possible to transform these second class contraints into first class.

FROM REALITY CONDITIONS TO FIRST CLASS CONSTRAINTS

The procedure to identify the reality conditions as second class constraints is as follows: the complex canonical variables are splitted into real and imaginary parts,

CP490, *Particles and Fields: Eighth Mexican School,* edited by J. C. D'Olivo et al.

each of which is taken as an independent new configuration variable. The corresponding momenta are subsequently defined from the action, leading to primary constraints. The real sector of the theory is next defined by introducing appropriate reality conditions in the form of additional primary constraints. This is possible because the original phase space has been extended. The whole set of constraints is next classified into first and second class, after imposing the conservation of the primary constraints upon evolution. Finally, one faces the problem of how to conveniently deal with the resulting second class constraints, which include the reality conditions.

The Ashtekar complex canonical variables are: (i) $\tilde{E}^{ai} := EE^{ai}$, with E^{ai} being the triad ($E^{ai}E^{b}{}_{i} := q^{ab}$, q^{ab} is the spatial three–metric), and $a, b, \ldots = 1, 2, 3$ are spatial indices, whereas $i, j, \ldots = 1, 2, 3$ are SO(3) internal indices. Also, $E := \det E_{bj}$ with E_{bj} being the inverse of E^{ai}. (ii) A_{ai} is the three–dimensional projection of the self-dual connection [1], with associated covariant derivative $\mathcal{D}_a\lambda_i = \partial_a\lambda_i + \epsilon_{ijk}A_a{}^j\lambda^k$ and $F_{ab}{}^i := \partial_a A_b{}^i - \partial_b A_a{}^i + \epsilon^i{}_{jk}A_a{}^j A_b{}^k$ is the corresponding curvature. In terms of these variables, the self-dual action of canonical GR is given by

$$S = \int dt d^3x \left\{ -i\tilde{E}^{ai}\dot{A}_{ai} - N\mathcal{S} - N^a\mathcal{V}_a - N^i\mathcal{G}_i \right\}, \tag{1}$$

where

$$\mathcal{S} := \epsilon_{ijk}\tilde{E}^{ai}\tilde{E}^{bj}F_{ab}{}^k, \quad \mathcal{V}_a := \tilde{E}^b{}_j F_{ab}{}^j, \quad \mathcal{G}_i := \mathcal{D}_a\tilde{E}^a{}_i\,, \tag{2}$$

are the constraints of the theory and N, N^a, N^i are Lagrange multipliers. Such constraints are first class and polynomial in the phase space variables. Let us denote by $\mathcal{R}$ any of them and by $\{\mathcal{R}\}$ the full set.

Notice that having 18 complex phase space variables $(A_{ai}, \tilde{E}^{bj})$, together with 7 complex first class constraints, $\{\mathcal{R}\}$, leaves us with 2 complex configurational degrees of freedom. Then, in order to recover the 2 real configurational degrees of freedom per point, further constraints are necessary. To this end, let us introduce the splitting

$$\tilde{E}^{ai} = \tilde{e}^{ai} + i\tilde{\mathcal{E}}^{ai}, \quad A_{bj} = \gamma_{bj} - iK_{bj}\,. \tag{3}$$

From now on, all the above thirty six degrees of freedom are taken as configuration variables in the action (1). Hence, the associated canonical momenta Π lead to the primary constraints $\Phi_{\mathcal{E}ai} = \Pi_{\mathcal{E}ai}$, $\Phi_\gamma{}^{ai} = \Pi_\gamma{}^{ai} + i\tilde{e}^{ai}$, $\Phi_K{}^{ai} = \Pi_K{}^{ai} + \tilde{e}^{ai}$, $\Phi_{eai} = \Pi_{eai}$, which as a set is denoted by $\{\Phi\}$. The coordinates of the total phase space are $Y_A = (\tilde{e}^{ai}, \tilde{\mathcal{E}}_{ai}, \gamma^{ai}, K_{ai}, \Pi_{eai}, \Pi_{\mathcal{E}ai}, \Pi_\gamma{}^{ai}, \Pi_K{}^{ai})$.

The reduction of the complex phase space $(A_{ai}, \tilde{E}^{bj})$ to a real one is achieved by means of the following reality conditions

$$\psi^{ai} := \tilde{\mathcal{E}}^{ai} = 0, \quad \chi_{ai} := \gamma_{ai} - f_{ai}(\tilde{e}) = 0, \tag{4}$$

which are subsequently taken as additional primary constraints. The constraints ψ^{ai} enforces the $\tilde{E}^{ai}$ to be real, and hence the corresponding three-metric. The constraints χ_{ai} ensures that, upon evolution, $\tilde{E}^{ai}$ keeps being real. Using the compatibility condition between a real torsion-free connection and the triad, the form of f_{ai} is chosen as $f_{ai} = \frac{1}{2}[\underset{\sim}{e}_{ai}\underset{\sim}{e}_{c}{}^{j}\epsilon_{jrs} - 2\underset{\sim}{e}_{aj}\underset{\sim}{e}_{c}{}^{j}\epsilon_{irs}]\tilde{e}^{dr}\partial_{d}\tilde{e}^{cs}$. Let us observe that χ_{ai} is not polynomial in $\tilde{e}^{bj}$.

The full set of primary constraints is $\{\ \{\mathcal{R}\},\ \psi,\ \chi, \{\Phi\}\ \}$, written in terms of the real variables Y_A. The evolution of the primary constraints does not introduce additional constraints. After redefining $\Phi_{eai} \rightarrow \Phi'_{eai} = \Phi_{eai} + \alpha_{aibj}\Phi_{\gamma}{}^{bj} + \beta_{ai}{}^{bj}\chi_{bj} + \eta_{aibj}\Phi_{K}{}^{bj}$, the Poisson brackets matrix for the subset $\{\Xi\} = \{\{\Phi'\},\ \psi,\ \chi\}$ reveals them as second class constraints. Besides having a simple form, it is a phase space independent, block diagonal matrix with non zero determinant.

To keep $\{\mathcal{R}\}$ as a first class set it is enough to redefine each element as $\mathcal{R}' = \mathcal{R} + \{\Phi_{\mathcal{E}bj}, \mathcal{R}\}\psi^{bj} + \{\Phi_{\gamma}^{bj}, \mathcal{R}\}\chi_{bj} + \{\Phi'_{ebj}, \mathcal{R}\}\Phi_{K}{}^{bj} - \{\Phi_{K}{}^{bj}, \mathcal{R}\}\Phi'_{ebj}$, so that they have zero Poisson brackets with the previous second class subset. This redefinition preserves the property $\{\mathcal{R}', \mathcal{Q}'\} \approx 0$, for any pair of constraints in $\{\mathcal{R}'\}$. In this way, there are no additional contributions to the set of primary constraints $\{\Upsilon\} := \{\mathcal{R}'\} \cup \{\Xi\}$, which includes the reality conditions. Counting the independent variables gives 2 real configurational degrees of freedom per space point, as it should be for real GR [3].

At this point Dirac's programme calls for the elimination of the second class constraints through the use of Dirac brackets. One might avoid such treatment of the second class constraints by transforming them into first class constraints. To achieve this, by means of the Batalin-Tyutin procedure [4,5], one adds a new canonical pair, $\{Q^{ai}, P_{bj}\} = \delta_{a}{}^{b}\delta_{i}{}^{j}\delta^{(3)}$, per original couple of second class constraints, i.e., the phase space is further enlarged with the new variables $\Psi_{\Xi} = (Q_{\mathcal{E}}{}^{ai}, Q_{\gamma ai}, Q_{e}{}^{ai}, P_{\mathcal{E}ai}, P_{\gamma}{}^{ai}, P_{eai})$. In the present case, the set of first class constraints replacing the former second class set is

$$\begin{aligned}
\bar{\psi}^{ai} &:= \tilde{\mathcal{E}}^{ai} + Q_{\mathcal{E}}{}^{ai}, \\
\bar{\Phi}_{\mathcal{E}ai} &:= \Pi_{\mathcal{E}ai} - P_{\mathcal{E}ai}, \\
\bar{\chi}_{ai} &:= \gamma_{ai} - f_{ai}(\tilde{e}) + Q_{\gamma ai}, \\
\bar{\Phi}_{\gamma}{}^{ai} &:= \Pi_{\gamma}{}^{ai} + i\tilde{e}^{ai} - P_{\gamma}{}^{ai}, \\
\bar{\Phi}_{K}{}^{ai} &:= \Pi_{K}{}^{ai} + \tilde{e}^{ai} + Q_{e}{}^{ai}, \\
\bar{\Phi}'_{eai} &:= \Phi_{eai} + \alpha_{aibj}\Phi_{\gamma}{}^{bj} + \beta_{ai}{}^{bj}\chi_{bj} + \eta_{aibj}\Phi_{K}{}^{bj} - P_{eai},
\end{aligned} \tag{5}$$

which reduces to the original set by setting $Q^{ai} = 0 = P_{bj}$. Let us denote any of the constraints in (5) by $\bar{\Xi}_{\Lambda}$. Any pair satisfies $\left\{\bar{\Xi}_{\Lambda}, \bar{\Xi}_{\Lambda'}\right\} = 0$; i.e. the set (5) is first class. Next, it is necessary to keep the set $\{\mathcal{R}'\}$ first class. This can be done by recalling that the Poisson brackets matrix among the constraints $\{\Xi\}$ is independent of the phase space variables and by following the method of [5]. Thus, one redefines $\mathcal{R}'$ as

$$\bar{\mathcal{R}}' \equiv \mathcal{R}'(Y - \bar{Y}) \ , \tag{6}$$

where

$$Y_A - \bar{Y}_A := \left\{ \tilde{e}^{ai} - Q_e{}^{ai}, \tilde{\mathcal{E}}^{ai} + Q_{\mathcal{E}}{}^{ai}, \gamma_{ai} + Q_{\gamma ai} + Q_e{}^{bj}\frac{\delta f_{ai}}{\delta \tilde{e}^{bj}}, K_{ai} + P_{eai} + iQ_e{}^{bj}\frac{\delta f_{bj}}{\delta \tilde{e}^{ai}}, \right.$$

$$\Pi_{eai} - P_{eai} - iQ_{\gamma ai} + P_\gamma{}^{bj}\frac{\delta f_{bj}}{\delta \tilde{e}^{ai}} + Q_e{}^{bj}\left(\frac{\delta^2 f_{ck}}{\delta \tilde{e}^{ai}\delta \tilde{e}^{bj}}\Phi_\gamma^{ck} + i\frac{\delta^2 f_{ai}}{\delta \tilde{e}^{ck}\delta \tilde{e}^{bj}}\Phi_K^{ck} + i\frac{\delta f_{bj}}{\delta \tilde{e}^{ai}}\right),$$

$$\left. \Pi_{\mathcal{E}ai} - P_{\mathcal{E}ai}, \Pi_\gamma{}^{ai} - P_\gamma{}^{ai} - iQ_e{}^{ai}, \Pi_K{}^{ai} \right\}. \tag{7}$$

The set (6) is in involution with $\{\bar{\Xi}_\Lambda\}$, i.e. $\left\{\bar{\mathcal{R}}', \bar{\Xi}_\Lambda\right\} = 0$. Hence, the final whole set of constraints is first class and contains an Abelian ideal: $\{\bar{\Xi}\}$. The non-Abelian sector is just given by $\{\bar{\mathcal{R}}'\}$. By construction, this sector preserves the structure of the first class algebra among the elements of $\{\mathcal{R}'\}$. Notice that the set $\{\mathcal{R}\}$ depends only on the configurational variables $(\tilde{e}, \mathcal{E}, \gamma, K)$. In this way, the most involved modifications to $\{\mathcal{R}'\}$, via Eq.(6), come from the terms that are proportional to the second class constraints. It is worth emphasizing that all the non-polynomiality of the constraints $\{\{\bar{\mathcal{R}}'\} \cup \{\bar{\Xi}\}\}$ arises only through one function, which is $f_{ai}(\tilde{e})$, appearing in the reality conditions (4).

The use of Dirac brackets, which is the standard way of eliminating the second class constraints, yields the expected real non polynomial form of the theory. For example, it leads to the Palatini canonical form in the case of pure gravity [1]. To explore an alternative preventing the use of Dirac brackets in the pure gravity case, we have implemented the conversion of the full set of second class constraints into a first class set, following the method of [5]. Thus, we have rewritten pure real gravity as a theory involving an alternative set of first class constraints, which, for example, has not been previously done starting from the Palatini formulation with second class constraints. However, their physical meaning, together with their usefulness in a quantum theory still needs to be clarified.

Partial support is acknowledged from CONACyT grant 3141P–E9608 and UNAM–DGAPA–IN100397. MM's postdoctoral fellowship is funded through the CONACyT of México, fellow number 91825.

REFERENCES

1. Ashtekar, A. Phys. Rev. Lett. **77**, 3288 (1986).
   Ashtekar, A. *Lectures on Non-Perturbative Canonical Gravity (Notes prepared in collaboration with R S Tate)*, Singapore: World Scientific, 1991.
2. Thiemann, T. *Class. Quantum Grav.* **15**, 839 (1998).
3. Morales–Técotl, H. A., Urrutia, L. F. and Vergara, J.D. *Class. Quantum Grav.* **13**, 2933 (1996).
4. Batalin, I. A. and Tyutin, I. V. *Int. J. Mod. Phys.* A **6**, 3255 (1991).
5. Amorim, R. and Das, A. *Mod. Phys. Lett.* A **9**, 3543 (1994).

Robustness of the Quantum Search Algorithm

B. Pablo-Norman and M. Ruiz-Altaba

Abstract. We find exact results for Grover's quantum search algorithm and analyze its behavior under noisy situations when no quantum correction codes are available. We compute how the algorithm slows down: it is still better than a classical one, provided the noise is smaller than some bound, which we also compute.

I INTRODUCTION

High reaction rates at hadronic colliders call for new ideas in triggering. Neural networks have been implemented in this setting with some success. We wish to explore the possibility of using quantum algorithms (presumably in classical computers) for this purpose. One crucial drawback of quantum algorithms implemented in quantum computers is their extreme dependence on the exact complex phase between various states. Quantum correcting codes can deal with simple situations, preventing to a large extent the loss of quantum coherence. Nevertheless, in an implementation on a classical computer, one would have to worry about the finite precision of these machines.

We thus analyze in this paper how a random gaussian noise, added to the output at each step of the algorithm, afects the recently proposed quantum search algorithm, i.e. a quantum procedure for finding a number in a phone book.

II GROVER'S QUANTUM SEARCH ALGORITHM

Classically, the only way to find a number in a random phone book (one not ordered alphabetically) is to search it entry by entry, checking each time whether it is the searched one. If it is, the search is over. If it is not, then continue. Thus, any classical algorithm (whether deterministic or probabilistic) will find the wanted number after $N/2$ steps, on the average. But this is only true when a classical algorithm is used. Recently, Grover found a quantum algorithm that requires only $O\left(\sqrt{N}\right)$ steps [1–4]. How does it work?

Suppose there are $N = 2^n$ entries in the phone book. Each of them can be represented by a quantum mechanical state of n spin-1/2 particles. Start with the initial state

CP490, *Particles and Fields: Eighth Mexican School,* edited by J. C. D'Olivo et al.

$$u_0 = 1/\sqrt{N} \begin{pmatrix} 1 \\ 1 \\ \vdots \\ 1 \end{pmatrix}. \tag{1}$$

Assume, for notational simplicity and without loss of generality, that the entry we are looking for is represented by the state $|\downarrow\downarrow \ldots \downarrow\rangle$, let B be the unitary transformation whose only action is invert the phase of the desired component, which in this case is

$$B = \begin{pmatrix} -1 & 0 & \cdots & 0 \\ 0 & 1 & 0 & \vdots \\ \vdots & 0 & \ddots & 0 \\ 0 & 0 & \cdots & 1 \end{pmatrix}. \tag{2}$$

The algorithm consists on the repeated action of the unitary transformation $X = DB$, where D is the diffusion matrix. Explicitly:

$$X = DB = \frac{2}{N} \begin{pmatrix} -1 + \frac{N}{2} & 1 & \cdots & 1 \\ -1 & 1 - \frac{N}{2} & 1 & \vdots \\ \vdots & 1 & \ddots & 1 \\ -1 & 1 & 1 & 1 - \frac{N}{2} \end{pmatrix}. \tag{3}$$

After m iterations of the unitary transformation X on u_0 the quantum state will be

$$u_m = X^m u_0 = \begin{pmatrix} A_m \\ B_m \\ \vdots \\ B_m \end{pmatrix}, \tag{4}$$

where the amplitudes are given by the recursion formula

$$\begin{pmatrix} A_{m+1} \\ B_{m+1} \end{pmatrix} = \begin{pmatrix} 1 - \frac{2}{N} & 2 - \frac{2}{N} \\ \frac{-2}{N} & 1 - \frac{2}{N} \end{pmatrix} \begin{pmatrix} A_m \\ B_m \end{pmatrix} = S \begin{pmatrix} A_m \\ B_m \end{pmatrix} = S^m \begin{pmatrix} \frac{1}{\sqrt{N}} \\ \frac{1}{\sqrt{N}} \end{pmatrix}, \tag{5}$$

The trick is that S can be diagonalized very easily, with eigenvalues $e^{\pm i\varphi}$ such that $\cos\varphi = 1 - \frac{1}{N}$ and therefore

$$A_m = \frac{1}{\sqrt{N}} \left(\cos(m\varphi) + \sqrt{N-1} \sin m\varphi \right) \tag{6}$$

$$B_m = \frac{1}{\sqrt{N}} \left(\cos(m\varphi) - \frac{1}{\sqrt{N-1}} \sin m\varphi \right) \tag{7}$$

Thus, the probability of finding the searched for state is:

$$P_m = \frac{1}{N}\left(\cos(m\varphi) + \sqrt{N-1}\sin m\varphi\right)^2, \tag{8}$$

with the change of variable $\varphi = 2\theta$, P_m can be written as [4]:

$$P_m = \sin^2(\theta(2m+1)), \tag{9}$$

This implies that P_m is periodic in m with period $\simeq \frac{\pi}{2}\sqrt{N}$, and reaches its maxima at

$$\theta(2m+1) = n\pi, n \text{ integer}, \tag{10}$$

For large N, the first maximum simplifies

$$m \sim \frac{\pi\sqrt{N}}{4}. \tag{11}$$

This exact result agrees with Grover's, in the sense of the existence of a number $m < \sqrt{N}$, such that after m iterations of the algorithm, if we measure the state of the system, we will find the searched one with a probability of at least 0.5.

III NOISY QUANTUM SEARCH ALGORITHM

Like all experimental devices, quantum computers will be subject to noise. Let us assume that the same Gaussian noise is present at each step of the algorithm, i.e. each time the unitary X matrix is applied. Even though some quantum correction codes have been developed [5,6], it is known that these codes work only if the noise is small enough. Moreover, it is not known whether these codes are subject to noise themselves, and if they are, whether they can still be useful. So, for simplicity, we assume that no quantum correction code is available, and study the effect of Gaussian noise in Grover's algorithm.

First, let us investigate the maximum noise the algorithm can put up with before it loses periodicity and, worse, the searched for amplitude is no longer enhanced. The size of the white noise is characterized by the standard deviation σ of its normal distribution. In numerical experiments, we found that the maximum noise the algorithm can allow before it breaks down is:

$$\sigma_{\max} \simeq \frac{4}{3N}. \tag{12}$$

Unfortunately, the amount of noise that the algorithm can handle is very small for large databases.

Secondly, assume that the noise is smaller than $\sigma_{\max}$. What happens to the number m of steps needed to reach $P_m \simeq 1$, i.e. to find almost certainly the searched–for state? Let us focus on the limiting case when the algoritm still works, with noise given by σ_{max}. In another numerical experiment we found that

$$m_{\text{max}}^{(\sigma_{\text{max}})} \simeq 2N^{\frac{3}{4}}. \tag{13}$$

Thus, the exponent of N increases from $\frac{1}{2}$ to $\frac{3}{4}$. The algorithm slows down but is still faster than a classical one (this is all for large N).

Recently, Grover's algorithm with $N = 4$ has been succesfully implemented experimentally [7]. Our explicit results 12 and 13 are evaluated for large N, so they do not apply to this case. Still, we can compute exactly the effect of white noise on the speed and robustness of the algorithm. These results will be presented elsewhere.

IV CONCLUSIONS

Grover's quantum search algorithm requires $\mathcal{O}\left(\sqrt{N}\right)$ steps, for a large database with N entries. It thus improves any classical algorithm, needing $\mathcal{O}(N)$ steps. Nevertheless, if noise is present, the algorithm slows down to $\mathcal{O}\left(N^{\frac{3}{4}}\right)$ steps, before breaking down completely. This breakdown occurs when the width of the white noise reaches $\frac{4}{3}N^{-1}$. For large N, consequently, the algorithm can withstand very little noise.

Acknowledgements. This work is supported in part by CONACYT 25504-E, DGAPA-UNAM IN103997. B.P.N. enjoys a scholarship from CONACYT.

REFERENCES

1. L.K. Grover, *Quantum computers can search rapidily by using almost any transformation*, Phys. Rev. Lett. vol. 80 (1998) 4329-4332.
2. L.K. Grover, *Quantum Mechanics helps in searching for a needle in a haystack*, Phys. Rev. Letters, vol. 78 (2) (1997) 325-328..
3. D. Pyo, J. Kim, *Quantum Database Searching by a Simple Query*, quant-ph/9708005.
4. M. Boyer, G. Brassard, P. Hoeyer / A. Tapp, *Tight bounds on quantum searching*, Proc., PhysComp 1996..
5. A. Steane, *Multi-particle interference and quantum error correction*, Proc. R. Soc. Lond. A 452 (1996) 2551-2577.
6. A.R. Calderbank and P.W. Shor, *Good quantum error correction codes exist*, Phys. Rev. A 54 (1996) 1098-1105.
7. I.L. Chuang, Lieren, M.K. Vandersypen, Xin-lan Zhou, D.W. Leung, S. Loyd, *Experimental realization of a quantum algorithm*, quant-ph/9801037.

Exotic sources of Ultra High Energy Cosmic Rays

Abdel Pérez-Lorenzana[1]

Department of Physics, University of Maryland, College Park, Maryland 20742, USA.

Abstract. Ultra High Energy Cosmic Rays events beyond the theoretical cutoff of 10^{19} eV have been detected by several experiments. Nevertheless we do not know any potential astronomical source which could accelerate particles to those energies. We discuss new ideas for the production of these events, the so called "top-down" mechanisms, which involve exotic relics of the early universe.

INTRODUCTION

More than 30 years ago, the Volcano Ranch experiment [1] reported the observation of one cosmic ray event with an energy above 10^{20} eV. Since then, several other experiments have observed, so far, up to 24 events around 10^{20} eV, and some hundreds above 10^{19} eV [2]. Nevertheless, the shock acceleration in the powerful astrophysical objets seems to be uncapable to produce these Ultra High Energy (UHE) cosmic rays (CR) [3], establishing a serious challenge to the theory. Along this short paper I will give a general overview of the current status of this problem and to comment on the open questions and how the particle physics could provide us with the answers. For a deeper and detailed discussion on this topic see ref. [4].

THE MYSTERY

The puzzle of the UHE CR may be summarized by a three parts problem: (i) the origin: how can the nature produce particles at such extremely high energies?; where do they come from?; (ii) the propagation: how may a UHE particle travel from their source to the earth? and (iii) their interactions with the atmosphere. Assuming that the last part is the best understood – even though the chemical composition is unknown and the statistics is still poor, since those UHE CR have

[1] On leave of absence from Departamento de Física, Centro de Investigación y de Estudios Avanzados del I.P.N. Apdo. Postal 14-740, 07000, México, D. F., México.

CP490, *Particles and Fields: Eighth Mexican School,* edited by J. C. D'Olivo et al.

a very low flux: one event per km^2 per century for energies above 10^{20} eV –, I will concentrate only on the other two aspects.

The problem of production. The energy spectrum of CR above 10 GeV presents three different regions characterized by a power law. The index changes from -2.7 to -3.0 around 3×10^{15} eV (the first "knee"), and it changes again to -3.3 around the second "knee" ($\sim 10^{17.5}$ eV), to increases to about -2.7 above 3×10^{18} eV (the "ankle"). The most of this spectrum is well explained with the first order Fermi shock acceleration mechanism at energetic objects such as supernova remanents [3]. Nevertheless, no matter how the particles (basically protons and heavy nuclei) are accelerated, there is an upper bound to the gained energy, since during the acceleration process, the gained energy competes with several other energy loss processes as the synchrotron radiation and the collisions with matter of the acceleration site. This upper bound may be estimated by the Hillas relation [5]

$$B_{\mu G} L_{Pc} > \frac{E}{10^{15} eV} \frac{1}{Z\beta} \tag{1}$$

where $B_{\mu G}$ is the magnetic field; L_{Pc} the size in parsec; and β is the speed of the scattering waves of the site. From this relationship the most of the astrophysical objects are ruled out as possible sources. Moreover, nuclei may disintegrate by collision in a dense ambient which also may rule out other candidates such as AGNs. Radio Galaxy Lobes and other similar objects remains as possible sources.

The case of photons and neutrinos as primary particles is harder than for protons and nuclei, since such particles are believed to be produced from the decaying processes of extremely high energy pions. Hence, a proton flux domination is the signature of the acceleration mechanism at those energies. Besides, the primary gamma ray hypothesis appears inconsistent with the temporal distribution of the Fly's Eye event [6], and also, the density profile of the Yakutsk event showed a large number of muons, which argues against this hypothesis [7]. On the other hand, the Fly's eye event occurred high in the atmosphere, whereas the expected event rate for early development of a neutrino induced air shower is down from that of an electromagnetic or hadronic interaction by six orders of magnitude.

It is possible that Gamma Ray Burst may be related with the origin of UHE photons and neutrinos, but those are vary rare and the possibility of detect a correlation of both events is extremely difficult [8].

The problem of propagation. If we assume there are enough candidate sites as to support the shock acceleration mechanism to produce UHE protons (or heavy nuclei), next step is to explain how these UHE particles travel through the cosmic microwave background (at $2.7^o K$) and reach the earth at energies above 10^{20} eV. Nucleons with energies above 7×10^{19} eV rapidly loss energy through photo pion production. This effect result in a cutoff in the energy spectrum (which seems to be absent), the Greisen-Zapsepin-Kuzmin (GZK) cutoff [9]. The typical attenuation length for a UHE proton produced by this effect is around 100 Mpc. A similar situation is present in the case of heavy nuclei, now photodisintegration is the limiting factor which produces a more stringent bound to the distance. The case of photons is not better since the mean free path for a 10^{20} eV photon before it annihilates on the microwave background into a e^-e^+ pair is of the order of 10 to 40 Mpc. All that means that the sources of the UHE CR should be very nearby, and since these UHE particles weakly interact with the galactic and extragalactic magnetic fields, they point directly to their sources. Nevertheless, there are not obvious astronomical sources within 100 Mpc and the distribution seems to be isotropic.

TOP-DOWN MECHANISMS, A WAY OUT

All difficulties in accelerating particles are avoided if the primaries are produced at so UHE. This called "top-down" scenarios involve relics of the early universe as topological defects (TD) and cold dark matter (CDM).

TD. Magnetic monopoles, (superconducting) cosmic strings, domain walls, textures, and various hybrid systems are predicted to form during the phase transition associated with the symmetry breaking of the Grand Unified Theory [10]. They are topologically stable, nevertheless, they can release part of their energy in the form of superheavy "X particles", with typical masses at the scale $\sim 10^{16}$ GeV, through several physical processes [11]. Cosmic strings may produce particles when a loop formed by collisions or self interactions collapses. Superconducting strings [12] produce X particles when the electric currents in the strings reaches critical values. Monopole-antimonopole pairs can form bound states (monopolonium) which spirals in and finally collapses. For an appropriate choice of the monopole density, this model is consistent with observations. Also monopoles connected by strings may produce X particles when the monopoles are pulled by the strings [13]. The X particles are usually unstable, and subsequently decay into leptons, quarks and gauge particles, then neutrinos and photons dominate the total flux in this scenario. TD predict an injection spectra which are considerably harder than shock acceleration spectra, although the predicted absolute flux levels are model dependent.

It has been also suggested that the primary particles could be relativistic massive magnetic monopoles, $M \lesssim 10^{10\pm1}\ GeV$ [14]. This is the less explored possibility.

CDM. Superheavy X particles produced at the end of the inflation and in the reheating phase could form part of the CDM. In order to account for UHE CR these relic particles should be long lived, $10^{10} yr \gtrsim \tau_X \gtrsim 10^{22} yr$, heavy, $m_X \gtrsim 10^{12}$ GeV, and weakly interacting. Several mechanism of production of such kind of particles have been identified, related with SUGRA, Quantum Gravity and the SUSY breaking sector [15]. These relic particles (and also monopolonia) are expected to be clustered in the Galactic Halo, therefore their decays and collision processes, which may produce UHE CR do not have the GZK cutoff [16]. Besides the photon domination of the flux the CDM scenario presents an anisotropy which could be at the level of 20% when the fluxes from the Galactic Center and Anticenter are compared [17], for the the asymmetric position of the sun into the galaxy.

CONCLUDING REMARKS

As we have discussed along the present work, the UHE CR problem could be a very promising window to test the particle physics that extent the Standard Model, which on the other hand seems to be far from being tested at the current accelerators. The production of primary particles in the top down scenarios predict clear signatures which could be observed in the surface detectors as the planed "Pierre Auger" Observatory [18].

Acknowledgments

This work was partially supported by CONACyT, México. I would like to acknowledge A. Zepeda for his encourage and advise.

REFERENCES

1. J.Linsley, *Phys. Rev. Lett.* **10**, 146 (1963).
2. For a review of the current experimental situation see: S.Yoshida and H.Dai, *J. Phys.* **G24**, 905 (1998), and references therein.
3. For a review see R. Blandford and D. Eichler, *Phys. Rep.* **154**, 1 (1987).
4. P.Bhattacharjee and G. Sigl, astro-ph/9811011.
5. A.M.Hillas, *Ann. Rev Astron. Astrophys.* **22**, 425 (1984).
6. F.Halzen, R.A.Vazquez, T.Stanev and V.P.Vankov, *Astropart. Phys.* **3**, 151 (1995).
7. N.N. Efimov *et. al.* (Yakutsk Collab.), *ICRR Symposium on Astrophysical Aspects of the Most Energetic Cosmic Rays.* ed. N. Nagano and F. Takahara, World Scientific pub. (1991); and *Proc. 22nd ICRC*, Dublin (1991).
8. E.Waxman, *Phys. Rev. Lett.* **75**,386 (1995). M.Vietry, *Astrophys. Journal* **453**, 883 (1995).
9. K. Greisen, *Phys. Rev. Lett.* **16**, 748 (1966); G.T. Zatsepin and V.A. Kuzmin, *Pisma Zh. Eksp. Teor. Fiz.* **4**, 114 (1966) [*JETP Lett.* **4** 78 (1966)];
10. See for example: A. Vilenkin and E.P.S. Shellard, *"Cosmic Strings and Other Topological Defects"*, Cambridge University Press (1994), and references therein.
11. P.Bhattacharjee, C.T.Hill and D.N. Schramm,*Phys. Rev. Lett.* **69**, 567 (1992). P.Bhattacharjee and G.Sigl, *Phys. Rev. D* **51**, 4079 (1995).
12. E. Witten, *Nucl. Phys. B* **249**,557 (1985).
13. V.Berezinsky, X. Martin and A. Vilenkin, *Phys. Rev. D* **56**, 2024 (1997).
14. N. A. Porter, *Nuovo Cim.* **16**, 958 (1960); T. W. Kephart and T. J. Weiler, *Astropart. Phys.* **4**, 271 (1996); *Nucl. Phys.* (Proc Suppl.) **51B**, 218 (1996).
15. See for example: K.Benakli, J. Ellis and D. Nanopoulos, hep-ph/9803333.
16. V.Berezinsky, M.Kachelriess and A.Vilenkin, *Phys. Rev. Lett.* **79**, 4302 (1997).
17. S. L. Dubovsky, P. G. Tinyakov, *JETP Lett.* **68**, 107 (1998).
18. The Pierre Auger Project Design Report, 2nd. edition, Fermilab, February 1997; also M. Boratav, astro-ph/9605087.

SUSY String-GUTs

Abdel Pérez–Lorenzana*[1], William A. Ponce*,†, and Arnulfo Zepeda*

* *Departamento de Física, Centro de Investigación y de Estudios Avanzados del I.P.N. Apdo. Post. 14-740, 07000, México, D.F., México.*
† *Departamento de Física, Universidad de Antioquia A.A. 1226, Medellín, Colombia.*

Abstract. We present the available list of Kac-Moody levels coming from Grand Unified Theories (GUTs), with the aim to look for new insights in the four dimensional supersymmetric (SUSY) strings.

String theory is the leading candidate for a unified theory of all the interactions, including quantum gravity and chiral supersymmetric (SUSY) gauge theories. When one loop string effects are included [1] they predict a unification of the gauge couplings at a scale $M_{string} \sim 3.6 \times 10^{17}$ GeV. Then the standard model (SM) particles plus the graviton are associated to string modes which are approximately massless compared to the huge string mass scale.

Unification of coupling constants is a necessary phenomenon in string theory. Specifically, at tree level, the gauge couplings $\alpha_i = g_i^2/4\pi$ of the three SM interactions are related at the string scale by [2]

$$\kappa_3\alpha_3 = \kappa_2\alpha_2 = \kappa_1\alpha_1 = \alpha_{string}, \tag{1}$$

where κ_i, $i = 1,2,3$ are the affine levels, or Kac-Moody levels, at which the group factor $U(1)_Y$, $SU(2)_L$, and $SU(3)_c$ is realized in the four-dimensional string, (g_1, g_2, g_3 are the gauge coupling constants of $U(1)_Y$, $SU(2)_L$ and $SU(3)_c$ respectively, and α_{string} is related to the string tension).

The starting point is the ten-dimensional heterotic string with gauge group $SO(32)$ or $E_8 \otimes E_8$ corresponding to an affine Lie algebra at level $\kappa = 1$. A standard compactification [3] leads to a four dimensional model with gauge group formed by a product of non-abelian gauge groups G_i realized at levels $\kappa_i = 1$, times $U(1)$ factors. Building string theories with non-abelian algebras at higher levels ($\kappa = 2, 3, \ldots$) is considerable more difficult than at level one, and new methods for compactification must be developed [4] (to produce levels beyond $\kappa = 3$ is a very cumbersome task). Now, the affine levels for abelian $U(1)$ factors can not be

---

1) On leave of absence at Department of Physics, University of Maryland, College Park, Maryland 20742, USA.

CP490, *Particles and Fields: Eighth Mexican School,* edited by J. C. D'Olivo et al.

determined from algebraic procedures and their values may be considered as free parameters in the four dimensional string [5].

Then, the compactification of the heterotic string to the four dimensional G_{SM} could be achieved at M_{string}, with $\kappa_2, \kappa_3 = 1, 2, \ldots n$, an integer number, and κ_1 a normalization free coefficient ($\kappa_1 > 1$ in order for the e_R to be in the massless spectrum of the four dimensional string [6]). The compactification to a four dimensional simple gauge group $G(= SU(5),\ SO(10),\ E_6$, etc.) has also been partially studied in the literature, with upper values for the integer κ levels calculated [7]. Also, string-GUTs with $SU(5) \subset SU(5) \otimes SU(5)$ and $SO(10) \subset SO(10) \otimes SO(10)$ at levels $\kappa_2 = \kappa_3 = 2$ have been presented in Ref. [4].

So far, almost the entire literature on four dimensional strings has been focused on the so called canonical values $\kappa_2 = \kappa_3 = 1$ and $\kappa_1 = 5/3$, featuring a Calabi-Yau inspired phenomenology.

In general the ten dimensional string can compactify directly to $G_{SM} = SU(3)_c \otimes SU(2)_L \otimes U(1)_Y$, the SM gauge group, or either it can compactify to a simple (or semisimple) gauge structure like SU(5), SO(10), etc. which may act as a gauge group for a grand unified theory (GUT). For this last case the Kac-Moody levels may take values different to the canonical ones.

But what are the values for the Kac-Moody levels allowed by the different GUTs? As far as we know this question has not been answer yet in a systematic way, and it is our aim to give in this note at least a partial answer to this question.

In the table we present the κ_i $i = 1; 2; 3$ values for most of the GUT groups in the literature, calculated using group theoretical analysis, by normalizing to a common value all the quantum numbers of the particles in each used representation, for each model. The **CANONICAL** entry is associated with the following nine groups: $SU(5)$ [8], $SO(10)$ [9], E_6 [10], $[SU(3)]^3 \times Z_3$ [11], $SU(15)$ [12], $SU(16)$ [13], $SU(8) \times SU(8)$ [14], E_8 [15], and $SO(18)$ [16]. The model $[\mathrm{SU}(3)]^4 \times Z_4$ is taken from Reference [17], $SU(5) \otimes SU(5)$ from [18], $SO(10) \otimes SO(10)$ from [19], $[SU(6)]^3 \times Z_3$ from [20], $[SU(6)]^4 \times Z_4$ from [21], E_7 from [22], $[SU(4)]^3 \times Z_3$ from [19], and $[SU(2F)]^4 \times Z_4$ (the Pati-Salam models for F families) from [23].

In the **CANONICAL** entry we have normalized the κ_i values for some groups to the SU(5) numbers; for example, the actual values for SO(10) are $\{\kappa_1; \kappa_2; \kappa_3\} = \{10/3; 2; 2\} = 2\{5/3; 1; 1\}$, and for SU(16) are $\{\kappa_1; \kappa_2; \kappa_3\} = \{20/3; 4; 4\} = 4\{5/3; 1; 1\}$. This normalization makes sense because the common factor can be absorbed in the string tension; besides, physical quantities such as $\sin^2 \theta_W$, M_{GUT}, etc., depend only on ratios of the κ_i values, for example [5]

$$\sin^2 \theta_W = \frac{\kappa_2}{\kappa_1 + \kappa_2}, \tag{2}$$

other normalizations have been used for other groups.

κ_3 can take only the values $1, 2, 3, 4$ for one family groups, or higher integer values for family groups. $\kappa_3 = 1$ when it is $SU(3)_c$ which is embedded in the GUT group G; $\kappa_3 = 2$ when it is the chiral color [24] $SU(3)_{cL} \times SU(3)_{cR}$ which is embedded in G, etc. For example $\kappa_3 = 4$ in SU(16) due to the fact that the color group in SU(16) is $SU(3)_{cuR} \times SU(3)_{cdR} \times SU(3)_{cuL} \times SU(3)_{cdL}$.

For some family groups κ_2 take the values $1, 2, \ldots F$ for $1, 2, \ldots F$ families. Indeed, the c_i values for the F family Pati-Salam models [23] $[SU(2F)]^4 \times Z_4$ are $\{\kappa_1; \kappa_2; \kappa_3\} = \{(9F-8)/3; F; 2\}$; and for $[SU(2F)]^3 \times Z_3 = SU(2F)_L \otimes SU(2F)_c \otimes SU(2F)_R \times Z_3$ (the $2F$ color vectorlike version of the Pati-Salam models [25]),are $\{\kappa_1; \kappa_2; \kappa_3\} = \{(6F-4)/3; F; 1\}$.

In general, $\kappa_{2(3)} = 1, 2, \ldots f$, where f is the number of fundamental representations of $SU(2)_L$ ($SU(3)_c$) contained in the fundamental representation of the GUT group. For example, $\kappa_2 = 4$ in $SU(16)$ because the 16 representation of $SU(16)$ contains four $SU(2)_L$ doublets; three for $(u, d)_L$ and one for $(\nu_e, e)_L$.

The group $[SU(4)]^3 \times Z_3$ in the Table is not the vector-like color version of the two family Pati-Salam model, but it is the one family model introduced in Ref. [19]. The group $[SU(6)]^4 \times Z_4$ in the Table could be the three family Pati-Salam model [23], or either the version of such a model without mirror fermions introduced in Ref. [21]. All models in the Table are realistic, except E_7 [22] which is a two family model with the right handed quarks in $SU(2)_L$ doublets (this is why $\kappa_1 < 1$ for this odd model).

From the table we see that all the values for κ_1 are integer multiple of 1/3, which is a necessary condition in order to avoid exotic electric charges in the four dimensional string. Indeed, the condition for having only standard electric charges after compactification reads [6]

$$\kappa_1 + \kappa_2 + \frac{4}{3}\kappa_3 = 0 \qquad \text{mod} \quad 4, \tag{3}$$

which is satisfied by all the entries in the table (in some entries the real values must be used instead of the normalized ones, (for example for $[SU(3)]^4 \times Z_4$ we have $\{\kappa_1; \kappa_2; \kappa_3\} = \{10; 6; 12\} = 6\{\frac{5}{3}; 1; 2\}$)

In this note most of the four dimensional string Kac-Moody levels which could be related to GUT theories are presented in the Table (a guide for string-GUT model builders). From the table we may visualize the wide spectrum available for the values κ_i, $i = 1, 2, 3$.

Concerning the actual value κ_1 in specific models, in the orbifold examples constructed up to now the value of κ_1 is never the canonical 5/3 and has in fact a tendency to be larger [4]. As a guide, the κ_1 value in an specific four dimensional string theory should point to the particular GUT group which unifies the low energy SM group.

So far, almost the entire literature on four dimensional strings has been focused on the canonical values $\kappa_2 = \kappa_3 = 1$, $\kappa_1 = 5/3$. As it is known, there are serious problems with the models constructed so far. Just to mention a few we have: the string-GUT problem [26], the doublet-triplet problem [4], the failure to produce a consistent low energy particle spectrum [4], etc. It may be feasible that the construction of four dimensional string theories with non-canonical κ_i values may

TABLE 1. κ_1, κ_2 and κ_3 values for most of the GUT models in the literature. $F = 1, 2, \ldots$ stand for the number of families in that particular model. The nine "canonical" groups are presented in the main text.

Group	κ_1	κ_2	κ_3
CANONICAL (9 groups)	5/3	1	1
$[SU(3)]^4 \times Z_4$	5/3	1	2
SU(5)⊗ SU(5), SO(10)⊗SO(10)	13/3	1	2
$[SU(6)]^3 \times Z_3$	14/3	3	1
$[SU(6)]^4 \times Z_4$	19/3	3	2
E_7	2/3	2	1
$[SU(4)]^3 \times Z_3$	11/3	1	1
$[SU(2F)]^4 \times Z_4$	$(9F-8)/3$	F	2
$[SU(2F)]^3 \times Z_4$	$(6F-4)/3$	F	1

ameliorate, or even cure some of the mentioned problems (in the model of Ref. [17], $M_{GUT} \geq M_{string}$, and the doublet triplet problem is not present at tree level).

κ_i, $i = 1, 2, 3$ values different from the canonical ones are in general related to the existence of non standard matter. That extra matter can have a mass at an intermediate scale, or either at the string-GUT scale. (most of the string-GUT models constructed so far contain non standard matter).

Finally let us say that the values attained by level κ are crucial in the string theory, since they fix at the string scale the electroweak mixing angle $\sin\theta_W$. Besides, they impose limits on possible representations allowed at low energies [4], and determine the conformal spin of the currents J which are forced to be in the spectrum because of charge quantization [6]. So, theories with different κ_i values must have quite different physical implications.

Acknowledgments

This work was partially supported by CONACyT, México.

REFERENCES

1. V.Kaplunovsky, Nucl. Phys. **B307**, 145 (1988); *ibid* **B382**, 436 (1992).
2. P.Ginsparg, Phys. Lett. **B197**, 139 (1987).
3. L.Dixon, J.Harvey, C.Vafa, and E.Witten, Nucl. Phys. **B261**, 620 (1985); **B274**, 285 (1986); L.E.Ibañez, J.Mas, H.P.Nilles, and F.Quevedo, Nucl. Phys. **B301**, 137 (1988).

4. A.Font, L.E.Ibañez, and F.Quevedo, Nucl. Phys. **B345**, 389 (1990); G.Aldazabal, A.Font, L.E.Ibañez, and A.M.Uranga, Nucl. Phys. **B452**, 3 (1995); **B465**, 34 (1996).
5. L.E.Ibañez, Phys. Lett. **B303**, 55 (1993); **B318**,73 (1993).
6. A.N.Schellekens, Phys. Lett. **B237**, 363 (1990).
7. J.Ellis, J.L.López, and D.V.Nanopoulos, Phys. Lett. **B245**, 375 (1990).
8. H.Georgi and S.L.Glashow, Phys. Rev. Lett. **32**, 438 (1974); H.Georgi, H.R.Quinn, and S. Weinberg, Phys. Rev. Lett. **33**, 451 (1974).
9. H.Georgi, in *Particles and Fields*-1974, edited by C.E.Carlson (American Institute of Physics, N.Y. 1975), p. 575; H. Frietzsch and P. Minkowsky, Ann. Phys. **93**, 193 (1975).
10. F.Gürsey, P.Ramond and P.Sikivie, Phys. Lett. **B60**, 177 (1975); S.Okubo, Phys. Rev. **D16**, 3528 (1977).
11. A. de Rújula, H.Georgi, and S. Glashow, in *Fifth Workshop on Grand Unification*, edited by K.Kang, H.Fried, and P.Frampton (World Scientific, Singapore, 1984), p. 88; K.S.Babu, X-G. He, and S. Pakvasa, Phys. Rev. **D33**, 763 (1986).
12. S.L.Adler, Phys. Lett. **B225**, 143 (1989); P.H.Frampton and B.H.Lee, Phys. Rev. Lett. **64**, 619 (1990); P.B.Pal, Phys. Rev. **D43**, 236 (1991); *ibid*, **D45**, 2566 (1992).
13. J.C.Pati, A.Salam, and J.Strathdee, Phys. Lett. **B108**, 121 (1982); N.G.Deshpande, E.Keith, and P.B.Pal, Phys. Rev. **D47**, 2893 (1993).
14. N. G. Deshpande and P.Mannheim, Phys. Lett. **B94**, 355 (1980); S. Day and M. K.Parida, Phys. Rev. **D52**, 518 (1995).
15. I.Bars and M.Gunaydin, Phys. Rev. Lett. **45**, 859 (1980); S.M.Barr, Phys. Rev. **D37**, 204 (1998).
16. F.Wilczek and A.Zee, Phys. Rev. **D25**, 553 (1982); J.Bagger *et al*, Nucl. Phys. **B258**, 565 (1985).
17. A.Pérez-Lorenzana and W.A. Ponce: "Unification above the string-GUT scale", hep-ph/9812402, submitted for publication.
18. A.Davidson and K.C.Wali, Phys. Rev. Lett. **58**, 2623 (1987); R.N.Mohapatra, Phys. Lett **B379**, 115 (1996).
19. P.Cho, Phys. Rev. **D48**, 5331 (1993).
20. A.H-Galeana, R.Martínez, W.A.Ponce, and A.Zepeda, Phys. Rev. **D44**, 2166 (1991); W.A.Ponce and A.Zepeda, Phys. Rev. **D48**, 240 (1993).
21. W.A.Ponce and A.Zepeda, Z. Physik **C63**, 339 (1994)
22. F.Gürsey and P.Sikivie, Phys. Rev. Lett. **36**, 775 (1976); Phys. Rev. **D16**, 816 (1977).
23. J.C.Pati and A.Salam, Phys. Rev. Lett. **31**, 661 (1973); V.Elias and S.Rajpoot, Phys. Rev. **D20**, 2445 (1979).
24. J.C.Pati and A.Salam, Nucl. Phys. **B50**, 76 (1979); P.H.Frampton and S.L.Glashow, Phys. Lett. **B190**, 157 (1987).
25. Unpublished results.
26. For a review see: K.R.Dienes, Phys. Rep. **287**, 447 (1997).

Pion Scattering Revisited

M. Ruiz-Altaba, J.L. Lucio* and M. Napsuciale*

Departamento de Física Teórica, Instituto de Física
Universidad Nacional Autónoma de México, A.P. 20-364, 01000 México, D.F.

**Instituto de Física, Universidad de Guanajuato*
Loma del Bosque 103, 37160 León, Guanajuato, México

Abstract. Chiral Ward identities lead to consistent accounting for the σ's width in the linear sigma model's Feynman rules. Reanalysis of pion scattering data at threshold imply a mass for the σ of 600^{+200}_{-100} MeV.

This short talk (by M.R-A) reviews our recent work on the linear sigma model [1,2], where full references can be found. At low energies, chiral perturbation theory is supposed to yield good agreement with strong interaction data. Unfortunately, chiral perturbation theory gives rather poor results on the scattering lengths of pion-pion scattering, which are relevant experimental quantities in the limit of zero momentum, that is to say, where chiral perturbation theory should work best.

A missing ingredient in the description at low energies of strong interactions is the σ field, in addition to the Goldstone bosons of chiral symmetry (the pions). A wide scalar resonance in the vicinity of 600 MeV exists, and can be identified naturally with the σ particle of the original *linear* σ–model.

What are the phenomenological consequences of the linear σ–model in $\pi\pi \to \pi\pi$ scattering at very low energies? The sole guiding principle is chiral symmetry, whose Ward identities allow us modify the various vertices to take into account the large width of the σ resonance.

The chiral symmetry breaking giving mass to the pions is soft, so that when we include the width Γ_σ of the σ in its propagator, we can exploit the chiral Ward identities to modify the vertices accordingly. The chiral Ward identities are satisfied by the resulting lagrangian (with parameters m_π, f_π, m_σ), from which we compute the amplitudes in the various isospin and angular momentum channels of experimental relevance. We use the expression for Γ_σ from the decay $\sigma \to \pi\pi$ to perform a simple and succesful one–parameter (m_σ) fit to data.

The field σ is very unstable: its tree–level width is

$$\Gamma\left(\sigma \to \pi\pi\right) = \frac{3m_\sigma^3}{32\pi f_\pi^2}(1-\varepsilon)^2\sqrt{1-4\epsilon}$$

CP490, *Particles and Fields: Eighth Mexican School,* edited by J. C. D'Olivo et al.

where we have introduced the convenient shorthand $\varepsilon = (m_\pi/m_\sigma)^2$. In strict analogy with the Higgs field in the standard model, the σ width Γ_σ grows very fast with its mass: $\Gamma_\sigma(350) = 65$, $\Gamma_\sigma(500) = 310$, $\Gamma_\sigma(650) = 785$, all in MeV. The effect of the width of the σ field is to modify its propagator from the usual $i\left(q^2 - m_\sigma^2\right)^{-1}$ to $\Delta_\sigma(q) = i\left(q^2 - m_\sigma^2 + i\Gamma_\sigma m_\sigma\,\theta(q^2 - 4m_\pi^2)\right)^{-1}$, where the step function ensures that the imaginary piece in the denominator appears only when the momentum of the propagator is above the kinematical threshold for σ decay.

Thus, in the physical process of $\pi\pi \to \pi\pi$ scattering, which we shall consider shortly, the propagator of the σ picks up the correction due to the width only in the s–channel, not in the u– nor the t–channels.

The crucial point is that, in the linear σ model, chiral symmetry is responsible for important cancellations which imply, notably, that the pion coupling is always derivative in the limit of soft pion momenta. Enforcing the chiral Ward identities on the vertices of the lagrangian implies that the latter pick up modifications related to the width Γ_σ. These vertex corrections depend on the kinematical variables (the incoming momenta) in a particular way, dictated by chiral symmetry. For instance, the $\sigma\pi^i\pi^j$ Feynman rule reads now

$$V_{\sigma\pi^i\pi^j} = \frac{-i}{f_\pi}\delta^{ij}\left(m_\sigma^2 - m_\pi^2 - i\Gamma_\sigma m_\sigma \theta(q^2 - 4m_\pi^2)\right)$$

where q^μ is the momentum of the σ.

We find also

$$V_{\pi^i\pi^j\sigma\sigma} = V_{\sigma\sigma\sigma}\Delta_\sigma(p_j)V_{\sigma\pi^i\pi^j}$$

where p_j is the momentum of a pion, so that $p_j^2 = m_\pi^2$ if it is on–shell. This equation *defines* the $\pi\pi\sigma\sigma$ vertex. Similarly, the chiral Ward identity satisfied by the π^4 Feynman rule is

$$V_{\pi^i\pi^j\pi^k\pi^\ell} = V_{\pi^k\pi^\ell\sigma}\Delta_\sigma(p_j)V_{\sigma\pi^i\pi^j} + V_{\pi^i\pi^k\sigma}\Delta_\sigma(p_k)V_{\sigma\pi^j\pi^\ell} + V_{\pi^i\pi^\ell\sigma}\Delta_\sigma(p_\ell)V_{\sigma\pi^j\pi^k}$$

Obviously, these relations hold at tree level before chiral symmetry breaking, that is to say, when $m_\pi = 0$, and also $\Gamma_\sigma = 0$. Powefully, they also hold when $m_\pi \neq 0$ and/or when $\Gamma_\sigma \neq 0$, to all orders in perturbation theory. This can be proved easily using the enormous advantage that the linear sigma model is a well–defined (renormalizable) field theory.

Since the vertex modifications ensure the preservation of exact chiral Ward identities, they also guarantee, for instance, that the pion couplings remain derivative as they should.

To illustrate the power of this implementation of chiral symmetry, we evaluate, at tree level, the amplitude for $\pi\pi$ scattering. Clearly, we do not expect the result to be the perfect answer, since the only resonance we will take into account is the σ. In particular, not taking into account the vector meson $\vec{\rho}^\mu$ is a rather bad approximation in the $I = 1$, $\ell = 1$ amplitude. Nevertheless, our results are in better agreement with experimental data than those of chiral perturbation theory. Let us emphasize that the kinematical region where we compare both predictions, namely at very low momenta, is precisely where chiral perturbation theory should be exact. This lends further support to the real existence of σ as a strong resonance.

At tree level, four diagrams contribute to $\pi\pi \to \pi\pi$: the four–pion contact term, and the exchange of a σ in the three s, t and u channels. Due to the structure of the Feynman rules dictated by chiral Ward identities, the width Γ_σ contributes, in the Born approximation, only to $T_0^{(0)}$.

The experimental knowledge of pion scattering near threshold is rather poor. The relatively badly measured scattering lengths and ranges are $a_0^{(0)}$, $b_0^{(0)}$, $a_0^{(2)}$, $b_0^{(2)}$, $a_1^{(1)}$, $a_2^{(0)}$ and $a_2^{(2)}$, These seven numbers come out of our computation with only m_σ as a free parameter.

An overall fit to these seven numbers gives $m_\sigma = 700^{+800}_{-150}$MeV. The χ^2 distribution is very flat towards increasing values of m_σ; $m_\sigma \geq 550$ Mev is the only useful information.

Of the seven numbers, if we eliminate the worst one ($a_1^{(1)}$ (presumably under strong influence from ρ exchange, which we do not take into account), then the fit improves and it yields $m_\sigma = 590^{+220}_{-90}$MeV. Nicely, the fit to only the scalar isoscalar values gives $m_\sigma = 525^{+85}_{-45}$MeV.

Overall, one may conclude that the data are consistent with a linear sigma resonance provided its mass is around 600 MeV (and thus its width also around 600 MeV). The errors on these numbers, from the pion data available, are substantial.

Although the low–energy moments $a_\ell^{(I)}$ and $b_\ell^{(I)}$ are the relevant quantities for us, what is actually measured is a momentum–dependent phase shift, which can be split in various isospin and angular momentum channels. From the analysis of the data available, we fit $m_\sigma = 550^{+450}_{-80}$MeV. Again the error on the heavy side is huge: the χ^2 distribution is very flat with increasing m_σ.

Exact unitarity is achieved iff

$$\operatorname{Im} T_\ell^{(I)} = \sqrt{\frac{s - 4m_\pi^2}{s}} \left|T_\ell^{(I)}\right|^2$$

from which the optical theorem can be derived. Since there are many other resonances in nature heavier than the σ, we should not worry much about possible unitarity violations at high momenta (say, above 1 GeV). It turns out that there is no problem with unitarity at center of mass momenta lower than the 400 MeV. Unfortunately, unitarity does not constrain m_σ from above in any meaningful way.

We have enhanced the linear sigma model by enforcing chiral Ward identities which take into account the (large) sigma width. We have found that low energy pion scattering data supports the existence of a wide σ field with mass around 600 MeV (actually $m_\sigma = 590^{+220}_{-90}$MeV), provided we exclude the datum in the vector isovector channel. The advantage of keeping the σ as a true resonance in the effective low energy theory of strong interactions is not only that its inclusion simulates more or less the results of chiral perturbation theory to one loop, but also, more crucially, that this opens the door to more industrious analyses of the whole scalar spectrum, including glueballs.

Acknowledgements. This work was supported in part by CONACYT through projects 3979P-E9608, 25504-E, and Cátedra Patrimonial II de Apoyo a los Estados, and by DGAPA–UNAM through IN103997.

REFERENCES

1. J.L. Lucio, M. Napsuciale and M. Ruiz-Altaba, *The Linear Sigma Model at Work: Succesful Postdictions for Pion Scattering*, `hep-ph/9903420`, and references therein.
2. M. Napsuciale, J.L. Lucio, G. Moreno and J. Toscano, these proceedings.

Study of the Process $e^+e^- \to W^+W^-$ in a Model with Four Majorana Neutrinos

W. J. C. Teves and R. Zukanovich Funchal

Instituto de Física da Universidade de São Paulo
05315-970 C.P. 66318 - São Paulo, SP
Brazil

Abstract. *We investigate the process $e^+e^- \to W^+W^-$, currently being studied at LEP, in the context of the simplest extension of the Stantard Model of electroweak interactions, where a singlet right-handed neutrino is added to the matter content of the model.*

INTRODUCTION

The Standard Model of electroweak interactions (SM) [1] has passed all the high-precision tests performed at LEP and SLC up to now. Nevertheless its neutrino sector is still very much open to theoretical speculation. Are neutrinos really massless even though there is no underlying principle of nature to prevent them to acquire mass ? The fact that neutrinos are the only known electrically neutral elementary fermions means that they could be Majorana particles. Why should they be considered Dirac particles ? We believe these questions will only eventually be answered by the confront of experimental data with theoretical assumptions.

In this work, we consider an extension of the standard electroweak model, where a singlet right-handed neutrino is added to the particle content of SM and study the possible consequences of this model to the process $e^+e^- \to W^+W^-$ as a function of the free mixing parameters. We have calculated the total cross-section considering on-shell W boson production at three level. This is a first attempt to estimate the maximal deviations from the SM that can be consistent with the LEP data and its consequences in terms of the model free parameters.

The excitation curve of W-pair production near threshold is dominated by the neutrino exchange t-channel diagram. As this increase depends strongly on the value of the W boson mass, M_W, and as the LEP experiments promise a very accurate determination of M_W by direct reconstruction of W bosons through their decay products as well as the scan of the production cross-section, we may hope that in the near future these data will impose very strong constraints in mixing in the leptonic sector.

CP490, *Particles and Fields: Eighth Mexican School,* edited by J. C. D'Olivo et al.

We do not discuss here the inclusion of finite-width effects for the off-shell W-pair production which is clearly very important and will be addressed in the near future.

BRIEF DESCRIPTION OF THE MODEL

In the minimal extension of the SM considered here, which we will call Minimal Model (MM), where one right-handed singlet neutral fermion is added to the particle content of the SM, the most general form of the neutrino mass term is [2]

$$\mathcal{L}_\nu^M = -\sum_{j=e,\mu,\tau} a_j \overline{\nu}'_{jL}\nu'_R - \frac{1}{2}M\overline{\nu}'^C_R\nu'_R + H.c., \tag{1}$$

where the primed fields are not yet the physical ones. The diagonalization of the neutrino mass matrix will result in four physical neutrinos fields ν_1, ν_2, ν_P and ν_F; the first two massless and the last two massive Majorana neutrinos with masses

$$m_P = \frac{1}{2}\left(\sqrt{M^2+4a^2} - M\right) \quad \text{and} \quad m_F = \frac{1}{2}\left(\sqrt{M^2+4a^2} + M\right), \tag{2}$$

where $a^2 = {a_e}^2 + {a_\mu}^2 + {a_\tau}^2$.

In terms of the physical fields the charged-current interactions ($\mathcal{L}^{CC}$) is proportional to the factor

$$CC = \left(\, \overline{\nu}_1\overline{\nu}_2\overline{\nu}_P\overline{\nu}_F \,\right)_L \gamma^\mu \Phi R \begin{pmatrix} e \\ \mu \\ \tau \\ 0 \end{pmatrix}_L W_\mu^+ + H.c., \tag{3}$$

where $\Phi = diag\,(1,1,i,1)$ and R is the matrix

$$\begin{bmatrix} c_\beta & -s_\beta s_\gamma & -s_\beta c_\gamma & 0 \\ 0 & c_\gamma & -s_\gamma & 0 \\ c_\alpha s_\beta & c_\alpha c_\beta s_\gamma & c_\alpha c_\beta c_\gamma & -s_\alpha \\ s_\alpha s_\beta & s_\alpha c_\beta s_\gamma & s_\alpha c_\beta c_\gamma & c_\alpha \end{bmatrix}. \tag{4}$$

In Eq. (4) c and s denote the cosine and the sine of the respective arguments. The angles α, β, e γ lie in the first quadrant and are related to the mass parameter as follows:

$$s_\alpha = \sqrt{\frac{m_P}{m_P + m_F}}, \quad s_\beta = \frac{a_e}{a}, \quad c_\beta s_\gamma = \frac{a_\mu}{a}, \quad c_\beta c_\gamma = \frac{a_\tau}{a}. \tag{5}$$

W-PAIR PRODUCTION IN THE MM

At the Born level the $e^+e^- \to W^+W^-$ process can take place not only via γ and Z^0 formation, but also through t-channel neutrino exchange. This last process dominates the counting rate at LEP 200 energies. The Higgs-exchange diagram which is suppressed by a factor m_e/m_W is completely negligible and can be omitted from our calculation.

If we set the convention $e^+(q_+,\kappa_+) + e^-(q_-,\kappa_-) \to W^+(p_+,\lambda_+) + W^-(p_-,\lambda_-)$, where the arguments are the momenta and helicities of incoming and outgoing particles, one can write the total three level helicity amplitude $\mathcal{M}_{\text{Born}}$ as

$$\mathcal{M}_{\text{Born}}\left(\kappa,\lambda_+,\lambda_-,s,t\right) = \sum_{\alpha=\nu,\gamma,Z^0} \mathcal{M}_{\alpha}\left(\kappa,\lambda_+,\lambda_-,s,t\right), \tag{6}$$

where $s = (q_+ + q_-)^2$ and $t = (q_+ - p_+)^2$ are the usual Mandelstam variables and $\mathcal{M}_\nu$, $\mathcal{M}_\gamma$ and $\mathcal{M}_{Z^0}$ can be calculated in accordance to the helicity amplitude prescription found in refs. [3–5]. We have neglected the electron mass, so the helicity of the positron will be opposite to that of the electron , i.e. $\kappa_- = -\kappa_+ = \kappa$. Imposing CP conservation instead of 36 amplitudes we only have to calculate 12 independent ones.

In the MM only the neutrino amplitude is modified with respect to the SM results. To calculate this amplitude in terms of the SM one we will write the electron neutrino (ν_e) eigenstate in terms of the physical neutrinos of the model. For this purpose, we use the lepton mixing matrix and then ν_e can be write as follows $|\nu_e\rangle = c_\beta\,|\nu_1\rangle - is_\alpha c_\beta\,|\nu_P\rangle + s_\alpha s_\beta\,|\nu_F\rangle$.

Therefore the invariant amplitude do ν_e in MM can be obtained directly from its expression in the SM by

$$\mathcal{M}_\nu^{\text{MM}}\left(\kappa,\lambda_+,\lambda_-,s,t\right) = \frac{e^2}{2s_W^2}\left[\frac{c_\beta^2}{t^2} + \frac{c_\alpha^2 s_\beta^2}{t_P^2} + \frac{s_\alpha^2 s_\beta^2}{t_F^2}\right]^{\frac{1}{2}} \mathcal{M}_1^\kappa(\lambda_+,\lambda_-)\delta_{\kappa-}, \tag{7}$$

where t_P and t_F are given by $t_i = t-(m_i)^2$, $\;i = P, F$ and the index MM indicates that the Minimal Model was used for its calculation. Also $\delta_{\kappa-}$ is equal to 1 for left-handed electrons or 0 for right-handed ones and the expression for $\mathcal{M}_1^\kappa(\lambda_+,\lambda_-)$ are the SM ones that can be found in ref. [3].

In this way we can now study the differences between the results of differential and total cross-sections for W^+W^- production in the SM and in the MM in terms of the model free parameters. Before doing so let us revise the constrains on masses and mixing angles of the MM imposed by the measured Z^0 invisible width [6]. Three mass regions can be considered for the massive neutrinos: **(1)** $m_P, m_F < \frac{M_Z}{2}$; **(2)** $m_P < \frac{M_Z}{2}$ and $\frac{M_Z}{2} < m_F < M_Z$ and **(3)** $m_P < \frac{M_Z}{2}$ and $m_F > M_Z$.

In the region **(1)**, as showed in ref. [6], there is a constraint on the masses, i.e. $m_F > 18.2\ m_P$. This implies for our present calculation no visible discrepancy from the SM cross-section even is one consider maximal mixing. In the case of

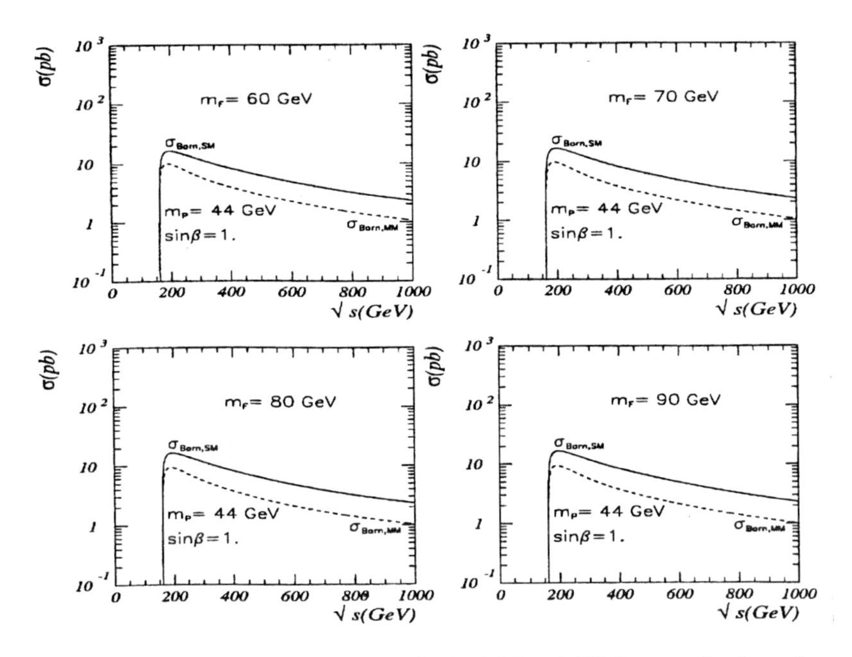

FIGURE 1. Region (2): Total cross-section in the MM and SM for several values of m_F.

region (2) all masses in the kinematic region are allowed. Here we have found, as exemplified by the plots in Fig. 1, that one expect that not all mixing values will be allowed by the LEP data. In region (3) the constraint is that the lightest neutrino (m_P) has been smaller than 9 GeV and again no visible discrepancy from the SM cross-section can be observed even is one consider maximal mixing.

CONCLUSIONS

In conclusion we have : **(i)** in the region **(1)** and **(3)** the total cross-section for on-shell W-pair production in MM is not distinguishable from the SM one. Therefore, in these regions, it seems that LEP data will not help much in constraining the model parameters. This may change when we perform the off-shell W-pair calculation; **(ii)** we can hope to experimentally constrain the MM parameters in region **(2)** and **(iii)** since the difference in the cross-section between the SM and the MM calculation vary dependent on $\sqrt{s}$ and the maximal discrepancy occurs near the W^+W^- pair production threshold because the t-channel dominates at this energy, we see that finite-width of the W will has really to be taken into account before any fit with experimental data can be performed.

ACKNOWLEDGMENTS

We thank Conselho Nacional de Desenvolvimento Científico e Tecnológico (CNPq) and Fundação de Amparo à Pesquisa do Estado de São Paulo (FAPESP) for financial support.

REFERENCES

1. S. L. Glashow, Nucl. Phys. **22**, 579 (1961); S. Weinberg, Phys. Rev. Lett. **19**, 1264 (1967); A. Salam, in *Elementary Particle Theory*, edited by N. Svarthom(Almquist and Wiksell, Stockholm, 1968); S. L. Glashow, J. Iliopoulos and L. Maiani, Phys. Rev. D **2**, 1285 (1970); M. Kobayashi e K. Maskawa, Prog. Theor. Phys. **49**, 652 (1973).
2. C. Jarlskog, Nucl. Phys. A, **518** 129(1990); Phys. Lett. B **241**, 579 (1990).
3. W. Beenakker and A. Denner, Int. Journ. of Phys. A **9**, 4837(1994).
4. K. Hagiwara and D. Zeppenfeld, Nucl. Phys. B **274**, 1(1986).
5. A. Denner Fortschr. Phys. **41**, 307 (1993).
6. C. O. Escobar, O. L. G. Peres, V. Pleitez and R. Z. Funchal, Phys. Rev. D **47**, 1747 (1993).

One particular approach to the non-equilibrium quantum dynamics

Eduardo S. Tututi† and Petr Jizba*

†*ECFM, UMSNH, Apdo. Postal 2-71, 58041, Morelia Mich., México*[1]
**DAMTP, University of Cambridge, Silver Street, Cambridge, CB3 9EW, UK*[2]

Abstract. We present a particular approach to the non-equilibrium dynamics of quantum field theory. This approach is based on the Jaynes-Gibbs principle of the maximal entropy and its implementation, throughout the initial-value data, into the dynamical equations for Green's functions. We use the ϕ^4 theory in the large N limit to show how our method works by calculating the pressure for a system which is invariant under both spatial and temporal translations.

INTRODUCTION: JAYNES-GIBBS PRINCIPLE

The objective of this talk is to present a novel approach to a non-equilibrium dynamics of quantum fields [1]. This approach is based on the Jaynes-Gibbs maximum entropy principle [2], which, in contrast to other approaches in use [3–6], constructs a density matrix ρ directly from the experimental/theoretical initial-time data (e.g. pressure, density of energy, magnetization, ionization rate, etc.). We illustrate our method on the ϕ^4 theory with the $O(N)$ internal symmetry in the large N limit, provided that the non-equilibrium medium in question is translationally invariant.

To start, we consider the following definition of expectation value of some dynamical operator A: $\langle A \rangle = \mathrm{Tr}(\rho A)$, where the trace is taken with respect to an orthonormal basis of *physical* states describing the ensemble at some initial time t_i. Let us consider the information (or Shannon) entropy $S[\rho] = -\mathrm{Tr}(\rho \ln \rho)$ [2]. According to the Jaynes-Gibbs principle, we have to maximize $S[\rho]$ subject to constrains imposed by the expectation value of certain experimental/theoretical observables: $\langle P_i[\Phi, \partial\Phi](x_1, \ldots) \rangle = g_i(x_1, \ldots)$, $\quad i = 1, \ldots n$, where the operators $P_i[\Phi, \partial\Phi]$, in contrast to thermal equilibrium, need not to be constants of the motion; space-time dependences are allowed. The maximum of the entropy determines the density matrix with the least informative content.

---

[1] Partially supported by CONACYT under grant I-29950 E and CIC-UMSNH.
[2] Fitzwilliam college

CP490, *Particles and Fields: Eighth Mexican School,* edited by J. C. D'Olivo et al.

$$\rho = \frac{1}{\mathcal{Z}(\lambda_i)} \exp\left(-\sum_{i=1}^{n} \int \prod_j d^4x_j \lambda_i(x_1, \ldots) P_i[\Phi, \partial\Phi] \right) , \tag{1}$$

where λ_i are the Lagrange multipliers to be determined. The 'partition function' $\mathcal{Z}$ is $\mathcal{Z}(\lambda_i) = \mathrm{Tr}\left\{\exp\left(-\sum_{i=1}^{n} \int \prod_j d^4x_j \lambda_i(x_1, \ldots) P_i[\Phi, \partial\Phi]\right)\right\}$. In the previous equations the time integration is not present at all (i.e. $g_i(\ldots)$ are especified at the initial time t_i). In case when the constraint functions $g_i(\ldots)$ are known over some gathering interval $(-\tau, t_i)$ the correspondent integration $\int_{-\tau}^{t_i} dt$ should be present in ρ. The Lagrange multipliers λ_i might be eliminated if one solves n simultaneous equations: $g_i = -\delta \ln \mathcal{Z} / \delta\lambda_i$. The solution can be formally written as $\lambda_i = \delta S_G[g_1, \ldots, g_n] \mid_{max} / \delta g_i$.

OFF-EQUILIBRIUM DYNAMICAL EQUATIONS

In this section we briefly introduce the off-equilibrium dynamical (or Dyson-Schwinger) equations. For simplicity we illustrate this on a single scalar field Φ coupled to an external source J described by the action $S'[\Phi] = S[\Phi] + \int J\Phi$. Associated with this action we have the functional equation of motion [1,5]:

$$\frac{1}{Z[J]} \frac{\delta S}{\delta \Phi} \left[\Phi_\alpha = -i \frac{\delta}{\delta J} \right] Z[J] = -J_\alpha , \tag{2}$$

with $Z[J] = \mathrm{Tr}\{\rho T_C \exp(i \int_C d^4x J(x)\Phi(x)\}$ being the generating functional of Green's functions. Here C is the Keldysh-Schwinger contour which runs along the real axis from t_i to t_f ($t_f > t_i$, t_f is arbitrary) and then back to t_i. In (2) we have associated with the upper branch of C the index "+" and with the lower one the index "−" (in the text we shall denote the indices $+/-$ by Greek letters α, β).

Let us define the classical field ϕ_α as the expectation value of the field operator in the presence of J: i.e. $\phi_\alpha = \langle \Phi_\alpha \rangle$. Defining the generating functional of the connected Green's functions as $Z[J] = \exp(iW[J])$, the two-point Green's function is $G_{\alpha\beta}(x, y) = -\frac{\delta^2 W}{\delta J_\alpha(x) \delta J_\beta(y)} = -i\langle T_C \Phi(x)\Phi(y)\rangle + i\langle\Phi(x)\rangle\langle\Phi(y)\rangle$. Eq.(2) is the first one of an infinite hierarchy of equations for Green functions. Further equations can be obtained from (2) by taking successive variations with respect to J. True dynamical equations are then obtained if one substitutes the physical condition $J = 0$ into equations obtained.

To reflect the effects of the density matrix in the Dyson-Schwinger equations it is necessary to construct the corresponding boundary conditions.[3] Using the cyclic property of the trace together with the Baker-Campbell-Hausdorff relation: $e^A B e^{-A} = \sum_{n=0}^{\infty} \frac{1}{n!} C_n$, (where $C_0 = B$ and $C_n = [A, C_{n-1}]$), and setting $A = \ln(\rho)$ and $B = \Phi(x_1)$ with $x_{10} = t_i$ we obtain the generalized KMS conditions:

[3] Let us remind that at equilibrium the corresponding boundary conditions are the Kubo-Martin-Schwinger (KMS) conditions.

$\langle\Phi(x_1)\cdots\Phi(x_n)\rangle = \langle\Phi(x_2)\cdots\Phi(x_n)\Phi(x_1)\rangle + \sum_{k=1}^{\infty}\frac{1}{k!}\langle\Phi(x_2)\cdots\Phi(x_n)C_k(x_1)\rangle$. So namely for the two-point Green function we have $G_{+-}(x,y) = G_{-+}(x,y) + \sum_{k=1}^{\infty}\frac{1}{k!}\mathrm{Tr}\{\rho\Phi(x)C_k(x)\}$. As an example of the latter relation we can choose the particular situation when $\rho = \exp(-\beta H)/\mathcal{Z}$, in which case we get the well known KMS condition: $G_{+-}(\mathbf{x};t,\mathbf{y};0) = G_{-+}(\mathbf{x};t-i\beta,\mathbf{y};0)$.

EXAMPLE: OUT-OF-EQUILIBRIUM PRESSURE

In order to apply our previous results let us consider the ϕ^4 theory with the $O(N)$ internal symmetry in the large N limit (also the Hartree-Fock approximation). It is well known that, in this limit only two-point Green's functions are relevant [1,6,7]. The Dyson-Schwinger equations for $G_{\alpha\beta}$ are automatically truncated and reduce to the Kadanoff-Baym equations [4]: $\left(\Box + m_0^2 + \frac{i\lambda_0}{2}G_{\alpha\alpha}(x,x)\right)G_{\alpha\beta}(x,y) = -\delta(x-y)(\sigma_3)_{\alpha\beta}$, where σ_3 is the Pauli matrix; λ_0 and m_0 are, respectively, the bare coupling and the bare mass of the theory. If the system is translationally invariant the Fourier transform solves the Kadanoff-Baym equations and the corresponding fundamental solution reads: $G_{\alpha\beta}(k) = \frac{(\sigma_3)_{\alpha\beta}}{k^2+\mathcal{M}^2+i\epsilon(\sigma_3)_{\alpha\beta}} - 2\pi i\delta(k^2+\mathcal{M}^2)f_{\alpha\beta}(k)$, where the (finite) $\mathcal{M}$ is $\mathcal{M}^2 = m_0^2 + i\frac{\lambda_0}{2}G_{++}(0)$. Function $f_{\alpha\beta}(k)$ must be determined through the generalized KMS conditions. Let us now choose the constraint to be used. Keeping in mind that we are interested in a system which is invariant under both spatial and temporal translations, we choose the constraint $g(\mathbf{k}) = \langle\tilde{\mathcal{H}}(\mathbf{k})\rangle$, where $\tilde{\mathcal{H}} = \omega_k a^{\dagger}(\mathbf{k})a(\mathbf{k})$, with $\omega_k = \sqrt{\mathbf{k}^2+\mathcal{M}^2}$ (notice that in the large N limit the Hamiltonian is always quadratic in the fields). The corresponding density matrix then reads

$$\rho = \frac{1}{\mathcal{Z}(\beta)}\exp\left(-\int\frac{d^3\mathbf{k}}{(2\pi)^3 2\omega_k}\beta(\mathbf{k})\tilde{\mathcal{H}}(\mathbf{k})\right), \tag{3}$$

with $\frac{\beta(\mathbf{k})}{(2\pi)^3 2\omega_k}$ being the Lagrange multiplier to be determined. According to the maximum entropy principle we find that $\beta(\mathbf{k})$ fulfils equation

$$g(\mathbf{k}) = \frac{V}{(2\pi)^3}\frac{\omega_k}{e^{\beta(\mathbf{k})\omega_k}-1}, \tag{4}$$

where V denotes the volume of the system. Eq.(4) can be interpreted as the density of energy per mode. Similarly as in the case of thermal equilibrium, $\beta(\mathbf{k})$ could be interpreted as "temperature" with the proviso that different modes have now different "temperatures".

The generalised KMS conditions in this case are $G_{+-}(k) = e^{-\beta(\mathbf{k})k_0}G_{-+}(k)$, and so the corresponding f_{++} reads: $f_{++} = [\exp(\beta(\mathbf{k})\omega_k)-1]^{-1}$. Let us now consider a particular system in which $g(\mathbf{k}) = \frac{V}{(2\pi)^3}\exp(\omega_k/\sigma)$. In this case σ is the physical parameter which, as we shall see below, can be interpreted as a "temperature" parameter. This particular choice corresponds to a system where the lowest frequency modes depart from equilibrium, while the high energy ones obey the Bose-Einstein distribution (typical situation in many non-equilibrium media, e.g. plasma heated up by ultrasound waves, hot fusion or ionosphere ionised by sun). In terms of the parameter σ the Lagrange multiplier may be written as $\beta(\mathbf{k}) = \frac{1}{\sigma} + \frac{1}{\omega_k}\sum_{n=1}^{\infty}\frac{(-1)^{n+1}}{n}\exp(-n\omega_k/\sigma)$. Notice that when $\omega_k \gg \sigma$, $\beta \sim \sigma^{-1}$, and we may see that f_{++} approaches to the Bose-Einstein distribution with temperature σ. However, when $\omega_k \sim \sigma$ the latter interpretation fails. Instead of the parameter σ, it may be useful to work with the expectation value of $\beta(\mathbf{k})$:

$$\langle\beta\rangle = \frac{\int d^3\mathbf{k}\,\beta(\mathbf{k})e^{-\omega_k/\sigma}}{\int d^3\mathbf{k}\,e^{-\omega_k/\sigma}} = \frac{1}{\sigma} + \frac{\sum_{n=1}^{\infty}\frac{(-1)^{n+1}}{n(n+1)}K_1((n+1)\mathcal{M}/\sigma)}{\mathcal{M}K_2(\mathcal{M}/\sigma)}, \tag{5}$$

where K_n is the Bessel function of imaginary argument of order n. An interesting feature of Eq.(5) is that it is actually insensitive to the value of $\mathcal{M}$ which is important if one wants to use $1/\langle\beta\rangle$ as a "temperature". The actual behaviour of $\langle\beta\rangle$ is depicted in Fig.1

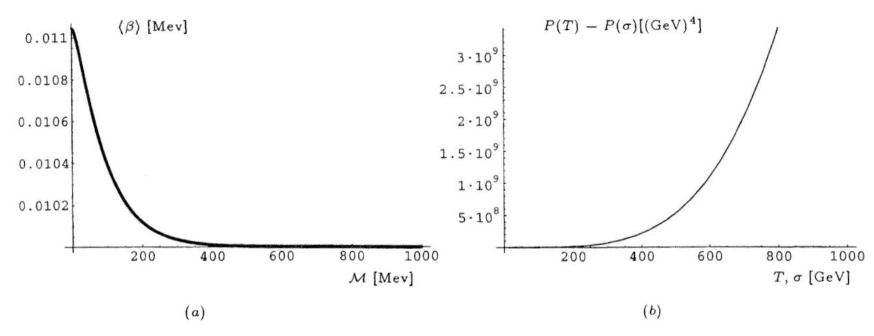

FIGURE 1. *In (a) we plot Eq.(5) at $\sigma = 100Mev$, while in (b) we plot the difference of equilibrium and non-equilibrium pressures for $m_r = 100MeV$.*

Let us now consider the renormalized expression for the expectation value of the energy momentum tensor [1]:

$$\langle\theta_{\mu\nu}\rangle_{\rm ren} = N\int\frac{d^dk}{(2\pi)^d}k_\mu k_\nu[G_{++}(k) - G(k)] - i\frac{Ng_{\mu\nu}\delta m^2}{4}\int\frac{d^dk}{(2\pi)^d}[G_{++}(k) + G(k)]\,,$$

with G being the usual $(T = 0)$ causal Green function and $\delta m^2 = \mathcal{M}^2 - m_r^2$ with m_r being the $(T = 0)$ renormalized mass. The pressure per particle, in the high "temperature" expansion (i.e. for large σ or small $\langle\beta\rangle$) for the system described by the density matrix (3) may be worked out either in terms of σ, using the Mellin transform technique [1,8]:

$$P(\sigma) = -\frac{1}{3N}\langle\theta^i{}_i\rangle_{\rm ren} = \frac{\sigma^4}{\pi^2} - \frac{\sigma^2\mathcal{M}^2}{2\pi^2} + \frac{\lambda_r}{8}\left(\frac{\sigma^2\mathcal{M}^2}{64\pi^4} - \frac{\sigma^4 3}{4\pi^4}\right) + \mathcal{O}\left(\ln(\mathcal{M}/\sigma);\lambda_r^2\right)\,,$$

or in terms of $1/\langle\beta\rangle$ using the Padé approximation [1]:

$$P(\langle\beta\rangle) = 0.0681122\,\langle\beta\rangle^{-4} - 0.0415368\,\langle\beta\rangle^{-2}\mathcal{M}^2 \;+ \lambda_r\,\left(-0.000647\,\langle\beta\rangle^{-4} + 0.0000164\,\langle\beta\rangle^{-2}\,\mathcal{M}^2\right) + \mathcal{O}\left(\mathcal{M}^2\ln(\mathcal{M}\langle\beta\rangle);\lambda_r^2\right).$$

It is interesting to compare the previous two results with the high-temperature expansion of the same system in thermal equilibrium [7]:

$$P(T) = \frac{T^4\,\pi^2}{90} - \frac{T^2\,\mathcal{M}^2}{24} + \frac{T\,\mathcal{M}^3}{12\pi} + \frac{\lambda_r}{8}\left(\frac{T^4}{144} - \frac{T^3\,\mathcal{M}}{24\pi} + \frac{T^2\,\mathcal{M}^2}{16\pi^2}\right) + \mathcal{O}\left(\ln\left(\frac{\mathcal{M}}{T4\pi}\right)\right).$$

Particularly, the leading "temperature" coefficients in the first two expansions approximate to a very good accuracy the usual Stefan-Boltzmann constant for scalar theory. The latter vindicates the interpretation of σ and $1/\langle\beta\rangle$ as temperatures for high energy modes. The behaviour of both $P(T)$ and $P(\sigma)$ are shown in Fig.1.

SUMMARY AND OUTLOOK

One of the main advantages of the Jaynes-Gibbs construction is that one starts with constraints imposed by experiment/theory. The constraints directly determine the density matrix with the least informative content (the least prejudiced density matrix which is compatible with all information one has about the system) and consequently the generalized KMS conditions for the Dyson-Schwinger equations. We applied our method on a toy model system ($O(N)$ $\lambda\phi^4$ theory), in the translationally invariant medium. The method presented, however, has a natural potential to be extensible to more general systems. Particularly to media where the translational invariance is lost. As an example we can mention systems which are in local thermal equilibrium. For such systems it is well known [2,8] that equilibrium β must be replaced by $\beta(\mathbf{x})$ (i.e. temperature which slowly varies with position). Obviously one may receive this result from the outlined Jaynes–Gibbs principle almost for free. Work on more complex systems is now in progress.

REFERENCES

1. Jizba P. and Tututi E.S., hep-th/9811236, to appear in *Phys. Rev.* D.
2. Jaynes E.T., *Phys. Rev.* **106**, 620 (1957); **108**, 171 (1957).
3. Calzetta E. and Hu B.L., *Phys. Rev.* D**37**, 2878 (1988).
4. Kadanoff L.P. and Baym G., *Quantum Statistical Mechanics*, New York, Benjamin, 1962.
5. Chou K.C., Su Z.B., Hao B.L., and Yu L., *Phys. Rep.* **118**, 1 (1985).
6. Éboli O., Jackiw R., and Pi S-Y., *Phys. Rev.* D**37** (1988).
7. Amelino-Camelia G. and Pi S.-A., *Phys. Rev.* D**47**, 2356 (1993).
8. Landsman N.P. and van Weert Ch.G., *Phys. Rep.* **145**, 141 (1987).

Fermion Damping and Reaction Rates in Hot Gauge Theories

Axel Weber[1]

Instituto de Ciencias Nucleares, UNAM,
Circuito Exterior C.U., A. Postal 70–543, 04510 México D.F., Mexico

Abstract. We examine the relation between the damping rate of a fermionic mode propagating in a hot QED or QCD plasma, and the corresponding reaction rate. We show that these two quantities should be equal provided the reaction rate is calculated using properly normalized wave functions for the mode in the medium. We use finite temperature cutting rules to identify the different terms in the expression for the damping rate with physical processes in the plasma.

In this contribution, we will describe some aspects of the physics of a hot QED or QCD plasma in thermal equilibrium. We will assume the temperature T to be sufficiently high, so that all (zero–temperature) masses can be neglected. Due to interactions with the particles present in the plasma, particles propagating through the medium are turned into collective modes. The poles of their propagators get thereby shifted from their vacuum positions to

$$\omega(p) = \omega_p(p) - \frac{i}{2}\,\gamma(p)\,, \tag{1}$$

where $P = (\omega, \vec{p})$ is the 4–momentum of the mode in the rest frame of the plasma, and $p \equiv |\vec{p}|$. The real part ω_p yields the dispersion relation of the mode, while $\gamma > 0$ is by definition the damping rate. Linear response theory then predicts that the mode, excited by an (arbitrarily weak) external current, gets exponentially damped with time scale γ^{-1} [1]. We assume throughout that $\gamma \ll \omega_p$ (this will be the case in hot gauge theories), so it makes sense to speak of a propagating mode.

We now turn our attention to the physical processes associated with the decay of the mode. As we indicate in Fig. 1 there are processes leading to the annihilation of the mode (rate $\Gamma^{>}$) and others leading to its creation (rate $\Gamma^{<}$). Compared to the vacuum (Fig. 1a) there are additional contributions in the plasma (Fig. 1b) due to the existence of real particles. The individual rates are calculated

---

[1] present address: Instituto de Física y Matemáticas, Universidad Michoacana de San Nicolás de Hidalgo, Edificio C–3 Cd. Universitaria, A. Postal 2–82, 58040 Morelia, Michoacán, Mexico, e–mail: axel@io.ifm.umich.mx

CP490, *Particles and Fields: Eighth Mexican School,* edited by J. C. D'Olivo et al.

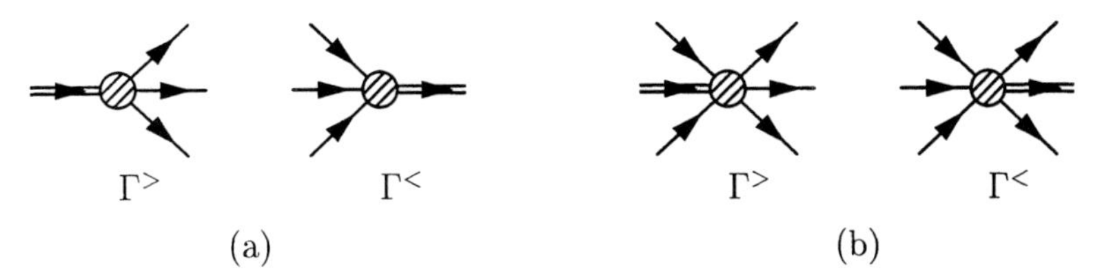

FIGURE 1. Typical contributions to the reaction rate of a particle (double line), at temperature zero (a) and at finite temperature (b).

by summing over all possible states of the other particles (single lines in Fig. 1), weighted by the appropriate statistical factors. The total reaction rate is then defined as $\Gamma = \Gamma^{>} + \Gamma^{<}$ (for fermions). Weldon [2] has shown that Γ^{-1} is the time scale for the considered mode to reach equilibrium, starting with a distribution infinitesimally out of equilibrium.

We conclude that the damping and reaction rates represent closely related concepts, and it is then natural to ask if $\gamma = \Gamma$, or else what the precise relation between these quantities is. This question has been answered in general by D'Olivo and Nieves [3]; they have shown that indeed $\gamma = \Gamma$ provided one uses properly normalized spinors in the calculation of the reaction rate of the collective mode, thus clearing up some confusion in the literature [4]. The continuous interest in the equality of damping and reaction rates is twofold: firstly, it can be used as a consistency check for the HTL formalism in hot gauge theories which is the currently accepted method to overcome the problems with perturbation theory at finite temperature. We will present the current status of the investigation in the sequel. Secondly, the calculation of $\gamma(p \neq 0)$ in QED plasmas is plagued by severe IR divergencies due to unscreened magnetic photons. This has recently led to claims that the decay of the fermionic modes is anomalous, namely faster than exponential [5]. In the present contribution, we will treat all the expressions formally, ignoring the IR divergency problem. Nevertheless, we consider the present study an important step towards the clarification of the issue of anomalous fermionic decay.

We will begin by summing up the argument of D'Olivo and Nieves to the equality of damping and reaction rates. The inverse propagator of the fermionic mode in the plasma rest frame is given by $iS^{-1}(P) = \not{P} - \Sigma(P)$ with the self energy Σ. For $\gamma \ll \omega_p$, the damping rate can be written to lowest order in $\mathrm{Im}\,\Sigma$ as [3]

$$\gamma = -\frac{1}{\omega_p}\,\bar{u}(\vec{p})\,\mathrm{Im}\,\Sigma(\omega_p, p)\,u(\vec{p})\,, \tag{2}$$

where the spinors $u(\vec{p})$ are the solutions of the Dirac equation in the medium, $\mathrm{Re}\,[iS^{-1}(\omega_p, \vec{p})]\,u(\vec{p}) = 0$. They have to be normalized as $u^{\dagger}(\vec{p})\,u(\vec{p}) = 2\,\omega_p\,Z_p$, where Z_p is the residue of the propagator pole and quantifies the one-particle contribution with frequency ω_p to the propagator.

To relate the expression in Eq. (2) to a reaction rate we make use of the finite–temperature cutting rules [6]. These are very similar to the well–known zero–temperature rules, which read schematically

$$\mathrm{Im}\, S_{fi} = -\frac{1}{2} \sum_{\substack{\text{different}\\ \text{cuts}}} \; \sum_{\substack{\text{intermediate}\\ \text{physical states } n}} \bar{S}_{fn}\, S_{ni} \tag{3}$$

for a connected diagram S_{fi}. At finite temperature, the statistical weights have to be included in the sum over intermediate states, and furthermore the "intermediate" particles can in fact be incoming *or* outgoing for every diagram (cf. Fig. 1) [6]. As a result of the application of these cutting rules to Eq. (2) one obtains $\gamma = \Gamma$, as D'Olivo and Nieves have shown for some particular cases.

Let us now calculate the damping rate from Eq. (2) for the case of hot gauge theories, where the presence of the two scales T and eT (e the coupling constant) leads to a breakdown of naïve perturbation theory. We will use the effective expansion proposed by Braaten and Pisarski [7], and independently by Frenkel and Taylor [8], which consists in including the hard thermal loop (HTL) contributions in the tree–level propagators and vertices with soft (order eT) external momenta. To lowest order in the effective expansion, the self energy for soft momenta (inside the light cone) turns out to be real. To next–to–leading order, the imaginary part of Σ receives contributions from two different diagrams [7],

$$\mathrm{Im}\,\Sigma = \mathrm{Im}\left[\qquad + \qquad \right], \tag{4}$$

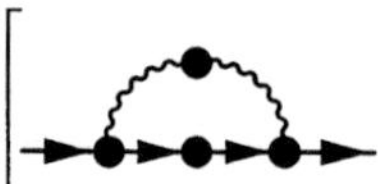

where we have denoted the effective propagators and vertices by full circles. As a consequence of the special Ward identities for the HTLs and the Dirac equation in the plasma, the damping rate γ in Eq. (2) can be shown to be gauge independent to this order [7].

The damping rate can be calculated explicitly by performing the frequency sum implied in the diagrams (4) and continuing analytically to real energies (in the imaginary time formalism). The result is [9]

$$\bar{u}(\vec{p})\, \mathrm{Im}\, \Sigma(\omega_p, p)\, u(\vec{p}) \;=$$

$$\pi e^2 \int \frac{\mathrm{d}^4 K}{(2\pi)^4} \frac{\mathrm{d}^4 P'}{(2\pi)^4}\, \delta^4(P - K - P')\left(e^{\omega_p/T} + 1\right) n_B(k_0) n_F(p_0') \tag{5}$$

$$\times \left\{ \bar{u}(\vec{p})\, \gamma^\mu \left[\rho_T(k_0, k)\, P^T_{\mu\nu} + \rho_L(k_0, k)\, P^L_{\mu\nu}\right] \right. \tag{6}$$

$$\times \left[\rho_+(p_0', p')\, \frac{1}{2}\,(\gamma_0 - \vec{\gamma}\cdot\hat{p}') + \rho_-(p_0', p')\, \frac{1}{2}\,(\gamma_0 + \vec{\gamma}\cdot\hat{p}')\right] \gamma^\nu u(\vec{p}) \tag{7}$$

$+$ HTL corrections to the bare vertices

$+$ contributions from the imaginary parts of the effective vertices $\Big\}$, where K and P' are the 4–momenta of the internal gauge boson and fermion, respectively, n_B and n_F are the usual Bose–Einstein and Fermi–Dirac distribution functions, continued to negative values of their arguments, $P^T_{\mu\nu}$ and $P^L_{\mu\nu}$ are the transverse and longitudinal projectors on the gauge boson momentum, $p_0 \equiv \omega_p$ and $\hat{p}' \equiv \vec{p}\,'/p'$. In the case of QCD, there is an additional factor of $C_F = 4/3$. The above expression already has the form of a reaction rate: the statistical factors in (5) combine to give the correct statistical weights to the states to be summed over [2], and the tensorial structures in square brackets in (6,7) can be interpreted as sums over the projectors corresponding to the different polarization states of the gauge boson and the fermion, respectively. The dispersion relations and the Z–factors for these modes are incorporated by the spectral functions ρ_i. In addition, there are contributions for spacelike momenta corresponding to Landau damping processes with internal gauge bosons or fermions.

The step that has not been concluded yet is the identification of the different terms appearing in the expression for γ with the possible processes entering Γ. However, the graphical application of the cutting rules yields the processes that should contribute to the given order, provided the HTL formalism is consistent. As an illustration, we show the results in QED for the following cuts:

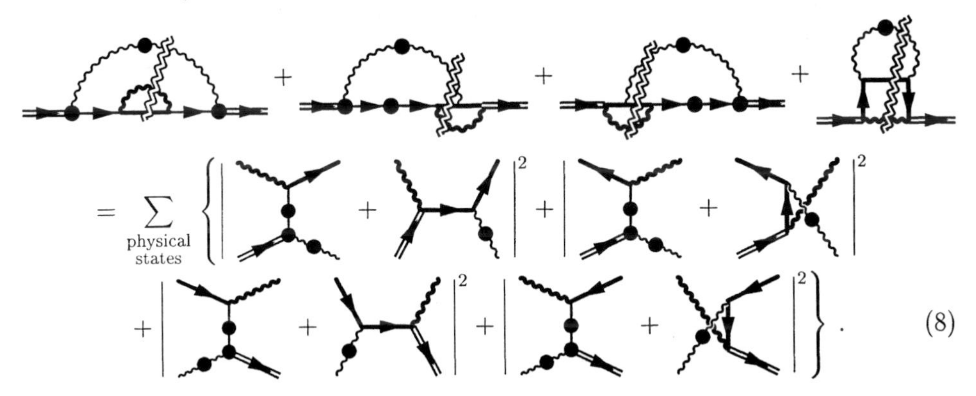

As before, the external soft fermion is denoted by a double line, the propagators of hard particles are marked with thick lines, and full circles denote effective quantities. The cut HTL contributions to the effective vertices and propagators are shown explicitly. For the terms that stem from cuts of the propagators the identification with the corresponding expressions in Eqs. (5–7) is rather straightforward, while the terms corresponding to cuts of the effective vertices are harder to identify. The latter task is the subject of present investigations.

ACKNOWLEDGEMENTS

I am indebted to my collaborators A. Ayala and J. C. D'Olivo at the Instituto de Ciencias Nucleares, with whom most of the work presented here was done. I am grateful to CONACyT for financial support under grant 3298P–E9608.

REFERENCES

1. see, for example, Le Bellac, M., *Thermal Field Theory*, Cambridge: CUP, 1996.
2. Weldon, H. A., *Phys. Rev.* **D28**, 2007 (1983).
3. D'Olivo, J. C., and Nieves, J. F., *Phys. Rev.* **D52**, 2987 (1995).
4. Thoma, M. H., and Gyulassy, M., *Nucl. Phys.* **B351**, 491 (1991);
   Altherr, T., Petitgirard, E., and del Rio Gaztelurrutia, T., *Phys. Rev.* **D47**, 703 (1993).
5. Blaizot, J. P., and Iancu, E., *Phys. Rev. Lett.* **76**, 3080 (1996),
   Takashiba, K., *Int. J. Mod. Phys.* **A11**, 2309 (1996);
   Boyanowsky, D., de Vega, H. J., Holman, R., and Simionato, M., "Dynamical renormalization group resummation of finite temperature infrared divergencies", preprint hep/ph–9809346.
6. Bedaque, P. F., Das, A., and Naik, S., *Mod. Phys. Lett.* **A12**, 2481 (1997);
   Landshoff, P. V., *Phys. Lett.* **B386**, 291 (1996).
7. Braaten, E., and Pisarski, R. D., *Nucl. Phys.* **B337**, 569 (1990).
8. Frenkel, J., and Taylor, J. C., *Nucl. Phys.* **B334**, 199 (1990).
9. Ayala, A., D'Olivo, J. C., and Weber, A., "The relation between damping and reaction rates of fermions in hot gauge theories", preprint hep–th/9809083.

List of Participants

ALARCÓN, Mariano	IFyM-UMSNH, México
ALDANA, Waleska	Univ. de Sn. Carlos, Guatemala
ALFARO, Adrián	ICN-UNAM, México
ALVAREZ, Erika	IF-UNAM, México
ARAIZA, Moises Elías	IF-BUAP, México
ARANDA, Jorge Isidro	UMSNH, México
AVILA, Manuel	FC-UAEM, México
AYALA, Alejandro	ICN-UNAM, México
BAÑADOS, Max	U. Zaragoza, España
BARREIRO, Julio	ICN-UNAM, México
BASHIR, Adnan	UMSNH-IF, México
BENITEZ, Fernando	ICN-UNAM, México
BUCIO, David	FC-UNAM, México
CAMACHO, Jaime	IF-UG, México
CARREÑO, Alexandra	ICN-UNAM, México
CARRILLO, Iván	IF-UNAM,México
CÁZARES, Federico	IF-UNAM, México
CERVERA, Pedro	FC-UNAM, México
CIFUENTES, Edgar	Universidad de San Carlos, Guatemala
CUAUTLE, Eleazar	CINVESTAV, México
DEL OSO, Alfredo	IF-UNAM, México
DÍAZ-CRUZ, Lorenzo	IF-BUAP, México
D'OLIVO, Juan Carlos	ICN-UNAM, México
ESPICHÁN, Jorge Abel	Universidade Estadual de Campinas, Brazil
ESPINOZA, Catalina	IF-UNAM, México
FÉLIX, Julián	IF-UG, México
FERNÁNDEZ, Arturo	FCFM-BUAP, México
FUENTES, Ivette	IF-UNAM, México
GALLEGOS, Armando	IF-UG, México
GARCÍA, Augusto	CINVESTAV, México
GARCÍA, José Luis	CINVESTAV, México
GARCÍA, Enrique	CINVESTAV, México
GARCÍA ZENTENO, J. Antonio	ICN-UNAM, México
GONZÁLEZ, Hernando	Univ. Nacional de Colombia, Colombia

GUPTA, Rajan	LANL, USA
GUZMÁN, Florencio	ESFM-IPN,México
HERNÁNDEZ, Orlando	FC-UNAM, México
HOLLIK, Wolfgang	Univ. Karlsruhe, Alemania
HERRERA, Simón	ESFM-IPN, México
JERÓNIMO, Gilberto	EFM-UMSNH, México
JIMÉNEZ, Uriel	FC-UNAM, México
JUÁREZ, Rebeca	ESFM-IPN, México
KIELANOWSKI, Piotr	CINVESTAV, México
KIRITSIS, Elias	CERN, Suiza
LEAL, José Luis	FC-UNAM, México
LEÓN, Gerardo	IFyM-UMSNH, México
LERMA, Sergio	ICN-UNAM, México
LINARES, Román	ICN-UNAM, México
LÓPEZ CASTRO, Gabriel	CINVESTAV, México
LÓPEZ, Celerino	UMSNH, México
LÓPEZ, Dennys A.	IF-BUAP, México
LÓPEZ, Rebeca	BUAP, México
MADRIZ-AGUILAR, José E.	IFyM-UMSNH, México
MARTÍNEZ, Juan Luis	ICN-UNAM, México
MARTÍNEZ, Mario I.	CINVESTAV, México
MATOS, Tonatiuh	CINVESTAV, México
MAZA, Marco A.	ICN-UNAM, México
MCLERRAN, Larry	Univ. of Minnesota, U.S.A.
MEDINA, Martín	UMSNH, México
MEGY GARCIN, E. Felipe	FQ-UNAM, México
MÉNDEZ, Héctor	CINVESTAV/Fermilab, México
MONDRAGÓN, Alfonso	IF-UNAM, México
MONDRAGÓN, Myriam	IF-UNAM, México
MORA, Gerardo	CINVESTAV, México
MORELOS, Antonio	IF-UASLP, México
MORENO, Enrique	IA-UNAM, México
MORENO, Gerardo	IF-UG, México
MORENO, Matías	IF-UNAM, México
NAPSUCIALE, Mauro	IF-UG, México
NELLEN, Lukas	IF-UNAM, México
NÚÑEZ, Carmen A.	IAFE, Argentina
PABLO NORMAN, Benjamín	IF-UNAM, México
PECCEI, Roberto	UCLA, U.S.A.

PÉREZ-LORENZANA, Abdel	CINVESTAV, México
PÉREZ CRUZ, Sergio	IF-UNAM, México
PONCE, William A.	Univ. de Antioquia, Colombia
RAFFELT, Georg	Max-Planck Institute, Alemania
RAMÍREZ, Cupatitzio	FCFM-BUAP, México
RAMÍREZ, Edgar H.	FC-UNAM, México
RAYA, Alfredo	IFyM-UMSNH, México
RIQUER, Verónica	IF-UNAM, México
RITTO, Pavel	CINVESTAV-UM, México
RODRÍGUEZ-JAUREGUI, Ezequiel	IF-UNAM, México
RODRÍGUEZ, Jairo A.	Univ. Nacional de Colombia, Colombia
RODRÍGUEZ, Marcos	Universidade Estadual Paulista, Brazil
RODRÍGUEZ, Simón	IF-UG, México
ROJAS, Ivett	FC-UNAM, México
ROMAN, Sergio	BUAP, México
ROSADO, Alfonso	IF-BUAP, México
ROUSSEAU, David	CERN, Suiza
RUIZ-ALTABA, Martí	IF-UNAM, México
SAAVEDRA, Oscar	INFN-Torino, Italia
SALAZAR, Humberto	IF-BUAP, México
SANTIAGO, Tania	FC-UNAM, México
SIGÜENZA, Pablo	IF-UNAM, México
SOD, Jordi	IF-UNAM, México
STEPHAN-OTTO, Christian	IF-UNAM, México
TAVARES, Gilberto	CINVESTAV, México
TEVES, Walter J.C.	Universidade de Sao Paulo, Brazil
TOLEDO, Genaro	CINVESTAV, México
TORRES, Manuel	IF-UNAM, México
TÚTUTI, Eduardo	IF-UNAM, México
URRUTIA, Luis F.	ICN-UNAM, México
VARGAS, Mauricio	Univ. Nacional de Colombia, Colombia
VENUGOPALAN, Raju	Brookhaven Nal. Lab., U.S.A.
VERGARA, José David	ICN-UNAM, México
VILLALOBOS, Manuel	FC-UNAM, México
VILLASEÑOR, Luis	IF-UMSNH, México
WEBER, Axel	ICN-UNAM, México
ZEPEDA, Arnulfo	CINVESTAV, México
ZOUPANOS, George	NTUA, Grecia

Author Index